W0042576

Computational
Microelectronics

Edited by
S. Selberherr

Arokia Nathan
Henry Baltes

Microtransducer CAD

Physical and Computational
Aspects

Springer-Verlag Wien GmbH

Prof. Dr. Arokia Nathan
Dept. of Electrical and Computer Engineering
University of Waterloo
Waterloo, Ontario N2L 3G1, Canada

Prof. Dr. Henry Baltes
Physical Electronics Laboratory
ETH Hoenggerberg
CH-8093 Zürich, Switzerland

This work is subject to copyright.
All rights are reserved, whether the whole or part of the material is concerned, specifically those of translation, reprinting, re-use of illustrations, broadcasting, reproduction by photocopying machines or similar means, and storage in data banks.

© 1999 Springer-Verlag Wien
Originally published by Springer-Verlag Springer-Verlag Wien New York in 1999
Softcover reprint of the hardcover 1st edition 1999

Typesetting: Thomson Press (India) Ltd., New Delhi 110 001

Printed on acid-free and chlorine-free bleached paper
SPIN: 10637710

With 137 Figures

ISSN 0179-0307
ISBN 978-3-7091-7321-3   ISBN 978-3-7091-6428-0 (eBook)
DOI 10.1007/978-3-7091-6428-0

To *Nandanee & Gabriella*

Preface

Semiconductor microtransducers have been investigated and developed for more than three decades while their numerical simulation has been underway for less than half that time. Integrated Si microtransducers are realized using microfabrication techniques similar to those for standard ICs. Unlike IC devices, however, microtransducers must interact with their environment, so their numerical simulation is considerably more complex. While the design of ICs aims at suppressing "parasitic effects", microtransducers thrive on the exploitation and optimization of the one or the other such effect. The challenging quest for physical models and simulation tools enabling microtransducer CAD is the topic of this book. The book is intended as a text for graduate students in Electrical Engineering and Physics and as a reference for CAD engineers in the microsystem industry.

The authors have been involved in microtransducer modeling since the days of the magnetic sensor modeling tool ALBERTINA, whose development was started in 1983 at the University of Alberta, Edmonton, Canada. We gratefully acknowledge our partners in those pioneering efforts: Prof. H. G. Schmidt-Weinmar and Prof. W. Allegretto.

Since then, many colleagues experienced in the art of numerical modeling of semiconductor IC devices have generously shared their valuable insights and tools. It is our pleasure to thank Prof. G. Baccarani, University of Bologna, Prof. S. G. Chamberlain, DALSA Inc., Prof. R. Dutton, Stanford University, Prof. W. Fichtner, ETH-Zürich, Prof. S. C. Jain, IMEC, Prof. D. J. Roulston, University of Waterloo, Prof. M. Rudan, University of Bologna, Prof. S. Selberherr, University of Vienna (whose book, Analysis and Simulation of Semiconductor Devices, initiated our numerical sensor simulation efforts), and last, but not least, Prof. S. D. Senturia of MIT for his pioneering efforts on CAD of MEMS.

We also would like to acknowledge the contributions to this book made by our colleagues, former and present research associates, and PhD students: Dr. K. Aflatooni, dpiX, Inc., Palo Alto; Dr. K. Benaissa, Texas Instruments, Richardson; Dr. T. Dravia, University of Waterloo; Dr. M. Ershov, University of Aizu; Dr. J. Funk, Boston Group, Zürich; Prof. J. G.

Korvink, University of Freiburg; M. Kulas, University of Waterloo; Dr. Y. Lu, IBM, San Jose; Prof. T. Manku, University of Waterloo; M. Nagata, Yamatake Corp.; Dr. N. O, DALSA Inc.; Prof. O. Paul, University of Freiburg, Germany; Prof. T. P. Pearsall, University of Washington; Dr. H. Pham, University of Waterloo; Q. Ma, University of Waterloo; Dr. C. Riccobene, AMD, Sunnyvale; M. Stevens, Nortel, Ottawa; and Dr. N. R. Swart, Analog Devices, Wilmington. Our sincere appreciation goes to Prof. R. Hornsey, University of Waterloo, and S. Taschini, ETH Zürich, for critical reading of the manuscript, and Prof. W. Huang, Prof. V. Karanassios, and Prof. A. Khandani, University of Waterloo, for stimulating discussions.

This text evolved from a series of courses offered to graduate students from Electrical Engineering and Physics; 1994, 1995, and 1996 at the University of Waterloo, 1995 at the ETH Zürich, where it received significant additions, and from a short course given at the CAD for MEMS Workshop, Zürich in 1997. Much of the material in the book can be presented in about 40 hours of lecture time. The text begins with an illustrative example to highlight the goals and benefits of microtransducer CAD. Chapter 2 summarizes the model equations describing electrical transport in semiconductor devices and microtransducers in the absence of external fields. Models treating the effects of the external radiant, magnetic, thermal, and mechanical fields on electrical transport are described systematically in Chapters 3 to 6. Chapter 7 presents an abridged version of solid structural and fluid mechanics. Here, we focus only on pertinent model equations and boundary conditions. Model equations and boundary conditions relevant to the various types of mechanical microactuation are discussed in Chapter 8. Chapter 9 concludes with a glimpse into simulation techniques for the mixed-signal microsystem, i.e., microtransducer plus circuitry. Where possible, we have supplemented the model equations with tables and/or graphs of process-dependent material data to enable the CAD engineer to carry out simulations even when reliable material models are not available.

This book would not have been written without the support of our institutions, ETH Zürich and the University of Waterloo, who granted sabbatical leaves, and provided hospitality, to both of us in 1995 and 1996. H. Baltes would like to thank Prof. G. Kovacs for the hospitality at Stanford University in 1996, which greatly benefited the book.

We are grateful to A. Nathan's wife, Nandanee, who not only prepared the manuscript by combining competence with devotion to the task, but also tolerated a living room and basement that were littered with books, papers, and manuscript pages over the grueling two-year period. As Mark Twain remarked in The Adventures of Huckleberry Finn (1885). "There ain't no more to write about, and I am rotten glad of it, because if I'd 'a' knowed what a trouble it was to make a book I wouldn't 'a' tackled it".

Waterloo and Zürich, October 1997   Arokia Nathan and Henry Baltes

Contents

Notation

A	electromagnetic vector potential
B	magnetic induction vector
$B(\Delta)$	Bernoulli function
Bi	Biot number
B_r	remnant flux density
B_s	saturation flux density
C	tensor of stiffness coefficients
D	electric displacement vector
D/Dt	substantial derivative
D_n, D_p	diffusion coefficient for (electrons, holes)
E	electric field vector
E	energy
E	Young's modulus
E_B	height of energy barrier
E_B	atomic binding energy
E_c, E_v	(conduction, valence) band edge
E_F	Fermi energy
E_{Fn}, E_{Fp}	quasi Fermi energy for (electrons, holes)
E_g	band gap energy
E_g^{eff}	effective band gap energy
E_i	intrinsic Fermi energy
E_p	phonon energy
F	force vector
$F_{1/2}$	Fermi integral of order 1/2
G	shear modulus
$G_j(r)$	scattering coefficients
G_l, G_t	(longitudinal, transverse) piezoresistance gauge factor
G_{opt}	optical generation rate
Gr	Grashof number
H	magnetic field strength vector
H	heat sources and sinks
H_c	coercivity
H_s	saturation field

I_0	intensity of incident radiant signal
I_T	intensity of transmitted radiant signal
$\mathbf{J}$	total electric current density vector
$\mathbf{J^e, J^q}$	(electric, heat) current density vector
$\mathbf{J_n^e, J_p^e}$	electric current density vector for (electrons, holes)
$\mathbf{J_n, J_p}$	electric current density vector for (electrons, holes)
$\mathbf{J_{nB}, J_{pB}}$	magnetic-field-dependent electric current density vector for (electrons, holes)
$\mathbf{J_{nf}, J_{pf}}$	stress- and magnetic-field-dependent electric current density vector for (electrons, holes)
$\mathbf{J_{n0}, J_{p0}}$	zero field electric current density vector for (electrons, holes)
$\mathbf{J_{n,p}^q}$	heat current density vector for (electrons, holes)
$\mathbf{J_{tot}^u}$	total energy flux density vector
$\mathbf{J^\varepsilon}$	strain flux density
K_j	magnetic-field-dependent galvanomagnetic transport coefficient
Kn	Knudsen number
L_c	characteristic length
$\mathbf{M}$	magnetization vector
M	bending moment
M_{easy}	easy axis of magnetization
M_j	magnetic-field-independent galvanomagnetic transport coefficient
$\mathbf{M_0}$	remnant magnetization vector
N_D, N_A	(donor, acceptor) dopant density (concentration)
N_c, N_v	effective density of states in (conduction, valence) band
N_q	Bose-Einstein phonon distribution function
N_T	total dopant density (concentration)
Nu	Nusselt number
$\mathbf{P}$	electric polarization vector
P	tensor of piezo-Hall coefficients
Pr	Prandtl number
Q_{Joule}	Joule heat
Q_{radin}	radiation-induced heat generation
Q_{cond}	heat transfer by conduction
Q_{fluid}	heat transfer to fluid
Q_{conv}	heat transfer by forced convection
Q_{radout}	radiative heat transfer
R	specific gas constant (8.31441 J/mol-K)
R	net generation-recombination rate
R, R_s	resistance, sheet resistance
Re	Reynold's number
R_{Hn}, R_{Hp}	Hall coefficient for (electrons, holes)
$(R\text{-}G)_{SRH}$	net Shockley-Read-Hall generation/recombination rate
R_{surf}	net surface generation/recombination rate
S	tensor of compliance coefficients
T	stress tensor
T	transmittance of radiant signal
T	temperature

T_∞	fluid's free stream temperature
a	acceleration of an interior point in a solid or fluid
b	damper to model internal friction
c	velocity of light in vacuum (2.99792458×10^8 m/s)
c	velocity of sound
c, c_{tot}	total heat capacity
c_n, c_p, c_L	electron, hole, and lattice contributions to total heat capacity
c_p, c_v	specific heat at constant (pressure, volume)
d	tensor of piezoelectric coefficients
$d\Gamma$	surface element
f^b	body force density
f^s	surface force
f^D	damping force
f^E	electrostatic force
f^M	magnetic body force density
f_n, f_p	Fermi distribution function for (electrons, holes)
f_0	carrier distribution function at equilibrium
g	edge factor
g	gravity
g_c, g_v	density of states in (conduction, valence) band
h	Planck constant (6.626176×10^{-34} Js)
h	heat transfer coefficient
$\mathbf{k}$	wave vector
k	Boltzmann constant (1.380662×10^{-23} J/K)
k	extinction coefficient
k	spring constant
m	mass
m^*	carrier effective mass
m_n^*, m_p^*	effective mass of (electron, hole)
m_l^*, m_t^*	(longitudinal, transverse) effective mass in conduction band
m_{hh}^*, m_{lh}^*	(heavy, light) hole effective mass
m_0	electron rest mass (9.109534×10^{-31} kg)
$\mathbf{n}$	unit normal vector
n	refractive index
n	electron density (concentration)
n_i	intrinsic carrier density (concentration)
n_{ie}	effective intrinsic carrier density (concentration)
p	pressure
p	hole density (concentration)
p	pyroelectric coefficient
p^E	electrostatic pressure
q	elementary charge ($1.6021892 \times 10^{-19}$ C)
q	local heat density
$\mathbf{r}$	position vector
r	scattering factor
r_{Hn}, r_{Hp}	Hall scattering coefficient for (electrons, holes)
r_{Hn0}	isotropic Hall scattering coefficient for electrons

u, v, w	x, y, and z components of the fluid's velocity field
u	displacement of an interior point in a solid
u	internal energy density
$\mathbf{v}$	carrier velocity vector
v	velocity of an interior point in a solid or fluid
v_F	Fermi velocity
v_∞	fluid's free stream velocity
w	diaphragm displacement
w	ionization energy
x, y, z	components of position vector
α	absorption coefficient
α	expansion coefficient
α_s	Seebeck coefficient
α_{sn}, α_{sp}	Seebeck coefficient in (n, p)-type material
χ_e	dielectric susceptibility
χ_m	magnetic susceptibility
χ^c, Λ^c	electron transport coefficients associated with stress gradient
χ^v, Λ^v	hole transport coefficients associated with stress gradient
ΔE_g	change in band gap energy
$\Delta E_c, \Delta E_v$	shift in (conduction, valence) band edge
δ	Kronecker delta
ε	strain tensor
ε	dielectric permittivity
ε_0	internal (residual) strain
ε_0	permittivity of free space ($8.854187818 \times 10^{-12}$ F/m)
ε_r	dielectric constant
ε^M	magnetostrictive strain
ε^T	thermal strain tensor
Φ	dissipation function
ϕ	velocity potential
ϕ_n, ϕ_p	quasi-Fermi potential for (electrons, holes)
Γ	integration surface or region
$\Gamma(x)$	gamma function
γ	Thomson coefficient
γ	ratio of specific heats
η	quantum efficiency
κ	torsion constant
κ_{fluid}	thermal conductivity of fluid
$\kappa, \kappa_{\text{tot}}$	thermal conductivity
$\kappa_n, \kappa_p, \kappa_L$	electron, hole, and lattice contributions to thermal conductivity
λ	wavelength
λ	eigenvalue
λ	coefficient of bulk viscosity
λ	magnetostriction coefficient
λ_l	longitudinal magnetostrictive strain
λ_s	saturation magnetostrictive strain
μ	attenuation coefficient

μ	coefficient of shear viscosity
μ_{Hn}, μ_{Hp}	Hall mobility for (electrons, holes)
μ_n, μ_p	drift mobility for (electrons, holes)
μ	permeability
μ_0	permeability of free space ($1.256637061 \times 10^{-6}$ H/m)
μ_r	relative permeability
ν	frequency
ν	Poisson ratio
ν	kinematic viscosity
ν	reluctivity
∇	nabla operator
Π	Peltier coefficient
π	tensor of piezoresistance coefficients
π_l, π_t	(longitudinal, transverse) piezoresistance coefficient
θ_D	Debye temperature
ρ	mass density
ρ	space charge density
ρ, ρ_s	resistivity, sheet resistivity
ρ_n, ρ_p	resistivity of (n, p)-type material
σ	stress tensor
σ	conductivity
σ	squeeze number
σ^M	Maxwell stress tensor
σ_n, σ_p	conductivity of (n, p)-type material
σ_0	internal (residual) stress
τ	carrier relaxation time
τ	torque
τ^M	magnetic couple
τ_{ac}	relaxation time for acoustic phonon (lattice) scattering
τ_{imp}	relaxation time for impurity scattering
τ_n, τ_p	relaxation time for (electrons, holes)
τ_0	energy independent relaxation time
ω	oscillation frequency
ω	resonant frequency
ψ	electrostatic potential

Relevant Conversion Units

$1 \text{ eV} = 1.6021892 \times 10^{-19}$ J

$1 \text{ J} = 1 \text{ kgm}^2/\text{s}^2$

In vacuum, $1 \text{ Oe} = 1 \text{ G} = 10^{-4} \text{ T} = 79.5775$ A/m

$1 \text{ T} = 10^4 \text{ G} = 1 \text{ Vs}/\text{m}^2$

$1 \text{ Pa} = 10 \text{ dyn}/\text{cm}^2 = 1 \text{ N}/\text{m}^2 = 1 \text{ kg}/\text{m-s}^2$

$1 \text{ atm} = 101.325$ kPa

$1 \text{ torr} = 1 \text{ mm Hg } (0°\text{C}) = 133.322 \text{ N}/\text{m}^2$

$1 \text{ bar} = 10^5 \text{ N}/\text{m}^2$

Introduction 1

1.1 Modeling and Simulation of Microtransducers

Modeling and simulation collectively describe the complex process of constructing models of a device, process, or system, and subsequently imitating its function on a computer [1].

Modeling involves development of a valid mathematical representation (model) of the particular system. In the case of microtransducers, this includes representation of the transducer effects (e.g., Hall effect) and the transducer material parameters (e.g., Hall mobility). There are three key steps in modeling: acquisition, simplification, and validation of the model (see Table 1.1). Acquisition is based on theory and/or measurement and results in differential equations and boundary conditions describing the underlying physical effects (e.g., carrier transport equations in the presence of magnetic field). Simplification involves assumptions that are suitably invoked to reduce the complexity of the full model equations. A model is considered valid when its predictions corroborate with experimental data (inductive method). Once validated, the model can predict results of further experiments and may finally replace them. Moreover, it can adequately reflect the behaviour of a real system by providing data that may be inaccessible by experiment, such as the distributions of physical quantities (electrical, thermal, mechanical, . . .) inside the device.

Simulation deals with the solution of the model and is most often numerical. The first essential step in simulation is to define the application domain for which the model is valid. Subsequently, through iterative or recursive steps, the model with a given set of input data is employed to produce a consistent solution, e.g., prediction of device performance under various operating conditions.

Microtransducers are miniaturized devices realized using microfabrication techniques established in the integrated circuit (IC) industry coupled with application-specific thin film deposition and micromachining technologies (see e.g., [2–8]). Microsensors or "input transducers"

Table 1.1. *The modeling and simulation process*

ACTIVITY	RESULT
Acquisition	Full effect equations and material data
Simplification	Reduced effect equations and material models
Validation	Experimentally confirmed model equations
Domain Definition	Established range of validity
Numerical Solution	Device function prediction and design optimization

convert an input physical (such as radiant, thermal, magnetic, mechanical) or chemical signal into an output electronic signal. Microactuators or "output transducers" convert an input electronic signal into an output physical or chemical signal. Devices that produce electrically-induced actuation in the mechanical domain are referred to as Micro Electro Mechanical Systems (MEMS), but the term MEMS has grown to encompass a broad family of micromachined sensors, actuators, and systems that exploit coupled electrical, mechanical, radiant, thermal, magnetic, and selected chemical effects [8]. Systems based on co-integration of MEMS with circuitry on the same chip are referred to as IMEMS. The variety of signal conversion effects in semiconductors and

IC-related materials effectively utilized in practical microsensor and microactuator structures are classified in Table 1.2. A selection of possible sensor and actuator structures are indicated in italics.

In contrast to semiconductor IC device modeling and simulation, which after thirty years of research has become a mature art [9–11], the field of microtransducer modeling and simulation is relatively young (see [12–21]). Microtransducers and IC devices come with very different design [22], and hence, computer-aided-design (CAD) requirements. The micro-transducer is designed to interact with the environment, while in the IC world devices are shielded from the environment. The design of ICs aims at the suppression of "parasitic" effects whereas many microtransducers thrive on optimizing the one or the other "parasitic" effect in a controlled manner for high transduction efficiency.

By design, the presence of a perturbing (physical or chemical) signal is meant to reduce the symmetry of device operation as much as possible. This makes the choice of appropriate model equations, physical and material parameters, as well as boundary and interface conditions very crucial. The various input signals (such as mechanical force, magnetic field, temperature, chemical concentration) interact with the inherent electrical transport phenomena in a complicated way. For example, the presence of flow, pressure (vacuum), or radiation, upsets the thermal balance of a microstructure; magnetic induction disturbs the carrier transport by the Lorentz force; incident radiation alters the generation recombination balance in photodetectors; mechanical stress modulates the electric conductivity, the capacitance of a microcavity or some mechanical resonant frequency.

Despite differences in design methodology, the progress in micro-transducer CAD will prove beneficial to the IC technology as it continues to move into the submicron and high frequency regimes, where modeling of "parasitic" effects based on interaction of mechanical, thermal, magnetic, and electrical signals becomes increasingly important. Table 1.3 summarizes the benefits which the IC technology may draw from microtransducer research; the area of packaging being of particular interest. Table 1.4 highlights further areas of CAD research which would mutually benefit both areas.

With the growing demand for microtransducers, simulation tools are becoming increasingly crucial for the study and design of devices, processes, and systems pertinent to sensors and actuators. The development of new microtransducers usually involves several design and fabrication cycles until the specifications are satisfied. Modeling and simulation can reduce the number of costly trial-and-error-based prototyping cycles. The savings in development effort is becoming increasingly more substantial with the growing degree of sophistication in the CAD models and the steadily declining costs of computer resources. In particular, CAD tools are

Table 1.2. *Signal conversion (transduction) effects in microsensors and microactuators* (sources: [2, 3])

PRIMARY \ SECONDARY	MECHANICAL	THERMAL	ELECTRICAL	MAGNETIC	RADIANT
MECHANICAL	Pneumatics Hydraulics Acoustics Resonance *cantilevers, bridges, diaphragms, gears, resonators, microfluidic devices*	Friction	Piezoelectricity Piezoresistivity Acoustoelectric effect Tribolelectric effect Tunneling Capacitive effect Inductive effect Electrohydrodynamics *pressure sensors, accelerometers, MEMS, liquid flow meters*	Magneto-elasticity	Photoelasticity Piezooptic effect Sagnac effect Doppler effect Evanescence *interferometers*
THERMAL	Thermal expansion Ciliary motion Shape memory effect Thermopneumatics *bimorph structures, microfluidic devices, shape memory alloys*	Heat conduction	Thermoresistance Seebeck effect Nernst effect Pyroelectricity Thermodielectric effect *resistors, thermocouples*	Thermo-magnetization effects	Thermo optical effects Radiation Thermoluminescence Incandescence *LCDs, interferometers*
ELECTRICAL	Inverse piezoelectricity Electrostatics Electrostriction Photostriction *MEMS, acoustic wave devices, optical elements*	Joule heating Peltier effect Thomson effect Bridgeman effect Lossy dielectric effect *resistors, electrical contacts, capacitors*	Electrical conduction *logic elements*	Biot-Savart's Law *driver coils*	Kerr effect Pockels effect Population inversion Electroluminescence *lasers, masers, plasmas*

MAGNETIC	Magnetostriction Magnetostatics *microfluidic devices, micropositioners*	Righi-Leduc effect Ettinghausen effect *cross-effects on Hall-based devices*	Hall and Suhl effects Magnetoresistance Photo-Hall effect Nernst effect *Hall devices, magnetotransistors, magnetodiodes, magnetoresistors*	Magnetic induction *flux microconcentrator*	Faraday effect Cotton-Mouton effect
RADIANT	Radiation pressure Photostriction *Golay cell*	Radiation heating *bolometer*	Photovoltaic effect Photoconductivity Photoelectric effect Photo-Hall effect Photodielectric effect *color sensors, radiation sensors*		Photoluminescence Cathodoluminescence Radioluminescence

Table 1.3. *Benefits to IC CAD drawn from microtransducer modeling research*

Microtransducer Efforts	IC Device Benefits
Modeling mechanical properties of microfabrication thin film materials retrieved using transducer test structures	Elastic moduli, fracture or yield point, thermomechanical coefficients
Modeling effects of mechanical strain on electronic transport parameters	Band structure, effective mass, mobility
Modeling thermal and thermo-electric properties of thin films retrieved based on transducer test structures	Thermal conductivity and specific heat Seebeck, Peltier, and Thomson coefficients
Simulation tools to predict thermal, electrothermal, mechanical, and thermomechanical behavior	Residual stress in composite structures stress distributions and concentrations in LOCOS and trench structures risk and fracture predictions thermomechanical analysis and packaging effects
Numerical discretization schemes for tensorial transport coefficients for improved flux conservation	Anisotropy in transport induced by stress or magnetic field
Coupling techniques for mixed signals in numerical device simulation	Interactions of mechanical, thermal, flow and heat transfer, electrical, magnetic signals
Field analysis techniques for capacitance and inductance including eddy currents, relevant at high frequencies	Signal routing and packaging
Models, including boundary conditions, of fluid flow and heat transfer	Packaging and its thermal and mechanical analysis

Table 1.4. *Further CAD requirements to benefit both microtransducers and ICs*

Microtransducers	IC Devices
Models to predict process-dependent thermal and mechanical properties of thin films	Thermal, electrothermal, and thermomechanical coefficients intrinsic stresses in thin films residual stresses in composite films
Modeling thin film growth/deposition and film's mechanical behavior	Visco-elastic material models for oxidation simulation
Modeling wet and dry etch micromachining processes	Etch simulators
Expedient modeling tools at circuit-level for mixed signal (mechanical, thermal, optical) interactions	Simulating strain- or electrothermally-induced offsets in matched pairs

indispensable whenever optimization with respect to two or more design variables is required.

1.2 Illustrative Example

We introduce the reader to the concept and purpose of microtransducer modeling and simulation in terms of an example. We begin with the specific case of an integrated hot-wire flow sensor (Section 1.2.1), then expand to a family of thermal sensors and actuators (Section 1.2.2) and finally discuss the goals and benefits.

1.2.1 Thermal Flow Sensor

Flow sensors fabricated using microfabrication technologies including micromachining offer several attractive features such as small size, low cost through batch fabrication, and most importantly, high resolution (small flow volume monitoring). For optimized device operation, design and process trade-offs are necessary between flow sensitivity, mechanical integrity, environmental durability, the required signal-to-noise ratio, and input power. The microsensor shown in Fig. 1.1 utilizes a thermally isolated, micromachined, thin silicon nitride membrane of high mechanical integrity with the desired electrical and thermal characteristics [23]. On the membrane there are two resistive heating and sensing elements (see Fig. 1.2) based on thin film platinum of high resistivity and high temperature coefficient of resistance. The mechanical integrity of the membrane and uniformity of the various materials are crucial to ensure reproducibility of

Fig. 1.1 Photomicrograph of flow microsensor chip [23] (courtesy of Yamatake Corp.)

Fig. 1.2 Photomicrograph of flow microsensor showing the resistance heating and sensing
elements [23] (courtesy of Yamatake Corp.)

electrical and thermal characteristics, and hence the manufacturing yield. In the presence of flow over the membrane, heat is transferred by forced convection from the heating element to the fluid and subsequently, from the fluid to the sensing element downstream. The temperature difference, and hence the output signal measured, across heating and sensing elements is proportional to the fluid velocity. Device operation and mechanism of flow measurement can be heuristically described by the transduction efficiency (flow sensitivity)

$$dV/df = (dV/dR) \cdot (dR/dT) \cdot (dT/dQ) \cdot (dQ/dv) \cdot (dv/df)$$

$$(1.1)$$

where V denotes output voltage, f the volume flow, R the electrical resistance, T the temperature, Q the heat flow, and v the velocity. Thus for a given flow configuration and velocity range, each of the terms in the above equation has to be carefully designed to optimize the flow sensitivity and operating power.

A key design issue in such structures is the exact location and separation of the resistive heating and sensing elements. If the spacing between elements is large, the heater has little influence on the device output response. Likewise, if the spacing is too small, the response is poor due to excessive thermal coupling. Based on these limits, it is clear that there is an optimum spacing for which the sensitivity is maximum for a given flow rate. These design aspects can best be addressed with microsensor simulations rather than by fabricating and testing a large variety of prototypes. The simulation involves numerically solving the coupled differential equations, governing electrical and thermal conduction, under appropriate boundary conditions that accurately account for

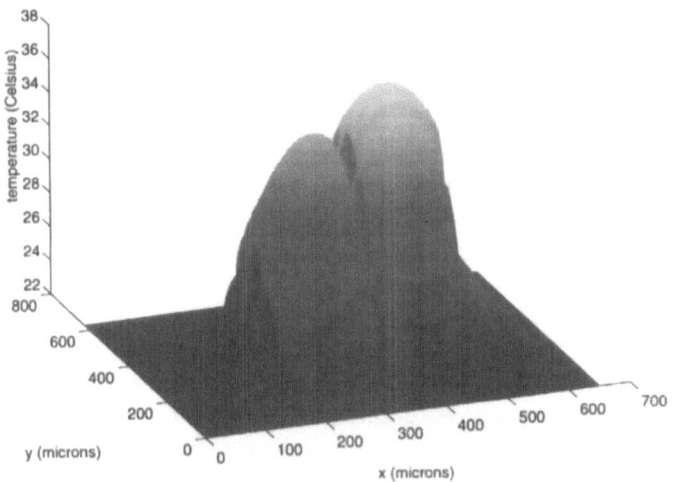

Fig. 1.3 Simulated temperature distribution on the sensor membrane for 5 m/s flow and 1.2 V operation [23]

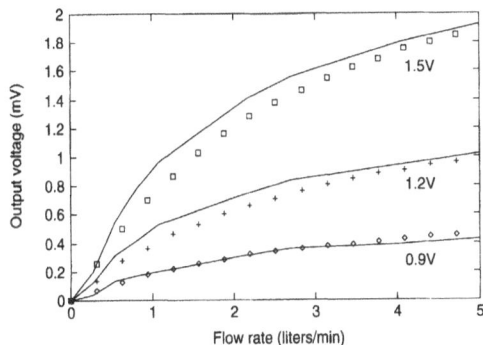

Fig. 1.4 Simulated and measured output response of flow sensor [25]: solid curve–simulation; symbol–measurement

heat transfer to, and from, the fluid by forced convection. As an example, Fig. 1.3 illustrates the simulated temperature distribution on the membrane of the microsensor (in Fig. 1.2) for a flow velocity of 5 m/s with the heating element at 1.2 V. Such a detailed two-dimensional temperature distribution with high spatial resolution can hardly be obtained by experimental means. In Fig. 1.4 the simulated flow sensor response for different operating voltages is compared with experimental data (validation). Figures 1.3 and 1.4 were obtained from simulations using the electrothermal simulator μ Therm developed for analysis of heat transport in thermal microstructures [24, 25].

1.2.2 Thermal Sensors and Actuators

Apart from the above flow detection example, thermal microstructures depending on a prevailing heat transfer mechanism can be employed to directly sense, or assist in sensing, quantities such as pressure (vacuum), gas species, AC power, and infra-red (IR) radiation, as well as to induce mechanical motion or vibration and IR imaging in actuation applications. For illustration, we describe in Fig. 1.5, a simplified generalization of the device structure shown in Figs. 1.1 and 1.2. The various components maintaining the thermal balance in the generic membrane microstructure shown in Fig. 1.5 constitute sources and sinks. The resistive heating element shown in the center of the membrane is provided with electrical power to generate Joule heat, Q_{Joule}. Alternatively, the resistor can also be heated by incident IR radiation, Q_{radin}. These denote heat sources while the various heat loss mechanisms shown denote sinks. The component Q_{cond} describes conductive heat loss through the membrane to the silicon substrate. In contrast to the heat loss mechanisms normally encountered in ICs, Q_{cond} in micromachined structures is dramatically lowered by virtue of thermal isolation. With the removal of the underlying (silicon) substrate, a huge conduction component directly below the membrane is eliminated resulting in low power (a few mW) device operation as well as low thermal time constant (a few ms). Consequently, the other heat sink mechanisms gain significance. For example, even in the absence of fluid motion, there is a non-trivial heat loss contribution, Q_{fluid} to the fluid above and below the membrane [24]. Part of Q_{fluid} may return to a cooler part of the membrane. This is significant when the temperature distribution on the membrane is highly nonuniform. The component Q_{conv} denotes heat transfer that is

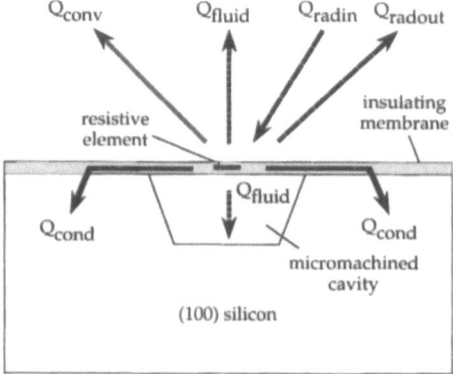

Fig. 1.5 Simplified general structure and thermal balance of the thermal microtransducer family

induced by forced convection such as in the presence of gas or liquid flow stream and Q_{radout} denotes the radiative heat loss component. The thermal balance for the microstructure reads:

$$Q_{Joule} + Q_{radin} = Q_{cond} + Q_{fluid} + Q_{conv} + Q_{radout}. \tag{1.2}$$

The component Q_{fluid} can be partitioned as $Q_{fluid} = Q_0 + Q_\infty$, where Q_0 and Q_∞ denote the heat loss components to cooler bodies in close proximity to, and far away from, the membrane, respectively.

Depending on the application, the one or the other components of heat transfer needs to be maximized to achieve the desired transduction efficiency (see Fig. 1.6). This is where simulation plays a key role. It

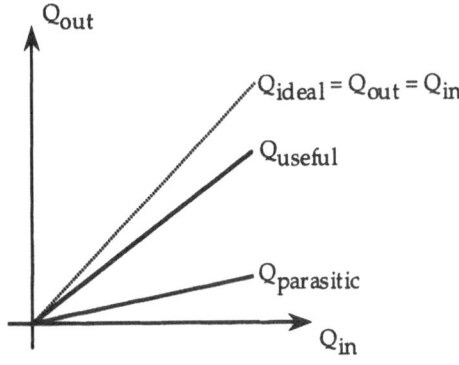

Q_{useful}	$Q_{parasitic}$	Application
Q_{conv}	$Q_{fluid} + Q_{cond} + Q_{radout}$	flow sensor
Q_{fluid}	$Q_{conv} + Q_{cond} + Q_{radout}$	vacuum sensor
Q_{cond}	$Q_{conv} + Q_{fluid} + Q_{radout}$	IR sensor thermal converter bimorph actuator (acoustic resonator)
Q_{radout}	$Q_{conv} + Q_{fluid} + Q_{cond}$	IR imager

Fig. 1.6 The parasitic and useful heat transfer components in different thermal micro-transducers; transduction efficiency is Q_{useful}/Q_{in} with $Q_{in} = Q_{radin}$ for the IR sensor and else $Q_{in} - Q_{Joule}$

provides insight into device operation for subsequent device optimization ($Q_{parasitic} \rightarrow 0$, $Q_{useful} \rightarrow Q_{ideal}$) with respect to the application in question.

Simple analytical models are usually the first choice and they indeed serve as valuable heuristic tools for trial device design, qualitative understanding of device behavior, and an order-of-magnitude estimate of the device sensitivity. Unfortunately, such models suffer from limited applicability since their solutions are valid only for the specific (simple) geometry and boundary conditions for which they have been derived. For example, if the device structure in Fig. 1.2 had a one-dimensional geometry (diaphragm thickness and width small relative to its length), a simple analytical solution of the heat conduction equation, reminiscent of the classical hot-wire anemometer, can be easily obtained. However, this is valid only for device operation in vacuum ($Q_{cond} = Q_{conv} = 0$) at relatively low temperature ($Q_{radout} = 0$). Closed form, Fourier series, solutions may be made possible, but require simplifying assumptions, which do not do justice to the physical nature of practical devices. Simplifying assumptions include geometrical regularity, material homogeneity, temperature uniformity, and periodic boundary conditions. Furthermore, these approaches do not permit simultaneous handling of all the heat loss components relevant for accurate prediction of output response. For example, Q_{fluid} can be significant under standard temperature and pressure. Thus for more general geometries and operating conditions, and for an indepth investigation of the underlying physical effects, their interaction, and the transduction mechanism(s) involved, it is imperative that the full system of equations be solved under suitable boundary conditions that adequately describe the relevant physical effects.

1.2.3 Goals and Benefits of Modeling and Simulation

One goal of modeling and simulation is to provide insight into the underlying transduction principle(s) by means of the distributions of various physical quantities in the device interior not readily accessible to characterization. For example, electrical measurements for the device shown in Fig. 1.2 can only yield an average value of resistance, and hence temperature, following measurement of the voltage across, or current through, the resistive element, based on known temperature coefficient of resistance. Knowledge of the heat flux from the diaphragm surfaces or the relative significance of the various heat dissipation components can only be determined numerically. Most importantly, simulation can predict the response of the device or its sensitivity to the perturbing signal, including cross sensitivity effects.

Output response optimization of pertinent device structures within the given framework of design, fabrication, and operating parameters is the next obvious goal. Here, we are interested in determining how design, fabrication, and operating parameters influence terminal characteristics (see [26]), and in particular, how they enhance or limit the device sensitivity or response with respect to the perturbing signal under consideration. As indicated in Fig. 1.6, we would like to maximize the heat transfer component of interest for that given application.

For example, in vacuum sensing applications (see Fig. 1.6), we maximize Q_{fluid}/Q_{Joule} and minimize the parasitic component, $Q_{parasitic} = Q_{cond} + Q_{conv} + Q_{radout}$, by suitable structural design. Within the fluid, we maximize Q_0/Q_{fluid}, so as to exploit the pressure-dependent mean free scattering length of the fluid. If $Q_{parasitic}$ and Q_∞ were zero, then we would have the ideal case where $Q_{fluid} = Q_0 = Q_{Joule}$, which would yield the maximum transduction efficiency. By design, we can minimize Q_{cond} with a high degree of lateral thermal isolation on the membrane, Q_{conv} by making the structure velocity insensitive (through a fluidic short circuit), and Q_{radout} by not operating the device at high temperatures (we assume $Q_{radin} = 0$ in this case).

On the other hand, in infra-red sensing applications (also in Fig. 1.6), the in-membrane conduction Q_{cond}/Q_{input} has to be maximized. Here, we would require, in addition, a thermal sensor (e.g., thermistor or thermopile) on the membrane for signal detection. Note that because the heat source is the absorbed radiation, $Q_{input} = Q_{radin}$ (while $Q_{Joule} = 0$), we have, in this case, a direct measure for transduction efficiency (i.e., the infrared sensitivity), viz., $\Delta Q_{cond}/\Delta Q_{input}$.

For flow sensing applications, we minimize $Q_{parasitic} = Q_{cond} + Q_{fluid} + Q_{radout}$, so as to maximize the power available for forced convection, i.e., $Q_{conv} = Q_{Joule}$. Note that because of the intrinsic presence of Q_{fluid}, differential configurations of the type shown in Fig. 1.2 require an optimal spatial separation between the resistive heating and sensing elements.

With thermal actuation (Fig. 1.6), such as the generation of thermal-based IR images, a highly thermally-isolated microhotplate arrangement is ideal because of high temperature uniformity and fast response time. Here, we maximize the radiative heat dissipation component, Q_{radout}. Device operation in vacuum yields $Q_{radout} = Q_{Joule}$, with $Q_{cond} = 0$ under conditions of high thermal isolation (and, of course, $Q_{radin} = 0$).

Regardless of device structure and application, the parasitic heat loss $Q_{parasitic}$ decreases the power available for transduction, thus reducing the strength of the output response. This subsequently degrades the minimum detectable measurand of interest. For example, if the on-membrane temperature gets to appreciable levels, the radiation emanating from the microstructure increases, thus increasing $Q_{parasitic}$, which reduces the available transduction (sensing) power.

In addition, modeling and simulation can facilitate the extraction of pertinent physical and material parameters from measurement data based on complex, realistic device geometries (rather than ideal limiting cases) where valid and accurate closed-form solutions of device behavior are difficult, if not impossible, to obtain [27–29]. For example, test structures reminescent of the shape shown in Fig. 1.5 are employed for retrieval of the thermo-physical properties of the membrane materials such as nitrides, oxides, polysilicon, through calibration of three-dimensional thermal simulation [29].

Finally, numerical modeling results can be used to refine or extend the application range of simple analytical formulas, and to establish design rules and scaling laws. Such formulas could serve as adequate initial guesses in iterative numerical schemes. Alternatively, simplified higher order models of device behavior for use in a CAD environment can be developed from numerous device simulations calibrated with pertinent experimental data [30]. Such approaches, which are popular in IC design, should be employed with caution in the case of microtransducers [31]. For example, the key design variable distinguishing any two MOSFETs is the aspect ratio, whilst for the microtransducer, no such description based on a single geometry parameter exists. Furthermore, a large number of these devices cannot be easily characterized in view of the physical interaction between the many variables involved in multiple transduction processes, which are not exclusively in the electrical domain. For example, thermal-based microsensors (Fig. 1.5) exploit signal transduction from thermal to electrical domain with the conversion process modulated by the rate of flow, pressure, or thermal conductivity (gas species).

The above example of thermal devices is just one family of microtransducers out of many that can be extracted from Table 1.2 with all sorts of effects and their couplings. Prediction of transduction efficiency is only one desirable simulation result. Others include resolution or signal-to-noise ratio, offset, temperature coefficient and other cross-sensitivities, impact of fabrication-induced mechanical stress and other fabrication tolerances, whose relative importance depends on the application specifications. Thus the world of microtransducer simulation, as it will unfold in the following chapters, is by far more complex than that of IC device simulation.

1.3 Progress in Microtransducer Modeling

The progress in the field of microtransducer modeling and simulation is summarized in Table 1.5. Here, we categorize progress by measurand and the corresponding devices or effects, along with a selected list of key papers and simulation tools, which have been developed specifically for, or are relevant to, microtransducer CAD. As it is typical for a new research area, academia has taken the lead. Examples include University of Alberta,

Table 1.5. *Progress in microtransducer modeling and simulation. Only selected publications are listed. Further references are given in the relevant chapters*

Measurand	Devices/Effects	References	Tools
Radiant	Photodiodes and colour sensors	[34–38]	HFIELDS [36, 37], PC3D [211], PISCES [212]
Magnetic	Hall devices and magnetotransistors	[39–66]	ALBERTINA [47], SOLIDIS [213]
	Flux microconcentrators	[67]	
	Piezo-Hall effects	[68, 69]	
Thermal	Flow microsensors, microhotplates, micro-Pirani gauges, thermal converters, and thermopiles	[70–83]	μTherm [24, 25], et3 [25], HFIELDS (see [37]), SOLIDIS [213], MICROTHERM [214], ANSYS [215], NASTRAN [216], ADINA [217]
Mechanical	Thermomagnetic effects	[84–85]	
	Transport modeling	[86–91]	
	Piezoresistive and capacitive pressure sensors, accelerometers, and microphones	[58, 82, 92–118]	SENSIM [95], SOLIDIS [213], ANSYS [215], NASTRAN [216], ADINA [217], CAPSIM [218], SESES [219, 220], ABAQUS [221], COSMOS [222], I-DEAS [223], FLOWERS [224], TPS10 [225], MARC [226]
Fluid Flow	Fluidic devices	[119–123]	FIDAP [227], FLUENT [228], FLOTRAN [229], FLOTHERM [230], PUSI [231]
	Fluidic damping	[124–127]	
Micro-actuation	Electrostatic	[58, 82, 118, 128–147, 246, 247]	MEMCAD [15], BEMMODULE [147], SOLIDIS [213], ANSYS [215], MAXWELL-2D [232], ALECSIS [233], MICROCOSM [234], IntelliSense [235], FASTCAP [236], EFCREL, EFDYN & EFCAD [237], CEDRAT [238], IES [239], WATCAP [248]
	Electrothermal and SMA	[148–152]	
	Magnetic	[153]	
	Acoustic and piezoelectric	[154–160]	
CAD Architecture and Structure Simulation		[161–176]	MEMCAD [15], OYSTER [161], CAEMEMS [165], PATRAN [240], Geomview [241], Pro/Engineer [242]
Process	Micromachining	[177–183]	ASEP [177], MICROCAD [243], SUPREM [244]
Packaging	Magnetic sensors, pressure sensors, and accelerometers	[60, 61, 184–190]	SOLIDIS [213], ANSYS [215]
Microsystems	Thermal, flow, magnetic, and chemical sensors, and microactuators	[20, 30, 191–210]	TRENDY [196], SPICE [245]

Table 1.6. *Evolution in microtransducer CAD*

Publications (■)

	radiant
	magnetic
	thermal
	mechanical
	micro actuators
	process fabrication
	packaging
	systems
	reviews

1970 1975 1980 1985 1990 1995 2000

Year

University of Bologna, Delft University of Technology, ETH Zürich (SOLIDIS), University of Michigan, MIT (MEMCAD), and University of Waterloo. Apart from the suppliers of general-purpose simulation tools, transducer-specific research contributions from industry have come from, e.g., NEC, Nova Sensors, IC Sensors, Toyota, and Yamatake-Honeywell.

Table 1.6 illustrates how microtransducer CAD has evolved over the years. Newly emerging areas are related to CAD of micro electro mechanical systems. Here, we are witnessing a large growth in publications, and sessions devoted to CAD are now in place at major international meetings. Future challenges lie in the areas of piezoelectric and magnetic micro-actuation, microfluidics, mixed-signal simulation at the microsystem-level where microsensors and microactuators are embedded in circuitry, and in microtransducer-specific process simulation.

This is the first comprehensive book on microtransducer CAD. Efforts to summarize the state of the art have been made hitherto only at the level of review papers or book chapters on specific topics:

• microsensor modeling–Baltes et al. [12], Nathan and Baltes [13].

- physical models for microsensor simulation – Nathan et al. [14].
- computer-aided-design of electromechanical actuators – Senturia et al. [15, 17].
- thermoelectromechanical sensors and actuators including material parameter extraction – Baltes et al. [18].
- computer-aided-design of microtransducers – Nathan [19].
- coupled modeling of microtransducer effects – Korvink and Baltes [20, 21].

A special issue dedicated to microtransducer modeling appeared in Sensors and Materials in 1994 [16].

The first workshop, including short courses, dedicated to microtransducer CAD was held in 1997 [32]. Thus the field is in its infancy, similar to the state IC device and process modeling was in 1979 when its first workshop was held [33]. By analogy, we expect a growing concerted effort aimed at microtransducer CAD in the years to come.

1.4 References

[1] Ziegler, B., *Theory of Modeling and Simulation*, Robert Krieger, 1984.
[2] Middelhoek, S., Audet, S. A., *Silicon Sensors*, New York: Academic Press, 1989.
[3] Grandke, T., Ko, W. H. (Eds.), *Sensors*, Vol. 1, Chapt. 1, Weinheim: VCH, 1989, pp. 1–16.
[4] Muller, R. S., Howe, R. T., Senturia, S. D., Smith, R. L., White, R. M. (Eds.), *Microsensors*, New York: IEEE Press, 1991.
[5] Sze, S. M. (Ed.), *Semiconductor Sensors*, New York: Wiley, 1994.
[6] Baltes, H., Future of IC Microtransducers, *Sensors and Actuators A*, 56 (1996), 179–192.
[7] Baltes, H., Paul, O., Korvink, J. G., Schneider, M., Bühler, J., Schneeberger, N., Jaeggi, D., Malcovati, P., Hornung, M., Häberli, A., von Arx, M., Mayer, F., Funk, J., IC MEMS Microtransducers, *Technical Digest*, IEEE IEDM, San Francisco, 1996, pp. 521–524.
[8] Proceedings, IEEE Micro Electro Mechanical Systems Conference, Nagoya, 1997.
[9] Engl, W. L., Dirks, H. K., Meinerzhagen, B., Device Modeling, *Proc. IEEE*, 71 (1983), 10–33.
[10] Selberherr, S., *Analysis and Simulation of Semiconductor Devices*, Vienna: Springer-Verlag, 1984.
[11] Baccarani, G., Rudan, M., Guerrieri, R., Ciampolini, P., Physical Models for Numerical Device Simulation, in: *Process and Device Modeling*, Engl, W. L. (Ed.), Amsterdam: North-Holland, 1986, pp. 107–158.
[12] Baltes, H., Allegretto, W., Nathan, A., Microsensor Modeling, in: *Simulation of Semiconductor Devices and Processes*, Vol. 3, Baccarani, G., Rudan, M. (Eds.), Tecnoprint, Bologna, 1988, pp. 563–577.
[13] Nathan, A., Baltes, H. P., Sensor Modeling, in: *Sensors*, Vol. 1, Grandke, T., Ko, W. H. (Eds.), Chapt. 3, Weinheim: VCH, 1989, pp. 45–77.
[14] Nathan, A., Baltes, H., Allegretto, W., Review of Physical Models for Numerical Simulation of Semiconductor Microsensors, *IEEE Trans. on CAD of ICAS*, 9 (1990), 1198–1208.

[15] Senturia, S. D., Harris, R. M., Johnson, B. P., Kim, S., Nabors, K., Shulman, M. A., White, J. K., A Computer-Aided Design System for Microelectromechanical Systems (MEMCAD), *IEEE J. of Microelectromechanical Systems*, 1 (1992), 3–14.

[16] Nathan, A. (Ed.), *Special Issue on Microsensor Modeling, Sensors and Materials*, Vol. 6, Nos. 2–4, 1994.

[17] Senturia, S. D., CAD for Microelectromechanical Systems, *Digest of Technical Papers*, Vol. 2, Transducers '95, Stockholm, 1995, pp. 5–8.

[18] Baltes, H., Korvink, J. G., Paul, O., Numerical Modelling and Materials Characterization for Integrated Micro Electro Mechanical Systems, in: *Simulation of Semiconductor Devices and Processes*, Vol. 6, Ryssel, H., Pichler, P. (Eds.), Vienna: Springer-Verlag, 1995, pp. 1–9.

[19] Nathan, A., Microtransducer CAD, *Proc. ESSDERC '96*, Baccarani, G., Rudan, M. (Eds.), Bologna, 1996, pp. 707–715.

[20] Korvink, J. G., Bächtold, M., Emmenegger, M., Paganini, R., Ruehl, R., Funk, J., Baltes, H., TCAD for MEMS, *Proc. ESSDERC '96*, Baccarani, G., Rudan, M. (Eds.), Bologna, 1996, pp. A5–A7.

[21] Korvink, J. G., Baltes, H., Microsystem Modeling, in: *Sensors Update*, Baltes, H., Göpel, W., Hesse, J. (Eds.), Chapt. 6, Weinheim: VCH, 1996, pp. 181–209.

[22] Senturia, S. D., Microsensors vs. Integrated Circuits: A Study in Contrasts, *Technical Digest*, IEEE IEDM, Washington, 1989, pp. 3–7.

[23] Nagata, M., Swart, N., Stevens, M., Nathan, A., Thermal Based Micro Flow Sensor Optimization Using Coupled Electrothermal Numerical Simulations, *Digest of Technical Papers*, Vol. 2, Transducers '95, Stockholm, 1995, pp. 447–450.

[24] Swart, N. R., *Heat Transport in Thermal Based Microsensors*, Ph.D. Dissertation, University of Waterloo, Waterloo, Ontario N2L 3G1, Canada, 1994.

[25] Stevens, M. E., *CMOS Electrothermal Microsensors for Flow and Pressure Measurements*, M.A.Sc. Dissertation, University of Waterloo, Waterloo, Ontario N2L 3G1, Canada, 1996.

[26] Gnudi, A., Ciampolini, P., Guerrieri, R., Rudan, M., Baccarani, G., Sensitivity Analysis for Device Design, *IEEE Trans. on CAD of ICAS*, CAD-6 (1987), 879–885.

[27] Briglio, D. R., Nathan, A., Baltes, H. P., Measurement of Hall Mobility in n-Channel Silicon Inversion Layer, *Can. J. Phys.*, 65 (1987), 842–845.

[28] Maseeh, F., Schmidt, M. A., Allen, M. G., Senturia, S. D., Calibrated Measurements of Elastic Limit, Modulus, and the Residual Stress of Thin Films Using Micromachined Suspended Structures, *Technical Digest*, IEEE Solid-State Sensor and Actuator Workshop, Hilton Head Is., 1988, pp. 84–87.

[29] Paul, O. M., Korvink, J., Baltes, H., Determination of Thermophysical Properties of CMOS Polysilicon, *Sensors and Actuators A*, 41–42 (1994), 161–164.

[30] Pham, H. H., Nathan, A., Compact MEMS-SPICE Modeling, *Sensors and Materials*, 10 (1998), 63–75.

[31] Swart, N., Nathan, A., Mixed-Mode Device-Circuit Simulation of Thermal-Based Microsensors, *Sensors and Materials*, 6 (1994), 179–192.

[32] *CAD for MEMS Workshop*, Zürich, Switzerland, March 16–18, 1997.

[33] Browne, B. T., Miller, J. J. H. (Eds.), *Numerical Analysis of Semiconductor Devices*, Proc. NASECODE I Conference, Dublin: Boole Press, 1979.

[34] Lundstrom, M. S., Schwartz, R. J., Gray, J. L., Transport Equations for the Analysis of Heavily Doped Semiconductor Devices, *Solid-State Electron.*, 24 (1981), 195–202.

[35] Gray, J. L., *Two-Dimensional Modeling of Silicon Solar Cells*, Ph.D. Dissertation, Purdue University, West Lafayette, USA, 1982.

[36] Ciampolini, P., Pierantoni, A., Baccarani, G., Efficient 3-D Simulation of Complex Structures, *IEEE Trans. on CAD of ICAS*, CAD-10 (1991), 1141–1149.

[37] Ciampolini, P., Pierantoni, A., Vecchi, M. C., Rudan, M., Application of General-Purpose Device Simulator to Analysis of Integrated Silicon Microsensors, *Sensors and Materials*, 6 (1994), 139–157.

[38] Mohajerzadeh, S., Nathan, A., Selvakumar, C. R., Numerical simulation of a *p-n-p-n* Color Sensor for Simultaneous Color Detection, *Sensors and Actuators A*, 44 (1994), 119–124.

[39] Mimizuka, T., Improvement of Relaxation Method for Hall Plates, *Solid-State Electron.*, 14 (1971), 107–110.

[40] Chwang, R., Smith, B. J., Crowell, C. R., Contact Size Effects on the Van der Pauw Method for Resistivity and Hall Coefficient Measurement, *Solid-State Electron.*, 17 (1974), 1217–1227.

[41] Andor, L., Baltes, H. P., Nathan, A., Schmidt-Weinmar, H. G., Carrier Transport in Semiconductor Magnetic Field Sensors, *Technical Digest*, IEEE IEDM, Washington, 1983, pp. 635–638.

[42] Baltes, H. P., Andor, L., Nathan, A., Schmidt-Weinmar, H. G., Two-Dimensional Numerical Analysis of a Silicon Magnetic Field Sensor, *IEEE Trans. Electron Devices*, ED-31 (1984), 996–999.

[43] Schmidt-Weinmar, H. G., Andor, L., Baltes, H. P., Nathan, A., Numerical Modeling of Silicon Magnetic Field Sensors: Magnetoconcentration Effects in Split-Metal-Contact Devices, *IEEE Trans. Magnetics*, MAG-20 (1984), 975–978.

[44] Andor, L., Baltes, H. P., Nathan, A., Schmidt-Weinmar, H. G., Numerical Modeling of Magnetic-Field-Sensitive Semiconductor Devices, *IEEE Trans. Electron Devices*, ED-32 (1985), 1224–1230.

[45] Nathan, A., Huiser, A. M. J., Baltes, H. P., Two-Dimensional Numerical Modeling of Magnetic Field Sensors in CMOS Technology, *IEEE Trans. Electron Devices*, ED-32 (1985), 1212–1219.

[46] Nathan, A., Andor, L., Baltes, H. P., Schmidt-Weinmar, H. G., Modeling of a Dual Drain NMOS Magnetic Field Sensor, *IEEE J. Solid-State Circuits*, SC-20 (1985), 819–821.

[47] Allegretto, W., Mun, Y. S., Nathan, A., Baltes, H. P., Optimization of Semiconductor Magnetic Field Sensors using Finite Element Analysis, *Proc. NASECODE IV Conf.*, Dublin: Boole Press, 1985, pp. 129–133.

[48] Baltes, H. P., Popovic, R. S., Integrated Semiconductor Magnetic Field Sensors, *Proc. IEEE*, 74 (1986), 1107–1132.

[49] Mun, Y., *Numerical Modeling of CMOS Magnetic Field Sensors by Finite Element Method*, M. Sc. Thesis, University of Alberta, Edmonton, Canada, 1986.

[50] Nathan, A., Allegretto, W., Baltes, H. P., Sugiyama, Y., Modeling of Hall Devices Under Locally Inverted Magnetic Field, *IEEE Electron Device Letts.*, EDL-8 (1987), 1–3.

[51] Nathan, A., Allegretto, W., Baltes, H. P., Sugiyama, Y., Carrier Transport in Semiconductor Detectors of Magnetic Domains, *IEEE Trans. Electron Devices*, ED-34 (1987), 2077–2085.

[52] Allegretto, W., Nathan, A., Baltes, H. P., Two-Dimensional Numerical Analysis of Silicon Bipolar Magnetotransistors, *Proc. NASECODE V Conf.*, Boole Press: Dublin, 1987, pp. 87–92.

[53] Nathan, A., Allegretto, W., Joerg, W., Baltes, H., Numerical Modeling of Bipolar Action in Magnetotransistors, *Digest of Technical Papers*, Transducers '87, Tokyo, 1987, pp. 519–522.

[54] Nathan, A., Allegretto, W., Baltes, H. P., Galvanomagnetic Transport in *p-n* Junctions, *Sensors and Materials*, 1 (1988), 1–6.

[55] Nathan, A., Maenaka, K., Allegretto, W., Baltes, H. P., Nakamura, T., The Hall Effect in Magnetotransistors, *IEEE Trans. Electron Devices*, ED-36 (1989), 108–117.

[56] Allegretto, W., Nathan, A., Baltes, H., Numerical Analysis of Magnetic-Field-Sensitive Bipolar Devices, *IEEE Trans. CAD of ICAS*, 10 (1991), 501–511.

[57] Riccobene, C., Wachutka, G., Baltes, H., Two-Dimensional Numerical Analysis of Novel Magnetotransistors with Partially Removed Substrate, *Technical Digest*, IEEE IEDM, San Francisco, 1992, pp. 513–516.

[58] Korvink, J., *An Implementation of the Adaptive Finite Element Method for Semiconductor Sensor Simulation*, Ph.D. Dissertation, ETH Zürich, No. 10143, 1993.

[59] Nathan, A., Bhatnagar, Y. K., Tang, D. D., Magnetic Field Bit Resolution of Integrated Circuit Polysilicon Hall Elements, *Digest of Technical Papers*, Transducers '93, Yokohama, 1993, pp. 896–899.

[60] Manku, T., Nathan, A., O, N., Aflatooni, K., Allegretto, W., Modeling of Encapsulation Stress Effects on Output Response of Hall Sensors, *Sensors and Materials*, 6 (1994), 225–234.

[61] Aflatooni, K., *Strained Silicon Hall Effect Devices*, M.A.Sc. Thesis, University of Waterloo, Waterloo, Ontario N2L 3G1, Canada, 1994.

[62] Riccobene, C., Wachutka, G., Bürgler, J. F., Baltes, H., Operating Principle of Dual Collector Magnetotransistors Studied by Two-Dimensional Simulation, *IEEE Trans. Electron Devices*, 41 (1994), 32–41.

[63] Riccobene, C., Wachutka, G., Baltes, H., Numerical Study of Structural Variants of Bipolar Magnetotransistors, *Sensors and Materials*, 6 (1994), 159–178.

[64] Riccobene, C., Gärtner, K., Wachutka, G., Baltes, H., Fichtner, W., First Three-Dimensional Numerical Analysis of Magnetic Vector Probe, *Technical Digest*, IEEE IEDM, San Francisco, 1994, pp. 727–730.

[65] Riccobene, C., Gartner, K., Wachutka, G., Baltes, H., Fichtner, W., Full Three-Dimensional Numerical Analysis of Multi-Collector Magnetotransistors with Directional Sensitivity, *Sensors and Actuators A*, 46–47 (1995), 289–293.

[66] Riccobene, C., *Multidimensional Analysis of Galvanomagnetic Effects in Magnetotransistors*, Ph.D. Dissertation, ETH Zürich, Diss. ETH No. 11077, 1995.

[67] Schneider, M., Korvink, J. G., Baltes, H., Magnetostatic Modeling of an Integrated Microconcentrator, *Digest of Technical Papers*, Vol. 2, Transducers '95, Stockholm, 1995, pp. 9–12.

[68] Nathan, A., Manku, T., Modeling the Piezo-Hall Effects in *n*-Doped Silicon Devices, *Appl. Phys. Letts.*, 62 (1993), 2947–2949.

[69] Allegretto, W., Nathan, A., Manku, T., Numerical Simulation of Piezo-Hall Effects in *n*-Doped Silicon Magnetic Sensors, in: *Simulation of Semiconductor Devices and Processes*, Vol. 5, Selberherr, S., Stipel, H., Strasser, E. (Eds.), Vienna: Springer-Verlag, 1993, pp. 377–380.

[70] Crary, S. B., Thermal Management of Integrated Microsensors, *Sensors and Actuators*, 12 (1987), 303–312.

[71] van Duyn, D. C., Munter, P. J. A., Finite-Element Modeling of Thermoelectric Materials and Devices, *Sensors and Actuators A*, 32 (1992), 413–418.

[72] van Duyn, D. C., Modeling and Simulation of Solid-State Transducers: The Thermal and Electrical Energy Domain, *Sensors and Actuators A*, 41–42 (1994), 268–274.

[73] Swart, N., Nathan, A., Design Optimization of Integrated Microhotplates, *Sensors and Actuators A*, 43 (1994), 3–10.

[74] Swart, N. R., Nathan, A., An Integrated CMOS Polysilicon Coil-Based Micro-Pirani Gauge with High Heat Transfer Efficiency, *Technical Digest*, IEEE IEDM, San Francisco, 1994, pp. 135–138.

[75] Allegretto, W., Shen, B., Lai, Z., Robinson, A. M., Numerical Modelling of Time Response of CMOS Micromachined Thermistor Sensor, *Sensors and Materials*, 6 (1994), 71–83.

[76] Dillner, U., Thermal Modeling of Multilayer Membranes for Sensor Applications, *Sensors and Actuators A*, 41–42 (1994), 260–267.

[77] Swart, N. R., Nathan, A., Reliability Study of Polysilicon for Microhotplates, *Technical Digest*, Solid-State Sensor and Actuator Workshop, Hilton Head Is., 1994, pp. 119–122.

[78] Allegretto, W., Shen, B., Haswell, P., Lai, Z., Robinson, A. M., Numerical Modeling of a Micromachined Thermal Conductivity Gas Pressure Sensor, *IEEE Trans. CAD of ICAS*, 13 (1994), 1247–1256.

[79] Nathan, A., Swart, N. R., Quasi Three-Dimensional Simulation of Heat Transport in Thermal-Based Microsensors, in: *Simulation of Semiconductor Devices and Processes*, Vol. 6, Ryssel, H., Pichler, P. (Eds.), Vienna: Springer-Verlag, 1995, pp. 30–33.

[80] Jaeggi, D., Funk, J., Häberli, A., Baltes, H., Overall System Analysis of a CMOS Thermal Converter, *Technical Digest*, Vol. 2, Transducers '95, Stockholm, 1995, pp. 112–115.

[81] Park, S., Kim, H., Kang, Y., Study of Flow Sensor Using Finite Difference Method, *Sensors and Materials*, 7 (1995), 43–51.

[82] Funk, J., *Modeling and Simulation of IMEMS*, Ph.D. Dissertation, ETH Zürich, Diss. ETH No. 11378, 1996.

[83] Mayer, F., Salis, G., Funk, J., Paul, O., Baltes, H., Scaling of Thermal CMOS Gas Flow Microsensors: Experiment and Simulation, *Proc. IEEE MEMS*, San Diego, 1996, pp. 116–121.

[84] Rudin, S., Wachutka, G., Baltes, H., Thermal Effects in Magnetic Microsensor Modeling, *Sensors and Actuators A*, 25–27 (1991), 731–735.

[85] Nathan, A., Manku, T., The Thermomagnetic Carrier Transport Equation, *Sensors and Actuators A*, 36 (1993), 193–197.

[86] Manku, T., Nathan, A., Electron Drift Mobility for Devices Based on Unstrained and Coherently Strained $Si_{1-x}Ge_x$ grown on <001> Silicon Substrate, *IEEE Trans. Electron Devices*, 39 (1992), 2082–2089.

[87] Manku, T., Nathan, A., Valence Energy-Band Structure for Strained Group-IV Semiconductors, *J. Appl. Phys.*, 73 (1993), 1205–1213.

[88] Manku, T., Nathan, A., Electrical Properties of Silicon Under Nonuniform Stress, *J. Appl. Phys.*, 74 (1993), 1832–1837.

[89] Nathan, A., Manku, T., Piezoresistance and the Drift-Diffusion Model in Strained Silicon, *Simulation of Semiconductor Devices and Processes*, Vol. 6, Ryssel, H., Pichler, P. (Eds.), Vienna: Springer-Verlag, 1995, pp. 94–97.

[90] Aflatooni, K., Nathan, A., Heat Transport Properties of Semiconductors Under Non-Uniform Stress, *Appl. Phys. Lett.*, 66 (1995), 1110–1111.

[91] Aflatooni, K., Hornsey, R., Nathan, A., Thermodynamic Treatment of Mechanical Stress Gradients in Coupled Electro-Thermo-Mechanical Systems, *Sensors and Materials*, 9 (1997), 449–456.

[92] Lee, K. W., *Modeling and Simulation of Solid State Pressure Sensors*, Ph.D. Dissertation, University of Michigan, Ann Arbor, USA, 1982.

[93] Suzuki, S., Yamada, K., Nishihara, M., Hachino, H., Minorikawa, S., Structural Analysis of a Semiconductor Pressure Sensor, *Proc.*, The 1st Sensor Symp., Japan, 1981, pp. 131–133.

[94] Lee, K. W., Wise, K. D., Accurate Simulation of High-Performance Silicon Pressure Sensors, *Technical Digest*, IEEE IEDM, 1981, pp. 471–474.

[95] Lee, K. W., Wise, K. D., SENSIM: A Simulation Program for Solid-State Pressure Sensors, *IEEE Trans. Electron Devices*, ED-29 (1982), 34–41.

[96] Suzuki, S., Yagi, Y., Optimum Design of Silicon Pressure Sensor by Nonlinear Finite Element Method, *Proc.*, The 2nd Sensor Symp., Japan, 1982, pp. 163–165.

[97]   Yamada, K., Nishihara, M., Shimada, S., Tanabe, M., Shimazoe, M., Matsuoka, Y.,
       Nonlinearity of the Piezoresistance Effect of *p*-Type Silicon Diffused Layers, *IEEE
       Trans. Electron Devices*, ED-29 (1982), 71–77.
[98]   Bin, T. Y., Huang, R. S., CAPSS: A Thin Diaphragm Capacitive Pressure Sensor
       Simulator, *Sensors and Actuators*, 11 (1987), 1–22.
[99]   Suzuki, K., Ishihara, T., Hirata, M., Tanigawa, H., Nonlinear Analysis of a CMOS
       Integrated Silicon Pressure Sensor, *IEEE Trans. Electron Devices*, ED-34 (1987),
       1360–1367.
[100] Barth, P. W., Pourahmadi, F., Mayer, R., Poydock, J., Peterson, K., A Monolithic
       Silicon Accelerometer with Integral Air Damping and Overrange Protection, *Tech-
       nical Digest*, IEEE Solid-State Sensor and Actuator Workshop, Hilton Head Is., 1988,
       pp. 35–38.
[101] Pourahmadi, F., Barth, P., Peterson, K., Modeling of Thermal and Mechanical
       Stresses in Silicon Microstructures, *Sensors and Actuators*, A21–A23 (1990),
       850–855.
[102] Zhang, Y., Crary, S. B., Wise, K. D., Pressure Sensor Design and Simulation Using the
       CAEMEMS-D Module, *Technical Digest*, IEEE Solid-State Sensor and Actuator
       Workshop, Hilton Head Is., 1990, pp. 32–35.
[103] Chau, K., Allegretto, W., Ristic, L., Simulation of Silicon Microstructures, *Sensors
       and Materials*, 5 (1991), 253–264.
[104] Tschan, T., de Rooij, N., Characterization and Modelling of Silicon Piezoresistive
       Accelerometers Fabricated by a Bipolar-Compatible Process, *Sensors and Actuators
       A*, 25–27 (1991), 605–609.
[105] Tschan, T., de Rooij, N., Bezinge, A., Analytical and FEM Modeling of Piezoresistive
       Silicon Accelerometers: Predictions and Limitations Compared to Experiments,
       Sensors and Materials, 4 (1992), 189–203.
[106] Bergqvist, J., Finite Element Modeling and Characterization of a Silicon Condenser
       Microphone with a Highly Perforated Backplate, *Sensors and Actuators A*, 39 (1993),
       191–200.
[107] Schellin, R., Mohr, R., A Monolithically-Integrated Transistor Microphone:
       Modeling and Theoretical Behaviour, *Sensors and Actuators A*, 37–38 (1993),
       666–673.
[108] Peizhong, H., Jianzhong, G., Finite Element Simulation of Thin-Film Strain Resis-
       tance, *Sensors and Actuators A*, 35 (1993), 239–241.
[109] Morikawa, T., Nonomura, Y., Tsukuda, K., Takeuchi, M., Hosono, A., 3-Dimensional
       Piezoresistive FEM Analysis of a New Combustion Pressure Sensor, *Digest of
       Technical Papers*, Transducers '93, Yokohama, 1993, pp. 598–601.
[110] Yamada, K., Kuriyama, T., FEM Analysis for Single-Chip Multiaxial Servo Accel-
       erometer, *Sensors and Materials*, 6 (1994), 211–223.
[111] Pourahmadi, F., Review of Modeling Silicon Microsensors and Actuators, *Sensors
       and Materials*, 6 (1994), 193–209.
[112] Lades, M., Frank, J., Funk, J., Wachutka, G., Analysis of Piezoresistive Effects in
       Silicon Structures Using Multidimensional Process and Device Simulation, in:
       Simulation of Semiconductor Devices and Processes, Vol. 6, Ryssel, H., Pichler,
       P. (Eds.), Vienna: Springer-Verlag, 1995, pp. 22–25.
[113] Ciampolini, P., Pierantoni, A., Rudan, M., A CAD Environment for the Numerical
       Simulation of Integrated Piezoresistive Transducers, *Sensors and Actuators A*, 46–47
       (1995), 618–622.
[114] Bonse, M. H. W., Mul, C., Spronck, J. W., Finite-Element Modeling as a Tool for
       Designing Capacitive Position Sensors, *Sensors and Actuators A*, 46–47 (1995),
       266–269.

[115] Kadar, Z., Bossche, A., Mollinger, J., Design of a Single-Crystal Silicon-Based Micromechanical Resonator Using Finite Element Simulations, *Sensors and Actuators A*, 46–47 (1995), 623–627.

[116] Benaissa, K., *Integrated Silicon Opto-Mechanical Sensors*, Ph.D. Dissertation, Electrical and Computer Engineering, University of Waterloo, Waterloo, Ontario N2L 3G1, Canada, 1996.

[117] Benaissa, K., Nathan, A., IC Compatible Optomechanical Pressure Sensors Using Mach-Zender Interferometry, *IEEE Trans. Electron Devices*, 43 (1996), 1571–1582.

[118] Funk, J. M., Korvink, J. G., Bühler, J., Bächtold, M., Baltes, H., SOLIDIS: A Tool for Microactuator Simulation in 3-D, *IEEE J. of Microelectromechanical Systems*, 6 (1997), 70–82.

[119] Thangaraj, D., Nathan, A., Two Dimensional Analysis of Incompressible Viscous Flow in Ducts Using a Rotated Difference Scheme, *Sensors and Materials*, 8 (1996), 13–22.

[120] Athavale, M. M., Yang, H. Q., Przekwas, A. J., Coupled Fluid-Thermo-Structures Simulation Methodology for MEMS Applications, *Digest of Technical Papers*, Transducers '97, Chicago, 1997, pp. 1043–1046.

[121] Olsson, A., Stemme, G., Stemme, E., Simulation Studies of Diffuser and Nozzle Elements for Valve-Less Micropumps, *Digest of Technical Papers*, Transducers '97, Chicago, 1997, pp. 1039–1042.

[122] Hsing, I.-M., Srinivasan, R., Harold, M. P., Jensen, K. F., Schmidt, M. A., Finite Element Simulation Strategies for Microfluidic Devices with Chemical Reactions, *Digest of Technical Papers*, Transducers '97, Chicago, 1997, pp. 1015–1018.

[123] Qiu, X. C., Hu, L., Masliyah, J. C., Harrison, D. J., Understanding Fluid Mechanics within Electrokinetically Pumped Microfluidic Chips, *Digest of Technical Papers*, Transducers '97, Chicago, 1997, pp. 923–926.

[124] Cho, Y.-H., Pisano, A. P., Howe, R. T., Viscous Damping Model for Laterally Oscillating Microstructures, *IEEE J. of Microelectromechanical Systems*, 3 (1994), 81–87.

[125] Zhang, X., Tang, W. C., Viscous Air Damping in Laterally Driven Microresonators, *Sensors and Materials*, 27 (1995), 415–430.

[126] Reuther, H. M., Weinmann, M., Fischer, M., von Münch, W., Aßmus, F., Modeling Electrostatically Deflectable Microstructures and Air Damping Effects, *Sensors and Materials*, 8 (1996), 251–269.

[127] Yang, Y.-J., Senturia, S. D., Numerical Simulation of Compressible Squeezed-Film Damping, *Technical Digest*, IEEE Solid-State Sensor and Actuator Workshop, Hilton Head Is., 1996, pp. 76–79.

[128] Price, R. H., Wood, J. E., Jacobsen, S. C., The Modeling of Electrostatic Forces in Small Electrostatic Actuators, *Technical Digest*, IEEE Solid-State Sensors and Actuators Workshop, Hilton-Head Is., 1988, pp. 131–135.

[129] Price, R. H., Wood, J. E., Jacobsen, S. C., Modeling Considerations for Electrostatic Forces in Electrostatic Microactuators, *Sensors and Actuators*, 20 (1989), 107–114.

[130] Johnson, B. P., Kim, S., Senturia, S. D., White, J., MEMCAD Capacitance Calculations for Mechanically Deformed Square Diaphragm and Beam Microstructures, *Digest of Technical Papers*, Transducers '91, San Francisco, 1991, pp. 494–497.

[131] Nabors, K., White, J., FastCap: A Multipole-Accelerated 3-D Capacitance Extraction Program, *IEEE Trans. CAD of ICAS*, 10 (1991), 1447–1459.

[132] Gilbert, J. R., Osterberg, P. M., Harris, R. M., Ouma, D. O., Cai, X., Pfajfer, A., White, J., Senturia, S. D., Implementation of a MEMCAD System for Electrostatic and Mechanical Analysis of Complex Structures from Mask Descriptions, *Proc. IEEE MEMS*, Fort Lauderdale, 1993, pp. 207–212.

[133] Sandmaier, H., Offereins, H. L., Folkmer, B., CAD Tools for Micromechanics, *J. Micromech. Microeng.*, 3 (1993), 103–106.

[134] Cai, X., Osterberg, P., Yie. H., Gilbert, J., Senturia, S., White, J., Self-Consistent Electromechanical Analysis of Complex 3-D Microelectromechanical Structures Using Relaxation/Multipole-Accelerated Method, *Sensors and Materials*, 6 (1994), 85–99.

[135] Ananthasuresh, G. K., Kota, S., Gianchandani, Y., A Methodical Approach to the Design of Compliant Micromechanisms, *Technical Digest*, IEEE Solid-State Sensor and Actuator Workshop, Hilton Head Is., 1994, pp. 189–192.

[136] Boyd, M. R., Crary, S. B., Giles, M. D., A Heuristic Approach to the Electromechanical Modeling of MEMS Beams, *Technical Digest*, IEEE Solid-State Sensor and Actuator Workshop, Hilton Head Is., 1994, pp. 123–126.

[137] Osterberg, P., Yie, H., Cai, X., White, J., Senturia, S., Self-Consistent Simulation and Modeling of Electrostatically Deformed Diaphragms, *Proc. IEEE MEMS*, Oiso, 1994, pp. 28–32.

[138] Gilbert, J. R., Legtenberg, R., Senturia, S. D., 3D Coupled Electro-Mechanics for MEMS: Applications of CoSolve-EM, *Proc. IEEE MEMS*, Amsterdam, 1995, pp. 122–127.

[139] Stewart, J. T., Finite Element Modeling of Microelectromechanical Structures for Sensing Applications, *Proc. SPIE*, 2642 (1995), pp. 194–205.

[140] Yie, H., Bart, S. F., White, J., Senturia, S. D., A Computationally Practical Approach to Simulating Complex Surface-Micromachined Structures with Fabrication Non-Idealites, *Proc. IEEE MEMS*, Amsterdam, 1995, pp. 128–133.

[141] Lefevre, Y., Lajoie-Mazenc, M., Sarraute, E., Camon, H., First Steps Towards Design, Simulation, Modeling and Fabrication of Electrostatic Micromotors, *Sensors and Actuators A*, 46–47 (1995), 645–648.

[142] Lee, J. S., Yoshimura, S., Yagawa, G., Shibaike, N., A CAE System for Micromachines: Its Application to Electrostatic Micro Wobble Actuator, *Sensors and Actuators A*, 50 (1995), 209–221.

[143] Bächtold, M., Korvink, J. G., Funk, J., Baltes, H., New Convergence Scheme for Self-Consistent Electromechanical Analysis of IMEMS, *Technical Digest*, IEEE IEDM, Washington, 1995, pp. 605–608.

[144] Gilbert, J. R., Ananthasuresh, G. K., Senturia, S. D., 3D Modeling of Contact Problems and Hysteresis in Coupled Electro-Mechanics, *Proc. IEEE MEMS*, San Diego, 1996, pp. 127–132.

[145] Funk, J., Korvink, J. G., Bachtold, M., Buhler, J., Baltes, H., Coupled 3D Thermo-Electro-Mechanical Simulations of Microactuators, *Proc. IEEE MEMS*, San Diego, 1996, pp. 133–138.

[146] Wang, P. K. C., Hadaegh, F. Y., Computation of Static Shapes and Voltages for Micromachined Deformable Mirrors with Nonlinear Electrostatic Actuators, *IEEE J. of Microelectromechanical Systems*, 5 (1996), 205–220.

[147] Bächtold, M., *Efficient 3D Computation of Electrostatic Fields and Forces in Microsystems*, Ph.D. Dissertation, ETH Zürich, Diss. ETH No. 12165, 1997.

[148] Funk, J., Korvink, J. G., Wachutka, G., Baltes, H., Electro-Thermo-Mechanical Field Analysis Using SESES, in: *Simulation of Semiconductor Devices and Processes*, Vol. 5, Selberherr, S., Stipel, H., Strasser, E. (Eds.), Vienna: Springer-Verlag, 1993, pp. 347–350.

[149] Korvink, J. G., Funk, J., Roos, M., Wachutka, G., Baltes, H., SESES: A Comprehensive MEMS Modelling System, *Proc. IEEE MEMS*, Oiso, 1994, pp. 22–27.

[150] Korvink, J. G., Funk, J., Baltes, H., IMEMS Modeling, *Sensors and Materials*, 6 (1994), 235–243.

[151] Ikuta, K., Shimizu, H., Two Dimensional Mathematical Model of Shape Memory Alloy and Intelligent SMA-CAD, *Proc. IEEE MEMS*, Fort Lauderdale, 1993, pp. 87–91.

[152] Krulevitch, P., Lee, A. P., Ramsey, P. B., Trevino, J. C., Hamilton, J., Northrup, M. A., Thin Film Shape Memory Alloy Microactuators, *IEEE J. of Microelectromechanical Systems*, 5 (1996), 270–282.

[153] Quandt, E., Seeman, K., Fabrication and Simulation of Magnetostrictive Thin Film Actuators, *Sensors and Actuators A*, 50 (1995), 105–109.

[154] Schwarzenbach, H. U., Lechner, H., Steinle, B., Baltes, H. P., Schwendimann, P., Calculation of Vibrations of Thick Piezoceramic Disk Resonators, *Appl. Phys. Lett.*, 38 (1981), 854–855.

[155] Langer, E., Selberherr, S., Markowich, P. A., Ringhofer, C. A., Numerical Analysis of Acoustic Wave Generation in Anisotropic Piezoelectric Materials, *Sensors and Actuators A*, 4 (1983), 71–76.

[156] Lerch, R., Finite Element Analysis of Piezoelectric Transducers, *Proc. IEEE Ultrasonics Symp.*, 1988, pp. 643–654.

[157] Lerch, R., Piezoelectric and Acoustic Finite Elements as Tools for the Development of Electroacoustic Transducers, *Siemens Forsch.-u. Entwickl.-Ber.*, Vol. 17, No. 6 (1988), pp. 283–290.

[158] Brand, O., *Micromachined Resonators for Ultrasound Based Proximity Sensing*, Ph.D. Dissertation, ETH Zürich, Diss. ETH No. 10896, 1994.

[159] Low, T. S., Guo, W., Modeling of a Three Layer Piezoelectric Bimorph beam with Hysteris, *IEEE J. of Microelectromechanical Systems*, 4 (1995), 230–237.

[160] Lim, Y.-H., Varandan, V. V., Varandan, V. K., Finite Element Modeling of the Dynamic Response of a MEMS Sensor, *Proc. SPIE*, 2642 (1995), pp. 233–240.

[161] Koppelman, G. M., OYSTER, a Three-Dimensional Structural Simulator for Micro-Electro-Mechanical Design, *Sensors and Actuators*, 20 (1989), 179–185.

[162] Maseeh, F., Harris, R. M., Senturia, S. D., A CAD Architecture for Micro-Electro-Mechanical Systems, *Proc. IEEE MEMS*, Napa Valley, 1990, pp. 44–49.

[163] Amster, R., Tavrow, L. S., Flynn, A. M., Intelligent CAD for Micromechanics, *Proc. Microsystems Conf.*, Berlin, 1990, Berlin: Springer-Verlag, 1990, pp. 23–28.

[164] Crary, S., Kota, S., Conceptual Design of Micro-Electro-Mechanical Systems, *Proc. Microsystems Conf.*, Berlin, 1990, Berlin: Springer-Verlag, 1990, pp. 17–22.

[165] Crary, S., Zhang, Y., CAEMEMS: An Integrated Computer-Aided Engineering Workbench for Micro-Electro-Mechanical Systems, *Proc. IEEE MEMS*, Napa Valley, 1990, pp. 113–114.

[166] Harris, R. M., Maseeh, F., Senturia, S. D., Automatic Generation of a 3-D Solid Model of a Microfabricated Structure, *Technical Digest*, IEEE Solid-State Sensor and Actuator Workshop, Hilton Head Is., 1990, pp. 36–41.

[167] Crary, S., Juma, O., Zhang, Y., Software Tools for Designers of Sensor and Actuator CAE Systems, *Digest of Technical Papers*, Transducers '91, San Francisco, 1991, pp. 498–501.

[168] Harris, R. M., Senturia, S. D., A Solution of the Mask Overlay Problem in Microelectromechanical CAD (MEMCAD), *Proc. IEEE MEMS*, Travemünde, 1992, pp. 58–62.

[169] Gilbert, J. R., Osterberg, P. M., Harris, R. M., Ouma, D. O., Cai, X., Pfajfer, A., White, J., Senturia, S. D., Implementation of a MEMCAD System for Electrostatic and Mechanical Analysis of Complex Structures from Mask Descriptions, *Proc. IEEE MEMS*, Fort Lauderdale, 1993, pp. 207–212.

[170] Gogoi, B., Yuen, R., Mastrangelo, C. H., The Automatic Synthesis of Planar Fabrication Process Flows for Surface Micromachined Devices, *Proc. IEEE MEMS*, Oiso, 1994, pp. 153–157.

[171] Poppe, A., Rencz, M., Szekely, V., CAD Framework Concept for the Design of Integrated Microsystems, *Proc. SPIE*, 2642 (1995), pp. 215–224.

[172] Lo, N. R., Pister, K. S. J., 3DμV – a MEMS 3-D Visualization Package, *Proc. SPIE*, 2642 (1995), pp. 290–295.

[173] Osterberg, P. M., Senturia, S. D., MEMBUILDER: An Automatic 3D Solid Model Construction Program for Microelectromechanical Structures, *Digest of Technical Papers*, Vol. 2, Transducers '95, Stockholm, 1995, pp. 21–24.

[174] Hasanuzzaman, M., Mastrangelo, C. H., MISTIC 1.1: A Process Compiler for Micromachined Devices, *Digest of Technical Papers*, Vol. 1, Transducers '95, Stockholm, 1995, pp. 182–185.

[175] He, Y., Harris, R., Napadenski, G., Maseeh, F., A Virtual Prototype Manufacturing Software System for MEMS, *Proc. IEEE MEMS*, San Diego, 1996, pp. 122–126.

[176] Nagler, O., Trost, M., Hillerich, B., Kozlowski, F., Efficient Design and Optimization of MEMS by Integrating Commercial Simulation Tools, *Digest of Technical Papers*, Transducers '97, Chicago, 1997, pp. 1055–1058.

[177] Buser, R. A., de Rooij, N. F., CAD for Silicon Anisotropic Etching, *Proc. IEEE MEMS*, Napa Valley, 1990, pp. 111–112.

[178] Sequin, C. H., Computer Simulation of Anisotropic Etching, *Digest of Technical Papers*, Transducers '91, San Francisco, 1991, pp. 801–806.

[179] DeLapierre, G., Anisotropic Crystal Etching: A Simulation Program, *Sensors and Actuators*, 31 (1992), 264–274.

[180] Hubbard, T. J., Antonsson, E. K., Emergent Faces in Crystal Etching, *IEEE J. of Microelectromechanical Systems*, 3 (1994), 19–28.

[181] Tabata, O., Effects of Etching Products and Diffusion on Silicon Anisotropic Etching, *Sensors and Materials*, Special Issue on CAD for MEMS, 10 (1998) (to appear).

[182] van Suchtelen, J., van Veenendaal, E., Nijdam, A. J., Elwenspoek, M., van Enckevort, W. J. P., Computer Simulation of Orientation-Dependent Etching of Silicon, presented at the CAD for MEMS Workshop, Zürich, 1997.

[183] Koide, A., Tanaka, S., Simulation of Three-Dimensional Etch Profile of Silicon During Orientation-Dependent Anisotropic Etching, *Proc. IEEE MEMS*, Nagoya, 1997, pp. 418–423.

[184] Senturia, S. D., Smith, R. L., Microsensor Packaging and System Partitioning, *Sensors and Actuators*, 15 (1988), 221–234.

[185] Fotheringham, G., Simulation Methods for Multi-Chip Modules, *Sensors and Actuators A*, 30 (1992), 157–165.

[186] Pourahmadi, F., Peterson, K., Package Design of Silicon Micromachined Sensors Using Finite Element Modeling, *Digest of Technical Papers*, Transducers '93, Yokohama, 1993, pp. 774–778.

[187] Lin, Y., Hesketh, P. J., Schuster, J. P., Finite-Element Analysis of Thermal Stresses in a Silicon Pressure Sensor for Various Die-Mount Materials, *Sensors and Actuators A*, 44 (1994), 145–149.

[188] Chin, S.-W., Rajan, S. D., Nagaraj, B. K., Mahalingam, M., Automated Design Tool for Examining Microelectronic Packaging Design Alternatives, *IEEE Trans. on Component, Packaging, and Manufacturing Technology*, 17 (1994), 76–82.

[189] Michel, B., Schubert, A., Dudek, R., Grosser, V., Experimental and Numerical Investigations of Thermo-Mechanically Stresses Micro-Components, *Microsystem Technology*, 1 (1994), 14–22.

[190] Koen, E., Pourahmadi, F., Terry, S., A Multilayer Ceramic Package for Silicon Micromachined Accelerometers, *Digest of Technical Papers*, Vol. 1, Transducers '95, Stockholm, 1995, pp. 273–276.

[191] Popovic, R. S., Numerical Analysis of MOS Magnetic Field Sensors, *Solid-State Electronics*, 28 (1985), 711–716.

[192] Caverly, R., Peck, E., A Finite-Element Model and Characterization of the *p-i-n* Magnetodiode at Microwave Frequencies, *Solid-State Electronics*, 30 (1987), 473–477.

[193] Swart, N. R., Nathan, A., Flow-Rate Microsensor Modelling and Optimization Using SPICE, *Sensors and Actuators A*, 34 (1992), 109–122.

[194] Swart, N., Nathan, A., Mixed-Mode Device-Circuit Simulation of Thermal-Based Microsensors, *Sensors and Materials*, 6 (1994), 179–192.

[195] Swart, N., Nathan, A., Coupled Electrothermal Modeling of Microheaters Using SPICE, *IEEE Trans. Electron Devices*, 41 (1994) 920–925.

[196] Mouthaan, T. J., Krabbenborg, B. H., Thermodynamic Analysis of Semiconductor Structures Using a Device Simulator and Lumped Circuit Modelling, *Sensors and Materials*, 6 (1994), 125–137.

[197] Auerbach, F. J., Meiendres, G., Müller, R., Scheller, G. J. E., Simulation of the Thermal Behaviour of Thermal Flow Sensors by Equivalent Electrical Circuits, *Sensors and Actuators A*, 41–42 (1994), 275–278.

[198] Massobrio, G. Martinoia, S., Grattarola, M., Use of SPICE for Modeling Silicon-Based Chemical Sensors, *Sensors and Materials*, 6 (1994), 101–123.

[199] Rombach, P., Langheinrich, W., Modelling of a Micromachined Torque Sensor, *Sensors and Actuators A*, 46–47 (1995), 294–297.

[200] Salim, A., Manku, T., Nathan, A., Modeling of Magnetic Field Sensitivity of Bipolar Magnetotransistors Using HSPICE, *IEEE Trans. on CAD of ICAS*, 14 (1995), 464–469.

[201] Veijola, T., Kuisma, H., Lahdenperä, J., Ryhänen, T., Equivalent-circuit Model of the Squeezed Gas Film in a Silicon Accelerometer, *Sensors and Actuators A*, 48 (1995), 239–248.

[202] Burstein, A., Kaiser, W. J., The Microelectromechanical Gyroscope – Analysis and Simulation Using SPICE Electronic Simulator, *Proc. SPIE*, 2642 (1995), pp. 225–232.

[203] Nathan, A., *Self-Consistent Network Synthesis for Mixed-Signal Simulations*, Int. Rep., No. 95/06, Physical Electronics Laboratory, ETH Zürich, Switzerland, 1995.

[204] Shie, J.-S., Chen, Y.-M., Ou-Yang, M., Chou, B. C. S., Characterization and Modeling of Metal-Film Microbolometer, *IEEE J. of Microelectromechanical Systems*, 5 (1996), 298–306.

[205] Mohajerzadeh, S., Nathan, A., Modeling Noise Correlation Behaviour in Dual-Collector Magnetotransistors Using Small Signal Equivalent Circuit Analysis, *IEEE Trans. Electron Devices*, 43 (1996), 883–888.

[206] Tilmans, H. A. C., Equivalent Circuit Representation of Electromechanical Transducers: I. Lumped-Parameter Systems, *J. Micromech. Microeng.*, 6 (1996), 157–176.

[207] Ando, S., Tanaka, K., Abe, M., Fishbone Architecture: An Equivalent Mechanical Model of Cochlea and its Application to Sensors and Actuators, *Digest of Technical Papers*, Transducers '97, Chicago, 1997, pp. 1027–1030.

[208] Voigt, P., Wachutka, G., Electro-Fluidic Microsystem Modeling Based on Kirchhoffian Network Theory, *Digest of Technical Papers*, Transducers '97, Chicago, 1997, pp. 1019–1022.

[209] Romanowicz, B., Lerch, Ph., Renaud, Ph., Fullin, E., de Coulon, Y., Simulation of Integrated Electromagnetic Device Systems, *Digest of Technical Papers*, Transducers '97, Chicago, 1997, pp. 1051–1054.

[210] Pham, H. H., Nathan, A., Circuit Modeling and SPICE Simulation of Mixed-Signal Microsystems, *Sensors and Materials*, Special Issue on CAD for MEMS, 10, No. 7 (1998) (to appear).

[211] *User's Guide*, Iowa State University Research Foundation (ISURF), Ames, IA 50011, Copyright 1985.

[212] *PISCES*, Integrated Circuits Laboratory (ICL), Department of Electrical Engineering, Stanford University, CA, USA. http://www-tcad.stanford.edu/tcad/org.html

[213] Korvink, J. G., *SOLIDIS Reference Manual 1.0*, Internal Report No. 95/01, Physical Electronics Laboratory, ETH Zürich, 1995. ISE Integrated Systems Engineering AG, Technopark Zürich, Technoparkstrasse 1, CH-8005 Zürich, Switzerland.

[214] Kriegl, W., Steiner, P., Folkmer, B., Lang, W., MICROTHERM: A Program for Thermal Modelling of Microstructures, *Sensors and Actuators A*, 46–47 (1995), 637–639.

[215] ANSYS Inc., 275 Technology Drive, Canonsburg, PA 15317, USA.

[216] *MSC/NASTRAN*, McNeal-Schwendler Corp., Los Angeles, CA, USA.

[217] *ADINA*, Adina R&D, Inc., 71 Elton Ave., Watertown, MA 02172, USA.

[218] Puers, B., Peeters, E., Sansen, W., CAD Tools in Mechanical Sensor Design, *Sensors and Actuators*, 17 (1989), 423–429.

[219] Schwarzenbach, H. U., Korvink, J. G., Roos, M., Sartoris, G., Anderheggen, E., A Micro Electro Mechanical CAD Extension for SESES, *J. Micromech. Microeng.*, 3 (1993), 118–122.

[220] Anderheggen, E., Korvink, J. G., Roos, M., Sartoris, G. E., Schwarzenbach, H. U., *SESES User Manual*, NM Numerical Modelling GmbH, Thalwil, Switzerland, 1993.

[221] *ABAQUS*, Hibbit, Karlsson, and Sorenson, Inc., 1080 Main Street, Pawtucket, RI 02860, USA.

[222] *COSMOS/M*, Structural Research Analysis Corp., Santa Monica, CA, USA.

[223] *I-DEAS*, Structural Dynamics Research Corp, Milford, OH., USA.

[224] *FLOWERS*, Inst. für Informatik, ETH, CH-8093 Zürich, Switzerland.

[225] *TPS10 Benutzerhandbuch*, T-Programm GmbH, Reutlingen, 11th Ed., 1989.

[226] *MARC*, MARC Analysis Research Corp., (see [142]).

[227] *FIDAP*, Fluid Dynamics International, Evanston, Illinois, USA.

[228] *FLUENT*, FLUENT Inc., Centerra Resource Park, 10 Cavendish Court, Lebanon, N.H. 03766-1442, USA.

[229] *FLOTRAN*, see, Ulrich, J., Zengerle, R., Static and Dynamic Flow Simulation of a KOH-Etched Microvalve Using the Finite Element Method, *Sensors and Actuators A*, 53 (1996), 379–385.

[230] *FLOTHERM*, see, Fotheringham, G., Simulation Methods for Multi-Chip Modules, *Sensors and Actuators A*, 30 (1992), 157–165.

[231] *PUSI*, see, Zengerle, R., Richter, M., Brosinger, F., Richter, A., Sandmaier, H., Performance Simulation of Microminiaturized Membrane Pumps, *Digest of Technical Papers*, Transducers '93, Yokohama, 1993, pp. 106–109.

[232] *Maxwell Solver*, Ansoft Corp., 4 Station Square, 660 Commerce Court Bldg., Pittsburgh, PA, USA.

[233] *ALECSIS*, Inst. of Prec. Eng., TU Vienna, Floragasse 7, 1040 Vienna, Austria.

[234] *MICROCOSM*, 201 Willesden Dr., Cary, NC 27513, USA.

[235] IntelliSense Corp., 16 Upton Dr., Wilmington, MA 01887, USA.

[236] Nabors, K., Kim, S., White, J., Senturia, S., *FastCap User's Guide*, Research Laboratory of Electronics, Department of Electrical Engineering and Computer Science, MIT, Cambridge, MA 02139, USA.

[237] *EFCREL, EFDYN, EFCAD*, see, Lefèvre, Y., Lajoie-Mazenc, M., Sarraute, E., Lamon, H., First Stop Towards Design, Simulation, Modeling and Fabrication of Electrostatic Micromotors, *Sensors and Actuators A*, 46–47 (1995), 645–648.

[238] CEDRAT S.A., 10 Chemin du Pré Carré, 38240 Meylan, France.

[239] *IES*, Integrated Engineering Software, 46-1313 Border Place, Winnipeg, Manitoba, R3H 0X4, Canada.

[240] *PATRAN*, PDA Engineering, Costa Mesta, CA, USA.

[241] *Geomview*, Software Development Group, Geometry Center, University of Minnesota, 1300 South Second Street, Suite 500, Minneapolis, MN 55454, USA. http://www.geom.umn.edu/welcome.html.

[242] *Pro/ENGINEER*, Parametric Technology, Waltham, MA, USA.

[243] Asaumi, K., Iriye, Y., Sato, K., Anisotropic-Etching Process Simulation System MICROCAD Analyzing Complete 3D Etching Profiles of Single Crystal Silicon, *Proc. IEEE MEMS*, Nagoya, 1997, pp. 412–417. MICROCAD, 3-D Etching Simulator, Fuji Research Institute Corp., URL http://www.fuji-ric.co.jp/crab/

[244] *SUPREM*, Integrated Circuits Laboratory (ICL), Department of Electrical Engineering, Stanford University, CA, USA. http://www-tcad.stanford.edu/tcad/org.html

[245] *SPICE*, Industrial Liaison Program, Research Software Catalog, EECS Department, University of California, Berkeley, USA. http://hera.eecs.berkeley.edu/~software/

[246] Pham, H. H., Nathan, A., A New Approach for Rapid Evaluation of the Potential Field in Three Dimensions, *Procdings of Royal Society London A*, 455 (1999), 1–39.

[247] Pham, H. H., Numerical Capacitance Extraction for Large Area Systems, Ph.D. Dissertation, University of Waterloo, Waterloo, Ontario NZL 3G1, Canada, 1998.

[248] Pham, H. H., Nathan, A., WATCAP: A New Simulation Engine for Interconnect Capacitance Extraction, 1st Canadian Workshop on RF IC Research and Development, Nov. 16, Ottawa, Canada, 1998.

2 Basic Electronic Transport

In contrast to very large scale integrated (VLSI) devices, microtransducers have relatively large dimensions and are not in the race to push the limits of feature size into the submicron regime. Thus with microtransducers, it is reasonable to assume a static picture for electrical transport in the device, whereby the mobile charge carriers are in equilibrium with the host lattice. This permits the use of the classical model comprising Poisson's equation, which relates the electrostatic potential and space charge in the device, and the electron and hole continuity equations, which account for charge conservation, with current density relations based on the drift-diffusion formulation. Effects of non-static transport have become very important in VLSI devices where the active device dimensions are reaching scales (nm) where the carrier transit time becomes comparable to the collision time.

In this chapter, we review the modeling equations necessary for simulation of electrical transport and relevant physical effects in crystalline and non-crystalline bulk and thin film materials pertinent to microtransducers. For clarity, only isothermal effects will be considered here. Non-isothermal effects and associated devices will be discussed in Chapter 5. The transport equation and physical models described here do not take into account effects of an external field such as radiant, magnetic field, temperature gradient, and strain. Effects of an external field will be discussed systematically in Chapts. 3 to 6 that follow.

The equations governing electrical behavior in semiconductor devices can be derived from Maxwell's equations, which in their general form are always valid. However, on their own they are incomplete for device analysis and require further relations drawn from the solid state theory of materials. Without explicit recourse to Maxwell's equations, we shall describe the formulation of the basic equations along with their limits of validity. A detailed derivation of the basic equations can be found in Refs. [1, 2].

2.1 Poisson's Equation

In microtransducer materials, the electric displacement, at sufficiently low electric fields and for a broad range of operating frequencies, can be expressed as [3, 4]

$$\mathbf{D} = \varepsilon_0\mathbf{E} + \mathbf{P} = \varepsilon_0\mathbf{E} + \varepsilon_0\chi_e\mathbf{E} = \varepsilon\mathbf{E}, \tag{2.1}$$

where ε_0 is the free-space permittivity, $\mathbf{E}$ is the applied electric field, $\mathbf{P}$ denotes the induced electric polarization, χ_e is the dielectric susceptibility, and $\varepsilon = \varepsilon_0(1 + \chi_e)$ is the material dielectric permittivity; the dielectric constant or relative permittivity being, $\varepsilon_r = (1 + \chi_e)$. To obtain a relation between the electric field and the electrostatic potential ψ, we introduce a vector potential, $\mathbf{A}$, *viz.*, $\mathbf{B} = \text{curl } \mathbf{A}$. This yields for the electric field

$$\mathbf{E} = -\partial\mathbf{A}/\partial t - \text{grad } \psi. \tag{2.2}$$

Thus the Maxwell's third equation

$$\text{div } \mathbf{D} = \rho, \tag{2.3}$$

where ρ denotes the space charge density (coul/cm^3) in the device, becomes

$$\text{div} \left(\varepsilon\partial\mathbf{A}/\partial t\right) + \text{div} \left(\varepsilon \text{ grad } \psi\right) = -\rho. \tag{2.4}$$

For a broad range of frequencies, the corresponding wavelengths are much larger than the dimensions of the transducer. Thus it is appropriate to assume a quasi-static condition [2] for the electric field which reduces Eq. (2.4) to

$$\text{div} \left(\varepsilon \text{ grad } \psi\right) = -\rho. \tag{2.5}$$

In most, if not all, cases the material structures considered are cubic or non-crystalline (isotropic) and thus the permittivity is a scalar quantity. Realistic simulations involve different materials and the permittivity is position-dependent (inhomogeneous) as accounted for by Eq. (2.5); in homogeneous cases, (2.5) reduces to

$$\text{div grad } \psi = -\rho/\varepsilon. \tag{2.6}$$

However, in anisotropic materials, such as in piezoelectric, pyroelectric, or strained materials, the orientation of the polarization vector can be different from that of the electric field (see Section 8.6). The dielectric susceptibility, permittivity, and dielectric constant become tensors of rank two [4]. This is illustrated in Table 2.1 for homogeneous materials. Typical values of the permittivity in transducer materials are shown in Table 2.2 in which the values for anisotropic materials shown denote principal components.

In piezoelectric and pyroelectric materials, there can be a polarization already present in the material induced by means of stress or temperature.

Table 2.1. *Compatible relations for isotropic and anisotropic materials*

Isotropic	Anisotropic	Property
$\mathbf{P} = \varepsilon_0 \chi_e \mathbf{E}$	$P_i = \varepsilon_0 \chi_{eij} E_j$	$\chi_{eij} = \chi_{eji}$
$\mathbf{D} = \varepsilon \mathbf{E}$	$D_i = \varepsilon_{ij} E_j$	$\varepsilon_{ij} = \varepsilon_{ji}$
$\varepsilon = \varepsilon_0 (1 + \chi_e)$	$\varepsilon_{ij} = \varepsilon_0 (\delta_{ij} + \chi_{eij})$	
ε_r	$\varepsilon_{rij} = \varepsilon_{ij} / \varepsilon_0$	$\varepsilon_{rij} = \varepsilon_{rji}$

Table 2.2. *Dielectric constants for commonly used materials in microtransducers; see Section 8.6 for other anisotropic materials. The deposition processes indicated for silicon nitride are: low pressure chemical vapor deposition (LPCVD) and plasma enhanced chemical vapor deposition (PECVD)*

Material	Value
Si	11.7
SiO$_2$	3.9
for most IC processes	
α-SiO$_2$ (quartz)	Principal components:
(see Section 8.6)	$\varepsilon_{r11} = \varepsilon_{r22} = 4.5$, $\varepsilon_{r33} = 4.6$
Si$_3$N$_4$ (LPCVD)	7
a-SiN$_x$:H (PECVD)	4 (Si rich) to 8 (N rich)
GaAs	12.5
Ge	16.1
3C-SiC	9.7
Diamond	5.5
Dry air	1

In this case, the electric displacement in (2.1) is modified to read

$$\mathbf{D} = \varepsilon_0 \mathbf{E} + \mathbf{P} + \mathbf{P}_i = \varepsilon \mathbf{E} + \mathbf{P}_i \qquad (2.7)$$

where $\mathbf{P}_i$ denotes the stress- or thermally-induced polarization in the material. This modifies the space charge term, ρ in (2.5) to include an additional charge term; div $\mathbf{P}_i$.

The space charge density, ρ in (2.6) and can generally be modeled as

$$\rho = q(p - n + N_D - N_A), \qquad (2.8)$$

where q is the elementary charge, p is the hole concentration (cm^{-3}), n is the electron concentration (cm^{-3}), N_D and N_A are fixed ionized dopant densities (cm^{-3}). The carrier densities, n and p, need to be modeled following suitable assumptions of the band structure of the material. The concentrations are highly nonlinear functions of the electrostatic potential, and they need to be solved using the electron and hole continuity equations for self-consistency. The dopant distributions are process-dependent and can be predicted using process simulators, e.g., SUPREM [5].

2.2 Continuity Equations

The continuity equations for electrons and holes stem directly from Maxwell's equations:

$$\text{div } \mathbf{J_n} = qR + q\frac{\partial n}{\partial t}, \tag{2.9}$$

$$\text{div } \mathbf{J_p} = -qR - q\frac{\partial p}{\partial t}. \tag{2.10}$$

Here, $\mathbf{J_n}$ (A/cm^2) is the electron current density due to electron (n) motion in the conduction band and $\mathbf{J_p}$ (A/cm^2) is the hole current density due to hole (p) motion in the valence band. Equations (2.9) and (2.10) take into account only the time-dependence of mobile carriers; the time-dependence of possible charged defects in the material has been assumed negligible in comparison. This may not be true in metastable amorphous materials, e.g., hydrogenated amorphous silicon (a-Si : H), where time-dependent defect creation can lead to shifts in device characteristics [6]. In this case, additional continuity equations are necessary to model carrier transport. In (2.9) and (2.10), R denotes the net recombination rate per unit volume (cm^{-3}s^{-1}) and accounts for all possible generation-recombination mechanisms present in the device; $R = 0$ implies thermal equilibrium, $R > 0$ denotes recombination, and $R < 0$ signifies generation, G. The various terms, *viz.*, $\mathbf{J_{n,p}}$, n, p, and R, in (2.9) and (2.10) have to be carefully modeled to maintain consistency with material- and device-physics considerations. Models for these terms are discussed in the sections that follow.

2.3 Carrier Transport in Crystalline Materials and Isothermal Behavior

The derivation process leading to a simple description of the transport relations is complex (see [1, 2]) and beyond the scope of this book. Thus only key steps, the physical relevance, and associated limits of validity will be outlined, along with pertinent relations which we will state without proof. However, what remains crucial in the context of microtransducer modeling is identification of the relevant terms in the transport relations that are modified in the presence of an external field such as radiant, magnetic field, temperature gradients, and mechanical strain.

2.3.1 Transport Relations

Given the Boltzmann Transport Equation (BTE) and the framework of assumptions underlying its validity [1, 2], a wealth of further assumptions are required to arrive at a relatively simple description of the current density relations. Key assumptions inherent in subsequent formulations shown below stem from considerations of microtransducer geometry and operating conditions. The relatively large active area of microtransducers yield charge propagation distances larger than the carrier mean free path length. Also carriers are in equilibrium with the lattice so that the carrier and lattice temperatures are the same. The following assumptions have a direct bearing on microtransducer-related simulations and cannot always be taken for granted:

- vanishing temperature gradients (grad $T = 0$); but grad $T \neq 0$ in the thermal sensor family;
- no Lorentz force ($B = 0$) effects; but $B \neq 0$ in magnetic sensors;
- parabolic band structure; this is not true when the material is strained (e.g., in pressure sensors and device packaging).

Given the various assumptions, the electron and hole current densities in the continuity equations (2.9) and (2.10), for not too large driving forces and in the absence of external fields, read [1, 2]:

$$\mathbf{J_n} = -q\mu_n n \operatorname{grad} \phi_n, \tag{2.11}$$

$$\mathbf{J_p} = -q\mu_p p \operatorname{grad} \phi_p, \tag{2.12}$$

where $\mu_{n,p}$ denote drift mobilities for electrons and holes, respectively, and $\phi_{n,p}$ the respective quasi-Fermi potentials. They are related to the associated Fermi energies as $E_{Fn,p} = E_i - q\phi_{n,p}$, where E_i denotes the intrinsic Fermi level. If the Fermi energies are sufficiently far from band edges, Boltzmann statistics can be employed to yield simple expressions for $\phi_{n,p}$ in terms of the carrier concentrations and the electrostatic potential ψ,

$$\phi_n = \psi - \frac{kT}{q} \log_e \left(\frac{n}{n_i} \right), \tag{2.13}$$

$$\phi_p = \psi + \frac{kT}{q} \log_e \left(\frac{p}{n_i} \right). \tag{2.14}$$

Here, k denotes Boltzmann's constant and T the temperature. The quasi-Fermi levels describe a system or a region in the device that is not in thermal equilibrium whereby a unique Fermi level can no longer be defined, $E_{Fn} \neq E_{Fp} \neq E_F$, and the Fermi levels differ by the applied voltage. In thermal equilibrium (i.e., no applied voltage and no current flow), the system has a unique Fermi level, $E_{Fn} = E_{Fp} = E_F$. In calculations of carrier concentrations, as seen by (2.13) and (2.14), only the difference in Fermi and electrostatic potentials is important. Thus absolute values are not relevant and hence, we can arbitrarily define the reference level. We define the quasi-Fermi potentials to be zero in the absence of an applied voltage. In (2.13) and (2.14), n_i represents the intrinsic concentration.

When the relations (2.13) and (2.14) are substituted into (2.11) and (2.12), we have a gradient in intrinsic concentration as a driving force. Practically, this can arise from spatial variations in band gap induced by heavy doping effects, temperature, or strain. This term can also be absorbed in the drift component by appropriately defining an effective electric field for the driving force [1, 2]. In the case of a uniform band gap, which typifies a system with no material inhomogeneities, no heavy doping effects, and no external fields, we have grad $n_i = 0$ and the current densities become

$$\mathbf{J_n} = -q\mu_n n \operatorname{grad} \psi + qD_n \operatorname{grad} n, \tag{2.15}$$

$$\mathbf{J_p} = -q\mu_p p \operatorname{grad} \psi - qD_p \operatorname{grad} p. \tag{2.16}$$

Here, $\mu_{n,p}$ denote the electron and hole mobilities, respectively, and $D_{n,p}$ denote the corresponding diffusion coefficients which can be related to the mobilities using Einstein's relations, viz., $D_{n,p} = \mu_{n,p}kT/q$. Equations (2.15) and (2.16) represent the classical drift-diffusion relations and are based on Boltzmann statistics for the carrier concentrations.

2.3.2 Carrier Concentrations

The free carrier concentrations are obtained by multiplying the available density-of-states functions with the Fermi distribution function and integrating over the energy space (see, e.g., [7]):

$$n = \int_{E_c}^{\infty} g_c(E) f_n(E) dE \tag{2.17}$$

$$p = \int_{-\infty}^{E_v} g_v(E) f_p(E) dE \tag{2.18}$$

Here, E_c and E_v denote conduction and valence band edges; $E_g = E_c - E_v$. In the integrand, $g_{c,v}(E)$ denote the energy-dependent density of available states for the conduction and valence bands, respectively, and $f_{n,p}$ are the Fermi distribution functions, viz.,

$$f_n(E) = \frac{1}{1 + \exp\left(\frac{E - E_{Fn}}{kT}\right)}, \tag{2.19}$$

$$f_p(E) = \frac{1}{1 + \exp\left(\frac{E_{Fp} - E}{kT}\right)}, \tag{2.20}$$

where $E_{Fn,p}$ are the Fermi energies. The density of available states, for a spherical parabolic band structure ($E \propto k^2$; with E denoting energy and k the wavenumber) following the simple description of effective mass ($1/m^* = 4\pi^2 \partial^2 E/h^2 \partial k^2$), read

$$g_c(E) = \frac{4\pi(2m_n^*)^{3/2}}{h^3}(E - E_c)^{1/2} \qquad \text{for } E > E_c, \qquad (2.21)$$

$$g_v(E) = \frac{4\pi(2m_p^*)^{3/2}}{h^3}(E_v - E)^{1/2} \qquad \text{for } E < E_v. \qquad (2.22)$$

Here, m_n^* and m_p^* denote the effective masses in the conduction and valence bands, respectively. In general, the bands are not parabolic, and most certainly not so when the material is mechanically strained. In this case, the simple approximation for the effective mass does not hold and we have two kinds of effective masses; one associated with the density of states function and the other with the free carrier concentration [9, 10]. The former is a function of carrier energy. The latter, needed in calculations of the carrier concentrations, is a function of both temperature and Fermi level. They can be calculated for Si based on the approximation shown below using values (taken from Ref. [8]) given in Table 2.3:

$$m_n^* = (m_l^* m_t^{*2})^{1/3}, \qquad (2.23)$$

$$m_p^* = (m_{lh}^{*3/2} + m_{hh}^{*3/2})^{2/3}. \qquad (2.24)$$

Here, m_l^* and m_t^* denote the respective electron longitudinal and transverse effective masses, and m_{lh}^*, and m_{hh}^* denote the respective light- and heavy-hole effective masses. Non-parabolicity in band structure will be addressed in Chapt. 6 when we look at the effects of mechanical strain on carrier transport. The density of states in lightly doped crystalline materials is zero for $E_v \leq E \leq E_c$. They become modified with heavy doping due to formation of band tail states, or in the presence of strain when the bands are no longer parabolic and isotropic. Substituting for $g_{c,v}(E)$ and $f_{n,p}(E)$ in

Table 2.3. *Effective mass values for electrons and holes in Si for calculation of carrier concentrations; m_0 denotes the free electron mass*

Effective mass	Value (m^*/m_0)
Longitudinal: m_l^*	0.98
Transverse: m_t^*	0.19
$m_n^* = (m_l^* m_t^{*2})^{1/3}$	0.33
Light holes: m_{lh}^*	0.16
Heavy holes: m_{hh}^*	0.49
$m_p^* = (m_{lh}^{*3/2} + m_{hh}^{*3/2})^{2/3}$	0.55

(2.17) and (2.18) with (2.19) to (2.22) yields

$$n = N_c \frac{2}{\sqrt{\pi}} F_{1/2}\left(\frac{E_{Fn} - E_c}{kT}\right), \tag{2.25}$$

$$p = N_v \frac{2}{\sqrt{\pi}} F_{1/2}\left(\frac{E_v - E_{Fp}}{kT}\right), \tag{2.26}$$

where

$$N_{c,v} = 2\left(\frac{2\pi kT m_{n,p}^*}{h^2}\right)^{3/2} \tag{2.27}$$

denote the effective density of states in conduction and valence bands, respectively, and $F_{1/2}$ is the Fermi integral of order $1/2$, viz.,

$$F_{1/2}(x) = \int_0^\infty \frac{\sqrt{y}}{1 + \exp(y - x)} \, dy. \tag{2.28}$$

Approximations to the Fermi integral for device simulation purposes are given in [1]. In a non-degenerate electron gas, if the Fermi level in the energy gap is several kT away from the band edges, viz., $(E_c - E_{Fn}) \gg kT$ and $(E_{Fp} - E_v) \gg kT$, we employ Boltzmann's statistics to obtain simplified expressions for the carrier concentrations:

$$n = N_c \exp\left(\frac{E_{Fn} - E_c}{kT}\right), \tag{2.29}$$

$$p = N_v \exp\left(\frac{E_v - E_{Fp}}{kT}\right). \tag{2.30}$$

Equations (2.29) and (2.30) can be written in terms of the intrinsic Fermi energy (E_i), the electrostatic potential, and quasi-Fermi potentials based on earlier definitions:

$$n = N_c \exp\left(\frac{E_i - E_c}{kT}\right) \exp\left(\frac{q(\psi - \phi_n)}{kT}\right), \tag{2.31}$$

$$p = N_v \exp\left(\frac{E_v - E_i}{kT}\right) \exp\left(\frac{q(\phi_p - \psi)}{kT}\right). \tag{2.32}$$

In intrinsic material, the quasi-Fermi and intrinsic levels line up under thermal equilibrium conditions, so $\phi_{n,p} = \psi = 0$. We can solve for the relative position of the intrinsic Fermi level by equating the electron and hole concentrations given by (2.31) and (2.32), viz.,

$$N_c \exp\left(\frac{E_i - E_c}{kT}\right) = N_v \exp\left(\frac{E_v - E_i}{kT}\right) \tag{2.33}$$

which yields

$$E_i - E_c = \frac{kT}{2} \log_e \left(\frac{N_v}{N_c}\right) - \frac{E_g}{2} = \frac{kT}{2} \log_e \left(\frac{m_p^*}{m_n^*}\right)^{3/2} - \frac{E_g}{2}. \qquad (2.34)$$

The intrinsic Fermi level is displaced by an amount of $-16.6\,\text{meV}$ (at room temperature) from the center of the gap due to the difference in the electron and hole effective masses; otherwise there would be unequal densities of thermally-generated electrons and holes in the conduction and valence bands, respectively, in the intrinsic material. Following the mass action law, $pn = n_i^2$, the intrinsic concentration can be computed as

$$n_i = \sqrt{N_c N_v} \exp\left(-\frac{E_g}{2kT}\right), \qquad (2.35)$$

which at room temperature yields a value of $1.48 \times 10^{10}\,\text{cm}^{-3}$ for Si. In general, the intrinsic concentration is a strong function of temperature and band gap which in turn depends on temperature, strain, and heavy doping effects. With Eqs. (2.34) and (2.35), the carrier concentrations, (2.31) and (2.32) which do not account for any band narrowing effects, can be reduced to:

$$n = n_i \exp\left[\frac{q(\psi - \phi_n)}{kT}\right], \qquad (2.36)$$

$$p = n_i \exp\left[\frac{q(\phi_p - \psi)}{kT}\right]. \qquad (2.37)$$

If Boltzmann's statistics does not apply, expressions for n and p remain as given in (2.25) and (2.26).

2.3.3 Doping-Induced Band Gap Narrowing

High impurity concentrations ($> 10^{18}\,\text{cm}^{-3}$) modify the band structure and correspondingly the electronic transport properties in the device. In a heavily doped material, there are rigid shifts in the conduction and valence band edges towards each other leading to an effective band gap reduction. Band gap lowering induced by heavy doping (or strain [11]) enhances the sensitivity of Si-based optical sensors to long wavelengths. The reduction in gap energy is a consequence of high carrier concentrations resulting from high doping or carrier injection induced electrically or optically [1, 2]. There is also formation of an impurity band and band tails which modify the density of states for electrons and holes. The density of states have infinite tails and despite their rapid decrease from band edges, they are non-zero in the gap region. Rather than using modified density of states functions along with Fermi statistics for calculation of carrier concentrations, it is more attractive, from a device modeling standpoint, to account for band gap narrowing effects through an effective intrinsic carrier concentration n_{ie} [1]. Thus (2.36) and (2.37) can be reexpressed as

$$n = n_{ie} \exp \left[\frac{q(\psi - \phi_n)}{kT} \right], \tag{2.38}$$

$$p = n_{ie} \exp \left[\frac{q(\phi_p - \psi)}{kT} \right]. \tag{2.39}$$

The associated error with this approach is less than 10% for total impurity concentrations less than $8 \times 10^{19} \, \text{cm}^{-3}$ and for materials with a doping compensation of less than 10% [1]. The approximation becomes poor at higher doping and compensation levels. The increase in intrinsic carrier concentration can be related to the band narrowing ΔE_g as [2]

$$n_{ie}^2 = n_i^2 \exp \left(\Delta E_g / kT \right). \tag{2.40}$$

Significant work has been pursued since the late seventies in the quest for a reasonably simple description for ΔE_g. The physical understanding of heavy doping effects has not been fully unraveled and models proposed to-date still yield discrepancies of varying degrees. Both theoretical and experimental investigations have been carried out, the latter being based on electrical and optical measurements [12–18]. For doping levels less than $4 \times 10^{19} \, \text{cm}^{-3}$ [2], reasonable agreement has been obtained between predicted values and experimental data based on electrical measurements. A widely used empirical model for band narrowing, first proposed by [14] following electrical measurements of the pn product in a bipolar transistor, takes the form

$$\Delta E_g = E_0 \left\{ \log_e \left(\frac{N_T}{N_0} \right) + \sqrt{\left[\log_e \left(\frac{N_T}{N_0} \right) \right]^2 + 0.5} \right\}. \tag{2.41}$$

Here, $E_0 = 9 \, \text{meV}$, $N_0 = 10^{17} \, \text{cm}^{-3}$, and N_T denotes the total impurity concentration. The band narrowing was found to be independent of temperature. A revised version of the above model has been recently proposed [19] based on new minority carrier mobility data [20–22]. The expression takes the same form as shown in (2.41) but with revised parameter values, $E_0 = 6.92 \, \text{meV}$ and $N_0 = 1.3 \times 10^{17} \, \text{cm}^{-3}$. The model appears to provide good agreement with recent electrical measurements [23–26].

The spatial dependence of impurity concentration correspondingly induces a spatial dependence of bandgap. Assuming Boltzmann's statistics applies, this can be accounted for with slightly modified current density relations [2,13,27]:

$$\mathbf{J_n} = -q\mu_n n \, \text{grad} \, (\psi + \Delta\psi_c) + qD_n \, \text{grad} \, n, \qquad (2.42)$$

$$\mathbf{J_p} = -q\mu_p p \, \text{grad} \, (\psi - \Delta\psi_v) - qD_p \, \text{grad} \, p, \qquad (2.43)$$

where, $\Delta\psi_{c,v}$ account for the effective shifts in band edges [13]. Assuming symmetric shifts in band edges, the carrier concentrations can be described as given in (2.38) and (2.39). Relations of the form (2.42) and (2.43) will be revisited in Sect. 6.5 when we discuss effects of stress variations in the material.

2.3.4 Temperature-Dependence of Band Gap Energy

The temperature-dependence of the band gap energy needs to be accounted for in simulation of thermal microsensors. It is also of importance in simulation of other junction-based microtransducers for prediction of the temperature coefficient (TC) of sensitivity. The band gap energy decreases with increasing temperature with a TC that is linear for most semiconductor materials. For Si, the dependence of gap energy on temperature can be modeled as (see [1])

$$E_g(T) = 1.1785 - 9.025 \times 10^{-5} \, T - 3.05 \times 10^{-7} \, T^2 \qquad (2.44)$$

where E_g and T are in units of eV and K, respectively. At 300 K, E_g is 1.12 eV in undoped Si with a TC that is approximately -241 ppm/K.

2.3.5 Carrier Mobility and Matthiessen's Rule

Modeling of mobility behavior and its dependence on doping concentration and temperature has a strong bearing on microtransducer design. For example, with thermal sensors, a reduced doping level is desirable to allow prevalence of lattice scattering so as to achieve a high temperature coefficient (TC). Here, the mobility degrades with increasing temperature because of increased thermally-induced lattice vibrations. On the other hand, for example, in mechanical sensors, a small TC is desirable, which is achieved at higher doping levels without compromising the magnitude of the piezoresistance effect.

The electron and hole drift mobilities in low-doped Si are determined by acoustic phonon and non-polar optical phonon scattering all of which are associated with the lattice. At higher doping concentrations, there is additional scattering due to ionized impurities as well as with other electrons and holes, themselves, the latter being termed as carrier-carrier scattering. In addition, there can be surface scattering effects such as those found in inversion layer devices (e.g., MOSFETs) and which lead to degradation of the mobility. Scattering with neutral impurities is important at temperatures below 77 K.

With acoustic phonons, energy and momentum conservation restrict scattering to long wavelength phonons. A long wavelength (longitudinal) acoustic displacement does not change carrier energy since all unit cells move by almost the same amount. Thus only the differential displacement, namely the strain, which is homogeneous in the material is of importance. Using deformation potential theory, the scattering relaxation time due to acoustic phonons is given by (see [2]):

$$\tau_{ac} = \frac{\rho u_l^2 h^4}{16\sqrt{2}\pi^3 m^{*3/2} E_{\text{eff}}^2 kT \sqrt{E}}, \tag{2.45}$$

where ρ is the material density, u_l is the longitudinal sound velocity in the material, E_{eff} is the effective deformation potential, m^* is the effective mass, and E is the carrier energy. As with acoustic phonons, scattering by optical phonons is restricted to long wavelength modes. However, the long wavelength optical displacement may affect the electronic energy directly. By defining an optical phonon coupling constant, the scattering relaxation time is given by [2]

$$\tau_{op} = \frac{\rho h^3 \omega_0}{4\sqrt{2}\pi^2 m^{*3/2} D_0^2 K} \left[N_q \sqrt{E + \frac{h\omega_0}{2\pi}} + (N_q + 1)\sqrt{E - \frac{h\omega_0}{2\pi}} \right]^{-1}, \tag{2.46}$$

where $D_0^2 K$ is the optical phonon coupling constant, E is the carrier energy, N_q is the Bose-Einstein phonon distribution for the qth mode with optical phonon frequency ω_0. The term in brackets represents the creation and annihilation of a phonon. Scattering by optical phonons is relatively weak below room temperature and becomes important only at higher temperatures [28].

The calculation of impurity scattering relaxation time follows directly from the theory of scattering of charged particles by Coulombic interaction [7, 28, 29]. The Coulombic potential associated with the nuclei of ionized impurities deflect the paths of electrons and holes. The scattering process can be considered elastic due to the large mass of impurity atoms and the deflection angles can be assumed small. The relaxation time for impurity scattering takes the form [2]

$$\tau_{\text{imp}} = \frac{16\sqrt{2m^*}\pi\varepsilon^2 E}{Z^2 q^4 N_T} g(x), \tag{2.47}$$

where ε is the material permittivity, Zq is the ionic charge, E is the carrier energy, and N_T is the total impurity concentration. In (2.47), $g(x)$ is a function which models the extent of charge screening in the scattering process. There are two descriptions for $g(x)$. In [30], the Coulombic potential is assumed unscreened and terminates at half the mean distance between neighboring impurity atoms. This yields [1, 2]

$$g(x) = [\log_e(1+x)]^{-1} \quad \text{with} \quad x = \left(\frac{4\pi\varepsilon E}{Zq^2N_T^{1/3}}\right)^2 \quad (2.48)$$

A more refined formulation for the screening function in (2.47) is given in [31]. The formulation takes into account the role of electrons and holes in the screening process,

$$g(x) = \left[\log_e(1+x) - \frac{x}{1+x}\right]^{-1} \quad \text{with} \quad x = \frac{32\pi^2 m^* \varepsilon kTE}{h^2 q^2(n+p)}. \quad (2.49)$$

The models shown do not account for the effects of heavy doping. They tend to overestimate the mobility for impurity concentrations greater than 10^{17} cm^{-3}. This can be corrected for by enhancing the scattering length [32, 33].

In the relaxation time approximation, the carrier mobility is computed from the first-order solution of the BTE as

$$\mu = \frac{q}{m^*}\langle\tau\rangle. \quad (2.50)$$

Here, $\langle\tau\rangle$ denotes the average relaxation time defined by the energy-dependent relaxation time $\tau(E)$ averaged over the equilibrium carrier distribution function $f_0(E)$

$$\langle\tau\rangle = \frac{\int E^{3/2}\tau(E)[\partial f_0/\partial E]dE}{\int E^{3/2}[\partial f_0/\partial E]dE}, \quad (2.51)$$

where an isotropic parabolic energy band structure is assumed (i.e., no heavy doping or strain).

However, the general practice in simulations is to combine the different scattering-limited mobility components using Matthiessen's rule. This rule was first proposed by Matthiessen for calculation of resistivity in metals (see, e.g., [34]):

$$\rho = \frac{m^*}{nq\tau} = \frac{m^*}{nq\tau_1} + \frac{m^*}{nq\tau_2} = \rho_1 + \rho_2. \quad (2.52)$$

Here, the relaxation times τ, τ_1, τ_2 are independent of wavevector and energy. Extension of the rule for semiconductor device analysis, $\mu^{-1} = \sum_i \mu_i^{-1}$ (where the μ_i denote the different scattering limited mobility components) is valid only when the scattering rates are (a) independent of energy, (b) have the same energy dependence, or (c) when one scattering rate is much larger than the other, i.e., when one or the other scattering mechanism predominates [10, 33, 35]. Realistically, the relaxation times for the different scattering mechanisms usually have different energy dependence, and there is usually no dominant scattering mechanism. Thus Matthiessen's rule generally does not hold. Its validity requires that

$$\left\langle \left(\sum_i \tau_i^{-1} \right)^{-1} \right\rangle = \left(\sum_i \langle \tau_i \rangle^{-1} \right)^{-1}. \tag{2.53}$$

Setting aside the possible errors that can arise from the models of the various scattering-limited mobility components themselves, the error in the approximation following (2.52) can be as large as 40% in Si [33] and 60% in GaInAs with an anomalous temperature dependence [35]. In Si, the main source of error stems from impurity scattering; for large enough electron energies, $\tau_{imp} \sim E^{3/2}$, whilst $\tau_{ac,op} \sim E^{-1/2}$. Strictly speaking, the mobility, following (2.50), should be computed using a sum of reciprocal relaxation times,

$$\mu = \frac{q}{m^*} \left\langle \left(\sum_i \tau_i^{-1} \right)^{-1} \right\rangle. \tag{2.54}$$

Modeling of the various scattering mechanisms and their interactions is, in general, a very complicated task. The models proposed to-date have been based on a combination of theoretical and empirical approaches. However, because of the inherently large scatter in the experimental data, they require adjustment of fitting parameters to yield reasonable agreement with measured and predicted mobility data for a given fabricated device over the operating range of interest. An excellent cross section of the various mobility models is reviewed in [1, 2]. In what follows, we only summarize models that have gained widespread usage based on either accuracy, consistency in terms of prediction, or simplicity of use.

A widely used mobility model, proposed by [36], which takes into account lattice (L) and impurity (I) scattering is given as:

$$\mu_{n,p}^{LI} = \mu_{n,p}^{min} + \frac{\mu_{n,p}^{L} - \mu_{n,p}^{min}}{1 + \left(N_T / N_{n,p}^{ref} \right)^{\alpha_{n,p}}}. \tag{2.55}$$

Here, N_T denotes the total impurity concentration. Lattice scattering accounts for acoustic and non-polar optical phonon scattering. Temperature dependence of mobility was presented by [37] using a model very similar to (2.55). Expressions for various parameters in (2.55), including their temperature dependence [1], is given in Table 2.4. Despite the large scatter

Table 2.4. *Model parameters, including temperature dependence, for majority carrier bulk mobility in Si*

Electrons	Holes	Units
$\mu_n^{\mathrm{min}} = 88 \left(\dfrac{T}{300\,\mathrm{K}} \right)^{-0.57}$	$\mu_p^{\mathrm{min}} = 54.3 \left(\dfrac{T}{300\,\mathrm{K}} \right)^{-0.57}$	cm^2/Vs
$\mu_n^{\mathrm{L}} - \mu_n^{\mathrm{min}} = 1252 \left(\dfrac{T}{300\,\mathrm{K}} \right)^{-2.33}$	$\mu_p^{\mathrm{L}} - \mu_p^{\mathrm{min}} = 407 \left(\dfrac{T}{300\,\mathrm{K}} \right)^{-2.33}$	cm^2/Vs
$N_n^{\mathrm{ref}} = 1.432 \times 10^{17} \left(\dfrac{T}{300\,\mathrm{K}} \right)^{2.546}$	$N_p^{\mathrm{ref}} = 2.67 \times 10^{17} \left(\dfrac{T}{300\,\mathrm{K}} \right)^{2.546}$	cm^{-3}
$\alpha_n = 1$	$\alpha_p = 1$	

in parameter values, there is reportedly good agreement with measured mobility values. Equation (2.55) with the parameter values shown in Table 2.4 is accurate within 13% for temperature and doping levels in the range; $(250–500)\,\mathrm{K}$ and $(10^{13}–10^{20})\,\mathrm{cm}^{-3}$, respectively [1]. At higher concentrations $(>10^{20}\,\mathrm{cm}^{-3})$, the model can be extended [38] to include terms that account for degradation in mobility which (2.55) fails to predict.

When concentrations of n and p become comparable to, or exceed, the impurity concentration, carrier-carrier scattering becomes important. This can be of significance in high injection devices or even in devices with large carrier concentrations that have been induced optically. Effects of carrier-carrier scattering (when $n \approx p \approx N_T$) can be accounted [39] by replacing N_T in (2.55) with

$$N_{T_{\mathrm{eff}}} = 0.34\,N_T + 0.66\,(n+p). \tag{2.56}$$

At high electric fields, the drift velocity of carriers saturates and this can be viewed in terms of a reduced effective mobility. This can be important, for example, in MOS-based magnetic sensors where high sensitivities can be achieved by virtue of high carrier velocity [40]. For purposes of simulations, this can be accounted for with an empirically-based field dependent effective mobility [1, 2]

$$\mu_{n,p} = \frac{\mu_{n,p}^0}{\left[1 + \left(\dfrac{\mu_{n,p}^0 E_{n,p}}{v_{n,p}^{\mathrm{sat}}} \right)^{\beta_{n,p}} \right]^{1/\beta_{n,p}}} \tag{2.57}$$

where $\mu_{n,p}^0$ denotes the low field drift mobility taking into account, for example, the scattering mechanisms described by (2.55), $\beta_n = 2$, $\beta_p = 1$, $v_{n,p}^{\mathrm{sat}}$ denote the electron and hole saturation velocities, and $E_{n,p}$ can be taken as the magnitude of the gradient in quasi-Fermi potentials for electrons and holes, respectively [1, 2]. The saturation velocities, which are temperature dependent, can be modeled as [41]

$$v_n^{sat} = 10^7 \left(\frac{T}{300\,K}\right)^{-0.87}, \tag{2.58}$$

$$v_p^{sat} = 8.37 \times 10^6 \left(\frac{T}{300\,K}\right)^{-0.52} \tag{2.59}$$

in the temperature range, (245–430) K. Mobility degradation due to surface scattering is most dominant in MOS devices where charge transport takes place at the Si/SiO_2 interface. Surface scattering can be attributed to interface roughness, interface charges, surface phonons, and other effects [1]. The nature and extent of the scattering mechanisms are not well understood. Models presently available have been constructed on an empirical basis. A reliable model, suggested in [42] and incorporated in the MINIMOS device simulator, accounts for interface scattering (S) in the following way:

$$\mu_{n,p}^{LIS} = \frac{\mu_{n,p}^{ref} + \left(\mu_{n,p}^{LI} - \mu_{n,p}^{ref}\right)[1 - f(y)]}{1 + f(y)\left(S_{n,p}/S_{n,p}^{ref}\right)^{\gamma_{n,p}}}, \tag{2.60}$$

where models and values for the various parameters are given in Table 2.5. The function $f(y)$ ensures a smooth transition from the Si/SiO_2 interface

Table 2.5. *Model parameters, including temperature dependence, for inversion layer carrier mobility in Si*

Electrons	Holes
$\mu_n^{ref} = \dfrac{638}{MR}\left(\dfrac{T}{300\,K}\right)^{-1.19}$ cm^2/Vs	$\mu_p^{ref} = \dfrac{106}{MR}\left(\dfrac{T}{300\,K}\right)^{-1.09}$ cm^2/Vs
$f(y) = \dfrac{2\exp\left[-(y/10\,nm)^2\right]}{1 + \exp\left[-2(y/10\,nm)^2\right]}$	
$S_n^{ref} = \dfrac{7 \times 10^5}{MT}$	$S_p^{ref} = \dfrac{2.7 \times 10^5}{MT}$
$\gamma_n = \dfrac{1.3}{MX}$	$\gamma_p = \dfrac{1}{MX}$
MR, MT, MX are fitting parameters; they can be assumed unity at 300 K	

$(y = 0)$ to the bulk. At the interface, (2.60) reduces to

$$\mu_{n,p}^{\text{LIS}} = \frac{\mu_{n,p}^{\text{ref}}}{1 + \left(S_{n,p}/S_{n,p}^{\text{ref}}\right)^{\gamma_{n,p}}} \tag{2.61}$$

The models presented above for bulk Si describe the majority carrier mobility. The same expressions can also be employed to model minority carrier mobility but only at not too large doping levels and not too low temperatures. At high doping levels or at low temperatures, the minority carrier mobility can be higher than its corresponding majority carrier value; for example, the value can be a factor of three higher at a doping concentration of 10^{20} cm^{-3} (see [21, 22]). Although there are independent model expressions describing minority carrier mobility, it is desirable from a numerical simulation standpoint, to have a single function that models both majority and minority carrier mobility taking into account the local scattering mechanisms, including temperature dependence. Such a model has been proposed in [21, 22], which provides good agreement with experiments as well as appears to provide a consistent prediction of band gap narrowing values (see Section 2.3.3).

2.3.6 Generation-Recombination

The generation-recombination term in Eqs. (2.9) and (2.10) is modeled according to material type and processes involved. Its role is significant in junction-based microtransducers, e.g., radiant, magnetic, thermal, and mechanical microsensors, notably under high injection conditions. Electron-hole recombination can take place either through a direct transition (such as in direct band semiconductors, e.g., GaAs) or through an indirect transition (such as in indirect band semiconductors, e.g., Si). In the latter, the recombination process involves an intermediate state arising from trap centers due to impurities and crystal imperfections. Direct transitions involve emission of a photon and carriers do not undergo a change in momentum. Indirect transitions involve a phonon to conserve crystal momentum [7].

The theory underlying trap-assisted recombination was developed by Shockley, Read and Hall and is known as the SRH theory. The processes involve

(1) transition of an electron from the conduction band to a trap;
(2) transition of a trapped electron to the conduction band;
(3) transition of a trapped electron to the valence band to neutralize a hole;
(4) transition of an electron to a trap from the valence band;

For traps of a single energy level, the four rates, under steady state conditions, lead to the following form (with slight change in notation) for the net recombination rate (see [1, 2])

$$(R - G)_{SRH} = \frac{pn - n_{ie}^2}{\tau_p(n + n_0) + \tau_n(p + p_0)} \tag{2.62}$$

where n_0 and p_0 are functions of the trap energy level,

$$n_0 = n_{ie} \exp\left[(E_t - E_i)/kT\right], \tag{2.63}$$

$$p_0 = n_{ie} \exp\left[(E_i - E_t)/kT\right]. \tag{2.64}$$

Here, E_t is generally a deep energy level impurity and can be assumed located at E_i. In deriving (2.62), it has been assumed that trap densities are larger than the concentration of excess carriers and that trapping times are negligible. The equation does not hold in transient situations if there is rapid change in carrier concentrations [2]. In this case, separate expressions for the recombination term are needed for electrons and holes. In (2.62), $\tau_{n,p}$ denote the electron and hole lifetimes, which are equal to the reciprocal capture probabilities [1, 2],

$$\tau_{n,p} = c_{n,p}^{-1} = \left[\sigma_{n,p} v_{th} N_t\right]^{-1} \tag{2.65}$$

where $\sigma_{n,p}$ are the electron and hole capture cross sections, v_{th} is the thermal velocity, and N_t the trap density. Additionally, the lifetimes are influenced by the impurity concentration due to doping-induced defect creation. This can be modeled as (see [1, 2])

$$\tau_{n,p} = \frac{\tau_{n,p}^0}{1 + \left(N_T/N_{n,p}^0\right)} \tag{2.66}$$

where N_T denotes the total impurity concentration. The rest are empirical coefficients [43]; $\tau_n^0 = 3.95 \times 10^{-4}$ s, $\tau_p^0 = 3.52 \times 10^{-5}$ s, and $N_{n,p}^0 = 7.1 \times 10^{15}$ cm^{-3}. Values of these coefficients, based on various published reports (see [1]), carry a very large scatter.

Generation-recombination mechanisms involving band-to-band transitions do not require assistance of any intermediate state but involve a photon. Here, we have:

(1) Transition of an electron from conduction to valence bands emitting photon energy. This is radiative electron-hole recombination and takes place in direct gap semiconductors.
(2) An electron in the valence band which absorbs an incident photon and moves to conduction band thereby creating an electron-hole pair. This is optical generation (band-to-band absorption) and takes place in both direct and indirect band semiconductors.

The net recombination rate is given as [1, 2]

$$(R - G)_{\text{opt}} = C_{\text{opt}}(pn - n_{ie}^2), \tag{2.67}$$

where C_{opt} denotes the capture rate. We shall look at band-to-band absorption when we deal with optical effects and associated optical sensors in Chapt. 3.

Another mechanism, important in heavily doped Si, is Auger recombination. Here, an electron moves from conduction to valence bands to recombine with a hole and transfers its excess energy to another electron in the conduction band or to another hole in the valence band. Based on qualitative understanding of underlying processes, Auger generation-recombination can be modeled as [1, 2]

$$(R - G)_{\text{AUGER}} = (c_n^A n + c_p^A p)(pn - n_{ie}^2), \tag{2.68}$$

where $c_{n,p}^A$ are Auger coefficients [44]; $c_n^A = 2.8 \times 10^{-31}\,\text{cm}^6/\text{s}$, $c_p^A = 9.9 \times 10^{-32}\,\text{cm}^6/\text{s}$ at room temperature. They are weakly dependent on temperature in the range, $(77-400)\,\text{K}$.

Another mechanism is impact ionization whereby there is creation of an electron-hole pair by a highly energetic electron in the conduction band or a highly energetic hole in the valence band. The generated electron then moves to the conduction band. Impact ionization is a generation process similar to Auger processes. The underlying difference is that impact ionization requires large current flow while Auger generation takes place in regions of high carrier concentration, with negligible current flow. A thorough review of this subject can be found in [1, 2]. Impact ionization is significant in high reverse biased photodiodes and in avalanche photo-diodes. The generation rate due to impact ionization can be modeled as

$$G_{\text{II}} = -(1/q)(\alpha_n|\mathbf{J_n}| + \alpha_p|\mathbf{J_p}|) \tag{2.69}$$

where $\alpha_{n,p}$ are the respective electron and hole ionization coefficients which are dependent on electric field and temperature, and $\mathbf{J_{n,p}}$ denote respective current densities. Expressions for the ionization rates in Si, based on theoretical and experimental investigations, are given as

$$\alpha_{n,p} = \alpha_{n,p}^\infty \exp\left[-\left(E_{n,p}^{\text{crit}}/E\right)^{\beta_{n,p}}\right]. \tag{2.70}$$

Depending on electric field strength E, α_n^∞ ranges from $(0.6\text{ to }1.0) \times 10^6\,\text{cm}^{-1}$, α_p^∞ from $(0.67\text{ to }22.5) \times 10^6\,\text{cm}^{-1}$, E_n^{crit} from $(1.2\text{ to }5.9) \times 10^6\,\text{V/cm}$, E_p^{crit} from $(1.4\text{ to }3.3) \times 10^6\,\text{V/cm}$, and $\beta_{n,p}$ take on values ranging from 1 to 2 (see [1]).

At device boundaries and interfaces, there can be surface recombination depending on the density of surface states and trapped interface charges. Surface recombination leads to leakage current. The generation-recombination rate at the device boundaries can be modeled along the same lines as the SRH theory [2]:

$$R_{\text{surf}} = \frac{pn - n_{ie}^2}{(n + n_0)s_p^{-1} + (p + p_0)s_n^{-1}} \delta(x). \tag{2.71}$$

Here, n_0 and p_0 are calculated using (2.63) and (2.64) but using the energy level value of interface traps, $s_{n,p}$ denote surface recombination velocities and $\delta(x)$ is the Dirac delta function where $x = 0$ denotes the surface. The surface recombination velocities can be defined in terms of the reciprocal lifetimes and thus be calculated using expression (2.65) but with the value of the bulk trap density N_t now replaced by the interface trap density N_{it}. A typical value of $s_{n,p}$ for an Si/SiO$_2$ interface is around $100 \, \text{cm/s}$ [1].

In simulations, the net rates for the above processes are added to yield the total generation-recombination rate.

2.4 Electrical Conductivity and Isothermal Behavior in Polycrystalline Materials

Apart from applications of polycrystalline silicon (poly-Si) in solar cells, VLSI interconnects, MOS and bipolar transistors, memory elements, and resistive elements [46], the material, as we will see in the Chapters that follow, has become a work-horse of microtransducers for a variety of sensor and actuator applications in view of its excellent electrical, thermoelectric, and mechanical properties. For example, thin film poly-Si is an integral part of the microsensor or microactuator whereby it is employed as: heating and/or sensing elements in thermal-based micro-sensors (see Chapt. 5); piezoresistors in mechanical microsensors (see Chapt. 6); and actuation elements in mechanical microactuators (see Chapt. 8).

In terms of structure, polycrystalline materials can be viewed as being composed of a chain of crystalline grains which are separated by grain boundary regions that are a few monolayers thick. The grain boundary regions, depending on processing conditions, can contain a large number of defects and impurities. In metals, these boundaries are very narrow and can be treated simply as scattering centers. However, in polycrystalline semiconductors, there can be considerable space charge formation at the boundary regions giving rise to a large perturbation in the Fermi level [45]. It is not within the scope of this section to review all polycrystalline semiconductor materials. Instead our discussion of carrier transport models will be focused on thin film poly-Si.

In comparison to crystalline silicon (c-Si), poly-Si has a reduced electrical conductivity which can be interpreted as being due to two effects

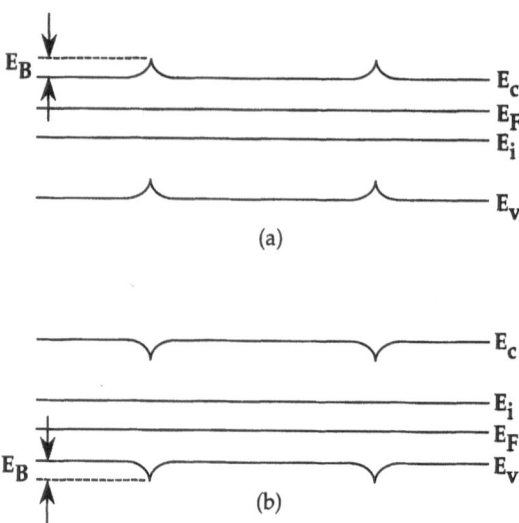

Fig. 2.1 Diagrams of the band energy in (a) *n*-type and (b) *p*-type poly-Si across grain boundaries

(see [46]); segregation of dopant atoms at the disordered grain boundary regions where they become electrically inactive thus resulting in a smaller number of free carriers for conduction, and trapping of free carriers (arising from substitutional dopant atoms within grains) by the high density of trapping sites at the grain boundaries thus immobilizing free carriers. The trapped carriers lead to charged depletion layers on both sides of the boundary, giving rise to local band bending and creation of a potential barrier (see Fig. 2.1) which impedes inter-grain carrier transport [47–49]. Dopant segregation, which is more common in n-type poly-Si doped with P or As, is unable to account for the temperature dependence of conductivity.

2.4.1 Doping-Dependence

Modeling of electrical conductivity in poly-Si is based on several simplifying assumptions mostly related to the material structure which is highly process-dependent [46]. The grains are assumed identical in size and orientation, have a band structure and other physical parameters similar to that of c-Si, and are much larger than the thickness of the grain boundaries. The traps in the grain boundary are assumed monovalent and occupy energy levels that peak at the midgap [49]. Dopant atoms are assumed to be uniformly distributed and fully ionized. Despite these assumptions and process dependence of material parameters, the models are able to provide results consistent with measurement data. They are adequate for simulation purposes when coupled with realistic material parameter values retrieved experimentally. For example, values of the grain size in the deposited material could be retrieved experimentally using transmission electron microscopy [50, 51].

Assuming the dominant transport mechanism is the thermionic emission of carriers over the potential barrier, the conductivity in poly-Si can be expressed as [47]

$$\sigma_{n,p} = \frac{q^2 L N_0}{\sqrt{2\pi k T m_{n,p}^*}} \exp\left(-\frac{E_B}{kT}\right),$$ (2.72)

where L is the grain size, E_B is the barrier height, $m_{n,p}^*$ is the effective mass based on a parabolic band structure in grains, and N_0 is the carrier concentration which can be assumed equal to the effective dopant concentration N remaining in the space charge neutral grain regions following dopant segregation [52]. Expression (2.72) shows that conductivity in poly-Si is an activated process; the activation energy can be approximated as being E_B. The average grain size, L depends on dopant concentration, dopant species, deposition conditions, and in particular, the deposition temperature. Higher deposition temperatures yield larger grain sizes. The grain size increases as film thickness increases [53].

The barrier height strongly depends on the doping concentration N and trap density N_t at the grain boundary [47–49]. The barrier height increases with increasing dopant concentration until all traps are completely filled. This takes place when a critical dopant concentration N^* is reached. For $N > N^*$, the barrier height decreases with increasing N for low trap densities or remains independent of N when trap densities are high. The critical concentration can be determined following an iterative solution of the equation [49]

$$N^* = \frac{8\varepsilon_{Si}}{q^2 L^2}\left[E_F - E_t + kT \log_e\left(\frac{2N_t}{N^*L} - 2\right)\right]$$ (2.73)

derived on basis of charge neutrality and assuming that the depletion layer width on each side of the grain boundary is $L/2$. At low trap density, the traps are filled before the formation of a space charge neutral region in the grain and $N^* \propto L^{-1}$ [49]. At high trap densities, the local bending of energy bands may be large enough to raise the trap energy level at the grain boundary above the Fermi level. This leads to formation of a neutral region before traps are filled, and N^* is proportional to L^{-1} and L^{-2} for small and large grains, respectively [49]. With integrated circuit poly-Si, which has a typical grain size of 100 nm, N^* is approximately 10^{17} cm^{-3} for a trap density, $N_t = 10^{12}$ cm^{-2} [46].

For dopant concentrations less than the critical concentration, $N < N^*$, the grains can be assumed fully depleted and the height of the energy barrier is given by [47–49]

$$E_B = \frac{q^2 L^2 N}{8\varepsilon_{Si}} \tag{2.74}$$

where we see an increase in the barrier height with increasing dopant concentration. The conductivity becomes

$$\sigma_{n,p} = \frac{q^2 L^2 N_{c,v} N}{2(N_t - LN)\sqrt{2\pi k T m^*_{n,p}}} \exp\left(-\frac{\frac{1}{2}E_g - E_t}{kT}\right) \tag{2.75}$$

and has an activation energy of $E_g/2$ for a trap distribution that is strongly peaked at or near the midgap ($E_t \approx 0$).

For doping concentrations larger than the critical concentration, $N > N^*$, the grains are only partially depleted and the barrier height can be obtained from an iterative solution [49] of the equation

$$E_B = E_F - E_t + kT \log_e\left(\frac{qN_t}{\sqrt{2\varepsilon N E_B}} - 2\right), \tag{2.76}$$

where E_F is defined relative to the intrinsic level. In the case of low and high trap densities, (2.76) reduces respectively to

$$E_B \approx \frac{q^2 L^2 N_t^2}{8\varepsilon N}, \tag{2.77}$$

$$E_B \approx \frac{1}{2}E_g - E_t + kT \log_e\left[\frac{qN_t\sqrt{N_D}}{N_{c,v}\sqrt{2\varepsilon E_B}}\right]. \tag{2.78}$$

Thus, at low trap densities, the conductivity, following (2.72) and (2.77), can be approximated as

$$\sigma_{n,p} = \frac{q^2 L N_0}{\sqrt{2\pi k T m^*_{n,p}}} \exp\left(-\frac{q^2 L^2 N_t^2}{8\varepsilon k T N}\right). \tag{2.79}$$

At high trap densities,

$$\sigma_{n,p} = \frac{qLN_{c,v}^2}{N_t}\sqrt{\frac{\varepsilon E_B}{\pi k T m^*_{n,p}}} \exp\left(-\frac{\frac{1}{2}E_g - E_t}{kT}\right), \tag{2.80}$$

where E_B is given as in (2.78).

In addition to the depletion layer potential barrier associated with trap states, the presence of a grain boundary barrier arising from the local disorder in crystallinity has been proposed to account for the variation in conductivity on the thermal history of, and dopant species in, the highly-doped material [50, 51]. The disordered boundary region is assumed to have a band gap ($\sim 1.6\,\mathrm{eV}$) which lies between that of crystalline and amorphous Si. Thus, the composite barrier ($\sim 10\,\mathrm{nm}$) between grains is wide at low energies and narrow at high energies. Inter-grain carrier transport can take place by thermionic emission over, and/or tunneling through, the barrier. The latter is predominant at lower temperatures and the former at high temperatures. At intermediate temperatures, both mechanisms are present at the same time and are collectively referred to as thermionic-field emission [46]. Here, the carriers are treated as having sufficient thermal energy to overcome the depletion layer barrier and tunnel across the grain boundary barrier. Assuming sufficiently large doping concentrations, so that the depletion layer barrier is much smaller than the grain boundary barrier, the average conductivity can be expressed in terms of a weighted combination of grain and grain boundary conductivities [50, 51]:

$$\sigma_{n,p} = L \left(\frac{L_g}{\sigma_g} + \frac{L_{gb}}{\sigma_{gb}} \right)^{-1}. \tag{2.81}$$

Here, L_g is the undepleted grain width, and L_{gb} is the composite barrier width, viz., $L_{gb} = 2W + W_{gb}$ where $W(= N_t/2N)$ is the depletion barrier width on one side of the boundary and W_{gb} is the grain boundary barrier width, and $L = L_g + 2W + W_{gb}$ is the grain size. The terms σ_g and σ_{gb} denote the effective grain and grain boundary conductivities, respectively. The conductivity in the grain region σ_g is modeled as in c-Si using a value of free carrier concentration that equals the effective doping density there. The carrier drift mobility in the grain regions is modeled along the lines shown for c-Si in Sect. 2.3.5. The low field (electric field at grain boundary $\ll kT$) grain boundary conductivity σ_{gb} in the linear region is given as [50, 51]

$$\sigma_{gb} = \frac{q^2 L_{gb} N_0}{\sqrt{2\pi kT m^*_{n,p}}} \exp\left(-\frac{E_B}{kT} \right) \left[\frac{\exp(-b_1)}{1 - c_1 kT} \cdot \frac{N_t + c_1 kTN W_{gb}}{N_t + N W_{gb}} \right], \tag{2.82}$$

where

$$b_1 = \frac{4\pi W_{gb}}{h} \sqrt{2m^*_{n,p}(q\varphi - E_B)}, \tag{2.83}$$

$$c_1 = \frac{4\pi W_{gb}}{h} \sqrt{2m^*_{n,p}(q\varphi - E_B)^{-1}}. \tag{2.84}$$

Here, $q\varphi$ denotes the grain boundary barrier energy defined relative to the conduction band. Values of various parameters in (2.82)–(2.84), obtained by fitting modeled and measured values from n-type (N ranging from $2 \times 10^{18}\,\mathrm{cm}^{-3}$ to $2 \times 10^{20}\,\mathrm{cm}^{-3}$), $0.5\,\mu\mathrm{m}$ thick samples annealed at different temperatures, are $N_t \sim 10^{12}\,\mathrm{cm}^{-2}$, $W_{gb} \sim 7\,\mathring{A}$, and φ_{gb} (i.e., grain boundary barrier height relative to the Fermi level) $\sim 0.66\,\mathrm{eV}$ yielding a grain boundary bandgap of $1.32\,\mathrm{eV}$.

From the above conductivity models, we can obtain an expression for the effective mobility in the material which describes the ease of current flow through grain boundaries; it is not representative of the underlying scattering events in poly-Si. Based on thermionic emission as the predominant carrier transport mechanism, Eq. (2.72) along with $\sigma = q\mu^{\mathrm{eff}} N_0$ leads to [46, 47]

$$\mu_{n,p}^{\mathrm{eff}} = \frac{qL}{\sqrt{2\pi kTm_{n,p}^*}} \exp\left(-\frac{E_B}{kT}\right). \qquad (2.85)$$

Because of the dependence of barrier height on doping, the effective mobility goes to a minimum at intermediate doping levels when N equals N^*. For $N > N^*$, the effective mobility increases because of the reduction in barrier height. However, at very high doping levels, ionized impurity scattering dominates and the conductivity behavior becomes analogous to c-Si where the mobility decreases with increasing dopant concentration. The mobility of poly-Si is approximately half that of c-Si and depends on the type of dopant species, e.g., n-type poly-Si doped with phosphorus yields $30\,\mathrm{cm}^2/\mathrm{Vs}$ while with arsenic, the value is $20\,\mathrm{cm}^2/\mathrm{Vs}$ due to increased impurity scattering [46].

The defect states in the grain boundaries not only act as traps but also as efficient recombination centers for minority carriers which are not impeded by the potential barrier [46, 54]. Due to the high defect density compared to c-Si, the minority carrier lifetime is correspondingly much lower. Assuming grains are cubic, a relatively simple expression for estimating the lifetime, based on treating the surface recombination centers as being uniformly distributed throughout the bulk material, reads [54]

$$\tau_{\mathrm{eff}} = \left(\frac{L}{6\sigma v N_{ss}}\right) \qquad (2.86)$$

where σ is the capture cross section, v is the thermal velocity of carriers, and N_{ss} is the density of surface recombination centers per unit of grain boundary area. Thus the lifetime varies linearly with grain size. Measurements yield $\tau_{\mathrm{eff}} = 5 \times 10^{-6}L$. For a grain size of $100\,\mathrm{nm}$, the effective lifetime is $50\,\mathrm{ps}$ for $\sigma = 2 \times 10^{-16}\,\mathrm{cm}^2$, $v = 10^7\,\mathrm{cm/s}$, yielding $N_{ss} = 1.6 \times 10^{13}\,\mathrm{cm}^{-2}$ [54].

2.4.2 Temperature-Dependence

The temperature-dependence of conductivity in poly-Si is dominated by the argument in the exponent (see, e.g., Eq. (2.72)) which carries the activation energy term. The conductivity increases with increasing temperature; the dependence being most pronounced in low doped samples. The rate of conductivity increase is determined by the activation energy [46]. The activation energy is high ($\sim E_g/2$) for doping levels up to the critical concentration, N^*. Beyond N^*, the activation energy decreases with increasing doping concentration. There is a weak temperature-dependence of the activation energy due to the temperature-dependence of the barrier height which stems from the shift in Fermi level with temperature [55]. Calculations show that

$$E_B \approx E_{B0}(1 + \gamma T),\tag{2.87}$$

where E_{B0} is the barrier height extrapolated to 0 K and $\gamma = 1.5 \times 10^{-3}/\text{K}$ [55].

The models for conductivity, as we have seen, are based on a variety of assumptions related to material structure which is dependent on fabrication process conditions [46]. For example, the trap density N_T depends on the structure of the grain boundary, which can vary from one grain to another as well as spatially along the boundary. Thus there is a corresponding variation in barrier height. The spatial distribution of dopant atoms may not be uniform due to segregation and they may not be fully ionized. In a given sample, there can be a fairly wide distribution in the size of grains; a statistical distribution of grain sizes in a given sample is necessary for accurate modeling. All these effects give rise to a highly non-uniform current density distribution over the film thickness due to grains being larger at the top than at the bottom. The current-voltage characteristics can be non-linear [56] as well as asymmetric. Thus, it may be more convenient, from a simulation standpoint, to retrieve an average value of the temperature-dependent conductivity from the measured terminal currents/ voltages over the temperature range of interest. Experimental characterization of temperature-dependence of conductivity, resistivity, sheet resistance, or resistance have been reported for different integrated circuit (IC) technologies and process conditions [57–63]. Values compiled from these sources are given in Table 2.6.

2.5 Electrical Conductivity and Isothermal Behavior in Metals

Thin metal films are becoming an integral part of microtransducers, and particularly, in thermal-based microsensors where they are designed to

Table 2.6. *Values of coefficients for modeling the average temperature-dependent behavior of conductivity (resistivity) in poly-Si*

Model Expression	Coefficient	Conditions
$\sigma(T) = a + bT + cT^2$	$a = 10^3/(\Omega\text{-cm})$ $b = 0.85/(\Omega\text{-cm-}^\circ\text{C})$ $c = 0.02/(\Omega\text{-cm-}^\circ\text{C}^2)$	Ref. [57], $0 \le T \le 200^\circ\text{C}$ gate poly: 1.2 μm CMOS, 18 Ω/sq.
$\rho(T) = \rho_0[1 + \alpha(T - T_0) + \beta(T - T_0)^2]$	$\alpha = -1.05 \times 10^{-3}/^\circ\text{C}$ $\beta = 1.4 \times 10^{-6}/^\circ\text{C}^2$	Ref. [59], $0 \le T \le 200^\circ\text{C}$, $T_0 = 23^\circ\text{C}$ capacitor poly: 1.5 μm CMOS, 1000 Ω/sq.
$\rho(T) = \rho_0[1 + \alpha(T - T_0)]$	$\alpha = 1.2 \times 10^{-3}/^\circ\text{C}$	Ref. [60], $50 \le T \le 250^\circ\text{C}$, $T_0 = 23^\circ\text{C}$ n-type, $\rho_0 = 10^{-3}\Omega\text{-cm}$
	$\alpha = 1.2 \times 10^{-3}/\text{K}$	Ref. [61], $300 \le T \le 800\,\text{K}$, $T_0 = 300\,\text{K}$ p-type (B: $10^{20}\,\text{cm}^{-3}$), $\rho_0 = 3.6 \times 10^{-3}\Omega\text{-cm}$
	$\alpha = 1.2 \times 10^{-3}/^\circ\text{C}$	Ref. [62], $0 \le T \le 200^\circ\text{C}$, $T_0 = 23^\circ\text{C}$ gate poly: 3 μm CMOS, 18 Ω/sq.
	$\alpha = 0.86 \times 10^{-3}/\text{K} \pm 3\%$	Ref. [63], $100 \le T \le 420\,\text{K}$, $T_0 = 300\,\text{K}$ $n+$ gate poly: 1.2 μm CMOS $\rho_0 = 0.85 \times 10^{-3}\Omega\text{-cm} \pm 7\%$
	$\alpha = -0.14 \times 10^{-3}/\text{K} \pm 3\%$	Ref. [63], $100 \le T \le 420\,\text{K}$, $T_0 = 300\,\text{K}$ $p+$ capacitor poly: 1.2 μm CMOS $\rho_0 = 5.8 \times 10^{-3}\Omega\text{-cm} \pm 7\%$
	$\alpha = 0.89 \times 10^{-3}/\text{K} \pm 3\%$	Ref. [63], $100 \le T \le 420\,\text{K}$, $T_0 = 300\,\text{K}$ $n+$ E^2 PROM poly: 2 μm CMOS $\rho_0 = 1.03 \times 10^{-3}\Omega\text{-cm} \pm 7\%$
	$\alpha = -4.4 \times 10^{-3}/\text{K} \pm 3\%$	Ref. [63], $100 \le T \le 420\,\text{K}$, $T_0 = 300\,\text{K}$ n gate poly: 2 μm CMOS $\rho_0 = 96 \times 10^{-3}\Omega\text{-cm} \pm 7\%$
	$\alpha = 0.83 \times 10^{-3}/\text{K} \pm 3\%$	Ref. [63], $100 \le T \le 420\,\text{K}$, $T_0 = 300\,\text{K}$ $n+$ gate poly: 2 μm CMOS $\rho_0 = 1.22 \times 10^{-3}\Omega\text{-cm} \pm 7\%$
	$\alpha = -0.59 \times 10^{-3}/\text{K} \pm 3\%$	Ref. [63], $100 \le T \le 420\,\text{K}$, $T_0 = 300\,\text{K}$ $p+$ gate poly: 2 μm CMOS $\rho_0 = 16.2 \times 10^{-3}\Omega\text{-cm} \pm 7\%$
	$\alpha = 0.54 \times 10^{-3}/\text{K} \pm 3\%$	Ref. [63], $100 \le T \le 420\,\text{K}$, $T_0 = 300\,\text{K}$ $n+$ gate poly: 1 μm CMOS $\rho_0 = 1.58 \times 10^{-3}\Omega\text{-cm} \pm 7\%$
	$\alpha = 0.49 \times 10^{-3}/\text{K}$	Ref. [63], $100 \le T \le 420\,\text{K}$, $T_0 = 300\,\text{K}$ $n+$ gate poly: 1 μm CMOS $\rho_0 = 2.25 \times 10^{-3}\Omega\text{-cm} \pm 7\%$

serve as low power, high precision, resistive heating and temperature sensing elements. Here, they are based on IC metallization or application-specific special thin film metals (e.g., Pt, NiCr, etc.).

Early interpretations of electrical carrier transport were based on the assumption that the free electrons in metals behave as an ideal gas which in thermal equilibrium obeyed Maxwell-Boltzmann statistics. While this gave satisfactory agreement with the Wiedemann-Franz law (see Chapt. 5), which yields a ratio of electrical-to-thermal conductivities that is the same for most metals, it resulted in inconsistencies with respect to thermal behavior. The classical theory (see, e.g., [7, 64]) could only account for the electronic component of the thermal conductivity and the specific heat. The limitations of the classical theory were resolved with the use of Fermi-Dirac statistics. Here, the electrical conductivity was explained in terms of the electrons at the Fermi level E_F and not on the total electron concentration in the metal. Since the density of states and hence, the electron concentration is highest around E_F, only a small electric field is required to substantially increase the electron concentration from E_F to higher states. Thus the associated electron energy is only slightly larger than E_F. The mean drift velocity of carriers, calculated from the electric field induced displacement of the Fermi (spherical) surface, can be approximated by the Fermi velocity, v_F which denotes velocity of electrons with energy E_F.

Following quantum mechanical considerations, the conductivity of metals in accordance to Ohm's law can be expressed as [7, 64]:

$$\sigma = \frac{1}{3} q^2 \tau v_F^2 \, n(E_F) \tag{2.88}$$

where τ denotes the relaxation time and $n(E_F)$ is the electron concentration at the Fermi level. The latter is proportional to the density of states which is the term that distinguishes the conductivity of different materials. The calculated values of the electron concentration [3] at the Fermi surface are shown in Table 2.7 for selected metals. For example, monovalent metals, such as Au, Ag, or Cu, have a density of states, and hence electron concentration, that is largest near the Fermi level which is the reason for their high conductivities. Variations in temperature do not lead to any appreciable change in the carrier concentration. However, the carrier scattering length, and hence relaxation time τ and conductivity, is temperature-dependent; its exact dependence being determined by the scattering mechanism that is prevalent.

Although expression (2.88) is not explicitly employed in simulations, it is stated to provide qualitative insight into conductivity behavior including temperature dependence. As in semiconductors, the relaxation time in (2.88) is determined by scattering due to phonons, other electrons, impurities (interstitial and substitutional), and mechanically-induced

Table 2.7. *Electronic parameters for selected metals at 300 K*

	E_F (eV)	$n(E_F)$ (10^{22} cm^{-3})	σ^L ($10^5/\Omega$-cm)	Θ_D (K)
Ag	5.5	5.9	6.21	225
Al	11.8	18.1	3.65	428
Au	5.5	5.9	4.55	165
Cr			0.78	630
Cu	7.0	8.5	5.88	343
Fe			1.02	470
Mo			1.89	450
Ni			1.43	450
Pd			0.95	274
Pt			0.96	240
Sn	9.4	14.5	0.91	200
Ti			0.23	420
W			1.89	400
Zn	11.0	13.1	1.69	327

defects (dislocations, vacancies, interstitials). The first two scattering mechanisms are dependent on temperature and the latter two can be assumed to be temperature-independent. We denote the corresponding scattering-limited conductivities as σ^L and σ^I, respectively. Assuming that scattering rates are independent, Matthiessen's rule yields for the conductivity (see [34]):

$$\frac{1}{\sigma} = \frac{1}{\sigma^L} + \frac{1}{\sigma^I}. \tag{2.89}$$

The temperature-dependent component, $1/\sigma^L$ in (2.89), also referred to as the ideal resistivity, is independent of impurities and defects when they are low in concentration. The temperature-independent component, σ^I, also referred to as the residual resistivity, can be retrieved by extrapolating the resistivity ($1/\sigma$) curve to absolute zero where $1/\sigma^L$ vanishes. The component σ^I decreases with increasing impurity content. For example, with alloys, the conductivity, in accordance to Matthiessen's rule, decreases linearly with increasing solute content. This increase in resistivity can be attributed to increased scattering due to variation in lattice parameters, differences in valence charges, and changes in the Fermi level due to presence of solute. For concentrated single phase alloys, σ^I can be modeled based on Nordheim's rule [64]:

$$\sigma^I = \left[\frac{\chi_A}{\sigma_A^I} + \frac{\chi_B}{\sigma_B^I} + C\chi_A\chi_B\right]^{-1}, \tag{2.90}$$

where $\chi_{A,B}$ are the fractional atomic compositions of the constituents in the binary system and C is a materials constant. Eq. (2.90) is valid only for alloys containing a non-transition metal. The conductivity of the alloy is a minimum when the constituents are equal in content.

At room temperature, the conductivity is dominated by scattering of conduction electrons with phonons. Although electron-electron scattering is present, it is negligible; the difference in scattering lengths is at least an order of magnitude [3]. At higher temperatures, the phonon density increases leading to increased scattering and decreased conductivity. In fact, beyond the Debye temperature (Θ_D), the phonon density increases linearly with increasing temperature and conductivity decreases approximately linearly. Values for Θ_D are shown in Table 2.7 for selected metals. At lower temperatures, Umklapp scattering of electrons with phonons prevails and at even lower temperatures, scattering with impurity atoms and defects dominates.

Models for the underlying scattering processes in metals are at present somewhat limited in terms of providing accurate estimates of the absolute conductivity. In any case, values of conductivity for a given fabrication

Table 2.8. *Values of coefficients for modeling the average temperature-dependent behavior of conductivity (resistivity) in selected metals*

Model Expression	Coefficient	Conditions
$\sigma(T) = a + bT + cT^2$	$a = 4.313 \times 10^5/(\Omega\text{-cm})$	Ref. [57], $0 \leq T \leq 200°C$
	$b = 9.493 \times 10^2/(\Omega\text{-cm-}°C)$	metal 1 and metal 2: 0.8 µm BiCMOS
	$c = 3.0/(\Omega\text{-cm-}°C^2)$	
$R_s(T) = a + bT + cT^2$	$a = 1.865\,\Omega/\text{sq.}$	Ref. [57, 66], $0 \leq T \leq 200°C$
(sheet resistance)	$b = 3.676 \times 10^{-3}\,\Omega/(\text{sq.K})$	Platinum
	$c = -0.677 \times 10^{-6}\Omega/(\text{sq.K}^2)$	
$R_s(T) =$	$\alpha = (2.96 \pm 0.1) \times 10^{-3}/\text{K}$	Ref. [63], $100 \leq T \leq 420\,K$, $T_0 = 300\,K$
$\quad R_{s0}\,[1 + \alpha(T - T_0)]$		metal 1: 1.2 µm CMOS
		$R_{s0} = (65 \pm 0.7) \times 10^{-3}\,\Omega/\text{sq.}$
	$\alpha = (3.01 \pm 0.1) \times 10^{-3}/\text{K}$	Ref. [63, 65], $100 \leq T \leq 420\,K$,
		$T_0 = 300\,K$
		metal 2: 1.2 µm CMOS
		$R_{s0} = (36 \pm 0.4) \times 10^{-3}\,\Omega/\text{sq.}$
	$\alpha = (4.38 \pm 0.09) \times 10^{-3}/\text{K}$	Ref. [63], $100 \leq T \leq 420\,K$, $T_0 = 300\,K$
		metal 1: 2 µm CMOS
		$R_{s0} = (44 \pm 0.2) \times 10^{-3}\,\Omega/\text{sq.}$
	$\alpha = (4.28 \pm 0.09) \times 10^{-3}/\text{K}$	Ref. [63], $100 \leq T \leq 420\,K$, $T_0 = 300\,K$
		metal 2: 2 µm CMOS
		$R_{s0} = (29 \pm 0.2) \times 10^{-3}\,\Omega/\text{sq.}$
	$\alpha = (3.29 \pm 0.1) \times 10^{-3}/\text{K}$	Ref. [63], $100 \leq T \leq 420\,K$, $T_0 = 300\,K$
		metal 1: 1 µm CMOS
		$R_{s0} = (59 \pm 0.5) \times 10^{-3}\,\Omega/\text{sq.}$
	$\alpha = (3.23 \pm 0.1) \times 10^{-3}/\text{K}$	Ref. [63], $100 \leq T \leq 420\,K$, $T_0 = 300\,K$
		metal 2: 1 µm CMOS
		$R_{s0} = (30 \pm 0.3) \times 10^{-3}\,\Omega/\text{sq.}$

process technology are readily measurable using well established test structures. Models for the average conductivity, resistivity, sheet resistance, or resistance retrieved from measurement data [57, 63, 65, 66] are given in Table 2.8 for selected metals. However, this does not rule out the need for accurate prediction of $\sigma(T)$ for a broad range of temperatures and fabrication process conditions, including new materials and alloys. This requires further theoretical work because empirical approaches can turn out to be costly to establish the complete matrix of data.

2.6 Boundary and Interface Conditions

Although electronic transport in microtransducers is inherently three-dimensional, most microtransducers can be reduced to two-dimensional rectangular structures to simplify the numerical and computational complexities. Most device structures have vanishing spatial derivatives of the pertinent variables (e.g., electrostatic potential and carrier concentrations) in the third dimension. In this section, we will only review boundary and interface conditions pertinent to electronic transport as described by the various model equations given in this chapter. We note, however, that some of the boundary conditions become inappropriate in the presence of an external field (e.g., magnetic field or strain). This will be indicated explicitly in subsequent chapters.

To illustrate the various boundary and interface conditions, we will consider a two-dimensional magnetic field sensor based on the bipolar transistor. A two-dimensional rectangular representation, for purposes of simulation, is depicted in Fig. 2.2. We shall adopt the elegant boundary decomposition representation and notation used in [1]. Consider device operation in the absence of an external field. Here, the simulation domain is described by ABCDEFGHIJKLMNA. The boundaries of the device, ∂D can be decomposed as

$$\partial D = \partial D_P \cup \partial D_A, \tag{2.91}$$

where

∂D_P = real (physical) boundaries, $viz.$, $\partial D_P = \partial D_O \cup \partial D_S \cup \partial D_I$,

∂D_O = ohmic contacts: BC, DE, FG, HI, JK,

∂D_S = Schottky contacts,

∂D_I = interfaces: AB, CD, EF, GH, IJ, KL,

∂D_A = artificial (non-physical) boundaries: LM, MN, NA.

No Schottky contacts are indicated in Fig. 2.2 as we have assumed a highly doped base; otherwise DE and HI become Schottky contacts. The

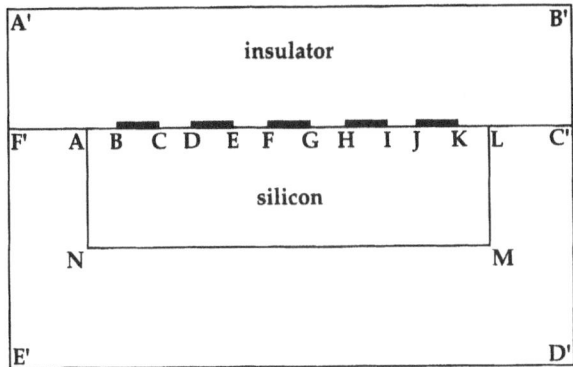

Fig. 2.2 Example of a bipolar magnetic sensor simulation domain for illustration of physical and artificial boundary conditions; FG denotes the emitter, DE, HI the bases, and BC, JK the collectors

interfaces denote Si/oxide boundaries where surface charges and surface recombination may be non-trivial. The artificial boundaries are non-physical and are introduced to reduce computational complexity. However, they have to be located judiciously to minimize physical non-plausibility of numerical solutions in the device active region.

Before dwelling on the models for the various boundary conditions, we identify, as depicted in Fig. 2.2, further artificial boundaries, A′B′C′D′E′ F′A′ and interfaces, LC′ and F′A. These have been introduced to account for the effects of magnetic field [67] or strain. In these cases, the standard condition on the electrostatic potential at boundaries and interfaces, AB, CD, EF, GH, IJ, KL, LM, MN, NA, may lead to solutions that are not meaningful in the device active region, by virtue of Hall or piezoresistance effects. Again, the extent of these boundaries is chosen based on strong physical and mathematical grounds. Alternatively, conditions for the electric field can be constructed [68, 69] that preserve the true behavior of electrostatic potential at these boundaries without the need for such artificial boundaries.

2.6.1 Ohmic Contacts

When a metal and a very heavily doped semiconductor are butted together, an ohmic contact is formed. Strictly speaking, at the interface there is formation of a potential barrier whose height and width depend on the materials involved and the dopant concentration [2]. When the semiconductor is heavily doped, the barrier width is very small and the probability of electrons tunneling through is close to unity. Thus the contact can be idealized by assuming space charge neutrality, *viz.*,

$p - n + N_D - N_A = 0$ there. Together with the mass action law; $pn = n_{ie}^2$, we obtain for the electrostatic potential and carrier concentrations at the contact [2]:

$$\psi_c = V_a + \frac{kT}{q} \sinh^{-1}\left(\frac{N_D - N_A}{2n_{ie}}\right),$$

$$n_c = \sqrt{\frac{(N_D - N_A)^2}{4} + n_{ie}^2} + \frac{N_D - N_A}{2},$$

$$p_c = \sqrt{\frac{(N_D - N_A)^2}{4} + n_{ie}^2} - \frac{N_D - N_A}{2}, \qquad (2.92)$$

where V_a denotes the applied voltage and the rest in usual notation. Equations (2.92) constitute Dirichlet conditions as the right-hand side of the equations are well prescribed. If the contact, instead of being ideal, has an associated internal contact resistance, then the potential drop across the contact can be accounted for as [2]

$$\psi_c = V_a + \frac{kT}{q} \sinh^{-1}\left(\frac{N_D - N_A}{2n_{ie}}\right) - \rho_c \mathbf{J} \cdot \mathbf{n}, \qquad (2.93)$$

where ρ_c is the contact resistivity and $(\rho_c \mathbf{J} \cdot \mathbf{n})$ is an ohmic term which contains the normal derivative of the electrostatic potential, $\partial\psi/\partial n$. This makes (2.93) a mixed boundary condition. In (2.93), $\mathbf{J}$ is the total current density through the contact, $viz.$, $\mathbf{J_n} + \mathbf{J_p}$, and $\mathbf{n}$ is a unit vector normal to the contact. If the resistance associated with contact is external, then we have an appropriate modification of the last term in (2.93)

$$\psi_c = V_a + \frac{kT}{q} \sinh^{-1}\left(\frac{N_D - N_A}{2n_{ie}}\right) - RI \qquad (2.94)$$

where R is the external resistor and I is the total current through the contact (see Eq. (2.95) below). Equation (2.94) is also a mixed boundary condition; both (2.93) and (2.94) create additional coupling between Poisson's and continuity equations [2].

When we have a current source at the contact, we have the relation

$$I = \int_{\Gamma_c} (\mathbf{J_n} + \mathbf{J_p}) \cdot \mathbf{n} \, d\Gamma, \qquad (2.95)$$

where the value specified for I denotes the value of the current source. Here, Γ_c denotes the contact surface and $d\Gamma$ a surface element. If $I = 0$, then we have a high impedance equipotential probe. In practice, I is set to some non-zero value to reflect the leakage current of the measurement circuit.

2.6.2 Schottky Contacts

Although the physics of Schottky contacts is complicated [8], accurate experimental characterization of key parameters, such as the barrier height and width, enables use of simplified models which may be adequate for simulation of contact behavior. With current flow, conditions at the contact based on a combination of thermionic emission and diffusion theories become [1, 39]

$$\psi_c = V_a + \frac{kT}{q} \sinh^{-1}\left(\frac{N_D - N_A}{2n_{ie}}\right) - \varphi_s,$$

$$\mathbf{J_n} \cdot \mathbf{n} = -q v_n (n - n_c)$$

$$\mathbf{J_p} \cdot \mathbf{n} = q v_p (p - p_c) \tag{2.96}$$

where φ_s is the Schottky barrier height which is a function of the materials constituting the contact, $v_{n,p}$ denote electron and hole thermionic recombination velocities, respectively, and n_c and p_c are given by (2.92). If $v_{n,p} = 0$, we have no current flow, viz., $\mathbf{J_{n,p}} \cdot \mathbf{n} = 0$. On the other hand, for infinite recombination velocities, relations (2.96) reduce to (2.92) for the carrier concentrations.

2.6.3 Insulators and Interfaces

In regions of the device containing insulating materials (e.g., oxide or nitride), only Poisson's equation needs to be solved [1, 2]:

$$\text{div}\,(\varepsilon_{ins}\,\text{grad}\,\psi) = -Q_{ins} \tag{2.97}$$

where ε_{ins} is the permittivity of the insulator and Q_{ins} is the trapped charge density in the insulator. At semiconductor-insulator interfaces, we assume continuity of electric potential. Here, we impose an additional condition [1,2] which follows from Gauss' law

$$\varepsilon_{sem}\left(\frac{\partial \psi}{\partial n}\right)_{sem} - \varepsilon_{ins}\left(\frac{\partial \psi}{\partial n}\right)_{ins} = -Q_{int}, \tag{2.98}$$

where ε_{sem} denotes the permittivity of the semiconductor and Q_{int} represents the interface charge density. At ideal interfaces, $Q_{int} = 0$ and we have conservation of displacement across the interface. With respect to current flow, we have [1, 2]

$$\mathbf{J_n} \cdot \mathbf{n} = q R_{surf}, \tag{2.99}$$

$$\mathbf{J_p} \cdot \mathbf{n} = -q R_{surf}, \tag{2.100}$$

where R_{surf} denotes the net recombination rate at the surface and takes the form shown in (2.71).

2.6.4 Outer Boundaries

Although we are only interested in behavior of the key variables within the active device region, suitable artificial boundaries enclosing this region are needed such that we can apply simple boundary conditions on its surface without compromising the true physical behavior of variables in the interior. As we saw earlier, in relation to Fig. 2.2, our artificial boundaries, in the absence of an external field, are LM, MN, NA, on which we impose

$$\mathbf{E} \cdot \mathbf{n} = 0, \tag{2.101}$$

$$\mathbf{J_n} \cdot \mathbf{n} = 0, \tag{2.102}$$

$$\mathbf{J_p} \cdot \mathbf{n} = 0. \tag{2.103}$$

For the condition on the normal electric field to hold at these boundary, we may need to extend our artificial boundaries in the presence of strain or magnetic field, since $\mathbf{E} \cdot \mathbf{n} \neq 0$, by virtue of the Hall or piezoresistance effects. Thus, the choice of location for these boundaries is very important for physical plausibility of numerical solutions in the device active region!

2.7 The External Fields – What Do They Influence?

- An optical/radiation field primarily affects the generation-recombination balance. The generation-recombination rate is a function of the absorption coefficient which is dependent on the material, the doping concentration, and wavelength of the incoming radiation.
- The thermal field affects all transport parameters starting with the effective mass, the gap energy, density of states, and subsequently, the carrier concentrations and mobilities.
- With a non-isothermal field, we have temperature gradient as a driving force of electrical current. Here, we have heat transport which needs to be simulated using the energy equation under appropriate boundary conditions. For large temperature differences, we have a position-dependent band gap in the material.
- The magnetic field makes the conductivity anisotropic by virtue of the Lorentz force. The conductivity becomes a tensor with off-diagonal components that are field-dependent. The artificial boundaries need to account for the possible presence of a Hall field.
- A strain field turns the various electronic transport parameters, such as effective mass and conductivity, from scalars to tensors, since

$1/m_{ij} \sim \partial^2 E_k/\partial k_i \partial k_j$. The band structure can no longer be assumed parabolic and isotropic. Depending on the strain field, the gap energy becomes position-dependent. A gradient in the strain field can be a driving force for electrical as well as thermal transport. The artificial boundaries need to account for the non-zero component of electric field due to the piezoresistance effect.

• All of the above effects can be simulated by coupling the associated partial differential equations; unfortunately, as yet, there is no straightforward description for the underlying coefficients!

2.8 References

[1] Selberherr, S., *Analysis and Simulation of Semiconductor Devices*, Vienna: Springer-Verlag, 1984.

[2] Baccarani, G., Rudan, M., Guerrieri, R., Ciampolini, P., Physical Models for Numerical Device Simulation, in: *Process and Device Modeling*, Engl, W. L. (Ed.), Amsterdam: North-Holland, 1986, pp. 107–158.

[3] Kittel, C., *Introduction to Solid State Physics*, 6th Ed., New York: Wiley, 1986.

[4] Nye, J. F., *Physical Properties of Crystals*, Oxford: Oxford University Press, 1957.

[5] *SUPREM*, Integrated Circuits Laboratory (ICL), Department of Electrical Engineering, Stanford University, CA, USA. http://www-tcad.stanford.edu/tcad/org.html

[6] Street, R. A., *Hydrogenated Amorphous Silicon*, Cambridge: Cambridge University Press, 1991.

[7] McKelvey, J. F., *Solid State and Semiconductor Physics*, New York: Harper & Row, 1966.

[8] Sze, S. M., *Physics of Semiconductor Devices*, New York: Wiley, 1981.

[9] Madarasz, F. L., Lang, J. E., Hemeger, P. M., Effective Mass for *p*-Type Si, *J. Appl. Phys.*, 52 (1981), 4646–4648.

[10] Manku, T., *Electronic Transport Properties of Strained and Relaxed $Si_{1-x}Ge_x$ Alloys*, Ph.D. Dissertation, University of Waterloo, Waterloo, Ontario N2L 3G1, Canada, 1993.

[11] Jain, S. C., *Germanium-Silicon Strained Layers and Heterostructures*, New York: Academic Press, 1994.

[12] Mock, M. S., Transport Equations in Heavily Doped Silicon, and the Current Gain of a Bipolar Transistor, *Solid-State Electronics*, 16 (1973), 1251–1259.

[13] van Overstraeten, R. J., de Man, H. J., Mertens, R. P., Transport Equations in Heavy Doped Silicon, *IEEE Trans. Electron Devices*, ED-20 (1973), 290–298.

[14] Slotboom, J. W., de Graaff, H. C., Measurements of Bandgap Narrowing in Si Bipolar Transistors, *Solid-State Electronics*, 19 (1976), 857–862.

[15] Slotboom, J. W., de Graaff, H. C., Bandgap Narrowing in Silicon Bipolar Transistors, *IEEE Trans. Electron Devices*, ED-24 (1976), 1123–1125.

[16] Slotboom, J. W., The *pn*-Product in Silicon, *Solid-State Electronics*, 20 (1977), 279–283.

[17] Mertens, R. P., van Meerenbergen, J. L., Nijs, J. F., van Overstraeten, R. J., Measurement of the Minority Carrier Transport Parameters in Heavily-Doped Silicon, *IEEE Trans. Electron Devices*, ED-27 (1980), 949–955.

[18] Tang, D. D., Heavy Doping Effects in *p-n-p* Bipolar Transistors, *IEEE Trans. Electron Devices*, ED-27 (1980), 563–570.

[19] Klaassen, D. B. M., Slotboom, J. W., de Graaff, H. C., Unified Apparent Bandgap Narrowing in *n*- and *p*-Type Silicon, *Solid-State Electronics*, 35 (1992), 125–129.

[20] Klaassen, D. B. M., A Unified Mobility Model for Device Simulation, *Technical Digest*, IEEE IEDM, San Francisco, 1990, pp. 357–360.

[21] Klaassen, D. B. M., A Unified Mobility Model for Device Simulation–I. Model Equations and Concentration Dependence, *Solid-State Electronics*, 35 (1992), 953–959.

[22] Klaassen, D. B. M., A Unified Mobility Model for Device Simulation–II. Temperature Dependence of Carrier Mobility and Lifetime, *Solid-State Electronics*, 35 (1992), 961–967.

[23] del Alamo, J., Swirhun, S., Swanson, R. M., Simultaneous Measurement of Hole Lifetime, Hole Mobility and Bandgap Narrowing in Heavily Doped *n*-Type Silicon, *Technical Digest*, IEEE IEDM, Washington, 1985, pp. 290–293.

[24] del Alamo, J., Swirhun, S., Swanson, R. M., Measuring and Modeling Minority Carrier Transport in Heavily Doped Silicon, *Solid-State Electronics*, 28 (1985), 47–54.

[25] Swirhun, S. E., Kwark, Y.-H., Swanson, R. M., Measurement of Electron Lifetime, Electron Mobility and Band-Gap Narrowing in Heavily Doped *p*-Type Silicon, *Technical Digest*, IEEE IEDM, Los Angeles, 1986, pp. 24–27.

[26] del Alamo, J., Swanson, R. M., Measurement of Steady-state Minority-Carrier Transport Parameters in Heavily Doped *n*-Type Silicon, *IEEE Trans. Electron Devices*, ED-34 (1987), 1580–1589.

[27] Marshak, A. H., van Vliet, K. M., Electrical Current in Solids with Position-Dependent Band Structure, *Solid-State Electronics*, 21 (1987), 417–427.

[28] Seeger, K., *Semiconductor Physics*, 3rd Ed., Berlin: Springer-Verlag, 1985.

[29] Ridley, B. K., *Quantum Processes in Semiconductors*, Oxford: Clarendon Press, 1988.

[30] Conwell, E. M., Weisskopf, V. F., Theory of Impurity Scattering in Semiconductors, *Phys. Rev.*, 77 (1950), 388–390.

[31] Brooks, H., Scattering by Ionized Impurities in Semiconductors, *Phys. Rev.*, 83 (1951), 879.

[32] Fiegna, C., Sangiorgi, E., Modeling of High Energy Electrons at the Microscopic Level, *IEEE Trans. Electron Devices*, 40 (1993), 619–627.

[33] Nathan, A., Ershov, M., Pearsall, T. P., Assessment of Matthiessen's Rule for Calculation of Carrier Mobility in Semiconductors, *Technical Report*, UW E&CE 96-10, University of Waterloo, Waterloo, Ontario N2L 3G1, Canada, 1996.

[34] Ziman, J. H., Electrons and Phonons, Oxford: Oxford University Press, 1960.

[35] Pearsall, T. P., Takeda, Y., Failure of Matthiessen's Rule in the Calculation of Carrier Mobility and Alloy Scattering Effects in $Ga_{0.47} In_{0.53}$ As, *Electron. Lett.*, 17 (1981), 573–574.

[36] Caughey, D. M., Thomas, R. E., Carrier Mobilities in Silicon Empirically Related to Doping and Field, *Proc. IEEE*, 52 (1967), 2192–2193.

[37] Arora, N. D., Hauser, J. R., Roulston, D. J., Electron and Hole Mobilities as a Function of Concentration and Temperature, *IEEE Trans. Electron Devices*, ED-29 (1982), 292–295.

[38] Masetti, G., Severi, M., Solmi, S., Modeling of Carrier Mobility Against Carrier Concentration in Arsenic-, Phosphorus-, and Boron-Doped Silicon, *IEEE Trans. Electron Devices*, ED-30 (1983), 764–769.

[39] Engl, W. L., Dirks, H. K., Meinerzhagen, B., Device Modeling, *Proc. IEEE*, 71 (1983), 10–33.

[40] O, N., Nathan, A., CCD-Based Magnetic Field Imaging, *IEEE Trans. Electron Devices*, 44 (1997), 1653–1657.

[41] Canali, C., Majni, G., Minder, R., Ottaviani, G., Electron and Hole Drift Velocity Measurements in Silicon and Their Empirical Relation to Electric Field and Temperature, *IEEE Trans. Electron Devices*, ED-22 (1975), 1045–1047.

[42] Selberherr, S., Hänsch, W., Seavey, M., Slotboom, J., The Evolution of the MINIMOS Mobility Model, *Solid-State Electronics*, 33 (1990), 1425–1436.

[43] Fossum, J. G., Computer Aided Numerical Analysis of Silicon Solar Cells, *Solid-State Electronics*, 19 (1976), 269–277.

[44] Dziewior, J., Schmid, W., Auger Coefficient for Highly-Doped and Highly Excited Silicon, *Appl. Phys. Letts.*, 31 (1977), 346–348.

[45] Pike, G. E., Seager, C. H., The dc Voltage Dependence of Semiconductor Grain-Boundary Resistance, *J. Appl. Phys.*, 50 (1979), 3414–3422.

[46] Kamins, T., *Polycrystalline Silicon for Integrated Circuit Applications*, Boston: Kluwer Academic Publishers, 1988.

[47] Seto, J. Y. W., The Electrical Properties of Polycrystalline Silicon Thin Films, *J. Appl. Phys.*, 46 (1975), 5247–5254.

[48] Kamins, T. I., Hall Mobility in Chemically Deposited Polycrystalline Silicon, *J. Appl. Phys.*, 42 (1971), 4357–4365.

[49] Baccarani, G., Riccò, B., Spadini, G., Transport Properties of Polycrystalline Silicon Films, *J. Appl. Phys.*, 49 (1978), 5565–5570.

[50] Mandurah, M. M., Saraswat, K. C., Kamins, T. I., A Model for Conduction in Polycrystalline Silicon–Part I: Theory, *IEEE Trans. Electron Devices*, ED-28 (1981), 1163–1171.

[51] Mandurah, M. M., Saraswat, K. C., Kamins, T. I., A Model for Conduction in Polycrystalline Silicon–Part II: Comparison of Theory and Experiment, *IEEE Trans. Electron Devices*, ED-28 (1981), 1171–1176.

[52] Mandurah, M. M., Saraswat, K. C., Helms, C. R., Kamins, T. I., Dopant Segregation in Polycrystalline Silicon, *J. Appl. Phys.*, 51 (1980), 5755–5763.

[53] Kamins, T. I., Cass, T. R., Structure of Chemically Deposited Polycrystalline Silicon Films, *Thin Solid Films*, 16 (1973), 147–165.

[54] Ghosh, A. K., Fishman, C., Feng, T., Theory of the Electrical and Photovoltaic Properties of Polycrystalline Silicon, *J. Appl. Phys.*, 51 (1980), 446–454.

[55] de Graaff, H. C., Huybers, M., de Groot, J. G., Grain Boundary States and the Characteristics of Lateral Polysilicon Diodes, *Solid-State Electronics*, 25 (1982), 67–71.

[56] Baccarani, G., Impronta, M., Riccò, B., I-V Characteristics of Polycrystalline Silicon Resistors, *Rev. Phys. Appl.*, 13 (1978), 777–782.

[57] Stevens, M. E., *CMOS Electrothermal Microsensors for Flow and Pressure Measurements*, M.A.Sc. Thesis, University of Waterloo, Waterloo, Ontario N2L 3G1, Canada, 1996.

[58] Volklein, F., Baltes, H., Thermoelectric Properties of Polysilicon Films Doped with Phosphorus and Boron, *Sensors and Materials*, 3 (1992), 325–334.

[59] Shen, B., Allegretto, W., Hu, M., Yu, B., Robinson, A. M., Lawson, R., Simulation and Physical Response of Negative TCR Polysilicon Micromachined Structures, *Canadian J. Phys.* (Suppl.), 74 (1996), S147–S150.

[60] Mastrangelo, C. H., Muller, R. S., Thermal Diffusivity of Heavily Doped Low Pressure Chemical Vapor Deposited Polycrystalline Silicon Films, *Sensors and Materials*, 3 (1988), 133–142.

[61] Mastrangelo, C. H., Yeh, J. H. J., Muller, R. S., Electrical and Optical Characteristics of Vacuum Sealed Polysilicon Microlamps, *IEEE Trans. Electron Devices*, ED-39 (1992), 1363–1375.

[62] Allegretto, W., Shen, B., Lai, Z., Robinson, A. M., Numerical Modeling of Time Response of CMOS Micromachined Thermistor Sensor, *Sensors and Materials*, 6 (1994), 71–83.

[63] Paul, O., von Arx, M., Baltes, H., Process-Dependent Thermophysical Properties of CMOS IC Thin Films, *Digest of Technical Papers*, Vol. 1, Transducers '95, Stockholm, 1995, pp. 178–181.

[64] Hummel, R. E., *Electronic Properties of Materials*, 2nd Ed., Berlin: Springer-Verlag, 1993.

[65] Paul, O., Baltes, H., Novel Fully CMOS-Compatible Vacuum Sensor, *Sensors and Actuators A*, 46–47 (1995), 143–146.

[66] Nagata, M., Swart, N., Stevens, M., Nathan, A., Thermal Based Micro Flow Sensor Optimization Using Coupled Electrothermal Numerical Simulations, *Digest of Technical Papers*, Transducers '95, Vol. 2, Stockholm, 1995, pp. 447–450.

[67] Allegretto, W., Nathan, A., Baltes, H., Numerical Analysis of Magnetic-Field-Sensitive Bipolar Devices, *IEEE Trans. CAD of ICAS*, 10 (1991), 501–511.

[68] Riccobene, C., Wachutka, G., Bürgler, J. F., Baltes, H., Operating Principle of Dual Collector Magnetotransistors Studied by Two-Dimensional Simulation, *IEEE Trans. Electron Devices*, 41 (1994), 32–41.

[69] Thangaraj, D., Nathan, A., The Discretization of Anisotropic Drift-Diffusion Equations, *Technical Report*, UW E&CE 97-11, University of Waterloo, Waterloo, Ontario N2L 3G1, Canada, 1997.

Radiation Effects on Carrier Transport 3

The model equations and boundary conditions reviewed in the preceding chapter describe electrical transport in semiconductor microtransducers in the absence of external fields. As summarized in Sect. 2.7, external fields appreciably alter carrier transport by introducing asymmetries in device operation. For example, radiation alters the generation-recombination rate, and hence, the electrical carrier transport in a semiconductor by virtue of its wavelength- and material-dependent absorption. In this chapter, we review the physical effects induced by radiant signals along with associated model equations relevant to simulation and subsequent optimization of optical microtransducers.

Radiant signals range from electromagnetic to nuclear particle radiation [1, 2] and can be classified as photons, charged-particles, and uncharged-particles. Photons are distinguished by their energy E (expressed in units of eV), frequency (ν) or wavelength (λ). Electromagnetic radiation in the energy range from 1 meV ($\lambda = 1.24$ mm) to 1 MeV ($\lambda = 1.24$ pm) includes infrared (IR) radiation, visible light, ultraviolet (UV) radiation, soft and hard X-rays, and gamma-rays. The spectrum of electromagnetic energies with corresponding frequencies and wavelengths are shown in Fig. 3.1 along with the associated relations in Table 3.1.

Semiconductor radiant sensors convert incident electromagnetic or particle radiation into an electrical output signal. Examples of radiant sensors include photoconductors, photocapacitors, phototransistors, p-i-n and Schottky diodes, and drift chambers (see, e.g., [1]). The energy associated with the radiant signal is partially or fully absorbed by the semiconductor, leading to generation of electron-hole (e-h) pairs. The rate of e-h pair generation is a measure of the intensity and/or energy of the incident radiation. Although the field of radiant sensing is rather mature, it is still a diverse research area and new device structures, materials, and fabrication process technologies continue to evolve. This evolution is posing new challenges in speed, imaging over large areas [4], detection in

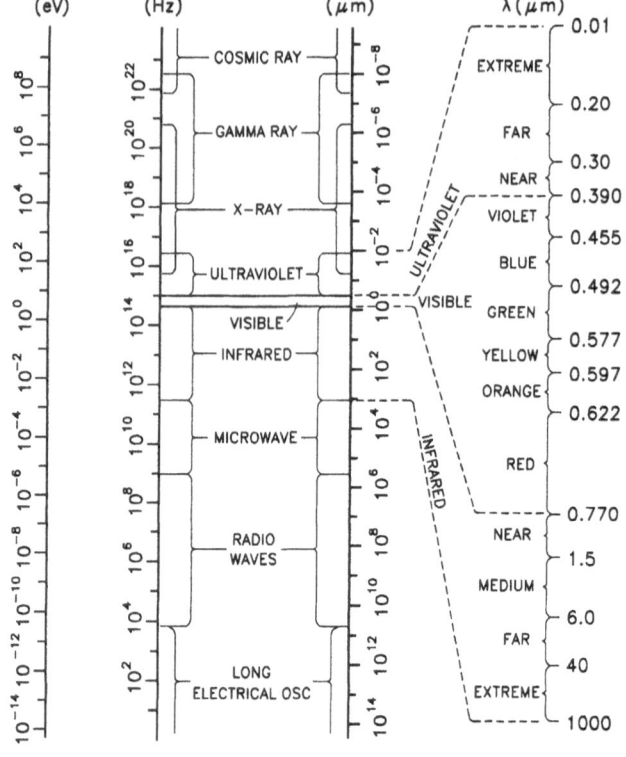

Fig. 3.1 The spectrum of radiant signals. Adapted from [3]

Table 3.1. *Relations for the photon energy*

Relation	Value
$E = h\nu = \dfrac{hc}{\lambda}$	$h = 6.626 \times 10^{-34}\,\text{Js} = 4.136 \times 10^{-15}\,\text{eVs}$
	$c = 2.998 \times 10^{8}\,\text{m/s}$
$E(\text{eV}) = \dfrac{1.24}{\lambda(\mu\text{m})}$	$1\,\text{eV} = 1.602 \times 10^{-19}\,\text{J}$

the UV [5] and IR range [6,7] using Si integrated circuit (IC) technology, and detection of high energy fields [8].

Radiant sensors come with a large number of design variants, involving different sensor materials, geometries, and detection schemes, which depend on the type and energy of radiation and the nature of its interaction [1]. Thus design trade-offs are necessary. This makes two- or three-dimensional numerical device optimization crucial to meet specifications of sensitivity, linearity over the required range, response time, noise level, and reliability with prolonged exposure (see [9–11]). In terms of level of complexity and effort, the simulation of radiant sensors at low photon energies is similar to ordinary semiconductor device simulation. At high photon energies, however, simple and accurate descriptions (models) for the interaction processes are difficult, if not impossible, to obtain for use in standard drift-diffusion simulators. Here, Monte Carlo schemes are invaluable for analysis and simulation of high energy coupled photon and electron transport as well as for computation of the associated interaction cross sections (see [12] and references therein).

Sections 3.1 to 3.6 describe principles and models underlying the interaction of radiation fields with IC-related materials and their effect on device operation. The exact interaction process depends on the energy of the incident radiation and the nature of the semiconductor material (*viz.*, band structure). A summary of model equations relevant to radiant sensor simulation is given in Sect. 3.7, followed by a numerical simulation example, in Sect. 3.8, of a p-n-p-n color sensor whose design has been optimized for efficient separation of the photogenerated electrons and holes, and hence, of the different wavelengths associated with the incident radiant signal.

We begin our discourse on interaction processes, and associated models, with a brief review of optical coefficients. These govern the wavelength-dependent optical characteristics, e.g., reflection and transmission of different materials such as semiconductors, insulators, and metals, at low photon energies. Reflection, transmission, and absorption properties are basic to computing the radiant signal intensity impinging on the sensor's active surface in the presence of single- or multiple-layers between the signal source and the active region. These layer(s), which can come with the fabrication process, e.g., passivation material, give rise to interference effects as a result of reflections at interfaces. Alternatively, they can, by design, be integrated with the sensor as interference filters for spectral selectivity [13–17].

3.1 Reflection and Transmission of Optical Signals

The refractive index of a material is generally a complex quantity, *viz.*,

$$N = n - ik, \tag{3.1}$$

where the real part n simply denotes the refractive index (since in ideal dielectrics, $N = n$) and k is the extinction coefficient. They can be related to the well-known electrical coefficients, namely the permittivity ϵ and conductivity σ, which are scalars for isotropic materials

[18, 19], *viz.*,

$$\varepsilon/\varepsilon_0 = n^2 - k^2, \tag{3.2}$$

$$\sigma = nk\nu\varepsilon_0, \tag{3.3}$$

respectively, where ε_0 denotes the permittivity of free space. The co-efficients are frequency-dependent. The use of *dc* values of ϵ and σ for estimating the material's optical properties is only valid at low frequencies. Relations (3.2) and (3.3) are not valid in anisotropic materials, in which ϵ and σ become tensors.

The refractive index and extinction coefficient can be obtained from reflectance and transmittance measurements of the material. The reflectance describes the ratio of the reflected intensity I_R to the incident intensity I_0 of the radiant signal, $R = I_R/I_0$, at the interface between two media. Similarly, the transmittance is the ratio of the transmitted to incident intensities, $T = I_T/I_0$. On transmission, the intensity of the radiant signal attenuates exponentially as it travels through the material [18], *viz.*,

$$I(x) = TI_0 \exp(-\alpha x), \tag{3.4}$$

where α is the absorption coefficient and x denotes the distance into the material from the location of incidence. The absorption coefficient is related to the extinction coefficient as [19]

$$\alpha = 4\pi k/\lambda \tag{3.5}$$

and its inverse $\lambda/4\pi k$ describes the penetration depth where the intensity degrades to $1/e$ of its initial value (TI_0).

3.1.1 Single- and Multi-Layer Thin Film Systems

The reflectance and transmittance characteristics of single-layer systems can be modeled quite accurately using simple formulas to account for both normal and oblique incidences of the radiant signal (see, e.g., [18, 19]). In particular, the reflectance associated with a radiant signal at normal incidence on a medium of index N_s can be computed as $R = [(N_0-N_s)/(N_0+N_s)][(N_0-N_s)/(N_0+N_s)]^*$. Here, N_0 is the index of the incidence medium and * denotes the complex conjugate. Correspondingly, the transmittance can be computed as $T = 4n_0n_s/[(N_0+N_s)(N_0+N_s)^*]$, where $n_{0,s}$ denote the real part of the complex indices, $N_{0,s}$. Now for a multi-layer thin film assembly sandwiched between source and sensor, optical interactions with each layer need to be taken into account to compute the intensity on the active surface of the sensor.

Transmission of the radiant signal through a multi-layer of q thin films (as depicted in Fig. 3.2), sandwiched between the incident medium (N_0)

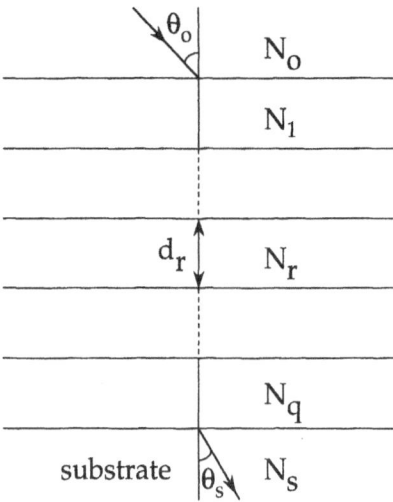

Fig. 3.2 Schematic of thin film assembly for modeling reflection and transmission characteristics. The assembly is sandwiched between substrate (N_s) and transparent medium (N_0); d_r denotes film thickness and N_r the complex refractive index

and the substrate or exit medium (N_s), can be estimated by computing the characteristic admittance matrix [20]. Here, the entire multi-layer thin film assembly is replaced by a single surface of an equivalent input admittance. To arrive at relatively simple expressions for the optical propagation coefficients of the system, we make the assumption that the associated materials are thin, isotropic, homogeneous, and form plane interfaces parallel to the substrate surface. The restriction on film thickness implies an optical path length difference that is less than the coherence length of the radiant signal. No thickness restrictions apply for the incident medium and the substrate. We further assume that the incident medium is transparent and that the incident radiant signal is a plane-polarized homogeneous plane wave. For an oblique angle of incidence θ_0, the reflectance, transmittance, and absorptance are given as [20]

$$R = \left(\frac{\eta_0 B - C}{\eta_0 B + C}\right)\left(\frac{\eta_0 B - C}{\eta_0 B + C}\right)^*, \tag{3.6}$$

$$T = \frac{4\eta_0 \, \mathrm{Re}\,(\eta_s)}{(\eta_0 B + C)(\eta_0 B + C)^*}, \tag{3.7}$$

$$A = \frac{4\eta_0 \, \mathrm{Re}\,(BC^* - \eta_s)}{(\eta_0 B + C)(\eta_0 B + C)^*} \tag{3.8}$$

and depend on the frequency, the angle of incidence, the polarization state of the incident wave, the thicknesses of the various layers, and the complex refractive indices of all layers involved. In (3.6) to (3.8), η_0 and η_s denote the tilted optical admittances for the incident (transparent) layer and the substrate (exit medium), respectively, and B and C denote elements that form the characteristic admittance matrix,

$$\begin{bmatrix} B \\ C \end{bmatrix} = \left(\prod_{r=1}^{q} \begin{bmatrix} \cos \delta_r & (i \sin \delta_r)/\eta_r \\ i\eta_r \sin \delta_r & \cos \delta_r \end{bmatrix} \right) \begin{bmatrix} 1 \\ \eta_s \end{bmatrix}. \qquad (3.9)$$

Here, η_r denotes the optical admittance of the rth layer in the thin film assembly. It takes the following form for transverse electric (TE) and transverse magnetic (TM) polarization:

$$s\text{-polarized plane wave (TE):} \qquad \eta_r = \mathscr{Y} N_r \cos \theta_r, \qquad (3.10)$$

$$p\text{-polarized plane wave (TM):} \qquad \eta_r = \mathscr{Y} N_r / \cos \theta_r, \qquad (3.11)$$

where $\mathscr{Y}$ denotes the characteristic optical admittance in free space, $(\epsilon_0/\mu_0)^{1/2}$, and θ_r can be determined from Snell's law

$$N_0 \sin \theta_0 = N_r \sin \theta_r = N_s \sin \theta_s \qquad (3.12)$$

with θ_0 and θ_s as denoted in Fig. 3.2. The term δ_r in (3.9) describes the phase shift experienced by the radiant signal as it traverses a distance d_r normal to the boundary. It is given by

$$\delta_r = (2\pi/\lambda) N_r d_r \cos \theta_r. \qquad (3.13)$$

In the case of normal incidence, the coefficients can be reduced to simple forms. If the incident radiant signal is non-polarized, one can assume an average value of the coefficients obtained for TE and TM.

In evaluating B and C in (3.9), the order of the layers have to be maintained in the multiplication. For example, if the layer q is next to the substrate, the characteristic matrix reads [20]

$$\begin{bmatrix} B \\ C \end{bmatrix} = [M_1][M_2] \dots [M_q] \begin{bmatrix} 1 \\ \eta_s \end{bmatrix}, \qquad (3.14)$$

where the M's indicate the matrices associated with each of the layers in the assembly.

3.2 Modeling Optical Absorption in Intrinsic Semiconductors

Optical absorption in intrinsic monocrystalline semiconductors is governed by the photoelectric effect. Generally, the absorption is high in the visible range, but as the wavelength approaches the infrared part of the electromagnetic spectrum, the material becomes transparent. The boundary between the absorption and transparent regions defines the fundamental absorption edge, whose wavelength (λ_e) and frequency (ν_e) can be approximated [18] as: $hc/\lambda_e = h\nu_e = E_g$, where E_g is the band gap energy of the semiconductor. When the energy of incident radiant signal exceeds the band gap energy of the material, $h\nu > E_g$ (or correspondingly when $\lambda < \lambda_e$ or $\nu > \nu_e$), interband transitions take place giving rise to band-to-band absorption. In this case, excitation of electrons and their transition from the valence to conduction band occurs, provided that unoccupied energy levels are available in the conduction band and that no energy states exist within the gap. The electron and hole then are free to conduct in the conduction and valence bands, respectively. The band gap, and hence, the band-to-band absorption threshold, is a function of parameters such as temperature and strain [21]. For example, increasing the temperature lowers the absorption threshold. This is discussed in Sect. 2.3.4. The effects of mechanical strain on gap energy are discussed in Chapt. 6. The maximum wavelength for which there is band-to-band absorption can be determined as: $\lambda_{\max} = hc/E_g^{\text{eff}}$, where E_g^{eff} is the effective band gap.

3.2.1 Band-to-Band Transitions

Band-to-band transitions can either involve only photons, or in addition, phonons (lattice vibration quanta) that are either generated or absorbed in the transition for conservation of crystal momentum [18]. The type of transition is determined by the energy band structure of the semiconductor. For example, in direct band semiconductors such as GaAs or InSb, the lowest and highest energy states in the conduction and valence bands, respectively, occur at the same value of the wave vector **k**. In indirect band semiconductors such as Si and Ge, the corresponding values of the wavevector are different and transitions are indirect, involving phonons for momentum conservation. The momentum (h/λ) associated with the incident photon is much smaller than the diameter of the Brillouin zone and is thus negligible in comparison to the momentum of the electron at optical frequencies [19]. Their momenta become comparable only at very high energy. Unlike photons, phonon energies are very small but since they have velocities in the acoustic and optical range, their momentum values are comparable to those of the electron. The various transitions for a selection of most commonly used semiconductors are illustrated in Fig. 3.3. There are considerable differences in absorption behavior between direct and indirect band semiconductors [21]. For example, in GaAs, since the conduction band minimum and valence band maximum occur at $\mathbf{k} = 0$, only vertical transitions (denoted as "d" in Fig. 3.3c) are allowed. The

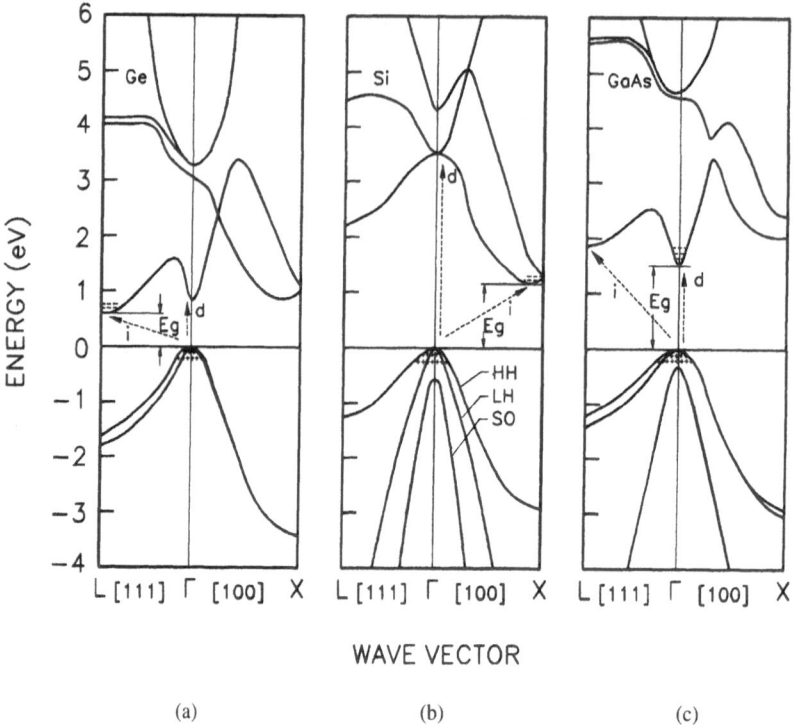

Fig. 3.3 Indirect (i) and direct (d) band-to-band transitions in (a) Ge, (b) Si, and (c) GaAs; E_g denotes band gap energy, and HH, LH, and SO the heavy hole, light hole, and spin orbit bands, respectively. Adapted from [3]

indirect transition (denoted as "i" in Fig. 3.3c) has a lower transition probability; it requires about 0.31 eV more than the direct transition and involves phonons for momentum conservation, since $\Delta \mathbf{k} \neq 0$. Direct transitions in Si and Ge do not correspond to the lowest value of energy or the absorption edge. Here, the valence band maximum takes place at $\mathbf{k} = 0$ but the conduction band minimum is at the edge of the first Brillouin zone. The direct transitions give rise to a steep increase in the absorption coefficient which takes place at 0.8 eV in Ge and 3.4 eV in Si (see Fig. 3.4). At energies below these values, transitions are indirect.

Since direct transitions do not involve phonons, the absorption is not strongly dependent on temperature (see Fig. 3.4); its temperature dependence stems from that of the band gap and Fermi energies. Indirect transitions, however, depend on the phonon distribution function governed by the Bose-Einstein statistics; $[e^{E_p/kT} - 1]^{-1}$. Thus the absorption has, in addition, an exponential temperature dependence which is prevalent when the probability of phonon absorption is high in the transition. The nature of

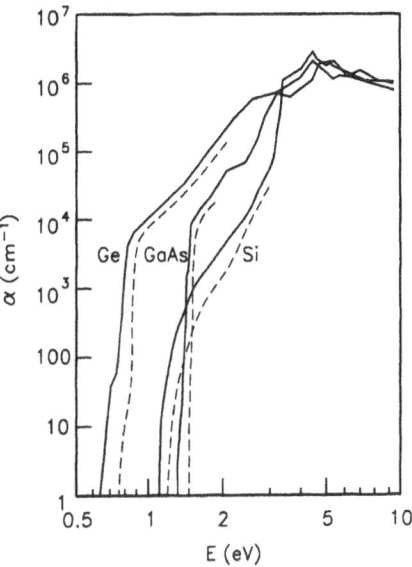

Fig. 3.4 The band-to-band absorption coefficient as a function of energy for intrinsic Si, Ge, and GaAs at 300K (solid curves) and 77K (dashed curves). Adapted from [3]

phonon participation in the transition is discussed next where the band-to-band absorption coefficient is modeled.

3.2.2 Absorption Coefficient

Optical absorption coefficients are needed for simulation of Si photo-detectors and color sensors. Apart from radiant sensing applications, these coefficients are also needed for retrieval of electronic parameters such as the minority carrier diffusion length. The band-to-band absorption coefficient in intrinsic semiconductors is proportional to: the probability of transition from initial to final energies; the density of occupied states at the initial energy; and the available density of states at the final energy [18]. Accounting for all possible phonon-assisted transitions between states separated by energy $h\nu$, the absorption coefficient in intrinsic Si is given as (see [22, 23])

$$\alpha_{ph} = \frac{8A^2}{\pi} \frac{\left(h\nu - E_g^{\text{eff}} + E_p\right)^2}{\exp\left(E_p/kT\right) - 1} I^+ + \frac{8A^2}{\pi} \frac{\left(h\nu - E_g^{\text{eff}} - E_p\right)}{1 - \exp\left(-E_p/kT\right)} I^-,$$

$$(3.15)$$

where

$$I^+ = \int_0^1 \frac{[x(1-x)]^{1/2} dx}{1 + \exp\left[E_F - x(h\nu - E_g^{\text{eff}} + E_p)\right]/kT} \tag{3.16}$$

and I^- is obtained from I^+ with the phonon energy E_p replaced by $-E_p$. Here, E_g^{eff} and E_F denote the effective band gap and Fermi energies, both of which are functions of temperature, and A^2 is a coefficient. Expression (3.16) is given in normalized form, where x is an integration variable. The first and second terms on the right-hand-side of (3.15) account for phonon absorption and emission, respectively, in the transition process. Phonon absorption is dominant at low energies (near the absorption threshold). This is where the temperature dependence of absorption is largest (see denominator of first term in relation 3.15). With decreasing temperature, the number of phonons decreases and so does the probability of phonon-absorption. Phonon-emission is dominant at higher energies. Here, the temperature dependence of absorption is weaker and is governed by the temperature dependence of the band gap and Fermi energies.

Because of its small departure from linearity, expression (3.15) can be approximated as [24]

$$\alpha_{\text{ph}} = \hat{A}^2 [(h\nu - E_g^{\text{eff}} + E_p)^2 I^+ + (h\nu - E_g^{\text{eff}} - E_p)^2 I^-]. \tag{3.17}$$

At high energies, the terms, $(h\nu - E_g^{\text{eff}} + E_p)^2$ and $(h\nu - E_g^{\text{eff}} - E_p)^2$, reduce approximately to $(h\nu - E_g^{\text{eff}})^2$ and the integrals, I^+ and I^-, become approximately equal. The coefficient $\hat{A}^2$ in (3.17) now takes a value slightly modified from A^2. To compute the absorption coefficient α_{ph} for the intrinsic material, we require values for the parameters E_g^{eff}, E_p, and $\hat{A}^2$ [23]. The band gap energy can be calculated using the models given in Sects. 2.3.3 and 2.3.4. For the phonon energy, only the contribution of transverse optical phonons is considered (see [23–26]). Transverse acoustic phonons are ignored since their contribution is important only near the threshold. For the optical phonon energy E_p, we can assume a value of 58 meV for Si [27]. The coefficient $\hat{A}$, retrieved from measurement data for intrinsic crystalline Si, is approximately 65 $\text{cm}^{-1/2}\,\text{eV}^{-1}$ [22].

3.3 Absorption in Heavily-Doped Semiconductors

The absorption behavior in heavily-doped semiconductors is governed not only by interband transitions, as in the intrinsic case, but also by intraband transitions. Intraband transitions are dominant at low photon energies and give rise to free carrier or infra-red absorption [18, 19]. Here, doping-generated free carriers are excited within the conduction or valence bands

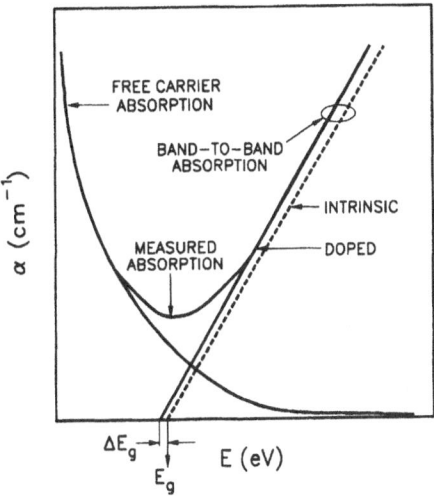

Fig. 3.5 Sketch of the absorption behavior in the visible range for doped and intrinsic crystalline semiconductors; $E_g - \Delta E_g = E_g^{\text{eff}}$

to higher energy levels provided that unoccupied states are available. Intraband transitions depend on the wavelength of incident radiation and the impurity concentration in the material. The doped material can become opaque to long wavelengths (e.g., radio waves). An illustration of the absorption behavior measured in doped Si (see, e.g., [18]) is sketched in Fig. 3.5. In the doped material, at energies below E_g^{eff}, the absorption is mainly due to free carriers and at energies larger than E_g^{eff}, both band-to-band and free carrier absorption processes overlap; the contribution of the latter decreasing rapidly with increasing energy [23–26].

3.3.1 Band-to-Band Absorption Coefficient

Unlike the intrinsic material where absorption is determined by transitions involving phonons, in the heavily-doped material additional transitions take place without phonon involvement [23]. Here, momentum is conserved by impurity or free carrier scattering, the latter at very high carrier concentrations. Thus, when computing the band-to-band absorption coefficient in heavily-doped Si, we need to take into account both phonon-assisted (α_{ph}) and impurity-assisted (α_i) transitions,

$$\alpha = \alpha_{\text{ph}} + \alpha_i. \tag{3.18}$$

Here, α_{ph} takes the same form as with the intrinsic material described by (3.15). The effect of impurity-assisted transitions, valid only for the doped material, can be modeled as [22]:

$$\alpha_i = \frac{8B_1^2}{\pi}(h\nu - E_g^{\text{eff}})^2 I^i. \tag{3.19}$$

The term I^i is evaluated at relatively large energies and is obtained from (3.16) with the the phonon energy, $E_p = 0$. The coefficient B_1^2 is independent of energy but a function of the dopant concentration. Apart from impurity-assisted transitions at high dopant concentrations, we must also consider the doping dependence of band gap energy and Fermi level, both of which lead to a lowering of the absorption threshold. Heavy doping leads to band gap narrowing (see Sect. 2.3.3) and the density of states at the threshold energy, which varies as $E_F^{1/2}$, is considerably larger than the intrinsic counterpart [23].

The effect of impurity-enhanced transitions can also be accounted for by using a simpler model [28]:

$$\alpha = B_2^2 \alpha_{\text{ph}}. \tag{3.20}$$

Here, the coefficient B_2^2 is similar to B_1^2, and α_{ph} is as given by (3.17). Equation (3.20), despite its simplicity, has provided good agreement in fitting selective absorption spectra with theory [29], and will form the basis for extraction of absorption coefficients presented in the following. By comparing the computed (α_{ph}) and measured (α) absorption [26] at high energies ($\sim 1.8\,\text{eV}$), the coefficient B_2^2 can be retrieved for different dopants and concentrations. At these energies, the free carrier absorption can be assumed negligible. Values of this coefficient retrieved for different doping levels have a large scatter, but on average, they can be represented as [23]

$$B_2^2(N_D) = 0.46, \tag{3.21}$$

$$B_2^2(N_A) = 0.8 + 5.0 \times 10^{-5}(N_A/N_0) - 3.2 \times 10^{-6}(N_A/N_0)^2$$
$$+ 5.7 \times 10^{-9}(N_A/N_0)^3 + 9.0 \times 10^{-10}(N_A/N_0)^4 \tag{3.22}$$

for arsenic- and boron-doped Si, respectively. Here, N_D and N_A denote the donor and acceptor concentrations, respectively, and $N_0 = 10^{18}\,\text{cm}^{-3}$. Following (3.21) and (3.22), relations for the band-to-band absorption coefficient retrieved as a function of energy, for arsenic- and boron-doped Si, are given as [23]

$$\alpha = a_0 + a_1(h\nu) + a_2(h\nu)^2 \tag{3.23}$$

with

$$a_0 = -18.4(N_D/N_0) + 5261.7,$$
$$a_1 = 23.2(N_D/N_0) - 9604.1,$$
$$a_2 = -6.4(N_D/N_0) + 4365.5,$$

and

$$\alpha = b_0 + b_1(h\nu) + b_2(h\nu)^2 \tag{3.24}$$

with

$$b_0 = 0.4(N_A/N_0)^2 - 44.0(N_A/N_0) + 5305.3,$$
$$b_1 = -0.6(N_A/N_0)^2 + 70.9(N_A/N_0) - 9574.0,$$
$$b_2 = 0.3(N_A/N_0)^2 - 28.2(N_A/N_0) + 4303.0,$$

respectively. Their behavior is illustrated in Fig. 3.6. The effects of doping are more pronounced in the latter, particularly, at high concentrations. In the case of arsenic, the absorption remains virtually independent of concentration for $N_D > 10^{19}\,\mathrm{cm}^{-3}$.

In the above models, we have assumed that there are no states within the band gap arising from presence of shallow impurity centers or excitons [21]. The latter describe electron-hole pairs bound by Coulomb interaction. Their presence is highly improbable at room temperature due to presence of phonons and free carriers generated by doping.

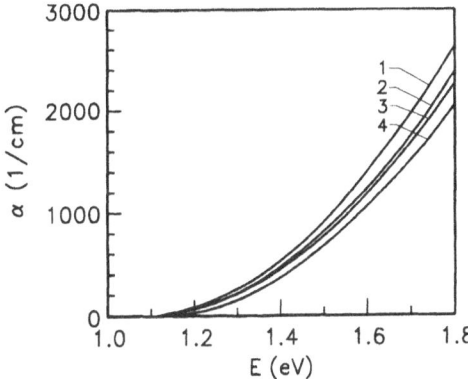

Fig. 3.6 Band-to-band absorption coefficients for Si at 300K as a function of energy for different densities of arsenic (As) and boron (B) [23]. Here, **1**, **2**, **3**, **4** denote 120 (B), 100 (B), 40 (As), 6 (As), respectively, following normalization of doping densities by $10^{18}\,\mathrm{cm}^{-3}$

3.3.2 Free Carrier Absorption Coefficient

Free carrier or infrared absorption due to intraband transitions is proportional to the number of free electrons and holes and the density of available states in the respective conduction and valence bands [18]. Free carrier absorption can be extracted, from the measured absorption spectra, by subtracting the computed band-to-band absorption coefficient described by Eqs. (3.23) and (3.24), taking into account the dependence of band gap narrowing and Fermi level shifts due to doping [23]. At large wavelengths and medium doping levels, the free carrier absorption coefficients vary as λ^2. This law, however, does not hold in cases of high doping or at higher energies for all doping levels. In fact, it has been indicated [30] that the discrepancy involved in extrapolating the free carrier absorption to high energies yields unreliable values of the band gap energy retrieved from the measured spectra for the doped material. Empirical relations retrieved for the free carrier absorption read [23]

$$\alpha = 664.0(N_D/N_{\text{ref}})(h\nu)^{-(1.4+N_D/N_{\text{ref}})}, \tag{3.25}$$

$$\alpha = 11.70(N_A/N_0)^{1/2}(h\nu)^{-2.0} \tag{3.26}$$

for arsenic- and boron-doped Si, respectively. Here, $N_{\text{ref}} = 40 \times 10^{18}/\text{cm}^3$.

3.4 Optical Generation Rate and Quantum Efficiency

With relations (3.23) and (3.24) for the band-to-band absorption coefficients, we can now compute the optical generation rate in the doped material, required in simulation of photodetectors and color sensors. For a constant absorption coefficient, the electron-hole pair generation with monochromatic radiation, takes the form (see, e.g. [18])

$$G_{\text{opt}}(x) = \eta\alpha I(x) = \eta\alpha T I_0 \exp(-\alpha x) \tag{3.27}$$

where η is the quantum efficiency, α is the absorption coefficient, and I_0 is the incident radiation intensity, cf. (3.4). In general, however, the incident radiation is not necessarily monochromatic, thus the quantum efficiency depends on the energy of the radiant signal, and so does the absorption coefficient. In addition, the latter need not be constant along the light path in view of the different levels of doping involved. In this case, the generation rate at position x in the sensor is found by summing the individual components at the different wavelengths,

$$G_{\text{opt}}(x) = \int_{\lambda_1}^{\lambda_2} \eta(\lambda)\alpha(x, \lambda)T(\lambda)I_0(\lambda) \exp[-\alpha(x, \lambda)x]d\lambda \tag{3.28}$$

where $\lambda_{1,2}$ denote the wavelength boundaries of the incident radiation contributing to electron-hole pair generation. Interference effects associated with transmission of the radiant signal through multilayers, sandwiched between source and sensor, are accounted for by the term $T(\lambda)$ as given by Eq. (3.7).

The quantum efficiency, $\eta(\lambda)$ in (3.28), defines the average number of electron-hole pairs produced per photon absorbed by the sensing material. For radiant signal energies near the absorption edge, the quantum efficiency is unity but it increases with increasing photon energy. To date, there are no reliable models for predicting this enhancement in quantum efficiency at higher energies where absorption of one photon results in creation of more than one electron-hole pair. A frequently used semi-empirical model, developed by Shockley [31], attempts to calculate the average ionization energy (w) of electron-hole pairs, viz., $w = 2.2\,E_i + rE_r$. Here, E_i is the energy gained by the ionized electron, E_r denotes the energy lost in the creation of a phonon ($\sim 63\,\text{meV}$ in Si), and r is the average number of phonons (~ 17.5 in Si) created in the process of one electron-hole pair generation. Dividing the photon energy absorbed by the ionization energy yields the quantum efficiency. Although Shockley's model yields a reasonable estimate of the energy threshold (where $\eta \geq 1$), the predicted increase in η values at higher energies is too abrupt. The shortcomings of the model are that it does not account for effects of band structure and is unrealistic in its assumed electron and phonon scattering rates [32, 33]. For simulation purposes, it may be more reliable to employ measured values of the quantum efficiency [34] (see Table 3.2). The threshold energy where η deviates from unity ranges from about 3.3 eV to 4.0 eV in Si. There is an apparent maximum at 5.3 eV and a minimum at 5.6 eV with corresponding η values of 1.4 and 1.35, respectively. Beyond 5.6 eV, η increases to 1.5 at 6.0 eV. The associated uncertainty in η at these energies is about 4% due to possible measurement errors. For Ge, at energies below 2.7 eV, η is unity and increases smoothly to a value of 2.0 at 5.0 eV [34].

Table 3.2. *Measured intrinsic quantum efficiency η in Si at 300 K in the energy range 1.2 eV to 6 eV* [34]

Energy	Quantum Efficiency
$1.2\,\text{eV} \leq E \leq 3\,\text{eV}$	1.00
3.3 eV to 4 eV	Deviation threshold
5.3 eV	1.40
5.6 eV	1.35
6.0 eV	1.50

3.5 Low Energy Interactions with Insulators and Metals

Insulators and metals have become an integral part of the microtransducer in a variety of applications. For example, as noted earlier in Sect. 3.1, a multi-layer assembly of dielectric thin films can be employed as anti-reflection coatings or as interference filters for spectral selectivity to ensure an optimal spectral distribution of the radiant signal incident on the detector [14]; thin film oxide/nitride dielectric layers can be employed in thermal sensor configurations for detection of IR signals [35]; and thin film metal layers are employed as optical reflectors in electrostatically-deflectable micromirrors [36, 37]. Calculation of the reflection and transmission characteristics of these thin film assemblies as given by Eqs. (3.6) to (3.8) requires knowledge of the associated optical coefficients, n and k which constitute the complex index.

3.5.1 Refractive Index and Extinction Coefficient

Insulators, because of their large gap energy, have an extremely low electron population in the conduction band. Thus there are no intraband transitions and correspondingly, no free carrier absorption. Interband transitions, and hence, band-to-band absorption, occur only at higher energies when the wavelength of the incident radiant signal approaches the UV range. This is with the exception of possible exciton absorption peaks which occur at energies slightly below the gap energy. Thus insulators are generally transparent from the IR to UV range. However, their absorption behavior, at certain wavelengths in the IR and UV regions, is strongly influenced by presence of impurities and defects in the material. For example, in the low energy (IR) region, phonon excitation can give rise to absorption peaks; in glass, including fused quartz, there is a pronounced absorption peak near 1.38 μm due to the presence of water or OH ions [19].

The wavelength-dependent refractive index and absorption behavior at room temperature in selected IC-compatible dielectric materials are shown in Figs. 3.7 and 3.8. These curves are from Ref. [38], which provides an excellent compilation of measured and theoretically extrapolated values gathered from various sources along with pertinent data retrieval procedures and associated accuracy. With silicon dioxide (Fig. 3.7), there is no absorption data shown in the 0.2 μm to 3.6 μm range. Here, the absorption is dominated by the presence of OH impurities which mask the intrinsic absorption. The optical characteristics shown for silicon nitride (Fig. 3.8) must be used with caution; the refractive index and absorption properties of the material strongly depend on the deposition method and temperature. Thus characterization of material composition of deposited

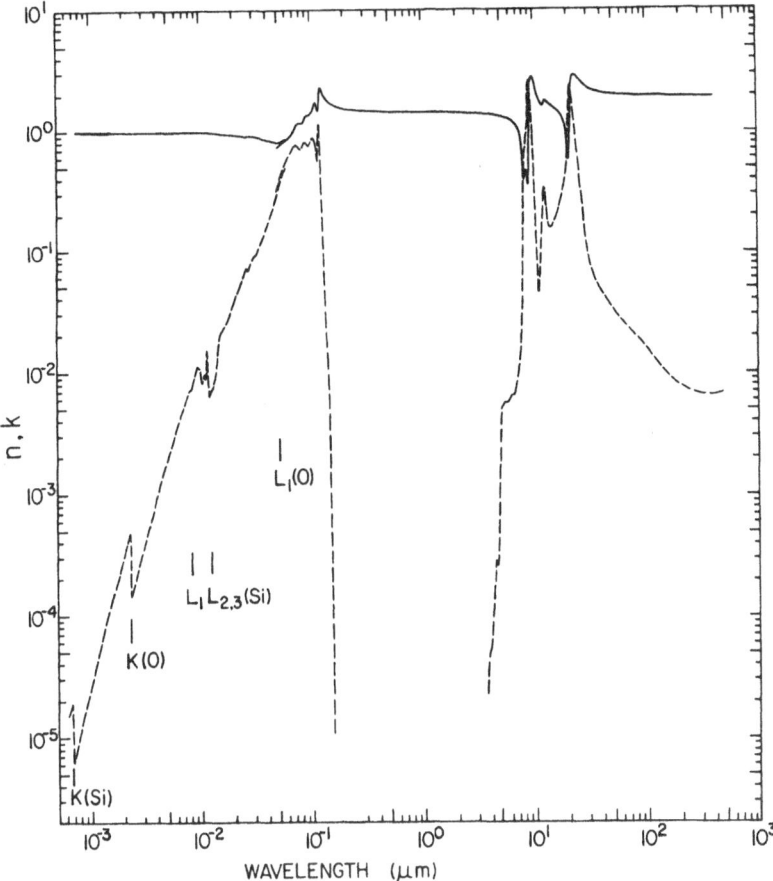

Fig. 3.7 Refractive index, n (solid curve), and extinction coefficient, k (dashed curve), as a function of wavelength for SiO_2. Also indicated is the onset of absorption of K- and L-shells. After [38]

films is crucial, particularly with low temperature processes such as plasma-enhanced chemical vapor deposition (PECVD). Some recent data of the refractive index for PECVD oxides and nitrides can be found in [39−41]. Values compiled for simulation purposes are given in Tables 3.3 and 3.4.

Absorption in metals is dominated by intraband transitions at relatively low photon energies and interband transitions at higher energies. However, unlike semiconductors and insulators, the physical mechanisms that govern optical properties of metals are rather involved [19]. At wavelengths in the far-IR region, metals are good reflectors and their optical properties in this region ($f \leq 10^{13}$ Hz) can be described by the well-known Hagen and

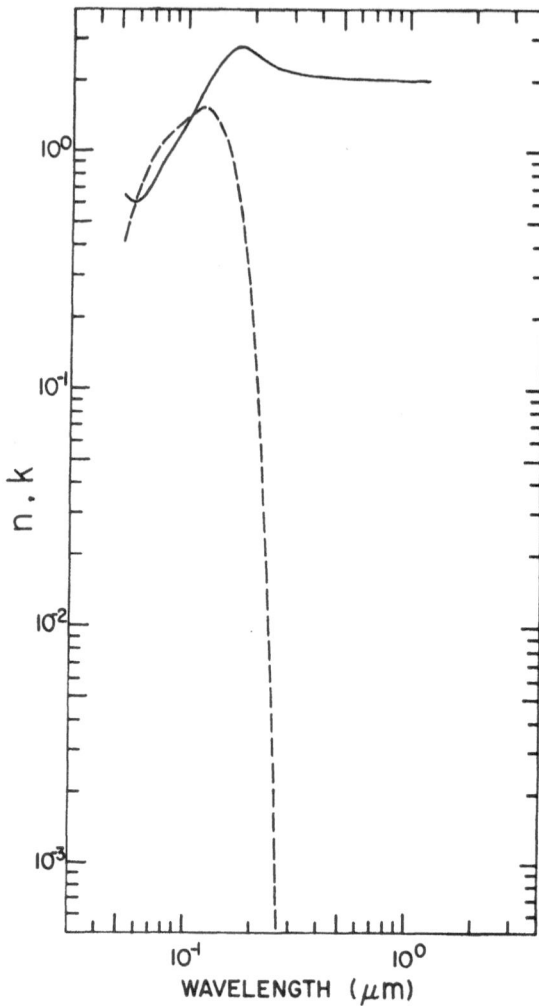

Fig. 3.8 Refractive index, n (solid curve), and extinction coefficient, k (dashed curve), as a function of wavelength for Si_3N_4. After [38]

Rubens relation based on continuum theory. At higher frequencies (in the near-IR and visible range), reflectivity decreases due to the presence of free electrons. Here, the reflectance is governed by Drude's law which takes into account the atomistic structure of the metal. At even higher frequencies, beyond the visible range, the reflectivity increases and then decreases again going through a series of peaks. The reflectivity behavior at these frequencies is described by the bound electron theory postulated by Lorentz.

Table 3.3. *Refractive index (real part) of PECVD oxide films at 633 nm wavelength for different process conditions* [39]

n SiO_2	Process conditions with $SiH_4 : N_2O = 1.5 : 40$ sccm, pressure = 2 Torr, Power = 10 W, power density = 10.4 mW/cm^2
1.340–1.442	No N_2 or He diluents
1.457–1.470	N_2 (2000 sccm), 300°C
1.451–1.463	He (2000 sccm), 250°C
1.444–1.469	He (2000 sccm), 300°C
1.469–1.481	He (2000 sccm), 350°C
1.469–1.476	He (3000 sccm), 300°C

Table 3.4. *Refractive index (real part) of PECVD nitride films at 633 nm wavelength for different process conditions* [40, 41]. Here, $x/y = 0.75$ is stoichiometric nitride. Hydrogen content varies from 32 to 39 atomic %

n a-Si$_x$N$_y$:H	x/y	Density (g/cm^3)	Process conditions NH$_3$/SiH$_4$ flow ratio
1.760	0.66	2.07	
1.766	0.63	2.11	
1.790			10
1.800			8
1.800	0.62	2.21	12
1.810			6
1.822	0.68	2.40	
1.830	0.68	2.42	
1.840			4
1.850	0.65	2.54	
1.870			2
1.909	0.67	2.85	
2.024	0.82	2.16	5

The refractive index and extinction coefficient for selected IC-compatible metals (Al, Cu, and W) are shown in Figs. 3.9 to 3.12 for wavelengths ranging from IR to soft X-ray regions. The curves are taken from refs. [42] and [43], which provide the numerical values and their scatter arising from limitations in measurement instrumentation and material differences in sample preparation, for a variety of thin film metals. Fig. 3.10 illustrates the reflectance as a function of wavelength for evaporated and sputter-deposited Al. Depending on deposition vacuum conditions, there is rapid formation of a very thin oxide (Al$_2$O$_3$) layer on the film surface which is highly transparent from about 0.18 µm to 6 µm. However, its effect on reflectance is strong at (shorter) wavelengths in the UV range.

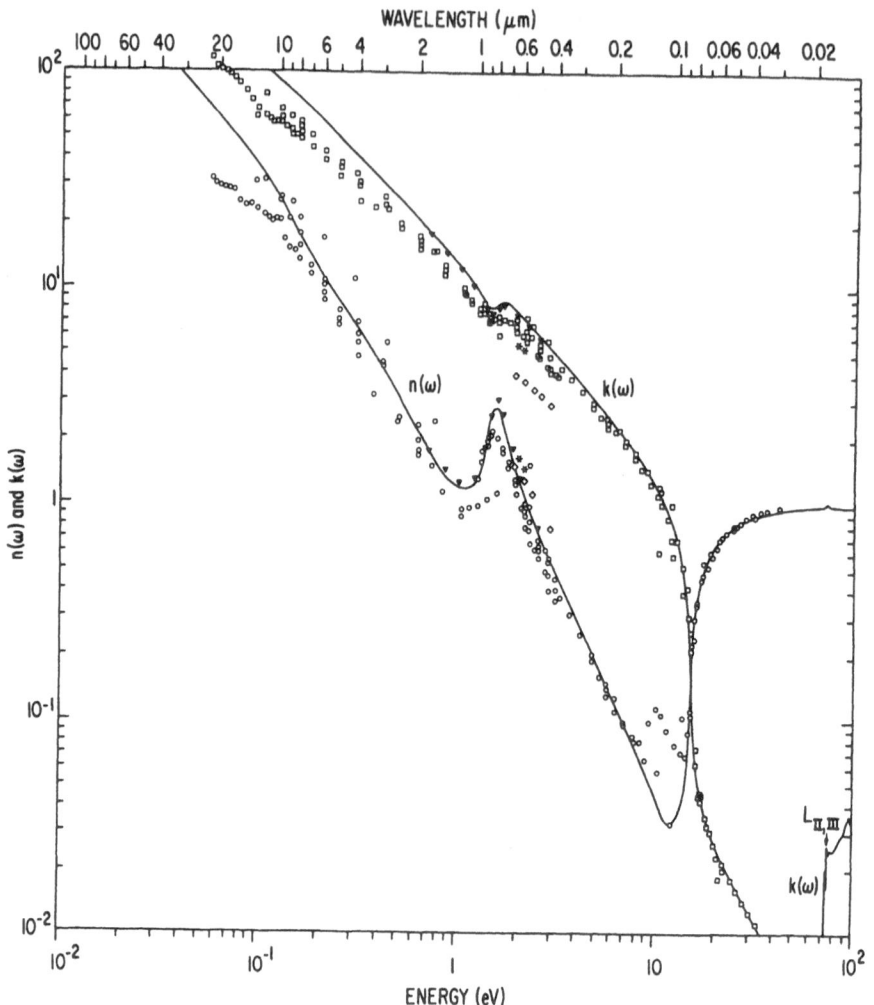

Fig. 3.9 The complex refractive index, $n(\omega) + ik(\omega)$, as a function of wavelength for (mostly evaporated) Al thin films; ω denotes frequency. After [42]

For example, at 121.6 nm [44], an oxide layer of thickness 17 Å can reduce reflectance by a factor of two. At these wavelengths, surface roughness is also an important consideration because of surface scattering as well as coupling to surface plasmons. The latter takes place at 10.6 eV for the Al-vacuum interface. For example, an rms surface roughness of 27 Å can lead to a reflectance degradation of 20% at 400 nm wavelength [45]. The film morphology, i.e., grain size and orientation, also affects the optical characteristics.

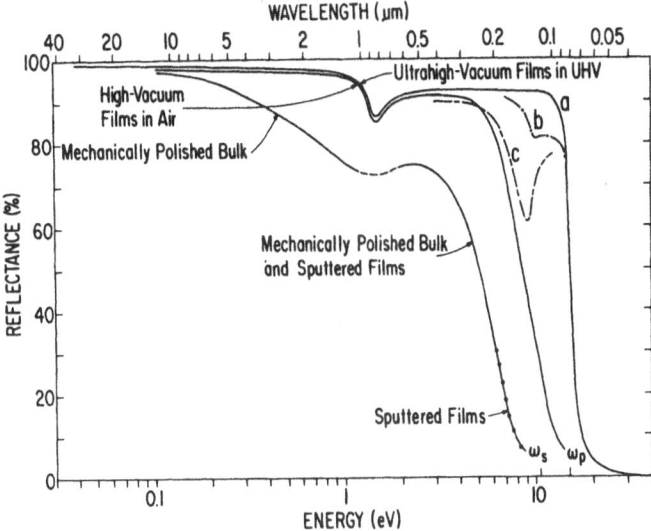

Fig. 3.10 Reflectance behavior of thin film Al at different processing conditions. Curves *b* and *c* indicate the effect of light scattering and coupling to surface plasmons; $\omega_{s,p}$ denote surface and bulk plasmon frequencies, respectively. After [42]

3.6 High Energy Interactions and Monte Carlo Simulations

At high radiant signal energies (X-rays: $100\,\mathrm{eV}$–$100\,\mathrm{keV}$, γ-rays: $100\,\mathrm{keV}$–$100\,\mathrm{MeV}$), interactions occur with atomic electrons in a variety of processes, notably, the photoelectric effect, Compton scattering, and pair production [1, 2, 46]. In Si, for example, the associated energy ranges are: $E < 50\,\mathrm{keV}$, $50\,\mathrm{keV} \leq E \leq 2\,\mathrm{MeV}$, and $E > 2\,\mathrm{MeV}$, respectively.

3.6.1 Photoelectric Effect, Compton Scattering, and Pair Production

We have seen the photoelectric effect at low energies (Sect. 3.2) where the incident photon transfers all its energy to outer shell electrons. The interaction process at high energies is similar except that all of the incident energy is now transferred to inner shell electrons; about 80% of the interactions involve the more tightly bound K-shell electrons and the remaining 20% involve L-shell electrons [46]. The energy of the ejected electron is given by $E_e = h\nu - E_B$, with E_B denoting the atomic binding energy. This places the atom in an excited state. Consequently, an electron from a higher energy level will fill the empty level in the K-shell, leading to

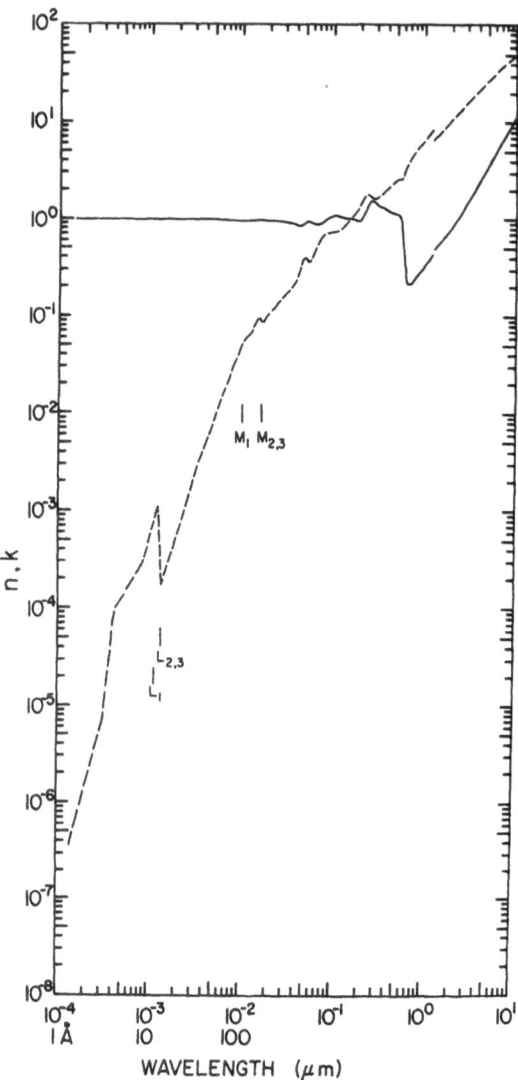

Fig. 3.11 Refractive index, n (solid curve), and extinction coefficient, k (dashed curve), as a function of wavelength for (mostly evaporated) Cu thin films. Also indicated is the onset of absorption of L- and M-shells. After [43]

emission of X-rays of energy equal to the electron transition energy. The X-rays get re-absorbed by the photoelectric effect or they generate Auger electrons, although the probability of the latter is small in comparison. Alternatively, they can escape from the material. The probability of escape can only be determined by using Monte Carlo simulations. The photo-electron from the K-shell gains significant kinetic energy as it settles to a

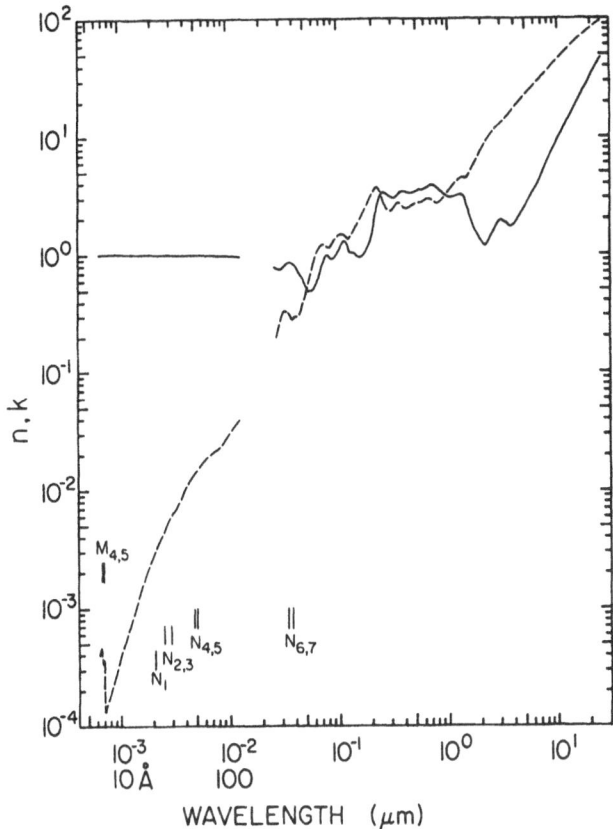

Fig. 3.12 Refractive index, n (solid curve), and extinction coefficient, k (dashed curve), as a function of wavelength for W (tungsten). Also indicated is the onset of absorption of M- and N-shells. After [43]

lower potential energy in the conduction band and creates further electron-hole pairs through a series of interaction processes. This, coupled with re-absorption of emitted X-rays, can lead to a quantum efficiency higher than unity despite the incident photon interacting with just one atom in the material. The ejected photo-electrons have an angular distribution in emission intensity; at low photon energies, the maximum is orthogonal to the direction of incident photons. With increasing energy, the maximum shifts to the direction of incident photons [46].

In Compton scattering, the incident photons scatter with electrons in the outermost shell. The scattering can be considered elastic because of the small binding energy of the outer shell electrons. The scattering process must include the momentum (h/λ) of the incident photon because of its high energy. In these interactions, the atom is only able to absorb some of the photon momentum. The entire energy of the primary photon cannot be transferred to the atom, thus yielding a secondary photon of reduced energy that is emitted along with the electron. The photon may experience, depending on its energy, a series of further Compton scattering events before getting re-absorbed by the photoelectric effect. Alternatively, it may escape from the medium. The scattered electron, as in photoelectric capture, can lead to generation of further electron-hole pairs, again yielding a quantum efficiency greater than unity.

In pair production, the incident photon interacts with the atomic nucleus to produce an electron-positron pair. The positron has the same mass as the electron but carries a positive charge q. Pair production takes place when the incident radiant energy is larger than the total rest mass of the electron-positron pair, viz., $2mc^2 = 1.022\,\text{MeV}$. Any excess energy associated with the radiant signal is converted to kinetic energy of the pair, $E_{pair} = h\nu - 1.022\,\text{MeV}$. Pair production, like Compton scattering, does not result immediately in the absorption of all the incident energy. The positron very quickly annihilates with another electron to form two photons, each of energy $0.511\,\text{MeV}$, which subsequently interact in Compton scattering or photoelectric capture, or escape from the material. Unlike the Compton interaction process, pair production, as in photoelectric capture, has an energy threshold for attenuation; its attenuation coefficient is zero for $h\nu \leq 1.022\,\text{MeV}$ [46].

In addition, there are a few other scattering mechanisms which, however, have very low interaction probability. For example, photons can be elastically scattered from bound and unbound electrons [47]; these being referred to as Rayleigh and Thompson scattering processes, respectively. The associated effects are generally small because the scattering is completely random and scattering in one direction tends to be canceled out by scattering in the opposite direction. Another interaction process is Delbruck scattering, which is present only at very high energies. Here, the incident photon excites the nucleus directly, which on de-excitation, emits one or more gamma rays whose total energy is equal to the energy of the initial photon.

3.6.2 Ionization Yield

Photoelectrons created from these high energy photon interactions undergo various elastic and inelastic collisions with other electrons and atoms. Excitation and ionization of atomic electrons dominate at lower energies ($< 1\,\text{MeV}$). Upon de-excitation of the atom, low energy photons are emitted which interact with, or escape from, the material. Bremsstrahlung dominates at higher electron energies. Here, there is photon emission when electrons are suddenly accelerated or decelerated by scattering with mostly the atomic nucleus of the material. The photons emitted have

Table 3.5. *The radiation ionization energy for selected materials* [48, 49]

Material	w (eV)
Si	3.65
Ge	3.0
GaAs	4.7
SiC	9
CdTe	4.4
HgI_2	4.2
C	17

continuous energies in the range of X-rays and γ-rays. Two parameters of notable importance in estimating the useful signal are the stopping power (dE/dx) and the ionization yield (dN/dx). The stopping power characterizes the mean energy loss of the electron per unit-path-length, which comprises of the total energy dissipated [46], *viz.*, $(dE/dx)_{total} = (dE/dx)_{e\&i} + (dE/dx)_{brem}$. Here, $(dE/dx)_{e\&i}$ and $(dE/dx)_{brem}$ are the energy dissipation components by excitation and ionization, and bremsstrahlung, respectively. The ionization yield per unit path-length for thin media, where dE is completely absorbed, can be estimated as $(dN/dx) = (dE/dx)w^{-1}$. Here, w is the radiation ionization energy and, as defined in Sect. 3.4, refers to the mean energy required to produce an electron-hole pair. Approximate values for w along the lines of Shockley's empirical model [31] are given in Table 3.5 for selected materials [48, 49].

3.6.3 Photon Attenuation Coefficients

All of the above photon and electron interaction processes can be described by cross sections from which the probability of occurrence can be calculated. However, the cross section data are difficult, if not impossible, to determine analytically, since the interactions are inherently random processes. Simple lumped approximations can be employed, in the case of photon transport, by the use of attenuation coefficients. These coefficients describe the fraction of photons which do not interact with the material and pass straight through. The term attenuation is generally associated with high energy radiant signals since the interaction probabilities are smaller than at low energies where the usual term is absorption. In the latter, the fraction of non-interacting photons is very small. Regardless of the interaction processes, each photon is eliminated individually. The energy of the non-interacting photons is a constant and so too is the interaction cross-section, which is given by the sum of the partial cross-sections associated with photoelectric capture (PE), Compton scattering (C), and pair production (PP). Thus the total attenuation coefficient is a sum of the partial coefficients [46]

$$\mu_T = \mu_{\text{PE}} + \mu_{\text{C}} + \mu_{\text{PP}} \tag{3.29}$$

and the number of photons across the beam at a given distance x in the material per unit area per unit time obeys

$$N(x) = N_0 \exp\left(-\mu_T x\right). \tag{3.30}$$

Here, N_0 denotes the number of photons incident on the material at $x = 0$. The term μ_T is a linear attenuation coefficient (units: cm^{-1}) and its value is assumed constant within a given material. Values for the attenuation coefficients can be retrieved from graphs or tables of measured data (see, e.g., [2, 46]).

3.6.4 Monte Carlo Simulations

Rigorous simulation of high energy interactions involves the solution of coupled photon and electron transport within the detector [50]. This can only be achieved using Monte Carlo methods. The Monte Carlo method is based on the study of individual photon and electron "histories" [47]. The history is a description of the passage of the photon or electron through the detector. The electrons produced in a photon interaction process must begin their own histories which are tracked in exactly the same manner as the original photon. But even with the Monte Carlo method, some degree of approximation is necessary to reduce the, otherwise impractical, amount of CPU time that is required for complete tracking of every interaction process associated with each photon and electron. For example, a 1 MeV electron will undergo 10^4 interactions as it slows down to 1 keV [50]. This large number of interactions for this process and similar numbers for other processes necessitates multiple scattering approximations where the outcomes, in terms of deflection and energy losses, for a large number of interactions are lumped into a single event. Table 3.6 summarizes key steps and associated physical processes in a typical simulation scenario [51]. Statistically meaningful results of detector efficiency, spectral sensitivity, and resolution can be simulated if enough photons, and corresponding electrons, are tracked through the detector. Although simulation accuracy is governed by the accuracy of models used in calculation of the interaction probability distributions, the limiting factor, in practice, almost always lies in CPU speed. Infinite CPU speed would allow use of arbitrarily complex models resulting in highly accurate results. Therefore a primary goal of the Monte Carlo user lies in finding the best compromise between execution time and accuracy, through pertinent modifications to the strict Monte

Table 3.6. *Monte Carlo simulation scenario of photon and electron transport*

Steps	Description
Source parameter selection	Six mutually independent parameters
	emission position: (x, y, z) or (ρ, θ, φ)
	emission velocity: $(v_x, v_y,$ energy$)$ or (v_x, v_y, v_z)
Path length selection	$R = -(1/\sigma) \ln \rho$
	$\sigma =$ interaction cross section that is dependent on
	position, direction, and energy
	$\rho =$ random number $(0 \leq \rho \leq 1)$
Photon interactions	Tabulation of interaction cross sections for photoelectric
	effect, Compton scattering, pair production, Rayleigh
	scattering, etc.
Electron interactions	Ionization and excitation, bremsstrahlung,
	stopping power, etc.
Interaction outcome	Energy loss, direction change, creation of photoelectrons
Scoring	Data retrieval during and after simulation
	data = photon and electron fluxes, energy deposited, etc.

Carlo method [51]. A list of Monte Carlo codes and code systems, including data bases of the various photon and electron interaction cross sections, compiled from Ref. [12] is summarized in Table 3.7. These codes specialize in photon and electron transport at (high) energies in excess of several keV. What appears to be lacking, however, are physical models for photon and electron interactions at lower energies, $E < 1$ keV. Low energy transport modeling may be a challenging task since the number of scattering events suffered by the electron is too small to be treated using the well-established multiple-scattering models.

Table 3.7. *Compilation of Monte Carlo codes and code systems, including databases, taken from Ref.* [12]

Codes/ Code Systems		Description	Reference
XCOM		Photon interaction cross sections	[52]
		attenuation coefficients	
ETRAN		Photon and electron transport	[53]
		(energy range: 1 keV to 1 GeV)	
	COMBIXD	Photon interaction cross sections	
	DATAPAC	Electron interaction cross sections	
	ZTRAN	Transport in heterogeneous multilayers	[54]
ITS		Steady state coupled photon-electron transport	[55]
EGS4		Coupled photon-electron transport	[56]
		(energy: a few keV to several TeV)	

3.7 Model Equations for Radiant Sensor Simulation

The model equations for radiant sensor simulation are essentially the same as those we have discussed in Chapt. 2; the key addition, however, lies in the incorporation of photo-generation terms in the equations. Extracting the key equations from Chapt. 2, we have: Poisson's equation (3.31), continuity equations for electrons (3.32) and holes (3.33) accounting for photogeneration G_{opt} from (3.27), and the current density relations for electrons (3.34) and holes (3.35)

$$\operatorname{div}(\epsilon \operatorname{grad} \psi) = -q(p - n + N_D - N_A), \qquad (3.31)$$

$$\operatorname{div} \mathbf{J_n} = qR - G_{opt} + q\partial n/\partial t, \qquad (3.32)$$

$$\operatorname{div} \mathbf{J_p} = -qR + G_{opt} - q\partial p/\partial t, \qquad (3.33)$$

$$\mathbf{J_n} = -q\mu_n n \operatorname{grad} \psi + qD_n \operatorname{grad} n, \qquad (3.34)$$

$$\mathbf{J_p} = -q\mu_p p \operatorname{grad} \psi - qD_p \operatorname{grad} p. \qquad (3.35)$$

Underlying assumptions to the above system of equations are discussed in Sects. 2.1, 2.2, and 2.3.1. Models for the physical parameters and

Table 3.8. *Model equations including physical parameters and boundary conditions, with reference to Chapter 2, for simulation of the broad range of radiant sensors*

Models	Reference	Description/validity
Governing Equations	(3.31)–(3.35)	
Physical Parameters		
Concentrations: n, p	(2.29), (2.30)	Non-degenerately doped
	(2.25), (2.26)	Degenerately-doped
n_{ie}	(2.40)	Band gap narrowing at high doping
Band edge shifts	(2.42), (2.43)	Spatially varying band gap
Generation-recombination: R	(2.62)	SRH trap-assisted recombination
	(2.68)	Auger recombination – heavy doping
	(2.69)	Impact ionization – high reverse bias
Photogeneration: G_{opt}	(3.28)	η values for 1.2 eV $\leq h\nu \leq$ 6 eV
Carrier mobility: $\mu_{n,p}$	Sect. 2.3.5	See Sect. 2.3.5 for model validity
Boundary Conditions		
Contacts	(2.92)	Ohmic, voltage source
	(2.93), (2.94)	Contact/external resistance
	(2.95)	Current source
	(2.96)	Schottky
	Photoelectric effects	Depends on metallization-type and $h\nu$
Insulators and interfaces	(2.97), (2.98)	Trapped and interface charges
	(2.99), (2.100)	Surface recombination
Outer boundaries	(2.101)–(2.103)	See Sect. 2.6.4 for physical plausibility

boundary conditions, and associated validity, pertinent to radiant sensor simulation are summarized in Table 3.8. The associated numerical procedures are described in Sect. 6.7.3.

3.8 Illustrative Simulation Example – Color Sensor

By way of a simulation example, we chose to illustrate the single-element p-n-p-n color sensor which permits simultaneous detection of the three primary colors [11]. Here, optimization of photoresponse and terminal characteristics for the different colors involves both process-and device-simulation. Process simulation aids in tailoring the doping profile to achieve the appropriate junction depths necessary for efficient separation of the different wavelengths (colors). The process simulation results shown are based on the SUPREM-IV software package [57]. Device simulation helps optimize bias conditions for high color discrimination, linearity in output response with photon flux, and to prevent possible parasitic p-n-p-n thyristor action at very high illumination levels. For this purpose, simulations were performed using the commercial device simulator, PISCES-2B [58]. Here, numerical results in terms of the electrostatic potential and output photoresponse (terminal characteristics) were obtained based on a simultaneous solution of the coupled system of Poisson's and the carrier (electron and hole) continuity equations, taking into account the photogeneration of electron-hole pairs using the wavelength- and doping-dependence of optical absorption.

Fig. 3.13 shows a schematic of the p-n-p-n device structure where terminals 1 to 4 denote the ohmic contacts to the various layers. The fabrication steps simulated using SUPREM-IV are summarized in Table 3.9, and the resulting doping profile is illustrated in Fig. 3.14. The device consists of three p-n junctions in series, all of which are reverse biased in order to deplete the entire device depth up to a thickness of $5\,\mu m$. The corresponding energy-band diagram is shown in Fig. 3.15, where we distinguish three regions. For very short wavelengths (blue light), most of the e-h pairs are generated in Region I. The electrons yield current I_1 and the holes drift to Region II to yield I_2. The thickness of Region I should be comparable to the penetration depth of the blue signal ($\sim 0.1\,\mu m$) to obtain reasonable discrimination between blue and green signals. With the red signal most of the e-h pair generation is in Region III. The electrons get collected at the backside contact to yield current I_4 whereas the holes diffuse in Region III to produce I_3. Thus it is clear from the above description of device operation, the layer thicknesses and bias conditions have to be optimized to yield terminal currents unique to the one or the other color component. The challenge here lies in the retrieval of the green signal from the terminal currents, I_2 and I_3.

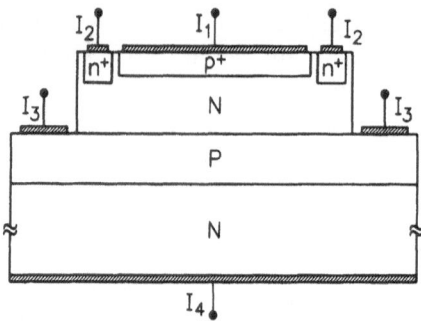

Fig. 3.13 Schematic cross section of p-n-p-n color sensor. I_1 to I_4 denote terminal currents

Table 3.9. *Fabrication steps of color sensor simulated using SUPREM-IV*

Device layer/region	Process conditions
n-Type Si-substrate	Doping concentration $= 1 \times 10^{15}$ cm^{-3}
Diffused p-layer	B diffusion at 925 °C for 20 mins
n-Epilayer	1.3–1.5 μm thick, 1100 °C
$n+$ Ohmic contacts to epilayer	P diffusion at 925 °C for 20 mins
$p+$ Top layer	B implantation, energy 10 keV and dose 5×10^{14} cm^{-2} low implantation energy needed for shallow junction crucial to blue light detection

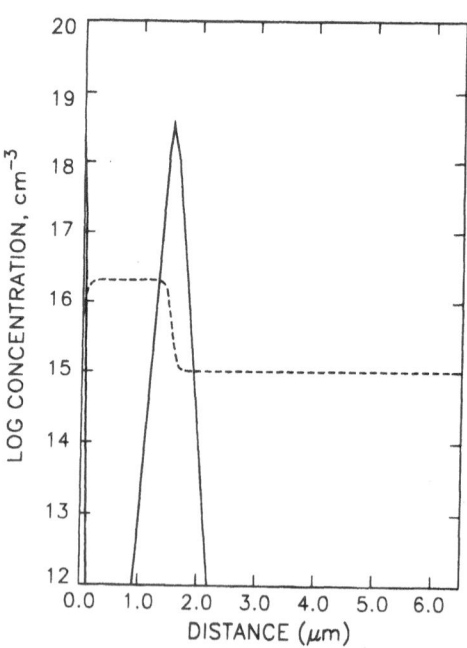

Fig. 3.14 Simulation results for the doping profile vs. distance from device surface. The solid and dashed curves indicate B and P concentrations, respectively. The B out-diffusion is due to high temperature treatment

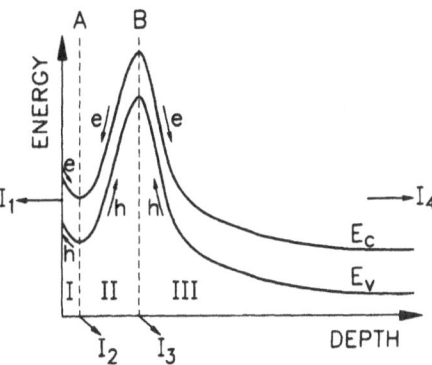

Fig. 3.15 Schematic of energy band diagram. The illumination is on the left side. Points A and B are located at $0.2\,\mu m$ and $1.5\,\mu m$, respectively

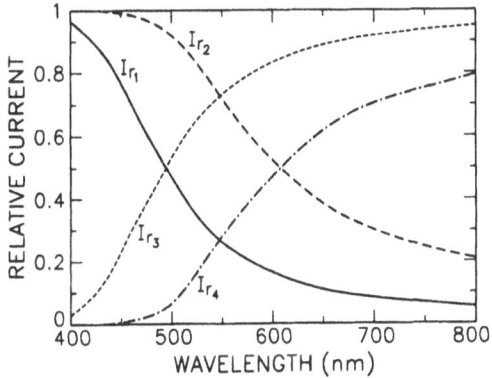

Fig. 3.16 The relative terminal currents, I_{rn}/I_{total} $(r = 1,\ldots,4)$ as a function of wavelength

   The simulated relative terminal currents are illustrated in Fig. 3.16 as a function of wavelength of the input radiant signal. Here, $I_{rn} = I_n/I_{total}$, where I_{total} is the total device current, and the bias voltages are 0 V, 1 V, -10 V, and 5 V, at terminals $n = 1$ to 4, respectively. The color components associated with the incident radiant signal can be retrieved through use of the following current ratios: $B = I_1/I_2$, $G = 2/[(I_2/I_3) + (I_3/I_2)]$, and $R = 1.25(I_2/I_3)$, which yield values of unity for B, G, and R signals at the respective wavelengths. The terminal photocurrents vary linearly with the intensity of the radiant signal in the range 10^{12}–10^{19} photons/cm^2 [11]. In the low intensity limit, the photoresponse is masked by the diode's reverse saturation current. At very high intensities, electron-hole pair generation in Region I becomes significant. This alters the potential distribution considerably (see Fig. 3.17) and the top transistor

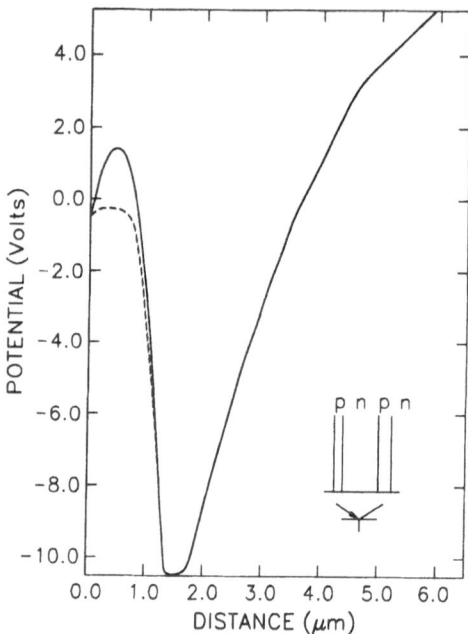

Fig. 3.17 Electrostatic potential profile over device depth for moderate (solid curve) and high (dashed curve) light intensities. At high intensities, the top p-n-p-n transistor becomes forward biased and current I_1 now flows in the opposite direction

structure (p-implant, n-epi, and p-diffused regions) becomes active allowing significant current through terminal 1 in the direction opposite to its normal operation. In this case, the device output currents are not representative of the input color and furthermore, the device risks the onset of possible parasitic p-n-p-n thyristor action. However, by ensuring that device design and operating bias is such that the total common base current transfer ratios of the n-p-n and p-n-p transistors are less than unity, the onset of thyristor action can be pushed to higher intensity levels thus achieving linearity over a wider range.

3.9 References

[1] Middelhoek, S., Audet, S. A., *Silicon Sensors*, London: Academic Press, 1989.
[2] Audet, S., Radiation Sensors, in: *Semiconductor Sensors*, Sze, S. M. (Ed.), New York: Wiley, 1994, pp. 271–329.
[3] Sze, S. M., *Physics of Semiconductor Devices*, New York: Wiley, 1981.
[4] Kanicki, J. (Ed.), *Amorphous and Microcrystalline Semiconductor Devices*, Norwood (MA): Artech House, 1992.
[5] Bolliger, D., *Integration of an Ultraviolet Sensitive Flame Detector*, Ph.D. Dissertation, ETH Zürich, No. 11359, Switzerland, 1995.

[6] Lenggenhager, R., *CMOS Thermoelectric Infrared Sensors*, Ph.D. Dissertation, ETH Zürich, No. 10744, Switzerland, 1994.

[7] Wood, R. A., Han, C.-J., Kruse, P. W., Integrated Uncooled Infrared Detector Imaging Array, *Technical Digest*, Solid-State Sensors and Actuators Workshop, Hilton-Head Is., 1992, pp. 132–135.

[8] Street, R. A., Antonuk, L. E., Amorphous Silicon Arrays Develop a Medical Image, *IEEE Circuits and Devices*, July 1993, pp. 38–42.

[9] Noorlag, D. J. W., Non-Linear Position Response in Position-Sensitive Photodetectors, *Sensors and Actuators*, 31 (1982/1983), 1–15.

[10] Vecchi, M. C., Rudan, M., Soncini, G., Numerical Simulation of Optical Devices, *IEEE Trans. CAD of ICAS*, CAD-12 (1993), 1557–1569.

[11] Moharjerzadeh, S., Nathan, A., Selvakumar, C. R., Numerical Simulation of a p-n-p-n Color Sensor for Simultaneous Color Detection, *Sensors and Actuators*, A44 (1994), 119–124.

[12] Jenkins, T. M., Nelson, W. R., Rindi, A. (Eds.), Monte Carlo Transport of Electrons and Photons, New York: Plenum Press, 1987.

[13] Wolffenbuttel, R. F., Operation of the Silicon Colour Filtering Element, *Sensors and Actuators*, 16 (1989), 13–23.

[14] Glass, A. S., Morf, R., Optimizing the Performance of Spectrally Selective Photodiodes by Simulated Annealing Techniques, *Sensors and Actuators*, A21–A23 (1990), 564–569.

[15] Verzellesi, G., Vecchi, M. C., Zen, M., Rudan, M., Optical Generation in Semiconductor-Device Analysis – A General Purpose Implementation, in: *Simulation of Semiconductors and Processes*, Vol. 4, Fitchner, W., Aemmer, D. (Eds.), (1991), pp. 57–64.

[16] Rossberg, D., Optical Properties of Integrated Infrared Sensor, *Digest of Technical Papers*, Vol. 2, Transducers '95, Stockholm, 1995, pp. 652–655.

[17] Bolliger, D., Popovic, R. S., Baltes, H., Integration of a Smart Selective UV Detector, *Digest of Technical Papers*, Vol. 2, Transducers '95, Stockholm, 1995, pp. 144–147.

[18] Smith, R. A., *Semiconductors*, 2nd Ed., Cambridge: Cambridge University Press, 1978.

[19] Hummel, R. E., *Electronic Properties of Materials*, 2nd Ed., Berlin: Springer-Verlag, 1993.

[20] Macleod, H. A., *Thin Film Optical Filters*, 2nd Ed., Bristol: Adam Hilger, 1986.

[21] Seeger, K., *Semiconductor Physics*, 3rd Ed., Berlin: Springer-Verlag, 1985.

[22] Haas, C., Infrared Absorption in Heavily Doped n-Type Germanium, *Phys. Rev.*, 125 (1962), 1965–1971.

[23] Jain, S. C., Nathan, A., Briglio, D. R., Roulston, D. J., Selvakumar, C. R., Yang, T., Band-to-Band and Free-Carrier Absorption Coefficients in Heavily Doped Silicon at 4K and at Room Temperature, *J. Appl. Phys.*, 69 (1991), 3687–3690.

[24] Vol'fson, A. A., Subashiev, V. K., Fundamental Absorption Edge of Silicon Heavily Doped with Donor or Acceptor Impurities, *Sov. Phys. Semicond.*, 1 (1967), 327–332.

[25] Balkanski, M., Aziza, A., Amzallag, E., Infrared Absorption in Heavily Doped n-Type Si, *Phys. Stat. Sol.*, 31 (1969), 323–330.

[26] Schmid, P. E., Optical Absorption in Heavily Doped Silicon, *Phys. Rev. B*, 23 (1981), 5531–5536.

[27] Macfarlane, G. G., Maclean, T. P., Quarrington, J. E., Roberts, V., Fine Structure in the Absorption-Edge Spectrum of Si, *Phys. Rev.*, 111 (1958), 1245–1254.

[28] Pankove, J. I., Aigrain, P., Optical Absorption of Arsenic-Doped Degenerate Germanium, *Phys. Rev.*, 126 (1962), 956–962.

[29] Wagner, J., Heavily Doped Silicon Studied by Luminescence and Selective Absorption, *Solid-State Electron.*, 28 (1985), 25–30.

[30] Pantelides, S., Selloni, A., Car, R., Energy-Gap Reduction in Heavily Doped Silicon: Causes and Consequences, *Solid-State Electron.*, 28 (1985), 17–24.

[31] Shockley, W., Problems Related to p-n Junctions in Silicon, *Czech. J. Phys. B*, 11 (1961), 81–121.

[32] Kane, E. O., Electron Scattering by Pair Production in Silicon, *Phys. Rev.*, 159 (1967), 624–631.

[33] Antoncik, E., di Coli, G., Farese, L., Radiation-Ionization Energy and Fano Factor in Semiconductors, *Radiation Effects*, 5 (1970), 1–13.

[34] Christensen, O., Quantum Efficiency of the Internal Photoelectric Effect in Silicon and Germanium, *J. Appl. Phys.*, 47 (1976), 689–695.

[35] Schneeberger, N., Paul, O., Baltes, H., Optimized Structured Absorbers for CMOS Infrared Detectors, *Digest of Technical Papers*, Vol. 2, Transducers '95, Stockholm, 1995, pp. 648–651.

[36] Buhler, J., Funk, J., Steiner, F.-P., Sarro, P. M., Baltes, H., Double Pass Metallization for CMOS Aluminium Actuators, *Digest of Technical Papers*, Vol. 2, Transducers '95, Stockholm, 1995, pp. 360–363.

[37] Sampsell, J. B., The Digital Micromirror Device and its Application to Projection Displays, *Digest of Technical Papers*, Transducers '93, Yokohama, 1993, pp. 24–27.

[38] Philipp, H. R., Silicon Dioxide (SiO_2) (Glass), in: *Handbook of Optical Constants of Solids*, Palik, E. D. (Ed.), Orlando: Academic Press, 1985, pp. 749–763. Also Silicon Nitride (Si_3N_4) (Noncrystalline), as above, pp. 771–774.

[39] Hsieh, S. W., Chang, C. Y., Hsu, S. C., Characteristics of Low-Temperature and Low-Energy Plasma-Enhanced Chemical Vapor Deposited SiO_2, *J. Appl. Phys.*, 74 (1993), 2638–2648.

[40] Kanicki, J., Wagner, P., Comparative Study of PECVD Nitride Films, in: *Proc. of the Symp. on Dielectric Films on Compound Semiconductors*, Kapoor, V. J., Hankins, K. T. (Eds.), *ECS Proc.*, 1987, pp. 261–274.

[41] Benaissa, K., *Integrated Silicon Opto-Mechanical Sensors*, Ph.D. Dissertation, University of Waterloo, Waterloo, Ontario N2L 3G1, Canada, 1996.

[42] Smith, D. Y., Shiles, E., Inokuti, M., The Optical Properties of Metallic Aluminium, in: *Handbook of Optical Constants of Solids*, Palik, E. D. (Ed.), Orlando: Academic Press, 1985, pp. 369–406.

[43] Lynch, D. W., Hunter, W. R., Comments on the Optical Constants of Metals and an Introduction to the Data for Several Metals, in: *Handbook of Optical Constants of Solids*, Palik, E. D. (Ed.), Orlando: Academic Press, 1985, pp. 275–367.

[44] Berning, P. H., Hass, G., Madden, R. P., Reflectance-Increasing Coatings for the Vacuum Ultraviolet and Their Applications, *J. Opt. Soc. Am.*, 50 (1960), 586–597.

[45] Endriz, J. G., Spicer, W. E., Study of Aluminum Films. I. Optical Studies of Reflectance Drops and Surface Oscillations on Controlled-Roughness Films, *Phys. Rev. B*, 4 (1971), 4144–4159.

[46] Tait, W. H., *Radiation Detection*, London: Butterworths, 1980.

[47] Debertin, K., Helmer, R. G., *Gamma- and X-Ray Spectrometry with Semiconductor Detectors*, Amsterdam: North-Holland, 1988.

[48] Tove, P. A., Review of Semiconductor Detectors for Nuclear Radiation, *Sensors and Actuators*, 5 (1984), 103–117.

[49] Halles, E. E., Goulding, F. S., in: *Handbook of Semiconductors*, Vol. 4, Moss, T. S., Hilsum, C. (Eds.), Amsterdam: North-Holland, 1981, pp. 799–825.

[50] Nahum, A. E., Overview of Photon and Electron Monte Carlo, in: *Monte Carlo Transport of Electrons and Photons*, Jenkins, T. M., Nelson, W. R., Rindi, A. (Eds.), New York: Plenum Press, 1987, pp. 3–20.

[51] Lux, I., Koblinger, L., *Monte Carlo Particle Transport Methods: Neutron and Photon Calculations*, Boca Raton: CRC Press, 1991.

[52] Berger, M. J., Hubell, J. H., *XCOM: Photon Cross Sections on a Personal Computer*, National Bureau of Standards Report NBSIR 87-3597, 1987.

[53] Ford, R. L., Nelson, W. R., *The EGS Code System: Computer Programs for the Monte Carlo Simulation of Electromagnetic Cascade Showers* (Version 3), Stanford Linear Accelerator Center Report SLAC-210, 1978.

[54] Seltzer, S. M., Berger, M. J., *Electron and Photon Transport in Multilayer Media: Notes on the Monte Carlo Code ZTRAN*, National Bureau of Standards Report NBSIR 84-2931, 1984. See also *Int. J. Appl. Radiat. Isot.*, 38 (1987), 349–364.

[55] Halbleib, J. A., Mehlhorn, T. A., *ITS: The Integrated Tiger Series of Coupled Electron/ Photon Monte Carlo Transport Codes*, Sandia National Laboratories Report SAND 84-0073, 1984. See also *Nucl. Sci. Eng.*, 92 (1986), 338–339.

[56] Nelson, W. R., Hirayama, H., Rogers, D. W. O., *The EGS4 Code System*, Stanford Linear Accelerator Center Report SLAC-265, 1985.

[57] *SUPREM*, Integrated Circuits Laboratory (ICL), Department of Electrical Engineering, Stanford University, CA, USA. http://www-tcad.stanford.edu/tcad/org.html

[58] *PISCES*, Integrated Circuits Laboratory (ICL), Department of Electrical Engineering, Stanford University, CA, USA. http://www-tcad.stanford.edu/tcad/org.html

4 Magnetic Field Effects on Carrier Transport

The domain of magnetic signals ranges from the very weak biomagnetic fields (~ 10 fT) to the very high fields associated with superconducting coils (~ 10 T) (see [1–6]). As a measure of the field strength $\mathbf{H}$, we use the related magnetic induction $\mathbf{B}$ whose unit is 1 tesla $= 1$ Vs/m^2 and is related to the field strength as: $\mathbf{B} = \mu_0 \mathbf{H}$ in vacuum, where μ_0 is the free space permeability. In this very large span of over 15 orders of magnitude in field strength, the lower limit of field strengths ($< 1\,\mu$T) requires relatively sophisticated detection devices and techniques [4], such as the flux-gate magnetometer, fiber optic magnetometer, nuclear magnetic resonance, and the superconducting quantum interference device, while the higher field strengths can be resolved by semiconductor magnetic sensors. Our discussion on the modeling issues will be restricted to the latter. Here, the signals are associated with geomagnetism (30–$60\,\mu$T), magnetic storage media (~ 1 mT), permanent magnets for contactless sensing (5–100 mT), and current carrying conductors (~ 1 mT at 10 A) [6]. These signals lend themselves to two categories of direct and indirect applications [1–3]. Direct applications include measurement of the geomagnetic field, reading of magnetic storage media, identification of magnetic patterns in cards and banknotes, and control of magnetic apparatus. In indirect applications, a non-magnetic signal is detected via the magnetic field which is used as an intermediate carrier. Examples include voltage-free current detection and watt-hour meters, and contactless sensors, based on mechanical displacement of a permanent magnet, for detection of linear or angular displacement and velocity.

Semiconductor magnetic sensors convert the magnetic signal into a useful electrical signal. Semiconductor magnetic sensors exploit galvanomagnetic effects due to the Lorentz force, $viz.$, $\mathbf{F} = -q(\mathbf{v} \times \mathbf{B})$ on mobile charge carriers. Here, q denotes the elementary charge and $\mathbf{v}$ the electron velocity. Note that while the field strength, $\mathbf{H}$ is the measurand, it is the magnetic induction, $\mathbf{B}$ that provides the sensor response; in semiconductors, the permeability $\mu_r \mu_0 \sim \mu_0$ in the relation, $\mathbf{B} = \mu_r \mu_0 \mathbf{H}$. Examples of semiconductor magnetic sensors include (see [6]): bulk, inversion layer, and heterojunction based Hall devices; bipolar junction magnetotransistors; magnetodiodes; carrier domain magnetometers; and magnetoresistors. This Section will, by and large, focus on simulation of Si magnetic sensors as they readily meet detection specifications for a large number of important applications in the range of μT and higher. Most importantly, they can be batch fabricated at low cost using Si IC technologies and offer co-integration of signal correction and conditioning circuitry for increased functionality [1]. Indeed co-integration with circuitry has now become the driving force of current research activity on magnetic sensors.

A magnetic field signal disturbs carrier transport in the device by virtue of the Lorentz force which reduces the symmetry of device operation [7–10]. Thus, by definition, the galvanomagnetic interaction has a vector character and realistic design of magnetic sensors can only be addressed by two-dimensional [10–26] and even three-dimensional [27–29] simulation. Critical design issues include:

- Assessment of the impact of packaging and application environment, e.g., mechanical stress [30–33] and temperature effects [34–37] on sensor performance.
- Geometry/structure optimization with respect to field orientation [18, 19, 26–29, 38] (e.g., components parallel ($B_\parallel$) or components perpendicular ($B_\perp$) to the chip surface, non-uniform B, or full vector **B**) and high spatial resolution.
- Prediction and optimization of signal level [39, 40] and sensitivity as well as signal-to-noise ratio or resolution through noise modeling [41, 42].
- Device optimization for high temporal resolution.
- Prediction of linearity and temperature coefficient of device output [43].
- Offset prediction based on process and lithography tolerances [44–46].
- Input and output impedance considerations for circuit integration.

Sects. 4.1 and 4.2 describe galvanomagnetic transport models, associated effects, and coefficients, including their dependence on magnetic and electric fields. Although the equations and underlying effects are valid for a large range of semiconductor magnetic sensors (see [5]), our discussions will focus primarily on models and/or data for the galvanomagnetic transport coefficients pertinent to the simulation of Si magnetic sensors. For practical reasons we deal with the non-degenerate material only, since the galvanomagnetic interaction is weak at degenerate concentrations. Our discussion of transport models and associated effects is restricted to isothermal conditions. Non-isothermal considerations are addressed in Chapt. 5. The effects of strain are not discussed here, but will be treated in Chapt. 6. A summary of model equations and boundary conditions is given in Sect. 4.3 with particular reference to different types of IC-compatible magnetic sensors. This is followed by a numerical simulation example, in Sect. 4.4, of a micromachined vector probe designed to detect both intensity and direction of the magnetic field.

4.1 Galvanomagnetic Transport Equation

Using microscopic transport theory, we will derive the general form of the carrier transport equation in the presence of a magnetic field. The derivation is based on a first-order perturbation of the electron (or hole) distribution function in the Boltzmann Transport Equation (BTE) for non-equilibrium conditions, to include higher order magnetic field terms. Here, we employ a series expansion of the magnetic field in the perturbation function to yield galvanomagnetic transport coefficients that are magnetic field independent. Magnetic field dependence of the transport coefficients will be treated in Sect. 4.2, whereas here a slightly different approach is employed whereby the series expansion of the magnetic field is contained within the coefficient terms. Our discussion throughout will be restricted to the weak field limit where the mobility-field product is small, $viz.$, $(\mu_{n,p}B)^2 \ll 1$ for all carrier energies. Because of mobility differences, the weak field limit has different implications in different materials [5]. For example, with n-type Si where $\mu_n = 0.14\,\text{m}^2/\text{Vs}$, B can be as high as 2.2 T. In contrast, with n-type InSb where $\mu_n = 8\,\text{m}^2/\text{Vs}$, the corresponding limit is 40 mT. Galvanomagnetic transport and associated coefficients in the high field limit, $viz.$, $(\mu_{n,p}B)^2 \gg 1$ can be found in [5]. The general derivation procedure of the transport equation is treated formally in Refs. [47, 48]. We review the solution procedure here, but for the sake of brevity, reproduce only the key steps and equations. We will retain terms containing temperature gradients during the derivation process, although our final result will be the isothermal galvanomagnetic transport equation. Thermomagnetic carrier transport will be treated formally in Sect. 5.4 using approaches based on microscopic theory [36, 47] as well as irreversible thermodynamics [35, 49], and notably Onsager's relations.

The solution to the BTE is based on the relaxation time approximation (see [47] for underlying assumptions). Using a first order expansion of the equilibrium distribution function f_0, we obtain the non-equilibrium distribution function

$$f\left[E(\mathbf{k}), E_F, T\right] = f_0[E(\mathbf{k}), E_F, T] - \frac{\partial f_0}{\partial E}\,\delta\varphi, \tag{4.1}$$

where $E(\mathbf{k})$ is the energy of the electron at wave vector $\mathbf{k}$, E_F is the Fermi level, $\delta\varphi$ is the first-order correction to the non-equilibrium distribution function, and T is the temperature. From here on, we only consider electron transport, the treatment for hole transport is analogous. Substituting (4.1) into the BTE and using the relaxation time approximation, we obtain a solution for the first-order correction, $\delta\varphi$. The BTE, under the relaxation time approximation, in steady state reads [47]

$$\frac{\mathbf{F}}{h/2\pi} \cdot \operatorname{grad}_{\mathbf{k}} f + \mathbf{v} \cdot \operatorname{grad}_{\mathbf{r}} f = -\frac{f - f_0}{\tau_n} \tag{4.2}$$

where τ_n is the relaxation time, $\mathbf{r}$ is the space vector, h is Planck's constant, $\mathbf{F}$ is the force on the electrons, and $\mathbf{v}$ is the group velocity of an electron wave packet. The force on the electron is given by the classical expression

$$\mathbf{F} = -q(\mathbf{E} + \mathbf{v} \times \mathbf{B}), \tag{4.3}$$

where q is the elementary charge, $\mathbf{E}$ is the electric field, and $\mathbf{B}$ the magnetic field. The group velocity of the electron is given by

$$\mathbf{v} = \frac{2\pi}{h} \operatorname{grad}_{\mathbf{k}} E(\mathbf{k}). \tag{4.4}$$

Equations (4.1) through (4.4) together with the Fermi distribution,

$$f_0 = \frac{1}{\exp\left[(E(\mathbf{k}) - E_F)/kT\right] + 1} \tag{4.5}$$

lead, after some manipulation [36], to the result

$$\delta\varphi = -\tau_n \mathbf{v} \cdot \left[q\mathbf{E} + \operatorname{grad}_{\mathbf{r}} E_F + \frac{E(\mathbf{k}) - E_F}{T} \operatorname{grad}_{\mathbf{r}} T \right]$$
$$+ \frac{2\pi q \tau_n}{h} (\mathbf{v} \times \mathbf{B}) \cdot \operatorname{grad}_{\mathbf{k}} \delta\varphi. \tag{4.6}$$

For the case of a free electron with an effective mass m_n^*, Eq. (4.6) can be further reduced by taking the gradient in $\mathbf{k}$-space of both sides, solving for $\operatorname{grad}_{\mathbf{k}} \delta\varphi$ and substituting it back into (4.6). In three iterations, we obtain

$$\delta\varphi = -\frac{\tau_n}{1 + \mathbf{s}^2} [\mathbf{v} \cdot \mathbf{F} + \mathbf{v} \cdot (\mathbf{s} \times \mathbf{F}) + (\mathbf{v} \cdot \mathbf{s})(\mathbf{s} \cdot \mathbf{F})], \tag{4.7}$$

where $\mathbf{s} = (q\tau_n/m_n^*) \mathbf{B}$ and $\mathbf{F}$ is the driving force, viz.,

$$\mathbf{F} = q\mathbf{E} + \operatorname{grad}_{\mathbf{r}} E_F + \frac{E(\mathbf{k}) - E_F}{T} \operatorname{grad}_{\mathbf{r}} T. \tag{4.8}$$

Expanding the magnetic field term, $\mathbf{s}$ in (4.7) as a series in ascending powers of $\mathbf{B}$ yields [36]

$$\delta\varphi = -\tau_n[(1 - \mathbf{s}^2)\mathbf{v} \cdot \mathbf{F} + (1 - \mathbf{s}^2)\mathbf{v} \cdot (\mathbf{s} \times \mathbf{F})$$
$$+ (\mathbf{v} \cdot \mathbf{s})(\mathbf{s} \cdot \mathbf{F})] + \cdots + O(B^4) \tag{4.9}$$

where we have neglected terms of order $\mathbf{B}^4$ and higher. Note that in going from (4.7) to (4.9), we have removed the magnetic field dependence of the transport coefficients. We will revisit this point in Sect. 4.2. With (4.9), we can now readily evaluate the magnetic current density in its usual form [47]

$$\mathbf{J}_{nB} = \frac{2\pi q}{h} \int \mathrm{grad}_{\mathbf{k}}\, E(\mathbf{k}) \frac{\partial f_0}{\partial E} \delta\varphi \frac{d^3\mathbf{k}}{4\pi^3}, \qquad (4.10)$$

where $d^3\mathbf{k}/4\pi^3$ is assumed to obey the classical density of states valid for a spherical parabolic band structure [50]

$$\frac{d^3\mathbf{k}}{4\pi^3} = \frac{1}{2\pi^2} \left(\frac{8\pi^2 m_n^*}{h^2}\right)^{3/2} E^{1/2} dE \qquad (4.11)$$

with the components of $\mathbf{v}$ being one third of the total velocity squared, viz., $v_i = (2E/3m_n^*)^{1/2}$. Since the electron concentration n is small relative to the degenerate concentration in practical semiconductor magnetic sensors, we can replace the Fermi distribution by the Boltzmann distribution. A further simplification is to describe the relaxation time by a power law [47, 50], $\tau_n = \tau_0 E^r$, where τ_0 is the energy-independent relaxation time, E is the electron energy, and r is a scattering factor which takes on the values of $-1/2$ for lattice scattering, $3/2$ for impurity scattering, and zero for neutral impurity scattering [50]. The scattering factor plays an important role in determining the influence of scattering processes on the transport coefficients [5] (see Sect. 4.2).

Substituting the expanded form of $\delta\varphi$, given in (4.9), into (4.10) and under isothermal conditions, $\mathrm{grad}_{\mathbf{r}}\, T = 0$, we obtain the following form for the magnetic-field-dependent electric current density [36]

$$\mathbf{J}_{nB} = -[M_1 + M_2 \mathbf{B} \times + M_3 \mathbf{B} \times \mathbf{B} \times - M_4 \mathbf{B} \cdot \mathbf{B}\mathbf{B} \times]\, \mathrm{grad}\, \phi_n, \qquad (4.12)$$

where ϕ_n denotes the Fermi potential for electrons and the second term on the right-hand-side signifies $\mathbf{B} \times (\mathbf{B} \times \mathrm{grad}\, \phi_n)$. The relation for the hole current density is analogous. In (4.12), the coefficients M_j do not depend on the magnetic field. Following the power law, $\tau_n = \tau_0(kT)^r$, the coefficients take the form

$$M_j = qn\left(\frac{q}{m_n^*}\right)^j \langle \tau_n^j \rangle = qn\left(\frac{q\tau_0}{m_n^*}\right)^j (kT)^{jr} \frac{\Gamma(5/2 + jr)}{\Gamma(5/2)}, \qquad (4.13)$$

where $j = 1, \ldots, 4$ denotes the coefficient suffix, n the electron concentration, and $\Gamma(x)$ is the gamma function. Relation (4.12) is analogous to the form shown in [47], but carries higher order terms in $\mathbf{B}$. The first term, M_1, is the electrical conductivity σ_n. From this we retrieve the electron drift mobility, which we express in the following notation [5] in order to highlight the averaging process over the different velocity-dependent components, *viz.*,

$$\langle \mu_n \rangle = \left(\frac{q\tau_0}{m_n^*} \right) (kT)^r \, \frac{\Gamma(5/2 + r)}{\Gamma(5/2)}. \tag{4.14}$$

Substituting (4.14) into (4.13) yields a simplified expression for the coefficients M_j

$$M_j = qn\langle \mu_n \rangle^j \, G_j(r) \equiv qn\langle \mu_n^j \rangle, \tag{4.15}$$

where $\langle \mu_n^j \rangle = (q/m_n^*)^j \langle \tau_n^j \rangle$ and the dimensionless scattering coefficients

$$G_j(r) = \frac{\Gamma^{j-1}(5/2)\, \Gamma(5/2 + jr)}{\Gamma^j(5/2 + r)}, \tag{4.16}$$

which are related to the scattering relaxation time. In general, they depend on the band structure, the nature of scattering mechanisms, the degree of degeneracy, and the statistics characterizing the velocity distribution of carriers [51]. The behavior of $G_j(r)$ as a function of r for different j values is illustrated in Fig. 4.1 [36]. If we compare the third- and first-order terms in magnetic field in (4.12), we see that $O(B^3)/O(B) \approx (\mu_n B)^2 G_4/G_2$. Realistically, $(\mu_n B)^2$ for Si is typically of the order of 10^{-4} at field strengths, $B \leq 100 \, \text{mT}$, while G_4/G_2 is of the order 10 (see Fig. 4.1). Thus the third term contributes less than 0.1% correction and it can be neglected. With these simplifications, Eq. (4.12) reads

$$\mathbf{J}_{nB} = -[M_1 + M_2 \mathbf{B} \times + M_3 \mathbf{B} \times \mathbf{B} \times] \,\text{grad}\, \phi_n. \tag{4.17}$$

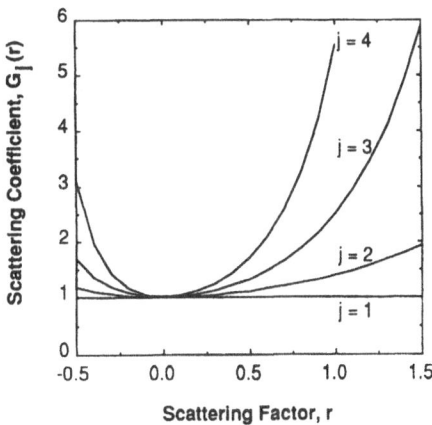

Fig. 4.1 Scattering coefficients $G_j(r)$ as a function of the scattering parameter, r for different scattering mechanisms

If we group the terms in (4.17) that are dependent on the scattering coefficients $G_j(r)$, the equation can be further reduced to the form

$$\mathbf{J}_{nB} = -q\mu_n n[1 + G_2(r)\mu_n \mathbf{B} \times + G_3(r)\mu_n^2 \mathbf{B} \times \mathbf{B} \times] \operatorname{grad} \phi_n,$$

(4.18)

where, for convenience, we have omitted the angular brackets on the mobility. The term $G_2(r)\mu_n$ is the Hall mobility and $G_2(r)$, as we will see in subsequent discussion, is the Hall scattering coefficient denoted as r_{Hn0} or as r_{Hn} when band anisotropies are taken into account (see Sect. 4.2). Taking the vector product of $\mathbf{B}$ with (4.18), a resubstitution of the term $(\mathbf{B} \times \mathbf{B} \times)$ back into Eq. (4.18) yields the widely used form of the isothermal galvanomagnetic electron transport equation [7–10]. With a slight change of notation, it reads

$$\mathbf{J}_{nB} + \mu_{Hn} \mathbf{J}_{nB} \times \mathbf{B} = -q\mu_n n \operatorname{grad} \phi_n,$$

(4.19)

where $\mu_{Hn} = G_2(r)\mu_n = r_{Hn}\mu_n$, denotes the Hall mobility. In arriving at (4.19), we have used the approximation, $G_2^2(r) \approx G_3(r)$ [36]. The error (see Fig. 4.2) involved in this approximation becomes insignificant in the low field limit, $(\mu_{Hn,p} B) \ll 1$. The equation for the magnetic field dependent hole current density is analogously obtained

$$\mathbf{J}_{pB} - \mu_{Hp} \mathbf{J}_{pB} \times \mathbf{B} = -q\mu_p p \operatorname{grad} \phi_p,$$

(4.20)

where μ_{Hp} is the hole Hall mobility, μ_p is the hole drift mobility, p is the hole density, and ϕ_p is the hole Fermi potential. Under the various assumptions stated in Sects. 2.3.1 and 2.3.2, the Fermi potentials $\phi_{n,p}$ in (4.19) and (4.20) can be expressed in terms of the electric potential and carrier densities [52] to yield the transport relations in terms of the classical

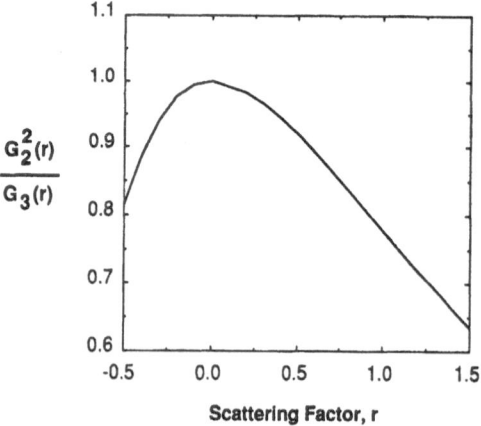

Fig. 4.2 Ratio of scattering coefficients $G_2^2(r)/G_3(r)$ as a function of the scattering parameter r

drift-diffusion formulation [7–10]:

$$\mathbf{J}_{nB} + \mu_{Hn}\mathbf{J}_{nB} \times \mathbf{B} = qD_n\left[\text{grad}\,n - n\,\text{grad}\,(q\psi/kT)\right], \qquad (4.21)$$

$$\mathbf{J}_{pB} - \mu_{Hp}\mathbf{J}_{pB} \times \mathbf{B} = -qD_p\left[\text{grad}\,p + p\,\text{grad}\,(q\psi/kT)\right], \qquad (4.22)$$

where $D_{n,p}$ denote the diffusion coefficients for electrons and holes, respectively, and ψ the electrostatic potential.

Eqs. (4.19) and (4.20) or (4.21) and (4.22) describe the basic magnetic field dependent electric current densities accounting for the effects of the Lorentz force on both drift and diffusion components. Other forms for the current density in the literature are merely variations of (4.19)–(4.22). The equations shown are exact within the framework of assumptions employed in the relaxation time approximation and the underlying band structure. Besides the above assumptions, the accuracy of solutions to these equations rests on validity of models employed for the various galvanomagnetic transport coefficients, such as, the conductivity, Hall coefficient, the Hall scattering coefficient, and Hall mobility. These coefficients, besides their usual dependence on temperature, doping, and band structure, can also be dependent on magnetic and electric fields. This is discussed in Sect. 4.2. But first, we deal with galvanomagnetic interactions, associated effects, and a review of pertinent coefficients.

4.1.1 Galvanomagnetic Effects

As specific manifestations of the Lorentz force acting on mobile electrons and holes, four basic effects, all of which are summarized in Eqs. (4.19) to (4.22), are usually assumed responsible for magnetic sensitivity in semiconductor magnetic sensors [1, 3]: Hall effect, carrier deflection, magnetoresistance, and magnetoconcentration. Only in very favorable special cases will any one of the effects prevail. Integrated magnetic sensors, depending on device geometry, fabrication process, and operating conditions, exploit the one or the other, or a combination, of these effects to produce a change in output response. A detailed discussion of the effects and precise forms of the associated analytical models is beyond the scope of this section. An elaborate account of these effects and models can be found in Refs. [5, 6]. Our focus here is on modeling issues pertinent to generic magnetic sensor structures and geometries; hence, our discussion of effects and analytical models will, where possible, be brief.

Some of the above effects can be best illustrated by assuming a specific (limiting) geometric and physical configuration whereby the magnetic field is perpendicular to the current flow, the components of current density and electric field in the direction of magnetic field can be ignored, and diffusion current components are negligible. This is illustrated in Table 4.1 for both

Table 4.1. *Analytical description of selected galvanomagnetic effects and associated relations*

$$\mathbf{B} = [0,0,B_z]; \mathbf{J_{nB}} = [J_{nx}, J_{ny}, 0]; \mathbf{J_{pB}} = [J_{px}, J_{py}, 0]; \mathbf{E} = [E_x, E_y, 0]; \text{grad } n = \text{grad } p = 0$$

$$\mathbf{J_{nB}} = \frac{\sigma_n}{1 + (\mu_{Hn}\mathbf{B})^2}(\mathbf{E} + \mu_{Hn}\mathbf{B} \times \mathbf{E}); \mathbf{J_{pB}} = \frac{\sigma_p}{1 + (\mu_{Hp}\mathbf{B})^2}(\mathbf{E} - \mu_{Hp}\mathbf{B} \times \mathbf{E})$$

Interaction	Geometry	n-Type	p-Type
Hall effect	$L \gg W$	$J_{ny} = 0$	$J_{py} = 0$
		$E_y/E_x = \sigma_n R_{Hn} B$	$E_y/E_x = \sigma_p R_{Hp} B$
		$\sigma_n = q\mu_n n$	$\sigma_p = q\mu_p p$
		$R_{Hn} = -r_{Hn}/(qn)$	$R_{Hp} = r_{Hp}/(qp)$
		$r_{Hn} = \mu_{Hn}/\mu_n$	$r_{Hp} = \mu_{Hp}/\mu_p$
		$\tan\Theta_{Hn} = \sigma_n R_{Hn} B = -\mu_{Hn}B$	$\tan\Theta_{Hp} = \sigma_p R_{Hp} B = \mu_{Hp}B$
Carrier deflection	$L \ll W$	$E_y = 0$	$E_y = 0$
		$J_{ny}/J_{nx} = -\sigma_n R_{Hn} B$	$J_{py}/J_{px} = -\sigma_p R_{Hp} B$
		$\tan\Theta_{Ln} = -\sigma_n R_{Hn} B = \mu_{Hn}B$	$\tan\Theta_{Lp} = -\sigma_p R_{Hp} B = -\mu_{Hp}B$
Mixed conduction (e.g., magneto-concentration (grad $n, p \neq 0$)		$\sigma = \sigma_n + \sigma_p$ $R_H = \dfrac{\sigma_n^2 R_{Hn} + \sigma_p^2 R_{Hp}}{(\sigma_n + \sigma_p)^2} = \dfrac{-[r_{Hn}(\mu_n/\mu_p)^2 n - r_{Hp}p]}{q[(\mu_n/\mu_p)n + p]^2}$	
Transverse magneto-resistance	$L \ll W$	$\dfrac{\sigma_n}{1 + (\mu_{Hn}B)^2}$	$\dfrac{\sigma_p}{1 + (\mu_{Hp}B)^2}$

n- and p-type materials along with pertinent relations. The Hall effect is predominant in extrinsic samples where the device active region is rectangular with an aspect ratio, $L \gg W$ [3, 12]. Here, L and W denote dimensions in the x- and y-directions, respectively, with the current electrodes at the short faces. In such geometries, $J_{ny} \approx 0$ and we get the build-up of a Hall field E_y which gives rise to a skew in the electric potential. Thus the equipotential and current lines are not orthogonal but are at the Hall angle, Θ_{Hn}. Carrier deflection and magnetoresistance prevail in device active regions with $L \ll W$, where current electrodes are now at the long faces. Here, we can assume that $E_y \approx 0$, which produces a rotation (deflection) in current flow by the Lorentz angle, Θ_{Ln} [6, 12]. The deflection in carrier-drift path leads to (transverse) magnetoresistance whereby there is an increase in resistivity regardless of field direction. In all of these cases, the magnitude of the effect responsible for sensor action is governed by the respective Hall or Lorentz angle. In its general form, the angle can be expressed as $\tan^{-1}(\sigma R_H B)$, where σ and R_H denote the conductivity and Hall coefficient, respectively, valid for mixed-conduction material. These coefficients are addressed in greater detail in Sect. 4.2. The $|\sigma R_H|$ product works out to $\tan^{-1}(\mu_{Hn,p}B)$ for n- and p-type material, respectively. For high sensitivity, we would like $(\mu_{Hn,p}B)$ to be large which is why semiconductors are preferred to metals and why n-type Si is preferred over the p-type counterpart.

The magnetoconcentration effect, also referred to as the Suhl effect [53], describes the compensation of the Lorentz force by concentration gradient. The effect dominates in diode-based structures at high injection levels where ambipolar diffusion creates comparable electron and hole concentrations on the same side of the sample. This leads to a magnetic field modulation of the sample resistivity. In regions where there is a large build-up of mobile charge, we see a localized decrease in base resistance leading to current crowding [14]. The reverse happens at the region where there is depletion of carrier concentration. Although there is a redistribution of mobile charge in regions where magnetoconcentration prevails, there is no noticeable effect of the magnetic field on the net space charge; $p(B) - n(B)$ [21, 22]. Since the effects on the electric potential are determined by the space charge, the local distribution of potential remains practically unperturbed. The Hall angle, $\Theta_{H,L} = \tan^{-1}(\sigma R_H B)$, should be virtually zero in this case of mixed conduction. Hence, there is no pronounced skew in the current flow lines and equipotential lines and indeed, they are virtually orthogonal [14].

4.2 Galvanomagnetic Transport Coefficients

The transport coefficients, *viz.*, conductivities $(\sigma_{n,p})$, Hall mobilities $(\mu_{Hn,p})$, Hall scattering coefficients $(r_{Hn,Hp})$, and Hall coefficients $(R_{Hn,Hp})$, which appear in the various galvanomagnetic effects as seen in Table 4.1, depend on temperature, carrier type and doping concentration, and band structure of the material. In addition, they also depend on the magnetic and electric fields. In this Section, we systematically review these coefficients and their dependencies, based on theoretical analysis and measurement data. But first, we revisit the derivation of the galvanomagnetic transport equation to recast it in a form such that we retain, in an explicit manner, the field-dependence of the coefficients.

4.2.1 Magnetic Field Dependence

In arriving at (4.12), we expanded the perturbation function, $\delta\varphi$ of the distribution function in ascending powers of the magnetic field to yield (4.9). This was then inserted into the current density equation, (4.10) which yielded coefficients M_i that were independent of the magnetic field.

Instead, if we use the unexpanded form of the perturbation function (4.7), we arrive at the following form of the isothermal electron current density [5, 47]

$$\mathbf{J}_{nB} = -[K_1 + K_2 \mathbf{B} \times + K_3 \mathbf{B} \mathbf{B} \cdot] \operatorname{grad} \phi_n, \tag{4.23}$$

where to distinguish coefficients, we have now switched notation from M_j to K_j. The latter now depend on the magnetic field and take the form [5]

$$K_j = qn \left(\frac{q}{m^*}\right)^j \left\langle \frac{\tau_n^j}{1 + (\mu_n \mathbf{B})^2} \right\rangle \tag{4.24}$$

These K_j are different from the earlier M_j (4.13) in that they now constitute energy-weighted averages over the magnetic field term, with the usual dependence of the relaxation time and drift mobility on energy. Eqs. (4.24) and (4.13) become identical for vanishing magnetic field. For $(\mu_n \mathbf{B})^2 \ll 1$, a series expansion of the denominator in ascending powers of $\mathbf{B}$ yields

$$\frac{1}{1 + (\mu_n \mathbf{B})^2} = \sum_{m=0}^{\infty} (-1)^m (\mu_n \mathbf{B})^{2m}. \tag{4.25}$$

Ignoring terms beyond those quadratic in $\mathbf{B}$ yields the following simplified form for the magnetic field dependent coefficients, in the weak field limit [5]:

$$K_1 = qn \left[\langle \mu_n \rangle - \langle \mu_n^3 \rangle \mathbf{B}^2 \right], \tag{4.26}$$

$$K_2 = qn \left[\langle \mu_n^2 \rangle - \langle \mu_n^4 \rangle \mathbf{B}^2 \right], \tag{4.27}$$

$$K_3 = qn \left[\langle \mu_n^3 \rangle - \langle \mu_n^5 \rangle \mathbf{B}^2 \right], \tag{4.28}$$

where $\langle \mu_n^j \rangle$ denotes the energy-weighted average value of the jth power of the carrier drift mobility. Following (4.15) it takes the form

$$\langle \mu_n^j \rangle = \langle \mu_n \rangle^j G_j(r), \tag{4.29}$$

where $\langle \mu_n \rangle^j$ is expression (4.14) to the jth power and $G_j(r)$ is as given in (4.16). Again, we note that the difference between (4.26) to (4.28) and (4.15) lies in the terms associated with $\mathbf{B}$ which have now been included in the averaging process. The above relations are strictly valid only for the non-degenerate material. As stated earlier, values of $G_j(r)$ depend on the energy distribution of the relaxation time. This in turn depends on the scattering processes which are themselves dependent on the temperature and doping level. Thus the $G_j(r)$ influence the transport coefficient through the scattering processes. The dependence of $G_j(r)$ on the scattering process is illustrated in Fig. 4.1 and corresponding numerical values are given in Table 4.2. The scattering coefficients have lower values in the lightly doped material where phonon scattering dominates. Its scattering relaxation time

Table 4.2. *Values of scattering coefficients for different scattering mechanisms*

Scattering coefficient	Acoustic phonon scattering $r = -1/2$	Ionized impurity scattering $r = 3/2$	Neutral impurity scattering $r = 0$
$G_2(r) = r_{Hn0}$ $r_{Hn} = r_{Hn0}\gamma$, where γ is given by (4.31)	1.18	1.93	1
$G_3(r)$	1.77	5.90	1
$G_4(r)$	4.16	19.14	1
$\beta = \dfrac{G_4(r)}{G_2(r)} - G_3(r)$	1.26	1.08	0
$\alpha = 1 - \dfrac{2G_3(r)}{G_2^2(r)} + \dfrac{G_4(r)}{G_2^3(r)}$	0.99	0.49	0

has a $E^{-1/2}$ dependence, where E denotes electron energy. The values of $G_j(r)$ increase with doping due to impurity scattering. Here, the scattering relaxation time has a $E^{3/2}$ dependence. In degenerate materials (such as in metals), the scattering coefficients are identically unity due to neutral impurity scattering. Here, the scattering relaxation time is energy independent.

With $j = 2$, we obtain the Hall scattering coefficient (*viz.*, $G_2(r) = r_{Hn0} = \langle \tau_n^2 \rangle / \langle \tau_n \rangle^2$) which relates the Hall mobility to drift mobility. Values of the Hall scattering coefficient are given in Table 4.2. They range from 1 to 1.93 depending on the nature of the scattering mechanism (see [54]). A compilation of values and their sources can be found in [5, 55]. The values of r_{Hn0} given in Table 4.2, although they account for the various scattering processes, are only valid for constant energy surfaces that are spherical, where the effective mass is independent of energy [5]. In the case of the conduction band in Si which is multivalleyed, the effective mass of an electron in any given valley depends on its direction of motion. Thus the effective mass is anisotropic and so too is its scattering probability. Assuming that the relaxation time in the longitudinal and transverse directions are equal, the band anisotropies can be accounted for by a factor γ in the following expression [5]:

$$r_{Hn} = r_{Hn0}\gamma. \tag{4.30}$$

Here, the anisotropy factor γ, following the analysis of Herring and Vogt [56] of the conduction band structure, takes the form:

$$\gamma = \frac{3\left(\dfrac{m_l^*}{m_t^*}\right)\left(2 + \dfrac{m_l^*}{m_t^*}\right)}{\left(1 + 2\dfrac{m_l^*}{m_t^*}\right)^2} \tag{4.31}$$

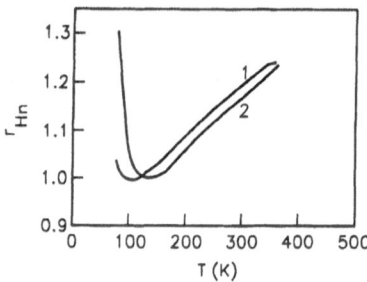

Fig. 4.3 The measured electron Hall scattering coefficient r_{Hn} as a function of temperature for donor (phosphorus) concentration of 8×10^{13} cm^{-3} (curve 1) and 9×10^{14} cm^{-3} (curve 2). Adapted from [57]

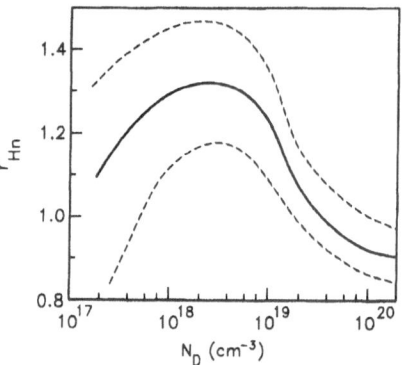

Fig. 4.4 Solid curve: the measured electron Hall scattering coefficient r_{Hn} as a function of donor (phosphorus) concentration at room temperature. The dashed curves denote the error range. Adapted from [58]

where $m_l^*/m_t^* \approx 5$ yielding $\gamma = 0.87$ for electrons in Si [5]. This leads to a 13% reduction in Hall electron mobility and the corresponding Hall coefficient when compared to values obtained with the isotropic scattering coefficient (r_{Hn0}) predicted in Table 4.2.

The measured behaviour of the Hall scattering coefficient as a function of temperature [5,57] and doping concentration is shown in Figs. 4.3 and 4.4. In relatively low doped material, the temperature dependence is governed by phonon and impurity scattering; the latter prevailing at lower temperatures (see curve 2 in Fig. 4.3). In the theoretical limit at low temperatures, Eq. (4.30) yields a value of r_{Hn} is 1.68 which comes from $r_{Hn0} = 1.93$ and $\gamma = 0.87$. The dependence of r_{Hn} on doping is shown in Fig. 4.4 at room temperature [5, 58]. At doping levels below 3×10^{18} cm^{-3}, we note an increase in r_{Hn} with doping due to impurity scattering. However, degeneracy effects set in at higher concentrations and r_{Hn}

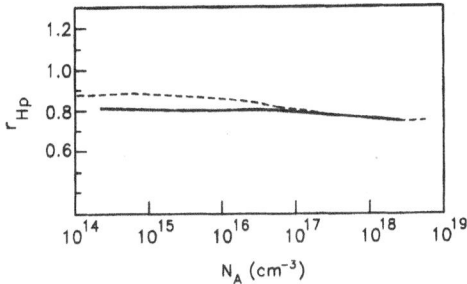

Fig. 4.5 The experimental (solid curve) and theoretical (dashed curve) hole Hall scattering coefficient r_{Hp} as a function of dopant concentration. Adapted from [59]

approaches the predicted value of 0.87; here, r_{Hn0} approaches unity, as in the case of metals, and $\gamma = 0.87$ [5].

The behavior of the Hall scattering coefficient in p-Si is not as straightforward to predict. Apart from the usual scattering effects accounted for by the scattering factors r, the non-parabolicity and anisotropy of the valence band, including the degeneracy of heavy and light hole bands at $\mathbf{k} = \mathbf{0}$, need to be taken into account (see [5,59]). As a result, the Hall scattering coefficient r_{Hp} for holes is smaller than that for electrons. Fig. 4.5 illustrates the doping dependence of the Hall scattering coefficient for holes in p-type Si [5, 59]. Its value remains at approximately 0.8 for a wide range of doping concentrations.

The conductivity implicitly determines the magnitude of the galvanomagnetic interaction, and hence, the sensor response as seen from the analytical relations given in Table 4.1. The magnetic field dependence of conductivity, $\sigma_n(B)$ can be retrieved from relation (4.23) where it corresponds to the first term, K_1. The expression takes the form [5]:

$$\sigma_n(B) = K_1 = qn\langle \mu_n \rangle \left(1 - \frac{\langle \mu_n^3 \rangle}{\langle \mu_n \rangle} B^2 \right). \tag{4.32}$$

After a grouping of the scattering coefficients, (4.32) can be written as

$$\sigma_n(B) = q\mu_n n (1 - \langle \mu_n \rangle^2 G_3(r) B^2) = q\mu_n n (1 - \mu_{Hn}^2 B^2), \tag{4.33}$$

where, for convenience, we dropped the angular brackets on the drift mobility. The term $q\mu_n n$ denotes the zero field ($B = 0$) conductivity, viz., K_1 evaluated at $B = 0$. In (4.33), we have used the approximation; $G_3(r) = G_2^2(r) = r_{Hn0}^2$ to yield an expression that reduces to the well known form of the magnetic-field-dependent conductivity. Indeed we employed this approximation earlier in arriving at the carrier transport relation (4.19). This is reasonable in the low field limit. The dependence of $\sigma_n(B)$ on temperature and doping is governed by the corresponding variations in the drift mobility, carrier concentration, and Hall mobility, $\mu_{Hn} = r_{Hn}\mu_n$. Models for the doping- and temperature-dependence of the drift mobility and carrier concentrations were reviewed in Chapt. 2, and that of the Hall scattering coefficient was addressed in the preceding discussion. An expression for the magnetic field dependence of the hole conductivity is analogously derived.

We now discuss the magnetic field dependence of the Hall mobility, μ_{Hn}. The Hall mobility explicitly determines the magnitude of the galvanomagnetic effect (see Table 4.1). In the equations, we have seen it appearing as a $(\mu_{Hn}B)$ product. This defines the Hall angle in extrinsic material. The magnetic field dependence of μ_{Hn} can be retrieved from coefficient K_2 in (4.23), since K_2 is the conductivity-mobility product, $\sigma_n(B)\mu_{Hn}(B)$. Thus $\mu_{Hn}(B) = K_2/K_1$, and its B-dependence stems from those of K_1 and K_2, as shown by relations (4.26) and (4.27). In non-degenerately doped material, we obtain, after some algebraic manipulation of K_2/K_1, the following reduced form for the Hall mobility for electrons [5]

$$\mu_{Hn}(B) = \mu_{Hn}[1 - \beta(\mu_{Hn}B)^2], \tag{4.34}$$

where

$$\beta = \frac{G_4(r)}{G_2(r)} - G_3(r) \tag{4.35}$$

is referred to as the non-linearity coefficient of the Hall mobility. Computed values of β are given in Table 4.2 for the different scattering mechanisms. The non-linearity coefficient is large for low doped material and vanishes when the carrier concentration becomes degenerate.

The Hall coefficient, like the conductivity, plays an important role in determining the magnitude of the galvanomagnetic interaction. Expressions for the zero-field Hall coefficient for n-type, p-type, and mixed conduction (e.g. intrinsic) materials are given in Table 4.1. With the relations we have already obtained for the magnetic field dependent conductivity (4.32) and Hall mobility (4.34), we can retrieve the dependence of the Hall coefficient on the magnetic field. The associated coefficients are K_1 and K_2. But first we identify the Hall coefficient as the ratio of the magnetic field dependent Hall mobility to the zero field conductivity. Its field-dependence stems from that of $\mu_{Hn}(B)$. Earlier, we saw that $\mu_{Hn}(B) = K_2/K_1$. Thus $\sigma_n = K_1[1 + (\mu_{Hn}B)^2] = K_1[1 + (K_2/K_1)^2B^2]$. Now we can state the expression for the magnetic-field-dependent Hall coefficient

$$R_H = \frac{K_2/K_1}{K_1[1 + (K_2/K_1)^2B^2]} \tag{4.36}$$

which can be reduced to [5]

$$R_{Hn}(B) = \frac{\mu_{Hn}}{\sigma_n}[1 - \alpha(\mu_{Hn}B)^2].$$ (4.37)

Here,

$$\alpha = 1 - \frac{2G_3(r)}{G_2^2(r)} + \frac{G_4(r)}{G_2^3(r)}$$ (4.38)

is the non-linearity coefficient of the Hall voltage which is also referred to as the material non-linearity coefficient. Values of α computed for the different scattering mechanisms are given in Table 4.2. Experimentally, a quadratic dependence of the Hall coefficient on magnetic field has been observed [5, 43],

$$R_{Hn}(B) = \frac{\mu_{Hn}}{\sigma_n}[1 - \alpha^* B^2]$$ (4.39)

in terms of an empirical non-linearity coefficient, α^*. Measured values of α^* and α, the latter retrieved from comparison of relations (4.37) and (4.39), are shown in Figs. 4.6 to 4.8 as a function of temperature and doping concentration [43]. Despite neglecting band anisotropies, which affect the Hall mobility in (4.39) via (4.30), the agreement between the retrieved and predicted values of α is satisfactory. For example, at low temperatures where impurity scattering prevails, the value of α retrieved is 0.4 (Fig. 4.6) which is close to the predicted value of 0.49 (Table 4.2). With

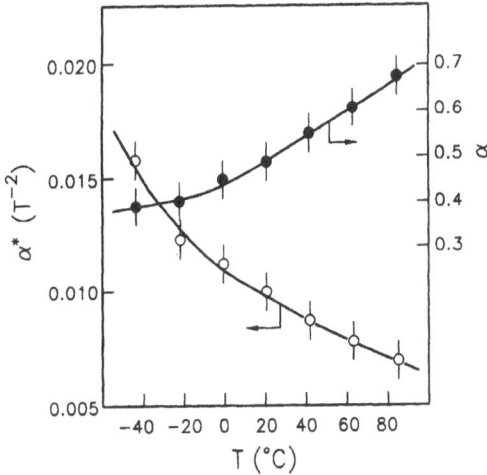

Fig. 4.6 Temperature-dependence of the non-linearity parameters in n-type Si with donor concentration, 10^{15} cm^{-3}; predictions (solid curve), measurements (symbols). Adapted from [43]

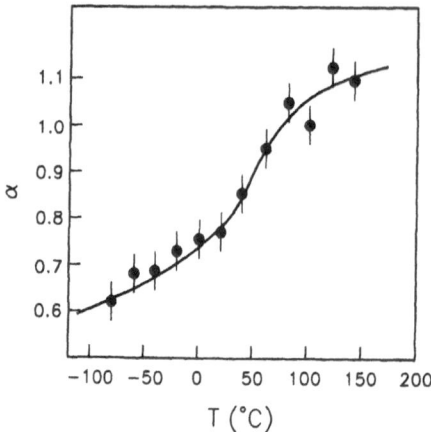

Fig. 4.7 Temperature-dependence of the non-linearity coefficient in n-type Si with donor (phosphorus) concentration, 1.7×10^{14} cm^{-3}; predictions (solid curve), measurements (symbols). Adaped from [43]

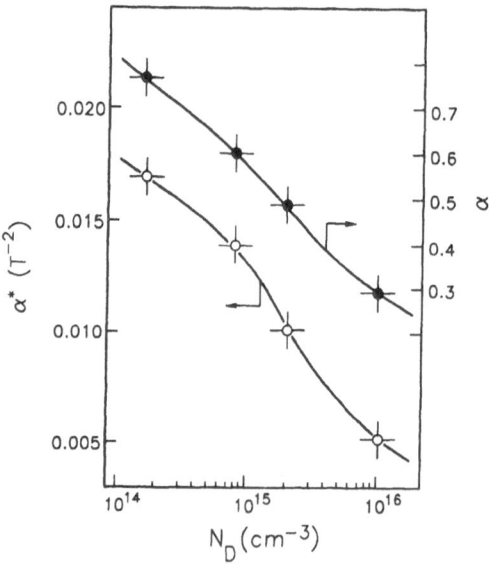

Fig. 4.8 Doping dependence of the non-linearity parameters in n-type Si at 20°C; predictions (solid curve), measurements (symbols). Adapted from [43]

increasing temperature, α increases showing the dominance of phonon scattering (Figs. 4.6 and 4.7). With increasing doping, α decreases due increased impurity scattering. Its behavior is consistent with the predictions given in Table 4.2. The temperature dependence of α at higher temperatures correlates well with that of the Hall scattering coefficient shown in Fig. 4.3. The temperature-dependence of the Hall coefficient, apart from the temperature-dependence of α, stems from that of the Hall mobility and conductivity.

The coefficient K_3 of the last term in the transport equation (4.23) describes the longitudinal magnetoresistance effect which arises when the magnetic field is parallel to the current density [5, 6]. In non-degenerate material, the coefficient can be approximated as: $K_3 = \sigma_n \mu_n^2 G_3(r)$. Following our earlier approximation, $G_3(r) \approx G_2^2(r) = r_{Hn0}^2$ used in conjunction with (4.19) for an isotropic conduction band effective mass, we see that the longitudinal magnetoresistance effect is described by $\sigma_n (\mu_{Hn} B)^2$.

4.2.2 Electric Field Dependence

Galvanomagnetic interactions and associated transport coefficients at high electric fields, have been a subject of theoretical and experimental investigations (see [5, 60–62]). However, studies have yet to yield fully convincing and consistent results. For example, the Hall scattering coefficient can no longer be assumed isotropic because of band anisotropies, which become even more enhanced at high electric fields. Furthermore, the Hall mobility decreases with increasing applied field. Experiments performed using conventional MOS inversion layer and bulk Hall voltage structures, where the induced Hall voltage is measured across probes located orthogonal to the direction of the applied electric field (see Fig. 4.9a), show a decrease in Hall mobility with increasing applied electric field. In fact, there is a saturation of the Hall field due to drift velocity saturation $(v_{d,\text{sat}})$ at high electric fields, viz., $|E_{H,\text{sat}}| \approx v_{d,\text{sat}} B$, leading to a

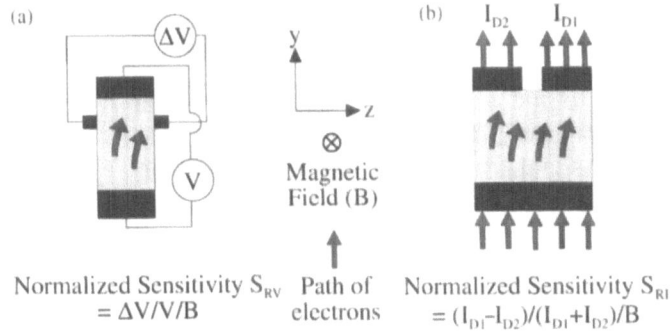

Fig. 4.9 Configurations of (a) Hall voltage and (b) Hall current sensors. Also indicated are their relative sensitivities, S_{RV} and S_{RI}

decrease in the Hall angle and hence, in the output sensitivity (S_{RV}). However, recent experiments with Si n-channel inversion layer Hall-current structures, show an opposite electric field dependence [63]. Here, the relative magnetic sensitivity (S_{RI}) increases with increasing applied field yielding sensitivity values comparable to devices fabricated using high mobility compound semiconductor materials. These effects appear to be stemming from anisotropies in the Hall mobility and scattering coefficients due to the multivalleyed conduction band structure. This behaviour in Hall mobility at high electric fields has been independently verified using the modified Einstein relation for interpretation of the measured diffusion coefficients [64, 65].

The effects of the electron effective mass (m_n^*) and momentum relaxation time (τ_n) anisotropies in the conduction band have been derived by Herring and Vogt [56]. For convenience of notation, we now discard the subscript n. Similar to the effective masses, viz., $m_l^*/m_t^* \approx 5$ as seen in (4.31), the value of τ depends on the direction of applied electric field and hence, conduction in the different valleys. The corresponding ratios (τ_l^*/τ_t^*) for the momentum relaxation time are approximately 0.7 for acoustic phonon scattering [66], 3.9 for impurity scattering [67] and 1 for intervalley scattering [66]. Because of the band anisotropy, electrons traveling in a direction with smaller mass have higher energies than electrons traveling in the direction with larger mass. Since scattering is energy dependent, the different valleys are limited by different scattering mechanisms. In addition, at very high electric fields, there may be hot electron induced valley repopulation [68, 69]. Denoting the proportion of electrons transferred from each of the four hot valleys by h, then the two cold valleys gain $(1 + 2h)$ times their original number of electrons, while the four hot valleys lose $(1 - h)$ times their original number of electrons. Thus the component of the drift mobility ($\mu_{di\alpha}$) from the ith valley in a direction α of the applied field is expressed [56] as

$$\mu_{di\alpha} = \frac{q\langle \Delta E \tau_{i\alpha} \rangle}{m_{i\alpha}^* \langle \Delta E \rangle}, \tag{4.40}$$

where E refers to the electron energy. The angular brackets represent Maxwellian averages with $\langle \Delta E \rangle \equiv \langle |E - E_b| \rangle$, where E_b is the band edge energy; $\langle \Delta E \rangle$ is equal to $(3/2)$ kT for cold electrons. The corresponding drift-Hall mobility product, in the weak magnetic field limit, reads [63]

$$\mu_{di\alpha} \mu_{Hi\beta} = \frac{q\langle \Delta E \tau_{i\alpha} \tau_{i\beta} \rangle}{m_{i\alpha}^* m_{i\beta}^* \langle \Delta E \rangle}. \tag{4.41}$$

Here, subscript β denotes the direction of the Lorentz force. Averaging the above expressions for the six valleys, we can determine the Hall

scattering coefficient as [63]

$$r_{Hn} = \frac{3\left[\frac{(1+2h)\langle\Delta E\tau_{lc}\tau_{tc}\rangle}{m_l^* m_t^*} + \frac{(1-h)\langle\Delta E\tau_{lw}\tau_{tw}\rangle}{m_l^* m_t^*} + \right.}{\left[\frac{(1+2h)\langle\Delta E\tau_{lc}\rangle}{m_l^*} + \frac{2(1-h)\langle\Delta E\tau_{tw}\rangle}{m_t^*}\right]\left[\frac{(1-h)\langle\Delta E\tau_{tw}\rangle}{m_t^*}\right.}$$

$$\frac{\left. + \frac{(1-h)\langle\Delta E\tau_{lw}\tau_{tw}\rangle}{m_t^{*2}}\right]}{\left. + \frac{(1+2h)\langle\Delta E\tau_{tc}\rangle}{m_t^*} + \frac{(1-h)\langle\Delta E\tau_{lw}\rangle}{m_l^*}\right]} \tag{4.42}$$

which turns out to be isotropic but dependent on electron energy. The subscripts c and w refer to the cooler (lower electron energy) and warmer (higher electron energy) valleys, respectively. Electrons in the cooler valleys are limited by impurity scattering while those in warmer valleys are limited by acoustic phonon scattering, since $\tau_{imp} \sim E^{3/2}$ and $\tau_{ac} \sim E^{-1/2}$.

The values of m_l^*/m_t^* and τ_l/τ_t enumerated earlier still do not contain sufficient information to allow use of Eqs. (4.40) through (4.42) to calculate precise values of the mobility components. In particular, values of τ_l/τ_t are not known and neither are their field dependencies; only their ratios have been determined. Nevertheless, the relations are useful for qualitative analysis of measured data [63] (see Fig. 4.10). In terms of (4.40) to (4.42), and with the aid of Fig. 4.9b, if $\tau_c(=\tau_{imp} \sim E^{3/2})$ increases by 40% and $\tau_h(=\tau_{ac} \sim E^{-1/2})$ decreases by 30% with the applied field (E_y), then μ_{nx} and μ_{Hnx} decrease by 0.74% on average. In contrast, μ_{ny} and μ_{Hny} increase

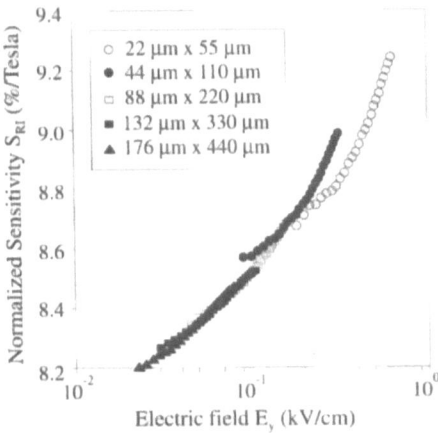

Fig. 4.10 Relative sensitivity S_{RI} as a function of applied electric field (E_y; see Fig. 4.9) in nMOS Hall current sensors for various channel geometries ($W \times L$)

by 12.4% on average. The relative sensitivity, which is mostly influenced by μ_{Hny}, increases by 22.3% and r_{Hn} increases by 19.2%. These predicted increases are consistent with the measured field dependence shown in Fig. 4.10.

The role of hot electron induced valley repopulation is almost negligible ($h < 0.4\%$) for the fields considered ($E_y < 1\,\text{kV/cm}$). Their effects become significant for fields larger than 1 kV/cm [68]. Here, all mobility components in Si are expected to increase, with those in the direction of the applied field increasing more rapidly than those in the direction of the Lorentz force. The Hall scattering coefficient is expected to decrease slightly along with the Hall coefficient.

4.3 Equations and Boundary Conditions for Magnetic Sensor Simulation

The system of model equations for simulation of magnetic sensors remains the same as that discussed in Chapt. 2. The key difference lies in the transport relations which describe the underlying galvanomagnetic interactions. Depending on the device in question, suitable approximations can be employed to reduce the system of equations to a much simpler numerical problem. Before discussing the different simulation scenarios and the associated boundary conditions, we first summarize the basic system of partial differential equations and transport relations. Restating for convenience the key equations from Chapt. 2, we have Poisson's Eq. (4.43) and the continuity equations for electrons (4.44) and holes (4.45):

$$\text{div}\,(\varepsilon\,\text{grad}\,\psi) = -q(p - n + N_D - N_A), \tag{4.43}$$

$$\text{div}\,\mathbf{J_{nB}} = qR + q\partial n/\partial t, \tag{4.44}$$

$$\text{div}\,\mathbf{J_{pB}} = -qR - q\partial p/\partial t, \tag{4.45}$$

where $\mathbf{J_{nB}}$ and $\mathbf{J_{pB}}$ denote the respective magnetic-field-dependent current density relations for electrons and holes. There are various forms of the current density relations that can be employed in the continuity equations. For convenience, we state below the different relations which we will employ in one form or another in subsequent chapters dealing with coupled energy fields. Restating (4.19) and (4.20), we have

$$\mathbf{J_{nB}} + \mu_{Hn}\mathbf{J_{nB}} \times \mathbf{B} = -q\mu_n n\,\text{grad}\,\phi_n, \tag{4.46}$$

$$\mathbf{J_{pB}} - \mu_{Hp}\mathbf{J_{pB}} \times \mathbf{B} = -q\mu_p p\,\text{grad}\,\phi_p \tag{4.47}$$

or (4.21) and (4.22) which are given explicitly in terms of drift-diffusion components

$$\mathbf{J_{nB}} + \mu_{Hn}\mathbf{J_{nB}} \times \mathbf{B} = qD_n\left[\operatorname{grad} n - n \operatorname{grad}\left(q\psi/kT\right)\right], \qquad (4.48)$$

$$\mathbf{J_{pB}} - \mu_{Hp}\mathbf{J_{pB}} \times \mathbf{B} = -qD_p\left[\operatorname{grad} p + p \operatorname{grad}\left(q\psi/kT\right)\right]. \qquad (4.49)$$

In some cases, it may be more suitable to express the above current densities in terms of the conductivity and Hall coefficient. Here, a substitution for the Hall mobility, for example, in (4.48) and (4.49) yields:

$$\mathbf{J_{nB}} - \sigma_n R_{Hn}\mathbf{J_{nB}} \times \mathbf{B} = qD_n\left[\operatorname{grad} n - n \operatorname{grad}\left(q\psi/kT\right)\right], \qquad (4.50)$$

$$\mathbf{J_{pB}} - \sigma_p R_{Hp}\mathbf{J_{pB}} \times \mathbf{B} = -qD_p\left[\operatorname{grad} p + p \operatorname{grad}\left(q\psi/kT\right)\right]. \tag{4.51}$$

We will illustrate how the above system of equations can be applied for analyses of different sensor types and field configurations. In particular, we consider unipolar devices [15–17, 63, 70] (e.g., Hall devices based on the inversion layer or bulk), including domain detector configurations [18, 19, 38] where the magnetic field can be highly spatially non-uniform, and bipolar devices [10, 25, 28] (e.g., magnetotransistors).

4.3.1 Unipolar Analysis

Unipolar analysis is by far the simplest since only a single carrier type needs to be considered. For convenience, as well as for practical reasons (since $\mu_n > \mu_p$), we restrict ourselves to n-type Hall devices and invoke the following assumptions:

- Magnetic field is perpendicular to the direction of current flow. This makes the detection configuration quasi-two-dimensional which implies that the components of current density and electric field in the direction of magnetic field can be ignored.
- Transport is governed only by electrons and generation-recombination in the bulk or inversion layer is negligible since these devices mostly operate under low level injection. Recombination at surfaces and other boundaries is also negligible.
- There is no space charge in the device active region although the device can be isolated by a space charge depletion layer through a reverse bias junction.
- Device is composed of uniform material. Thus concentration gradients and diffusion currents are negligible. However, this does not imply that the conductivity is homogeneous. In spatially varying magnetic field configurations, the conductivity is inhomogeneous. In addition, the conductivity is tensorial due to anisotropies induced by magnetic field as well as large electric fields. Stress-induced anisotropies in conductivity

and Hall coefficient, by virtue of piezoresistance and piezo-Hall effects, will be dealt with in Chapt. 6.

Following the above assumptions, the system of Eqs. (4.43) through (4.45) can be reduced to a single equation and the galvanomagnetic effects of interest are best described by the electron continuity equation (4.44) along with the pertinent transport relation (4.48) or (4.50). Here, the current density in the two-dimensional system, with $\mathbf{B} = [0, 0, B_z(x, y)]$ and $\mathbf{J_{nB}} = [J_{nx}, J_{ny}, 0]$, can be integrated with respect to z (as depicted by the device cross sections in Fig. 4.11) over an appropriate conduction layer thickness, $viz.,$

$$\mathbf{J_{nB}}(x, y) = \int_{thickness} \mathbf{J_{nB}}(x, y, z)dz, \tag{4.52}$$

$$Q(x, y) = q \int_{thickness} n(x, y, z)dz \tag{4.53}$$

to yield

$$\mathbf{J_{nB}}(x, y) + \mu_{Hn} \mathbf{J_{nB}}(x, y) \times \mathbf{B} = \mu_n Q(x, y)\mathbf{E}(x, y). \tag{4.54}$$

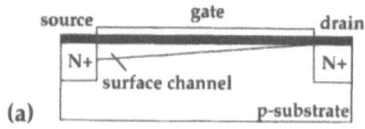

(a)

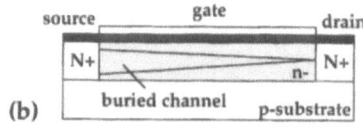

(b)

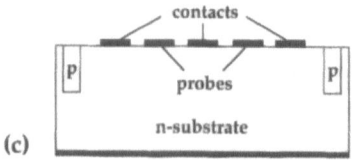

(c)

Fig. 4.11 Two-dimensional model of device active area for (a) surface channel MOS Hall device (b) buried channel MOS Hall device (c) bulk Hall device

Thus the current continuity equation, under steady state conditions, now becomes:

$$\text{div } \mathbf{J_{nB}}(x, y) = \text{div } [\beta \text{ grad } U(x, y)] = 0, \tag{4.55}$$

where

$$\beta = \frac{\mu_n}{1 + [\mu_{Hn} B_z(x, y)]^2} \begin{bmatrix} 1 & \mu_{Hn} B_z(x, y) \\ -\mu_{Hn} B_z(x, y) & 1 \end{bmatrix}. \tag{4.56}$$

Note that (4.55) now contains a new solution variable U, which is a scalar function [15, 16] whose form depends on the sensor under consideration (e.g., surface channel [63, 70] and buried channel [71] MOS Hall devices, or bulk Hall devices [72, 73], the cross sections of which are shown in Fig. 4.11).

With surface channel MOSFETs operating in the linear region (Fig. 4.11a), $Q(x, y)$ is the mobile surface charge density in the channel if the integration with respect to thickness extends over the channel depth. It is usually assumed that this surface charge density (Q_n) is only a function of the channel potential, $V(x, y)$. Defining [15, 16]

$$U(x, y) = \int_{V_s}^{V(x, y)} Q_n(V')dV' \tag{4.57}$$

with V_s being the source voltage and V' denoting an integration variable, we obtain

$$\text{grad } U(x, y) = Q_n(x, y) \text{ grad } V(x, y), \tag{4.58}$$

where [63]

$$Q_n(x, y) = C_{ox}[\alpha V(x, y) - V_{GS} + V_T], \tag{4.59}$$

$$U(x, y) = C_{ox}\left[V_{GS} - V_T - \frac{\alpha kT}{q} - \frac{\alpha V(x, y)}{2} \right] V(x, y). \tag{4.60}$$

Here, C_{ox} is the gate oxide capacitance per unit area, α is a factor used in the BSIM model [74] that accounts for bulk bias effects, V_{GS} is the gate-source voltage, and V_T is the threshold voltage.

With buried channel MOSFETs (Fig. 4.11b), the spatial relationship between the free charge density and the voltage distribution in the channel can be developed along the same lines. However, unlike MOSFETs, different conduction modes are possible depending on the bias applied. For example, surface conduction occurs in accumulation mode while buried channel conduction occurs in depletion mode. The latter is preferred from a standpoint of high mobility (and hence, high magnetic sensitivity) and low noise due to the reduced surface scattering and oxide trapping. Following an analytical solution of Poisson's equation [63, 75], the free charge density in the buried channel can be expressed as

$$Q_n(x, y) = qN_D(z_n - z_s),$$ (4.61)

where N_D denotes the buried layer doping concentration, and z_s and z_n define the top and bottom boundaries of the conduction channel, respectively, *viz.*,

$$z_s = \left[\frac{2\varepsilon_{Si}}{qN_D}(V(x, y) + V_{bi} - V_{GS} + V_{FB}) + \left(\frac{\varepsilon_{Si}t_{ox}}{\varepsilon_{ox}}\right)^2\right]^{1/2} - \frac{\varepsilon_{Si}t_{ox}}{\varepsilon_{ox}},$$ (4.62)

$$z_n = z_i - \left[\frac{2\varepsilon_{Si}}{qN_D}\left(\frac{N_A}{N_A + N_D}\right)(V(x, y) + V_{bi} - V_{BS})\right]^{1/2}.$$ (4.63)

Here, V_{bi} denotes the built-in voltage,

$$V_{bi} = \frac{kT}{q} \log_e \frac{N_A N_D}{n_{ie}^2},$$ (4.64)

z_i the buried layer depth, N_A the substrate doping, V_{BS} the substrate-source voltage, ε_{Si} the permittivity of Si, ε_{ox} the permittivity of oxide, t_{ox} the oxide thickness, V_{FB} the flat-band voltage, and n_{ie} the intrinsic concentration. A substitution of (4.61) to (4.64) into (4.57) yields an exceedingly cumbersome expression for U (see [63]).

In bulk Hall devices (Fig. 4.11c), $Q(x, y)$ is simply the approximately constant electron density multiplied by the linear dimension of the active device region in the direction of magnetic induction. In this case, we can define [15, 73]

$$U(x, y) = Q(x, y)V(x, y).$$ (4.65)

The differential equation (4.55), in its stated form, allows simulation of two-dimensional MOS and bulk Hall magnetic sensors and accommodates the anisotropies and inhomogeneities in transport coefficients arising from the magnetic field [18, 19] and high electric fields [63]. If there are no spatial variations in mobility and magnetic field distribution in the device, (4.55) reduces to

$$\text{div grad } U(x, y) = 0.$$ (4.66)

In this case, there is no magnetic field term in the governing differential equation. The magnetic field influences the solution of (4.66) through the boundary conditions at the insulating boundaries, i.e., boundaries between the electrodes (see Fig. 4.9). At these boundaries, imposing a zero normal current to the boundary yields [15]

$$[\mathbf{n} + \beta(x, y) \times \mathbf{n}] \operatorname{grad} U(x, y) = 0 \qquad (4.67)$$

where $\mathbf{n}$ is the normal to the boundary. This boundary condition contains the magnetic field through β, see (4.56). At contact regions where we have a bias voltage, as indicated in Fig. 4.9, the applied potential is known. Hence, from (4.60) or (4.65), the value of U is also known at those contacts [15],

$$U = U|_{V=V_a}, \qquad (4.68)$$

where V_a denotes the applied bias voltage. If biased by a current source (I_{appl}), the solution of (4.66) needs to fulfill an additional equation (see Sect. 2.6.1)

$$\int_{\Gamma_c} [\mathbf{n} + \beta(x, y) \times \mathbf{n}] \operatorname{grad} U(x, y) d\Gamma = I_{appl}, \qquad (4.69)$$

which contains the normal derivative of U. This makes (4.69) a mixed boundary condition. Here, Γ_c denotes the contact region and the rest in usual notation. With voltage-output Hall devices (Fig. 4.9a), the high impedance Hall probe regions are treated as equipotential with unknown potentials. Here, we integrate the normal component of the current density through the probe to yield

$$\int_{\Gamma_p} [\mathbf{n} + \beta(x, y) \times \mathbf{n}] \operatorname{grad} U(x, y) d\Gamma = I_{leakage}, \qquad (4.70)$$

where Γ_p denotes the probe region and $I_{leakage}$ constitutes the current drawn from the device by the instrumentation circuit. With a good instrumentation circuit, $I_{leakage}$ is negligible.

4.3.2 Bipolar Analysis

In bipolar analysis, we need to consider the system of coupled differential equations in its entirety, comprising (4.43) through (4.45), to account for both electron and hole transport, in addition to the electric potential. Depending on the choice of solution variables, we need to include the appropriate form of the magnetic-field-dependent current density relations (4.46, 4.47), (4.48, 4.49) or (4.50, 4.51). The system of model equations are solved subject to Dirichlet, homogeneous Neumann boundary conditions, and interface conditions described in what follows. For convenience, Fig. 4.12 replicates Fig. 2.2 in Chapt. 2 for illustration of the bipolar

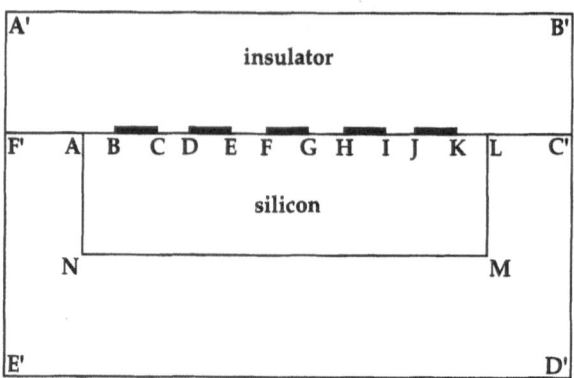

Fig. 4.12 Two-dimensional model of a bipolar magnetotransistor for illustration of Neumann boundary conditions

magnetotransistor. At ohmic contact regions (BC, DE, FG, HI, and JK), the electrostatic potential and carrier concentrations are prescribed by the usual Dirichlet boundary conditions. These are identical to the conditions employed in standard device modeling and are described in Sect. 2.6.1. Interface conditions are standard, and if required, can be extracted directly from Sect. 2.6.3. The exception lies in the Neumann boundary conditions for the electric potential at insulating (AB, CD, EF, GH, IJ, KL) and artificial (LM, MN, NA) boundaries since the presence of a magnetic field could result in a significant Hall field at these boundaries. Thus the standard condition $\mathbf{E} \cdot \mathbf{n} = 0$ employed in zero B cases may be physically invalid. Hence, conditions justified by physical or mathematical plausibility have to be imposed. We will consider both approaches.

Following physical reasoning [25, 29], we can assume that the Lorentz force induces a charge layer (ρ_{ind}) at the boundary which is counteracted by the Hall electric field. In the case of n-type material, the charge induced is given as

$$\rho_{\text{ind}} = -\varepsilon \, \mathbf{E_H} \cdot \mathbf{n} = \varepsilon R_{Hn} (\mathbf{J_{nB}} \times \mathbf{B}) \cdot \mathbf{n}, \qquad (4.71)$$

where ε denotes the material permittivity and $\mathbf{n}$ the outward normal unit vector along the insulating boundary. For an ideal interface (i.e., no interface charges), (4.71) yields the boundary condition for Poisson's equation

$$\text{grad } \psi \cdot \mathbf{n} = -\rho_{\text{ind}}/\varepsilon, \qquad (4.72)$$

where ρ_{ind} contains the magnetic field. In the bipolar material, the total charge induced is due to both electron and hole currents, and hence, the normal component of the potential at the boundary will be governed by the Lorentz force acting on both electron and hole drift and diffusion current

Table 4.3. *Summary of models pertinent to magnetic sensor simulation. Equations shown come from Chapt. 2 and this chapter*

Models	Reference	Description/Validity/Application
Governing PDE(s) & Transport Relations	(4.66) with (4.48) or (4.50)	2-D unipolar electron transport (nMOS or *n*-type bulk Hall devices)
	(4.66) with (4.49) or (4.51)	2-D unipolar hole transport (pMOS or *p*-type bulk Hall devices)
	(4.55)	2-D unipolar with anisotropic and inhomogeneous coefficients (Hall devices, domain detectors)
	(4.43)–(4.45) with (4.46, 4.47) or (4.48, 4.49) or (4.50, 4.51)	2-D/3-D coupled electron and hole transport (npn/pnp magnetotransistors, vector probes, magnetodiodes)
Physical Parameters		
Hall scattering coefficient: r_H	Table 4.2	r_{Hn0} : no band anisotropies
	(4.30)	r_{Hn} : with band anisotropies
	(4.42)	r_{Hn} : E-field dependence
	Fig. 4.5	r_{Hp}
Conductivity: σ	Table 4.1	unipolar and mixed conduction
	(4.33)	B-field dependence
Hall mobility: μ_{Hn}	(4.34)	B-field dependence
	(4.40),(4.41)	E-field dependence
Hall coefficient: R_H	Table 4.1	unipolar and mixed conduction
	(4.37), (4.39)	B-field dependence
Concentrations: n, p	(4.59)	Q_n : 2D surface channel MOS Hall
	(4.61)	Q_n : 2D buried channel MOS Hall
	(2.29), (2.30)	bipolar magnetic sensors
n_{ie}	(2.40)	band gap narrowing at high doping
Band edge shifts	(2.42), (2.43)	spatially varying band gap
Generation-recombination: R	(2.62), (2.68), (2.69)	SRH, Auger, and impact ionization
Drift mobility: $\mu_{n,p}$	Sect. 2.3.4	see Section 2.3.4 for model validity
Boundary Conditions		
Unipolar		
Contacts	(4.68), (4.69)	voltage source, current source
Probes	(4.70)	Hall voltage output
Insulating boundaries	(4.67)	
Bipolar		
Contacts	(2.92)	ohmic, voltage source
	(2.93), (2.94)	contact/external resistance
	(2.95)	current source
Insulating boundaries & interfaces	(2.97), (2.98)	trapped and interface charges
	(2.99), (2.100)	surface recombination
	(4.73)	Poisson's equation
	(2.102), (2.103)	continuity equations
Outer boundaries	(2.101)	see Sect. 2.6.4 & Sect. 4.3.2 for physical & mathematical plausibility

components, $viz.$, $(\mathbf{J}_{\mathrm{nB}} + \mathbf{J}_{\mathrm{pB}})$. This yields (see [76])

$$\mathrm{grad}\,\psi \cdot \mathbf{n} = -\frac{\rho_s}{\varepsilon} + (\sigma_n + \sigma_p)R_H(\mathbf{B} \times \mathrm{grad}\,\psi) \cdot \mathbf{n}$$

$$+ \frac{\sigma_n}{\sigma_n + \sigma_p}\,\mathrm{grad}\,\zeta_n \cdot \mathbf{n} - \frac{\sigma_p}{\sigma_n + \sigma_p}\,\mathrm{grad}\,\zeta_p \cdot \mathbf{n}$$

$$- \frac{\sigma_n^2 R_{Hn}}{\sigma_n + \sigma_p}(\mathbf{B} \times \mathrm{grad}\,\zeta_n) \cdot \mathbf{n} \tag{4.73}$$

$$+ \frac{\sigma_p^2 R_{Hp}}{\sigma_n + \sigma_p}(\mathbf{B} \times \mathrm{grad}\,\zeta_p) \cdot \mathbf{n},$$

where $\zeta_n = \psi - \phi_n$, $\zeta_p = \phi_p - \psi$, and ρ_s accounts for possible presence of surface and interface charges at the insulating boundary. It is interesting to note that relation (4.73), for vanishing magnetic field ($\mathbf{B} = 0$), does not yield the standard condition, $\mathrm{grad}\,\psi \cdot \mathbf{n} = 0$. Thus even in the absence of the external field, the standard condition $\mathrm{grad}\,\psi \cdot \mathbf{n} = 0$ has to be employed with caution since it does not account for the influence on potential by concentration gradients (contained in $\mathrm{grad}\,\zeta_{n,p}$). Again, this clearly demonstrates how device boundaries have to be judiciously located to justify use of standard boundary conditions. Relation (4.73) is also applicable in the presence of a stress field; in this case, regardless of the magnetic field, the transport coefficients (e.g., conductivities, Hall coefficients) have tensorial descriptions (see Chapt. 6).

Alternatively, an additional non-physical boundary (A′B′C′D′E′F′A′ in Fig. 4.12) can be introduced around the device and Poisson's equation is solved over the entire domain with the condition $\mathbf{E} \cdot \mathbf{n} = 0$ applied at the boundary [10, 21–23]. Here, the actual boundary condition on ψ at the physical boundaries (AB, CD, EF, GH, IJ, KL, LM, MN, NA) is treated as unknown. However, this would entail extending the outer boundaries far enough so that the potential in the device active region remains independent of small changes in their physical location. This justifies the mathematical plausibility of the boundary condition on ψ.

The boundary conditions for the continuity equations are identical to standard device modeling discussed in Sect. 2.6. With reference to Fig. 4.12, the solution process is restricted to within the device domain, ALMNA, with the conditions, $\mathbf{J}_{\mathrm{n,p}} \cdot \mathbf{n} = 0$ applied at these boundaries. If there is surface recombination at these boundaries, we employ (2.99) and (2.100).

For quick reference, we summarize in Table 4.3 modeling equations, physical parameters, and boundary conditions, pertinent to the simulation of unipolar and bipolar magnetic sensors. The numerical procedures are given in Sect. 6.7.3.

4.4 Illustrative Simulation Example – Micromachined Magnetic Vector Probe

For multi-dimensional position detection associated with motion of a permanent magnet or measurement of the geomagnetic fields for navigation, the detection of both intensity and direction of the magnetic field is required. This can be achieved with a magnetic vector probe realized using an appropriate configuration of unipolar Hall plates or with multi-collector magnetotransistors (see [6]). The latter constitutes the example we have chosen to illustrate magnetic sensor simulation. The magnetotransistor has a high spatial resolution achieved by a lower inter-action volume and high magnetic field resolution due to high sensitivity and low noise.

Fig. 4.13 shows a three-dimensional view of the vector probe [26] indicating the arrangement of two orthogonal dual-collector pnp magneto-transistors with a common emitter. The device structure is reminiscent of the suppressed sidewall injection magnetotransistor [77]. The arrangement shown allows detection of the two spatial magnetic field components parallel to the chip surface. There are a variety of issues related to magnetic probe design which make numerical simulations mandatory. It is important to identify the transduction principles underlying device sensitivity, including sensitivity coupling or cross sensitivity, to the different magnetic field components for subsequent device optimization. For example, the sidewall suppression guard ring has been identified, through numerical simulations, to be a crucial feature that determines the degree of coupling in output collector currents [28, 29]. Thus an optimal shape of the sidewall suppression guard ring is essential to minimize cross sensitivity. We have to

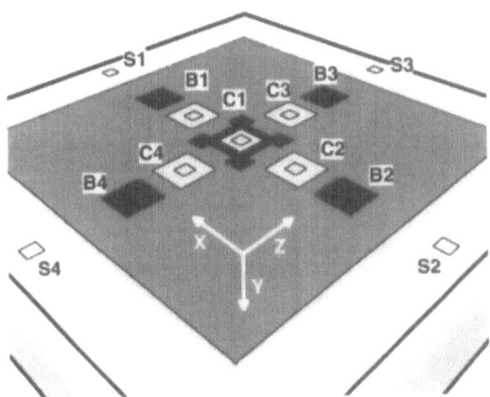

Fig. 4.13 Three-dimensional model of pnp (magnetotransistor) magnetic vector probe. Source [29]

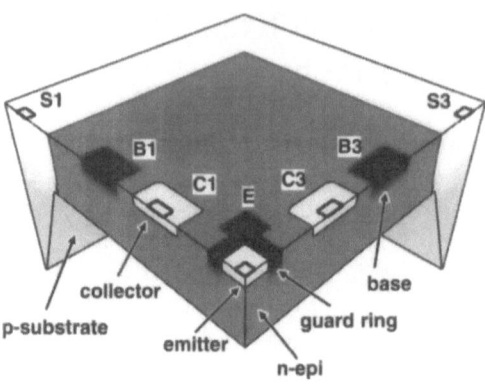

Fig. 4.14 A quadrant of the magnetotransistor illustrating the micromachined substrate for elimination of parasitic (substrate) current. Source [29]

predict the role of parasitic effects associated with substrate currents and its effect on device sensitivity. Again, extensive numerical simulations [26] have shown that elimination of the parasitic current does not compromise device sensitivity. In fact in the structure shown (Fig. 4.13), the parasitic (pnp) transistor to the substrate has been eliminated by micromachining the substrate below the n-epi base region. The resulting device structure illustrating the location of the micromachined cavity is shown in Fig. 4.14 (due to symmetry, only one quadrant of the device is shown). All of the above design considerations can only be addressed with three dimensional simulations. A superposition of two-dimensional analyses may lead to

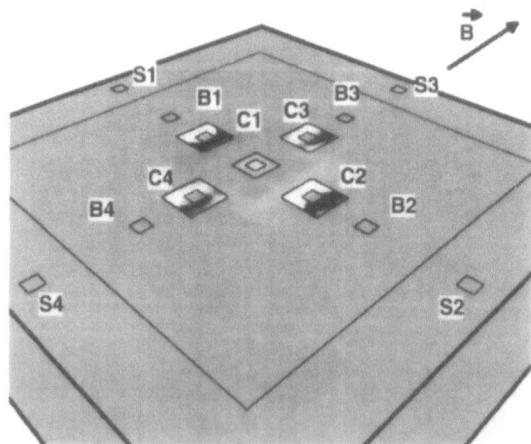

Fig. 4.15 Change in hole current density in pnp magnetotransistor at $B = 0.1$ tesla. The bright areas indicate a large change. Source [29]

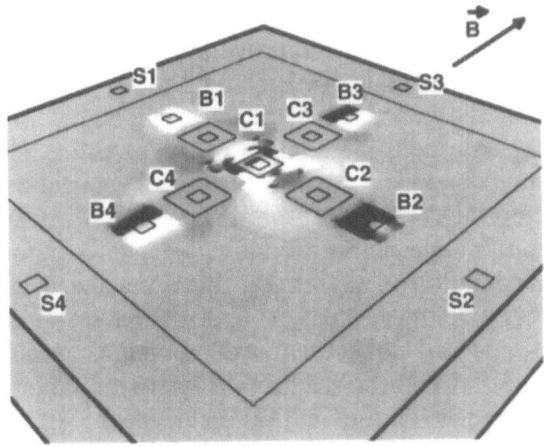

Fig. 4.16 Change in electron current density in pnp magnetotransistor at $B = 0.1$ tesla. The bright areas indicate a large change. Source [29]

Table 4.4. *Relative sensitivities $\Delta I/I$ in percent per tesla of micromachined vector probe for different magnetic field orientations. Source [29]*

$|\mathbf{B}| = 0.1$ [T]	$S_{RI}(\mathbf{x})[\%/\mathrm{T}]$	$S_{RI}(\mathbf{z})[\%/\mathrm{T}]$
$\mathbf{B} \parallel \mathbf{x}$	0.43	3.56
$\mathbf{B} \parallel \mathbf{y}$	0.00	0.00
$\mathbf{B} \parallel \mathbf{z}$	3.56	0.43
$\mathbf{B} \parallel \mathbf{y, z}$	3.55	0.59
$\mathbf{B} \parallel \mathbf{x, y, z}$	4.16	2.98

erroneous results, and more importantly, it rules out cross sensitivity effects due to the lack of dimensional coupling in the governing equations and boundary conditions.

Simulation results [29] of magnetic field modulated collector (hole) and base (electron) current densities on the device surface are shown in Figs. 4.15 and 4.16, along with the computed magnetic sensitivities in Table 4.4. As usual, we observe the expected asymmetry in collector currents. The current density in C2 is higher relative to C1 which yields a differential output collector current as the useful signal. The magnetic field orientation in this case is $(\mathbf{B} \parallel \mathbf{z}) = 0.1$ T. However, now comes the important question concerning the underlying galvanomagnetic effect(s) responsible for this asymmetry. If we look closely at the collector current distribution, we see that there is not only deflection of holes at the base region but there is also differential hole injection at the emitter due to a Hall voltage build up at the emitter-base junction. The latter effect, although small in magnitude, has been a source of significant controversy in the past (see, e.g., [23]). More importantly, is the Hall voltage induced by minority or majority carrier transport in the base? This becomes evident in Fig. 4.16, which illustrates the magnetic field modulation of base current. Here, due to the large base current, majority carrier (electron) deflection in the base becomes non-negligible. This perturbs the potential distribution, particularly at the p^+ emitter edges and surrounding n^+ guard regions, leading to a magnetic field modulation of the emitter-base voltage. Consequently this induces differential hole injection at the opposite halves of the emitter. This additional contribution to collector current imbalance can be significant particularly in these structures. Hence, in terms of underlying galvano-magnetic effects, sensor action in the vector probe structure is due to the combined minority (hole) and majority (electron) carrier deflection. The relative sensitivity to different field components is shown in Table 4.4; here, $S_{RI}(\mathbf{x}) = [(I_{C1} - I_{C2})/(I_{C1} + I_{C2})]/\mathbf{B}$ and $S_{RI}(\mathbf{z}) = [(I_{C3} - I_{C4})/(I_{C3} + I_{C4})]/\mathbf{B}$. As expected, we note no device response when $\mathbf{B}\|\mathbf{y}$, which in part can be attributed to the optimized design of the guard ring geometry. We also observe a small degree of coupling in collector currents due to the common emitter. For example, the sensitivity components do not superpose as seen from the values for $\mathbf{B}\|\mathbf{z}$ and $\mathbf{B}\|\mathbf{y}$ with $\mathbf{B}\|(\mathbf{y}, \mathbf{z})$.

4.5 References

[1] Baltes, H. P., Popovic, R. S., Integrated Semiconductor Magnetic Field Sensors, *Proc. IEEE*, 74 (1986), 1107–1132.

[2] Middelhoek, S., Audet, S. A., *Silicon Sensors*, London: Academic Press, 1989.

[3] Baltes, H., Nathan, A., Integrated Magnetic Sensors, in: *Sensors*, Grandke, T., Ko, W. H. (Eds.), Chapt. 7, Weinheim: VCH, 1989, pp. 195–215.

[4] Lenz, J. E., A Review of Magnetic Sensors, *Proc. IEEE*, 78 (1990), 973–989.

[5] Popovic, R. S., *Hall Effect Devices*, Bristol: Adam Hilger, 1991.

[6] Baltes, H., Castagnetti, R., Magnetic Sensors, in: *Semiconductor Sensors*, Sze, S. M. (Ed.), Chapt. 5, New York: Wiley, 1994, pp. 205–269.

[7] Baltes, H., Allegretto, W., Nathan, A., Microsensor Modeling, in: *Simulation of Semiconductor Devices and Processes*, Vol. 3, Baccarani, G., Rudan, M. (Eds.), Bologna: Tecnoprint, 1988, pp. 563–577.

[8] Nathan, A., Baltes, H., Sensor Modeling, in: *Sensors*, Grandke T., Ko, W. H. (Eds.), Chapt. 3, Weinheim: VCH, 1989, pp. 45–47.

[9] Nathan, A., Baltes, H., Allegretto, W., Review of Physical Models for Numerical Simulation of Semiconductor Microsensors, *IEEE Trans. on CAD of ICAS*, 9 (1990), 1198–1208.

[10] Allegretto, W., Nathan, A., Baltes, H., Numerical Analysis of Magnetic-Field-Sensitive Bipolar Devices, *IEEE Trans. CAD of ICAS*, 10 (1991), 501–511.

[11] Andor, L., Baltes, H. P., Nathan, A., Schmidt-Weinmar, H. G., Carrier Transport in Semiconductor Magnetic Field Sensors, *Technical Digest*, IEEE IEDM, Washington, 1983, pp. 635–638.

[12] Baltes, H. P., Andor, L., Nathan, A., Schmidt-Weinmar, H. G., Two-Dimensional Numerical Analysis of a Silicon Magnetic Field Sensor, *IEEE Trans. Electron Devices*, ED-31 (1984), 996–999.

[13] Schmidt-Weinmar, H. G., Andor, L., Baltes, H. P., Nathan, A., Numerical Modeling of Silicon Magnetic Field Sensors: Magnetoconcentration Effects in Split-Metal-Contact Devices, *IEEE Trans. Magnetics*, MAG-20 (1984), 975–978.

[14] Andor, L., Baltes, H. P., Nathan, A., Schmidt-Weinmar, H. G., Numerical Modeling of Magnetic-Field-Sensitive Semiconductor Devices, *IEEE Trans. Electron Devices*, ED-32 (1985), 1224–1230.

[15] Nathan, A., Huiser, A. M. J., Baltes, H. P., Two-Dimensional Numerical Modeling of Magnetic Field Sensors in CMOS Technology, *IEEE Trans. Electron Devices*, ED-32 (1985), 1212–1219.

[16] Nathan, A., Andor, L., Baltes, H. P., Schmidt-Weinmar, H. G., Modeling of a Dual Drain NMOS Magnetic Field Sensor, *IEEE J. Solid-State Circuits*, SC-20 (1985), 819–821.

[17] Allegretto, W., Mun, Y. S., Nathan, A., Baltes, H. P., Optimization of Semiconductor Magnetic Field Sensors Using Finite Element Analysis, *Proc. NASECODE IV Conf.*, Dublin: Boole Press, 1985, pp. 129–133.

[18] Nathan, A., Allegretto, W., Baltes, H. P., Sugiyama, Y., Modeling of Hall Devices Under Locally Inverted Magnetic Field, *IEEE Electron Device Letts.*, EDL-8 (1987), 1–3.

[19] Nathan, A., Allegretto, W., Baltes, H. P., Sugiyama, Y., Carrier Transport in Semiconductor Detectors of Magnetic Domains, *IEEE Trans. Electron Devices*, ED-34 (1987), 2077–2085.

[20] Allegretto, W., Nathan, A., Baltes, H. P., Two-Dimensional Numerical Analysis of Silicon Bipolar Magnetotransistors, *Proc. NASECODE V Conf.*, Boole Press: Dublin, 1987, pp. 87–92.

[21] Nathan, A., Allegretto, W., Joerg, W., Baltes, H., Numerical Modeling of Bipolar Action in Magnetotransistors, *Digest of Technical Papers*, Transducers '87, Tokyo, 1987, pp. 519–522.

[22] Nathan, A., Allegretto, W., Baltes, H. P., Galvanomagnetic Transport in p-n Junctions, *Sensors and Materials*, 1 (1988), 1–6.

[23] Nathan, A., Maenaka, K., Allegretto, W., Baltes, H. P., Nakamura, T., The Hall Effect in Magnetotransistors, *IEEE Trans. Electron Devices*, ED-36 (1989), 108–117.

[24] Riccobene, C., Wachutka, G., Baltes, H., Two-Dimensional Numerical Analysis of Novel Magnetotransistors with Partially Removed Substrate, *Technical Digest*, IEEE IEDM, San Francisco, 1992, pp. 513–516.

[25] Riccobene, C., Wachutka, G., Bürgler, J. F., Baltes, H., Operating Principle of Dual Collector Magnetotransistors Studied by Two-Dimensional Simulation, *IEEE Trans. Electron Devices*, 41 (1994), 32–41.

[26] Riccobene, C., Wachutka, G., Baltes, H., Numerical Study of Structural Variants of Bipolar Magnetotransistors, *Sensors and Materials*, 6 (1994), 159–178.

[27] Riccobene, C., Gartner, K., Wachutka, G., Baltes, H., Fichtner, W., First Three-Dimensional Numerical Analysis of Magnetic Vector Probe, *Technical Digest*, IEEE IEDM, San Francisco, 1994, pp. 727–730.

[28] Riccobene, C., Gartner, K., Wachutka, G., Baltes, H., Fichtner, W., Full Three-Dimensional Numerical Analysis of Multi-Collector Magnetotransistors with Directional Sensitivity, *Sensors and Actuators A*, 46–47 (1995), 289–293.

[29] Riccobene, C., *Multidimensional Analysis of Galvanomagnetic Effects in Magnetotransistors*, Ph.D. Dissertation, ETH Zürich, No. 11077, Switzerland, 1995.

[30] Hälg, B., Popovic, R. S., How to Liberate Integrated Sensors from Encapsulation Stress, *Sensors and Actuators*, A21–A23 (1990), 908–910.

[31] Nathan, A., Manku, T., Modeling the Piezo-Hall Effects in n-Doped Silicon Devices, *Appl. Phys. Letts.*, 62 (1993), 2947–2949.

[32] Manku, T., Nathan, A., O, N., Aflatooni, K., Allegretto, W., Modeling of Encapsulation Stress Effects on Output Response of Hall Sensors, *Sensors and Materials*, 6 (1994), 225–234.

[33] Aflatooni, K., *Strained Silicon Hall Effect Devices*, M.A.Sc. Thesis, University of Waterloo, Waterloo, Ontario N2L 3G1, 1994.

[34] in't Hout, S. R., Middelhoek, S., High Temperature Silicon Hall Sensor, *Sensors and Actuators*, A37–A38 (1993), 26–32.

[35] Rudin, S., Wachutka, G., Baltes, H., Thermal Effects in Magnetic Microsensor Modeling, *Sensors and Actuators A*, 25–27 (1991), 731–735.

[36] Nathan, A., Manku, T., The Thermomagnetic Carrier Transport Equation, *Sensors and Actuators A*, 36 (1993), 193–197.

[37] Paul, O., Baltes, H., Measuring Thermogalvanomagnetic Properties of Polysilicon for the Optimization of CMOS Sensors, *Digest of Technical Papers*, Transducers '93, Yokohama, 1993, pp. 606–609.

[38] Nathan, A., Bhatnagar, Y. K., Tang, D. D., Magnetic Field Bit Resolution of Integrated Circuit Polysilicon Hall Elements, *Digest of Technical Papers*, Transducers '93, Yokohama, 1993, pp. 896–899.

[39] Schneider, M., Korvink, J. G., Baltes, H., Magnetostatic Modeling of an Integrated Microconcentrator, *Digest of Technical Papers*, Vol. 2, Transducers '95, Stockholm, 1995, pp. 9–12.

[40] Nathan, A., Baltes, H., Integrated Silicon Magnetotransistors – High Sensitivity or High Resolution?, *Sensors and Actuators*, A21–A23 (1990), 780–785.

[41] Nathan, A., Baltes, H., Briglio, D. R., Doan, M. T., Noise Correlation in Dual-Collector Magnetotransistors, *IEEE Trans. Electron Devices*, ED-36 (1989), 1073–1075.

[42] Moharjerzadeh, S., Nathan, A., Modeling Noise Correlation Behaviour in Dual-Collector Magnetotransistors Using Small Signal Equivalent Circuit Analysis, *IEEE Trans. Electron Devices*, 43 (1996), 883–888.

[43] Popovic, R. S., Hälg, B., Nonlinearity in Hall Devices and Its Compensation, *Solid-State Electron.*, 31 (1988), 1681–1688.

[44] van den Boom, J. M., Kordic, S., Offset Reduction in Hall Plates: Simulations and Experiments, *Sensors and Actuators*, 18 (1989), 179–193.

[45] Munter, P. J. A., A Low-Offset Spinning-Current Hall Plate, *Sensors and Actuators*, A21–A23 (1990), 743–746.

[46] O, N., Nathan, A., Extracting MOS Parameter Variations from Dual-Drain MOSFET Offset, *IEEE Trans. Electron Devices*, 44 (1997), 1084–1090.

[47] Madelung, O., *Introduction to Solid State Theory*, Berlin: Springer-Verlag, 1978.

[48] Madelung, O., Halbleiter, in: *Encyclopedia of Physics*, Vol. XX, Flügge, S. (Ed.), Berlin: Springer, 1957, pp. 1–243.

[49] Callen, H. B., *Thermodynamics*, New York: Wiley, 1960.

[50] Sze, S. M., *Physics of Semiconductor Devices*, New York: Wiley, 1981.

[51] Beer, A. C., *Galvanomagnetic Effects in Semiconductors*, New York: Academic Press, 1963.

[52] Selberherr, S., *Analysis and Simulation of Semiconductor Devices*, Vienna: Springer-Verlag, 1984.

[53] Suhl, H., Shockley, W., Concentrating Holes and Electrons by Magnetic Fields, *Phys. Rev.*, 75 (1949), 1617–1618.

[54] Norton, P., Braggins, T., Levinstein, H., Impurity and Lattice Scattering Parameters as Determined from Hall and Mobility Analysis in n-Type Silicon, *Phys. Rev. B*, 8 (1973), 5632–5653.

[55] Landolt-Bornstein, Numerical and Functional Relationships in Science and Technology, in: *Semiconductors*, Vol. III/17a, Berlin: Springer, 1982.

[56] Herring, C., Vogt, E., Transport and Deformation-Potential Theory for Many-Valley Semiconductors with Anisotropic Scattering, *Phys. Rev.*, 101 (1956), 944–961.

[57] Okta, E., Sakata, M., Temperature Dependence of Hall Factor in Low Compensated *n*-Type Silicon, *Jpn. J. Appl. Phys.*, 17 (1978), 1795–1804.

[58] del Alamo, J. A., Swanson, R. M., Measurement of Hall Scattering Factor in Phosphorus-Doped Silicon, *J. Appl. Phys.*, 57 (1985), 2314–2317.

[59] Lin, J. F., Li, S. S., Linares, L. C., Teng, K. W., Theoretical Analysis of Hall Factor and Hall Mobility in *p*-Type Silicon, *Solid-State Electron.*, 24 (1981), 827–833

[60] Kachlishvili, Z. S., Galvanomagnetic and Recombination Effects in Semiconductors in Strong Electric Fields, *Phys. Stat. Sol.* (a), 33 (1976), 15–51.

[61] Heinrich, H., Kriechbaum M., Galvanomagnetic Effects of Hot Electrons in *n*-Type Silicon, *J. Phys. Chem. Solids*, 31 (1970), 927–938.

[62] Kordic, S., *Offset Reduction and Three-dimensional Field Sensing with Magnetotransistors*, Ph.D. Dissertation, Delft University of Technology, The Netherlands, 1987.

[63] O, N., *Magnetic Sensor Arrays with CCD Readout*, Ph.D. Dissertation, University of Waterloo, Waterloo, Ontario N2L 3G1, Canada, 1996.

[64] Brunetti, R., Jacoboni, C., Nava, F., Reggiani, L., Bosman, G., Zijlstra, R. J. J., Diffusion Coefficient of Electrons in Silicon, *J. Appl. Phys.*, 52 (1981), 6713–6722.

[65] Nougier, J.-P., Fluctuations and Noise of Hot Carriers in Semiconductor Materials and Devices, *IEEE Trans. Electron Devices*, 41 (1994), 2034–2047.

[66] Long, D., Myers, J., Scattering Anisotropies in *n*-Type Silicon, *Phys. Rev.*, 120 (1960), 39–44.

[67] Samoilovich, A. G., Ya Korenblit, I., Dakhovskii, I. V., Anisotropic Electron Scattering by Ionized Impurities, *Soviet Phys–Doklady*, 6 (1962), 606–608.

[68] Jacobini, C., Canali, C., Ottaviani, G., AlberigiQuaranta, A., A Review of Some Charge Transport Properties of Silicon, *Solid-State Electron.*, 20 (1977), 77–89.

[69] Asche, M., Gribnikov, Z. S., Ivastchenko, V. M., Kostial, H., Mitin, V. V., Conductivity and Transverse Fields in *n*-Si for Currents in the {110} and {100} Planes, *Phys. Stat. Sol. (B): Basic Research*, 114 (1982), 429–438.

[70] Lau, J., Ko, P. K., Chan, P. C. H., Modelling of Split-Drain Magnetic Field-Effect Transistor (MAGFET), *Sensors and Actuators A*, 49 (1995), 155–162.

[71] O, N., Nathan, A., CCD-Based Sensor Array for Magnetic Pattern Recognition, *Technical Digest*, IEEE IEDM, Washington, 1995, pp. 167–170.

[72] Popovic, R. S., The Vertical Hall-Effect Device, *IEEE Electron Device Letts.*, EDL-5 (1984), 357–358.

[73] Huiser, A. M. J., Baltes, H., Numerical Modeling of Vertical Hall-Effect Devices, *IEEE Electron Device Letts.*, EDL-5 (1984), 482–484.

[74] Sheu, B. J., Scharfetter, D. L., Ko, P. K., Jeng, M.-C., BSIM: Berkeley Short-Channel IGFET Model for MOS Transistors, *IEEE J. Solid-State Circuits*, SC-22 (1987), 558–566.

[75] van der Tol, M. J., Chamberlain, S. G., Potential and Electron Distribution Model for the Buried-Channel MOSFET, *IEEE Trans. Electron Devices*, ED-36 (1989), 670–689.

[76] Thangaraj, D., Nathan, A., The Discretization of Anisotropic Drift-Diffusion Equations, *Technical Report*, UW E & CE 97-11, University of Waterloo, Waterloo, Ontario N2L 3G1, Canada, 1997.

[77] Ristic, L., Baltes, H., Smy, T. J., Filanovsky, I., Suppressed Sidewall Injection Magnetotransistor with Focussed Injection and Carrier Double Deflection, *IEEE Electron Device Letts.*, EDL-8 (1987), 395–397.

5 Thermal Non-Uniformity Effects on Carrier Transport

Physical properties of semiconductor materials and devices are sensitive to variations in temperature, whether generated from the ambient or internally in a device or integrated circuit (IC). While the variations in temperature associated with the ambient can be treated as uniform (isothermal) relative to device dimensions, internal heat generation is highly localized giving rise to a temperature gradient, which constitutes a non-isothermal signal. Various methods can be employed for detection of thermal signals. For measurement of ambient temperature, we can employ the highly predictable and stable temperature dependence of the base-emitter voltage V_{BE} of a bipolar junction transistor. Together with co-integrated biasing, signal correction, and amplification circuitry, they provide an output voltage or current that is proportional to absolute temperature (PTAT) [1, 2]. On-chip temperature gradients or non-isothermal signals transduced by physical signals, not necessarily from the thermal domain (see [3, 4]), can be detected using thermoelectric or thermoresistive effects. Our discussion of modeling issues will be restricted to non-isothermal signals and related microtransducers; models pertinent to isothermal signals are reviewed in Chapt. 2.

Thermal microtransducers can be realized using IC microfabrication, including industrial CMOS or bipolar, technologies coupled with additional post-processing application-specific thin film deposition and bulk/surface micromachining steps. The latter serves to thermally isolate the micro-transducer from the underlying substrate, thus reducing parasitic heat losses. This reduces input power requirements associated with in situ thin-film-resistive (Joule) heating elements. Advantages of thermal micro-transducers over their conventional counterparts lie in their small size, low operating power and temperature, and fast thermal response time, due to the favorable thermal scaling laws. Most importantly, thermal micro-transducers can be integrated with control, signal processing, and correction circuitry for increased functionality and cost-effectiveness as compared to conventional transducers. The broad family of thermal microtransducers can be classified in terms of application areas: flow sensors [5–13], pressure (vacuum) sensors [14–23], chemical (reaction enthalpy) sensors based on microhotplates [24–26], AC power converters [27–29], (infrared) radiation sensors [30–33], thermoresistive radiation sources [34–36], and thermomechanical (bimorph) microactuators including microresonators [37–39]. The last category of thermal microactuators is discussed in Sect. 8.4. A review of recent progress in thermal microsensors can be found in [4].

Despite the overwhelming advantages associated with use of IC microfabrication technologies, the microsensor designer is faced with a constrained design space that stems from the restricted selection of materials, material thicknesses, and associated physical properties. In view of this restriction in design latitude, extensive optimization of transduction efficiency and reliability is indispensable. Here, numerical simulations play a vital role. Analytical models, including quasi-analytical schemes based on Fourier series solutions, require symmetrical device geometries and boundary conditions (see, e.g., [40–43]), thus imposing restrictions on their applicability range. Thermal microsensors can be complex in design with aperiodic boundary conditions, which must account for the strong thermal interactions with the ambient.

Numerical simulations provide insight into electrothermal interactions and the subsequent heat transport responsible for the temperature distribution in the device [20, 44–48]. This allows the designer to suppress or eliminate, by design, parasitic heat loss components to achieve high transduction efficiency (see [49] and Sect. 1.2). Subsequent optimization of device output response or sensitivity with respect to device structure/geometry and operating conditions is the next obvious step [13, 50–52]. For example, optimized operation of a flow sensor requires design trade-offs between flow sensitivity, mechanical integrity, environmental durability, the required signal-to-noise ratio, and the input power. A key consideration in thermal microsensor design is device reliability which can be undermined by irreversible changes in material state and eventually lead to catastrophic failure caused by excessively high temperatures at localized regions (hot spots) [53, 54]. This takes place even at relatively low electrical current densities through a microheater. Localized elevations in temperature can be alleviated by selective placement of (temperature-leveling) metallization layers over the thin film heater region [26, 55]. Uniformity in temperature is also an essential requirement for microhotplates in chemical sensing applications. Last, but not least, reliable simulation of thermal microsensors requires accurate description of underlying thermophysical properties/data (e.g., thermal conductivities) of the materials involved. Numerical simulations can facilitate the acquisition of such data from measurements using complex characterization structures [56] where valid and closed form solutions of device behavior are difficult, if not impossible, to obtain.

Sections 5.1 and 5.2 describe electrothermal effects and transport models. Transport coefficients and their dependence on fabrication process conditions are given in Sect. 5.3. Although the equations and underlying effects are valid for a broad range of conducting and insulating materials, our focus will be restricted to models and/or data pertinent to microfabrication materials compatible to IC technologies. The models and material data, although discussed in the context of microtransducer simulations, are also applicable for non-isothermal IC device simulation (see [57, 58]); however, the transport model is only valid within the framework of the drift-diffusion approximation. Section 5.4 describes the extension of the galvanomagnetic transport model, discussed in Chapt. 4, to include non-isothermal effects. Here, electro-thermo-magnetic effects are briefly reviewed along with associated modeling equations and transport coefficients. Thermal interactions of the microtransducer with the ambient is a crucial design consideration as it determines the device transduction efficiency. The mechanisms and models underlying heat transfer to the ambient are reviewed in Sect. 5.5. Section 5.6 summarizes model equations and boundary conditions for different types of microsensors comprising the thermal microsensor family. This is followed by a numerical simulation example, in Sect. 5.7, of a low power CMOS micro Pirani gauge designed for a heat transfer efficiency to the ambient that is in excess of 99%, for measurement of pressure (vacuum) with potential applications for gas identification.

5.1 Non-Isothermal Effects

Under non-isothermal conditions, the presence of a temperature gradient causes additional coupling between electrical and heat transport in semiconductors and metals. The coupling gives rise to a number of electrothermal effects (see [59]), notably, the Seebeck, Peltier, and Thomson effect, whose magnitude depends on the material-type as well as electrical and thermal operating conditions. The effects and their interactions can be understood in terms of microscopic transport theory or irreversible thermodynamics. The latter, by virtue of Onsager's reciprocal relations, makes apparent the nature of electrothermal coupling and the interdependence of the underlying coefficients. Microscopic transport theory, on the other hand, provides insight into the dependence of these coefficients on physical and material parameters. In Sect. 5.1.1, we review electrothermal effects following the linear theory of irreversible thermodynamics. We then turn to microscopic transport theory in Sect. 5.1.2, but only in the context of the Wiedemann-Franz law and associated inaccuracies arising from its application to semiconductors.

5.1.1 The Seebeck, Peltier, and Thomson Effects

In the thermodynamics of irreversible processes, the local entropy production is described by the sum of products of relevant fluxes (e.g., electrical current, heat current) and associated driving forces (e.g., potential gradient, temperature gradient). The fluxes and driving forces can be assumed to be linearly related [60]:

$$\mathbf{J}_k = \sum_j L_{jk}\mathbf{F}_j. \tag{5.1}$$

Here, L_{jk} are kinetic coefficients, which are functions of local intensive parameters (electric potential, temperature) as well as magnetic field (see Sect. 5.4), $\mathbf{F}_j$ denotes the jth driving force, and $\mathbf{J}_k$ the kth component of flux density. In arriving at (5.1), we have assumed that the driving forces acting on the system are small, i.e., the thermodynamic system deviates only slightly from equilibrium. Here, Onsager's relations are a powerful tool. The relations state the reciprocity between the jth component of the driving force and the kth component of the flux, and vice-versa [60]:

$$L_{jk} = L_{kj}. \tag{5.2}$$

The kinetic coefficients are scalars in isotropic situations, but become tensors in the presence of magnetic field or stress. For an electrothermal system, which is homogeneous (uniform and non-degenerate doping) and isotropic (absence of magnetic field or stress), the dynamical equations for the individual (electrical and heat) flux densities, after various substitutions, read [3]:

$$\mathbf{J}^e = -\sigma \operatorname{grad} \psi + \sigma \alpha_s \operatorname{grad} T, \tag{5.3}$$

$$\frac{1}{T}\mathbf{J}^q = \frac{\sigma\Pi}{T} \operatorname{grad} \psi - \left(\frac{\kappa + \sigma\alpha_s\Pi}{T}\right) \operatorname{grad} T. \tag{5.4}$$

Here, $\mathbf{J}^e$ and $\mathbf{J}^q$ denote the electrical and heat current densities, respectively, σ the electrical conductivity, T the temperature, ψ the electric potential, α_s the Seebeck coefficient, Π the Peltier coefficient, and κ the thermal conductivity. The Seebeck coefficient is associated with the open circuit potential (see Fig. 5.1a) that forms when two materials (a and b) are butted to form a junction with a temperature difference between the junction and the free ends, viz.,

$$\operatorname{grad} \psi = (\alpha_{sa} - \alpha_{sb}) \operatorname{grad} T. \tag{5.5}$$

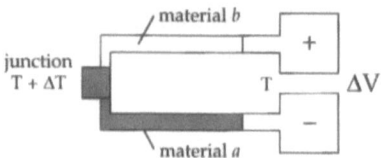

Fig. 5.1a Schematic of the Seebeck effect

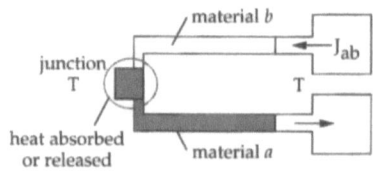

Fig. 5.1b Schematic of the Peltier effect

Here $\alpha_{sa,b}$ denote the respective Seebeck coefficients of the materials. The Peltier effect describes the generation or absorption of heat at a junction of two different materials when current flows through the junction in the absence of any temperature gradient. The Peltier heat (absorbed or released) is proportional to the magnitude of the electrical current flow through the junction. According to Onsager's theorem, the coefficients of the cross terms are equal, *viz.*, $\sigma\alpha_s = \sigma\Pi/T$. This yields Kelvin's relation which connects the Peltier and Seebeck coefficients

$$\Pi = \alpha_s T. \tag{5.6}$$

In the absence of a temperature gradient, i.e., grad $T = 0$ in (5.3) and (5.4), $\mathbf{J^e} = -\sigma\,\mathrm{grad}\,\psi$ and $\mathbf{J^q}/T = -(\sigma\Pi/T)\,\mathrm{grad}\,\psi$, which yields

$$\mathbf{J^q} = -\Pi\,\mathbf{J^e} \tag{5.7}$$

indicating that a heat current accompanies the flow of electrical current. The Peltier effect can give rise to asymmetries in temperature distribution [3, 47] in resistively heated thermal microstructures due to cooling (and heating) at the current carrying contacts. This asymmetry can be reduced by placing the contacts very close together (thermal short-circuiting). The Thomson effect is complementary to the Peltier effect; it also describes the absorption or release of heat that accompanies the flow of electric current, but now in the presence of a temperature gradient. The Thomson heat can be described as

$$\mathbf{J^q} = \gamma\,\mathbf{J^e} \cdot \mathrm{grad}\,T, \tag{5.8}$$

where γ denotes the Thompson coefficient. The Thompson coefficient is related to the Seebeck coefficient via the second Kelvin relation

$$\gamma = T \frac{\partial \alpha_s}{\partial T}. \tag{5.9}$$

5.1.2 Wiedemann-Franz Law

For microscopic descriptions for the electrothermal coefficients, we follow [61]. In the relaxation time approximation, the generalized description for the electric and heat current densities are

$$\mathbf{J}^{\mathbf{e}} = -M_{00} \, \text{grad} \, \psi - M_{01} \left(\frac{\text{grad} \, T}{T} \right), \tag{5.10}$$

$$\mathbf{J}^{\mathbf{q}} = -M_{01} \, \text{grad} \, \psi - M_{02} \left(\frac{\text{grad} \, T}{T} \right), \tag{5.11}$$

which are analogous to Eqs. (5.3) and (5.4) but stated in slightly different form. Here, the coefficients M_{ik} are given as

$$M_{ik} = \frac{-q}{3\pi^2} \left(\frac{8\pi^2 m^*}{h^2} \right)^{3/2} \int_0^\infty E^{3/2} \frac{\partial f_0}{\partial E} \left[\frac{q\tau(E)}{m^*} \right]^{i+1} \left[\frac{E(\mathbf{k}) - E_F}{q} \right]^k dE, \tag{5.12}$$

where q denotes the elementary charge, m^* the effective mass, h the Planck's constant, $E(\mathbf{k})$ the electron energy at wave vector $\mathbf{k}$, f_0 the Fermi distribution function, $\tau(E)$ the energy dependent relaxation time, and E_F the Fermi level. Here, we identify the coefficient M_{00} in (5.10) with the more familiar electrical conductivity (σ), M_{02}/T with the thermal conductivity (κ), and $-M_{01}/(M_{00}T)$ with the Seebeck coefficient. A more detailed description of the Seebeck coefficient is given in Sect. 5.3.1.

Accurate calculation of the coefficients M_{ik} requires a rigorous description of the band structure, $E(\mathbf{k})$ and the various scattering mechanisms basic to the relaxation time, $\tau(E)$. Eqs. (5.10) and (5.11) describe the electrical and heat transport for one particle species only, *viz.*, electrons or holes. In semiconductors, we have to account for both species of particles, particularly in the case of mixed (ambipolar) conduction. Accordingly, the electron and hole transport need to be accounted for by appropriate descriptions for the conduction and valence band structures described by $E(\mathbf{k})$. From the band structure evolve other microscopic transport parameters such as effective mass, relaxation time, and drift mobility. In addition, with heat transport in semiconductors, conduction by the lattice (phonons) has to be taken into account. The full system of equations, accounting for electron, hole, and phonon contributions to electrical and heat transport, is given in Sect. 5.2.

In metals, the integrand in (5.12) is evaluated at $E = E_F$ along with a higher order approximation for the Fermi distribution function to account for electrothermal interactions [61]. Upon evaluation of the electrical and thermal conductivites using (5.12), we state the Wiedemann-Franz law

$$\frac{\kappa}{\sigma T} = \frac{M_{00}}{M_{02}T^2} = \frac{\pi^2}{3}\left(\frac{k}{q}\right)^2,$$ (5.13)

which yields the Lorenz number, $L = \kappa/\sigma T = 2.443 \times 10^{-8} W\Omega/K^2$. The Wiedemann-Franz law is valid only when the relaxation time approximation applies, i.e., when the band structure and scattering processes are isotropic. Nevertheless, the agreement with the Lorenz number is reasonable for most metals at room temperature. However, the Wiedemann-Franz law when applied to semiconductors leads to large discrepancies, thus limiting its validity.

Consider a non-degenerate semiconductor. In this case, we can replace the distribution function, f_0 in (5.12), with the Boltzmann distribution function. Furthermore, we describe the energy-dependent relaxation time by a power law, viz., $\tau(E) = \tau_0 E^r$. Here, τ_0 denotes the energy independent relaxation time and r is a scattering coefficient whose value is $-1/2$ for acoustic phonon scattering or $3/2$ for ionized impurity scattering. With these assumptions, the integral in (5.12) reduces to the following form [61]:

$$\int_0^\infty E^j \exp(-E/kT)dE = (kT)^{j+1}\Gamma(j+1),$$ (5.14)

which yields for the Lorenz number

$$L = \frac{\kappa}{\sigma T} = (r + 5/2)\left(\frac{k}{q}\right)^2$$ (5.15)

Here, we now have $(r + 5/2)$ instead of $\pi^2/3$ in (5.13) and hence a different value of the Lorenz number. This value (5.15), however, differs from the measured values for semiconductors. The discrepancy arises because we only accounted for the electronic part of the thermal conductivity. Because of the relatively low carrier concentration in semiconductors, the lattice component of heat conduction is dominant and must be considered. The experimental Lorenz number includes the contribution of the electrons (κ_n) and that of the lattice (κ_L), viz., $L = (\kappa_n + \kappa_L)/[\sigma_n T]$, which holds for unipolar conduction. In the case of bipolar conduction, we include hole transport, which yields $L = (\kappa_n + \kappa_p + \kappa_L)/[(\sigma_n + \sigma_p)T]$. In practice, regardless of the doping level in semiconductors, the lattice component of heat conduction is usually orders of magnitude higher than the electronic contribution. Thus, $\kappa_{n,p} \ll \kappa_L$ in semiconductors as opposed to $\kappa_n \gg \kappa_L$ in metals (see Sect. 5.3.2).

5.2 Electrothermal Transport Model

We consider a non-degenerate bipolar material and assume that the carriers (electrons and holes) are in thermal equilibrium with the lattice. This allows us to describe carrier concentrations in terms of Boltzmann statistics with identical carrier and lattice temperatures. We assume that the material is not subject to magnetic or mechanical fields; this permits a scalar description for both electrical and thermal transport coefficients. We further assume that there is no interaction of carriers with optical fields. Interactions with the magnetic field and associated thermomagnetic coefficients are discussed in Sect. 5.4. A reduced form of the model equations for the unipolar system, pertinent to thermomechanical actuation, is described in Sect. 8.4.

5.2.1 Governing Equations

The above assumptions lead to the following system of electrical and heat transport equations, associated with electrons and holes, to include temperature gradient as a driving force [62]

$$\mathbf{J_n^e} = -\sigma_n(T) \operatorname{grad} \phi_n + \sigma_n(T)\alpha_{sn}(T) \operatorname{grad} T, \tag{5.16}$$

$$\mathbf{J_p^e} = -\sigma_p(T) \operatorname{grad} \phi_p - \sigma_p(T)\alpha_{sp}(T) \operatorname{grad} T, \tag{5.17}$$

$$\mathbf{J_n^q} = -\alpha_{sn}(T)T\mathbf{J_n^e} - \kappa_n(T) \operatorname{grad} T, \tag{5.18}$$

$$\mathbf{J_p^q} = \alpha_{sp}(T)T\mathbf{J_p^e} - \kappa_p(T) \operatorname{grad} T. \tag{5.19}$$

Here, $\mathbf{J_{n,p}^e}$ denote the respective electric current densities associated with electrons and holes, $\mathbf{J_{n,p}^q}$ the corresponding heat current densities, $\phi_{n,p}$ the quasi-Fermi potentials, $\sigma_{n,p}(T)$ the temperature-dependent electrical conductivities, $\alpha_{sn,p}(T)$ the temperature-dependent Seebeck coefficients, and $\kappa_{n,p}(T)$ the temperature-dependent thermal conductivities. The quasi-Fermi potentials $\phi_{n,p}$ are related to the electrostatic potential ψ and the carrier concentrations via Boltzmann relations (see Sect. 2.3.1). Thanks to Onsager's theorem, there are only three independent coeffcients in the above system, viz., electrical conductivity, thermal conductivity, and the Seebeck coefficient. The Seebeck coefficient is a coupling term which determines electrical transport in the presence of a temperature gradient and heat transport in the presence of a potential gradient.

Equations (5.16) and (5.17) have to be supplemented with the carrier continuity equations for electrons and holes along with Poisson's equation for the electric potential, as given in Sects. 2.1 and 2.2. That completes the description of electrical transport in the electrothermal system.

To obtain an equivalent description for heat transport, we compute the total energy current density [62]

$$\mathbf{J}_{\text{tot}}^{\mathbf{u}} = \phi_n \mathbf{J}_{\mathbf{n}}^{\mathbf{e}} + \phi_p \mathbf{J}_{\mathbf{p}}^{\mathbf{e}} + \mathbf{J}_{\mathbf{n}}^{\mathbf{q}} + \mathbf{J}_{\mathbf{p}}^{\mathbf{q}} - \kappa_L \, \text{grad} \, T, \tag{5.20}$$

where κ_L denotes the thermal conductivity associated with the lattice. Equation (5.20) comprises the energy density (i.e., product of the potential and current density) and heat flux associated with each carrier, along with the lattice (or phonon) contribution. The divergence of (5.20) yields the total energy balance

$$\frac{\partial u}{\partial t} + \text{div} \, \mathbf{J}_{\text{tot}}^{\mathbf{u}} = 0, \tag{5.21}$$

where u denotes the total energy density.

Substitution for the various terms in (5.21) yields the well-known conduction equation which governs heat transport within electrothermal microstructures

$$\text{div} \left(\kappa_{\text{tot}} \, \text{grad} \, T \right) = -H + c_{\text{tot}} \frac{\partial T}{\partial t}, \tag{5.22}$$

where κ_{tot} accounts for the electron (κ_n), hole (κ_p), and lattice (κ_L) contributions, viz., $\kappa_{\text{tot}} = \kappa_n + \kappa_p + \kappa_L$, and correspondingly, c_{tot} is the total heat capacity which accounts for electron (c_n), hole (c_p), and lattice contributions (c_L), viz., $c_{\text{tot}} = c_n + c_p + c_L$. The dominant contributions to both of these coefficients come from the lattice (see Sects. 5.3.2 and 5.3.3). At the limit of non-degeneracy (e.g., doping concentrations $\sim 10^{19} \, \text{cm}^{-3}$); $\kappa_{n,p}/\kappa_L < 1\%$ and $c_{n,p}/c_L < 1\%$ (see [62]). The term H in (5.22) signifies heat generation. Reverting to the standard notation for the electric current densities, $\mathbf{J}_{\mathbf{n,p}}$, the generic description for H takes the following form [62]:

$$H = \mathbf{J}_{\mathbf{n}}^2/\sigma_n + \mathbf{J}_{\mathbf{p}}^2/\sigma_p + qR \left[T \left(\frac{\partial \phi_n}{\partial T} - \frac{\partial \phi_p}{\partial T} \right) - (\phi_n - \phi_p) \right]$$

$$- T \left(\frac{\partial \phi_n}{\partial T} - \alpha_{sn} \right) \text{div} \, \mathbf{J}_{\mathbf{n}} - T \left(\frac{\partial \phi_p}{\partial T} + \alpha_{sp} \right) \text{div} \, \mathbf{J}_{\mathbf{p}}$$

$$+ T(\mathbf{J}_{\mathbf{n}} \cdot \text{grad} \, \alpha_{sn} - \mathbf{J}_{\mathbf{p}} \cdot \text{grad} \, \alpha_{sp}), \tag{5.23}$$

where the partial derivative $\partial/\partial T$ is evaluated at constant n and p. Here, R denotes the generation-recombination rate. In most thermal microtransducers, the electrical time constant is several orders of magnitude smaller than the corresponding thermal time constant, and thus transient analysis is only significant in terms of heat transport. Hence, we only need to consider the steady-state form of the carrier continuity equations. This greatly simplifies not only electrical transport analysis but also the heat generation term, (5.23). Replacing the divergence of current densities with the generation-recombination term, the steady-state form of (5.23) reduces to

$$H = \mathbf{J}_n^2/\sigma_n + \mathbf{J}_p^2/\sigma_p + qR[\phi_p - \phi_n + T\alpha_{sn} + T\alpha_{sp}]$$
$$+ T(\mathbf{J}_n \cdot \mathrm{grad}\,\alpha_{sn} - \mathbf{J}_p \cdot \mathrm{grad}\,\alpha_{sp}). \tag{5.24}$$

The first two terms in (5.24) signify the well-known Joule heat associated with the electron and hole current densities. The third term describes the heat exchange associated with recombination of electrons and holes. The fourth term describes the collective effect of Peltier and Thomson heat. The Peltier heat accounts for the exchange of heat flux accompanying the flow of electron and hole currents between materials of differing Seebeck coefficients. It thus constitutes a heat current discontinuity at material interfaces. The Thomson heat is related to the exchange of heat flux when the electron and hole current densities flow in a temperature gradient. The Thomson heat is significant only when the Seebeck coefficient is strongly temperature-dependent. The system of equations, (5.16), (5.17), and (5.22), supplemented with Poisson's equation and carrier continuity equations, are valid for non-degenerate n- and p-type semiconductor material in the absence of magnetic or mechanical fields. The presence of these fields turns both electrical and thermal transport coefficients from scalars to tensors. The system of equations are applicable not only to a wide range of thermal microsensors, but also to IC device simulation in the non-isothermal regime (see, e.g., [57, 58]). The system of equations are solved subject to appropriate boundary conditions at contacts, interfaces, and outer boundaries.

5.2.2 Boundary Conditions

Conditions for electrical transport are a straightforward and consistent adaptation of the isothermal case seen in Sect. 2.6. At contacts, space charge neutrality and the mass action law lead to the following conditions for the electrostatic potential (ψ_c) and carrier concentrations (n_c, p_c):

$$\psi_c = V_a + \frac{kT}{q}\,\sinh^{-1}\left(\frac{N_D - N_A}{2n_{ie}^2(T)}\right), \tag{5.25}$$

$$n_c = \sqrt{\frac{(N_D - N_A)^2}{4} + n_{ie}^2(T)} + \frac{N_D - N_A}{2}, \tag{5.26}$$

$$p_c = \sqrt{\frac{(N_D - N_A)^2}{4} + n_{ie}^2(T)} - \frac{N_D - N_A}{2}. \tag{5.27}$$

Here, V_a denotes the applied voltage, $N_{D,A}$ the respective donor and acceptor donor concentrations, and n_{ie} the effective intrinsic concentration, whose value is influenced by the temperature at the contact. If the contact has an associated internal resistance, the drop in the electrostatic potential across the contact can be accounted for through incorporation of an additional term, $\rho_c(\mathbf{J_n} + \mathbf{J_p}) \cdot \mathbf{n}$, to Eq. (5.25); see Sect. 2.6.1. This makes it a mixed boundary condition. Here, ρ_c denotes the contact resistivity, $\mathbf{n}$ the unit vector normal to the contact, and $\mathbf{J_{n,p}}$ are given by (5.16) and (5.17), respectively. If the contact is biased by a constant current source (I_0), the contact becomes a floating equipotential, whose value is then determined by solving

$$\int_{\Gamma_c} (\mathbf{J_n} + \mathbf{J_p}) \cdot \mathbf{n}\, d\Gamma = I_0, \tag{5.28}$$

which makes (5.28) a mixed boundary condition. Here, $\mathbf{J_{n,p}}$ are given by (5.16) and (5.17), respectively, Γ_c denotes the (contact) integration surface, $d\Gamma$ the surface element, and I_0 the value of the constant current source. Often the current is set to some non-zero value to reflect the leakage current of the measurement circuit. Relation (5.28) is a necessary condition in simulation of thermopiles in order to determine the open-circuit thermoelectric voltage.

At conductor-insulator interfaces and at outer device boundaries, we impose the standard conditions described in Sect. 2.6.3. The key difference here is that the current densities now contain the temperature gradient as described by (5.16) and (5.17).

The conditions for the heat transport at isothermal surfaces are analogous to those for electrical transport. At regions with a prescribed temperature T_0, we have $T = T_0$. Regions where only the heat flux q_0 is prescribed constitute floating isothermal regions and the temperature is then computed from

$$\int_{\Gamma} \kappa_{\text{tot}}(T)(\text{grad}\, T \cdot \mathbf{n}) d\Gamma = q_0 \tag{5.29}$$

in the usual notation. Relation (5.29) is analogous to the floating equipotential condition (5.28).

At interfaces and device boundaries, the conditions on temperature must be consistent with the total energy balance. Here, the energy current density,

$$\mathbf{J_{tot}^u} = -\kappa_{\text{tot}}(T)\, \text{grad}\, T + (\phi_n - T\alpha_{sn})\, \mathbf{J_n} + (\phi_p + T\alpha_{sp})\, \mathbf{J_p} \tag{5.30}$$

obtained from substitution of relations (5.18) and (5.19) into (5.20), must satisfy the continuity condition,

$$\left(\mathbf{J}_{tot}^{u} \cdot \mathbf{n}\right)_1 = \left(\mathbf{J}_{tot}^{u} \cdot \mathbf{n}\right)_2 \tag{5.31}$$

at an interface. Here, the subscripts 1 and 2 denote the associated materials at the interface. For a conductor-conductor interface, (5.30) and (5.31) yield

$$\left[-\kappa_{tot}(T)\,\text{grad}\,T + (\phi_n - T\alpha_{sn})\mathbf{J_n} + \left(\phi_p + T\alpha_{sp}\right)\mathbf{J_p}\right]_{c1} \cdot \mathbf{n}$$
$$= \left[-\kappa_{tot}(T)\,\text{grad}\,T + (\phi_n - T\alpha_{sn})\mathbf{J_n} + \left(\phi_p + T\alpha_{sp}\right)\mathbf{J_p}\right]_{c2} \cdot \mathbf{n}, \tag{5.32}$$

where subscripts c_1 and c_2 denote the associated conductor regions at the interface. For an ideal conductor-insulator interface, (5.30) and (5.31) yield

$$\kappa_c(\text{grad}\,T \cdot \mathbf{n})_c - \kappa_i(\text{grad}\,T \cdot \mathbf{n})_i = (\phi_n - T\alpha_{sn})\mathbf{J_n} \cdot \mathbf{n}$$
$$+ \left(\phi_p - T\alpha_{sp}\right)\mathbf{J_p} \cdot \mathbf{n}, \tag{5.33}$$

where the subscripts c and i denote the conductor and insulator regions, respectively.

At device boundaries exposed to an ambient fluid, there is a choice between one of the following conditions depending on the respective (conductor or insulator) region at the interface:

$$\kappa_c(T)(\text{grad}\,T \cdot \mathbf{n})_c = h(T_\infty - T) + (\phi_n - T\alpha_{sn})\mathbf{J_n} \cdot \mathbf{n}$$
$$+ \left(\phi_p + T\alpha_{sp}\right)\mathbf{J_p} \cdot \mathbf{n} \tag{5.34}$$

or

$$\kappa_i(T)(\text{grad}\,T \cdot \mathbf{n})_i = h(T_\infty - T). \tag{5.35}$$

Here, h denotes the effective heat transfer coefficient (resulting from heat loss to the fluid by conduction and convection) and T_∞ the fluid free-stream temperature (see Sect. 5.5).

The electrothermal system, (5.16), (5.17), and (5.22), solved under apppropriate boundary and interface conditions collectively described by (5.25) through (5.35), supplemented with pertinent models/data for the various coefficients (described in Sect. 5.3 that follows), yields the temperature distribution in the device. Under certain conditions of device structure, materials, and operating conditions, the model can be simplified to reduced forms (see Sect. 5.5.5 on quasi three-dimensional analysis and Sect. 8.4.1 on thermomechanical actuators). Models for the heat transfer coefficients associated with conduction, convection, and radiation are discussed in Sect. 5.5 along with the electrothermal computational procedure in Sect. 5.6.

5.3 Electrical and Thermal Transport Coefficients

Accurate description of thermoelectric and thermal parameters such as the Seebeck coefficient, thermal conductivity, and specific heat is important for reliable simulation and optimization of thermal microtransducers. It is also becoming important in design and thermal management of large-scale integrated circuits where reduction in feature size has led to increased power density, and hence, to increased local heat density. This section provides parameter models and/or material data for crystalline and polycrystalline semiconductors, metals, and insulators, pertinent to IC technologies.

5.3.1 The Seebeck Coefficient in Semiconductors and Metals

In Sect. 5.1.1, we identified the coupling of electrical and heat transport and the interdependence of the underlying coefficients. For microscopic insight and to obtain estimates of these coefficients, including their dependence on physical and material parameters, we turn to microscopic transport theory. Here, we need to solve the Boltzmann transport equation (BTE) under non-equilibrium conditions following a first-order perturbation of the carrier distribution function and taking into account the external field. An excellent treatment of the BTE solution procedure including derivation of various transport coefficients is given in [61]. These results were adopted in Sect. 4.1 where we reviewed the derivation of the galvanomagnetic transport equation and associated coefficients.

In what follows, we describe a procedure to arrive at a relatively simple model for the Seebeck coefficient in crystalline semiconductors. Here, rather than using the equations given in [61] for the kinetic coefficients as a starting point, we begin with the more familiar drift-diffusion form of the transport relations, which are obtained from solution to the BTE under the usual approximations (see [63]). But there is one key difference, we assume a non-vanishing temperature gradient; grad $T \neq 0$. This leads to the following description for the non-isothermal electron current density

$$\mathbf{J_n} = q\mu_n n\mathbf{E} + q \operatorname{grad}(nD_n), \tag{5.36}$$

where q denotes the elementary charge, μ_n the electron mobility, n the electron concentration, $\mathbf{E}$ the electric field, and D_n the electron diffusion coefficient. Equation (5.36), as we will see later, reduces to the well-known form given earlier by (5.16) in terms of the quasi-Fermi potential, ϕ_n. The relation for the hole current density is analogous. Equation (5.36) can be rewritten as

$$\mathbf{J_n} = q\mu_n n\mathbf{E} + qD_n \operatorname{grad} n + qn \operatorname{grad} D_n. \tag{5.37}$$

The diffusion coefficient is related to mobility by Einstein's relation, $D_n = \mu_n kT/q$, and following the relaxation time approximation, the mobility can be expressed as $\mu_n = q\tau_n/m_n^*$. Here, m_n^* denotes the effective mass for electrons and τ_n the relaxation time (described by the power law [61], $\tau_n = \tau_0 E^r$, where τ_0 is the energy-independent relaxation time), E the electron energy, and r the scattering factor. The factor, r takes on the values of $-1/2$ for acoustic phonon (lattice) scattering, $3/2$ for impurity scattering, and zero for neutral impurity scattering. Assuming the free electron energy, $E = kT$, the gradient of diffusion coefficient, following the above approximations, can be expressed as

$$\mathrm{grad}\, D_n = \frac{(r+1)}{T} D_n \,\mathrm{grad}\, T. \tag{5.38}$$

For not too large a temperature gradient, we can assume the system to be in equilibrium locally, and the free electron distribution function is described by Boltzmann statistics

$$n(T) = N_c \exp\left(\frac{E_{Fn} - E_c}{kT}\right). \tag{5.39}$$

Here, E_{Fn} denotes the quasi-Fermi level, E_c the conduction band edge, and N_c the effective density of states in the conduction band

$$N_c = 2\left(\frac{2\pi kT m_n^*}{h^2}\right)^{3/2} \tag{5.40}$$

in the usual notation. From Eqs. (5.39) and (5.40), we can easily evaluate the concentration gradient in (5.37),

$$\begin{aligned}\mathrm{grad}\, n = {} & \mathrm{grad}\, N_c \exp\left(\frac{E_{Fn} - E_c}{kT}\right) \\ & + N_c \exp\left(\frac{E_{Fn} - E_c}{kT}\right) \mathrm{grad}\left(\frac{E_{Fn} - E_c}{kT}\right),\end{aligned} \tag{5.41}$$

where

$$\mathrm{grad}\, N_c = \frac{3}{2}\frac{N_c}{T}\,\mathrm{grad}\, T \tag{5.42}$$

and

$$\mathrm{grad}\left(\frac{E_{Fn} - E_c}{kT}\right) = \frac{\mathrm{grad}\,(E_{Fn} - E_c)}{kT} - \frac{(E_{Fn} - E_c)}{kT^2}\,\mathrm{grad}\, T. \tag{5.43}$$

Substitution of Eqs. (5.38), (5.41), (5.42), and (5.43) into (5.37) yields

$$\mathbf{J_n} = q\mu_n n\mathbf{E} - \mu_n nk\left[\frac{5}{2} + r - \frac{(E_{Fn} - E_c)}{kT}\right] \text{grad } T$$
$$+ \mu_n n \text{ grad } (E_{Fn} - E_c). \tag{5.44}$$

To evaluate the Seebeck coefficient, we impose the open circuit condition, $\mathbf{J_n} = 0$, and with $\mathbf{E} = -\text{grad } \psi$, (5.44) yields

$$\text{grad } \psi = -\frac{k}{q}\left[\frac{5}{2} + r - \frac{(E_{Fn} - E_c)}{kT}\right] \text{grad } T + \frac{1}{q} \text{grad } (E_{Fn} - E_c), \tag{5.45}$$

where the first term on the right-hand side signifies the Seebeck coefficient,

$$\alpha_{sn} = -\frac{k}{q}\left[\frac{5}{2} + r - \frac{(E_{Fn} - E_c)}{kT}\right]. \tag{5.46}$$

The second term does not contribute to the Seebeck effect. It signifies a material property, independent of the operating or measurement conditions, which stems from variations in the doping level or variations in the material. In uniformly doped homogeneous samples, this term vanishes since grad $n = 0$. In fact, it can be shown that

$$\text{grad } (E_{Fn} - E_c) = \frac{kT}{n} \text{ grad } n. \tag{5.47}$$

Following (5.46) and (5.47), the non-isothermal current density, (5.44) can be written as

$$\mathbf{J_n} = -\sigma_n \text{ grad } \psi + qD_n \text{ grad } n + \sigma_n\alpha_{sn} \text{ grad } T \tag{5.48}$$

with $\sigma_n = q\mu_n n$, which is equivalent to (5.16). Following (5.39), the Seebeck coefficient (5.46) can also be expressed as

$$\alpha_{sn} = -\frac{k}{q}\left[\frac{5}{2} + r + \log_e \frac{N_c}{n}\right] \tag{5.49}$$

or

$$\alpha_{sn} = -\frac{k}{q}\left[\frac{5}{2} + r + \log_e \frac{2(2\pi kT m_n^*/h^2)^{3/2}}{N_D}\right] \tag{5.50}$$

by employing (5.40) and assuming complete ionization, $n = N_D$. The corresponding expression for the Seebeck coefficient in p-type material is

$$\alpha_{sp} = \frac{k}{q}\left[\frac{5}{2} + r + \log_e \frac{N_v}{p}\right] \tag{5.51}$$

or

$$\alpha_{sp} = \frac{k}{q}\left[\frac{5}{2} + r + \log_e \frac{2\left(2\pi kT\, m_p^*/h^2\right)^{3/2}}{N_A}\right] \qquad (5.52)$$

where N_v denotes the effective density of states in the valence band and N_A the acceptor dopant density.

In intrinsic samples or samples with multiple (conduction/valence) bands, the Seebeck coefficient can be computed using the relation [64]:

$$\alpha = \frac{\sum_i \alpha_i \sigma_i}{\sigma_i}, \qquad (5.53)$$

where the sum is taken over both carrier types in all bands.

The computed Seebeck coefficient using model Eqs. (5.49) and (5.51) varies approximately linearly with the scattering parameter, r. For an n-type Si sample with a doping concentration of $10^{16}\,\mathrm{cm}^{-3}$, $|\alpha_{sn}|$ increases from $0.64\,\mathrm{mV/K}$ to $0.82\,\mathrm{mV/K}$ as r increases from -0.5 to 1.5. The dominant contribution to α_{sn} stems from the $\log_e(N_c/n)$ term. As the doping concentration increases from $10^{14}\,\mathrm{cm}^{-3}$ to $10^{19}\,\mathrm{cm}^{-3}$, $|\alpha_{sn}|$ decreases from $1.27\,\mathrm{mV/K}$ to $0.28\,\mathrm{mV/K}$ due to increased impurity scattering. Thus lower doped samples yield a larger Seebeck coefficient, however, the associated temperature coefficient is also larger. The Seebeck coefficient degrades with increasing temperature due to increased lattice scattering.

The agreement in model predictions and measurement data is acceptable at temperatures above 200 K and for the moderately doped material. However, this is not true at low temperatures and for intrinsic or low doped materials. In particular, at low temperatures, there is a large anomaly in the Seebeck coefficient [64] which is attributed to phonon drag effects. These are not accounted for in the models (5.49) and (5.51), since the reduced form of the non-isothermal current density, Eq. (5.36), which served as our starting point, followed from the assumption that phonons are in equilibrium with the mobile carriers. However, in the presence of a temperature gradient, the system is no longer in thermal equilibrium. There is a preferential flow of phonons from the hot to the cold regions rather than random scattering. Thus mobile carriers are dragged along by the net flow of phonon energy, as in viscous flow, forcing an accumulation of carriers at cold regions. The phonon-electron scattering can be dramatic at low temperatures particularly in pure (undoped) material. Effects of phonon diffusion result in measured values of the Seebeck coefficient much larger than the values predicted by (5.49) [65] (see Fig. 5.2). This makes the Seebeck coefficient dependent on device dimensions.

If we assume that the phonons transfer all of their energy to electrons, the resulting phonon drag contribution to the Seebeck coefficient is $\rho c_p/qn$

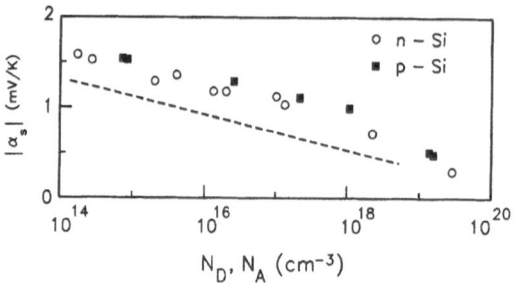

Fig. 5.2 Variation of Seebeck coefficient $|\alpha_s|$ with doping concentration for n- and p-type silicon at room temperature. Dashed line signifies the computed Seebeck coefficient, which does not account for phonon drag contribution. Adapted from [65]

[66]. Here, ρ denotes the material density and c_p the specific heat at constant pressure. For example, under this extreme condition of total energy transfer, an n-type sample of $10^{18}\,\text{cm}^{-3}$ electrons with $\rho c_p = 2$ Ws/K-cm^3 has a phonon drag contribution of the order of 10 V/K! In practice, however, only a small fraction of the phonon energy is transferred to electrons; the transfer in energy is greater for low doped samples than it is for samples with high doping. The phonon drag contribution rises as T^3 and reaches a maximum around 50 K. Subsequently it falls as T^{-1}. Including the phonon drag contribution, expressions (5.50) and (5.52) for the Seebeck coefficient now become [3]

$$\alpha_{sn} = -\frac{k}{q}\left[\frac{5}{2}+r+\log_e\frac{2\left(2\pi kT\,m_n^*/h^2\right)^{3/2}}{N_D}+\varphi_n\right], \qquad (5.54)$$

$$\alpha_{sp} = \frac{k}{q}\left[\frac{5}{2}+r+\log_e\frac{2\left(2\pi kT\,m_p^*/h^2\right)^{3/2}}{N_A}+\varphi_p\right] \qquad (5.55)$$

where

$$\varphi_n = \begin{cases} 0, & \text{for highly doped Si at 300 K,} \\ 5, & \text{for lowly doped Si at 300 K,} \end{cases} \qquad (5.56)$$

$$\varphi_p = \begin{cases} 0, & \text{for highly doped Si at 300 K,} \\ 100, & \text{for lowly doped Si at 100 K.} \end{cases} \qquad (5.57)$$

The measured values of the Seebeck coefficient for n- and p-type Si are shown in Fig. 5.2 as a function of doping concentration, and in Figs. 5.3 and 5.4 as a function of temperature. With the exception of the phonon drag contribution, which is dominant only at lower temperatures in low doped

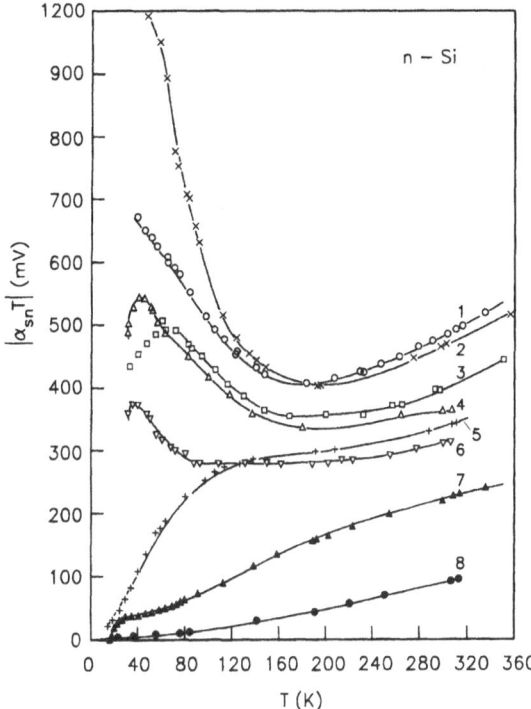

Fig. 5.3 Measured values of the Seebeck coefficient-temperature product, $|\alpha_{sn}T|$ as a function of temperature for n-type silicon at various doping levels. Here, **1**: $N_D = 2.75 \times 10^{14}\,\mathrm{cm}^{-3}$, $N_A = 10^{14}\,\mathrm{cm}^{-3}$; **2**: $N_D = 3.7 \times 10^{14}\,\mathrm{cm}^{-3}$, $N_A = 9 \times 10^{13}\,\mathrm{cm}^{-3}$; **3**: $N_D = 2.6 \times 10^{15}\,\mathrm{cm}^{-3}$, $N_A = 5 \times 10^{14}\,\mathrm{cm}^{-3}$; **4**: $N_D = 2.2 \times 10^{16}\,\mathrm{cm}^{-3}$, $N_A = 1.5 \times 10^{15}\,\mathrm{cm}^{-3}$; **5**: $N_D = 1.1 \times 10^{18}\,\mathrm{cm}^{-3}$, $N_A = 10^{18}\,\mathrm{cm}^{-3}$; **6**: $N_D = 1.3 \times 10^{17}\,\mathrm{cm}^{-3}$, $N_A = 2.2 \times 10^{15}\,\mathrm{cm}^{-3}$; **7**: $N_D = 2.2 \times 10^{18}\,\mathrm{cm}^{-3}$; **8**: $N_D = 2.7 \times 10^{19}\,\mathrm{cm}^{-3}$. Adapted from [67]

material, the behavior of the measured Seebeck coefficient is consistent with model predictions, (5.50) and (5.52), in terms of its variation with doping and temperature. For both n- and p-type material, the Seebeck coefficient decreases with increasing doping and increasing temperature. The effects associated with the latter being more pronounced in low doped samples.

Polycrystalline materials have a Seebeck coefficient that is lower than their crystalline counterpart. This is not surprizing since the mobility, and hence, diffusivity of mobile carriers is degraded by scattering at grain boundaries, as evident from the comparably higher material resistivity. Here, carrier transport is due to thermionic emission and tunneling across the grain boundary barrier which stems from presence of defect states. Furthermore, the contribution from phonons to the Seebeck coefficient is reduced due to the absence of a well-defined crystal structure in the

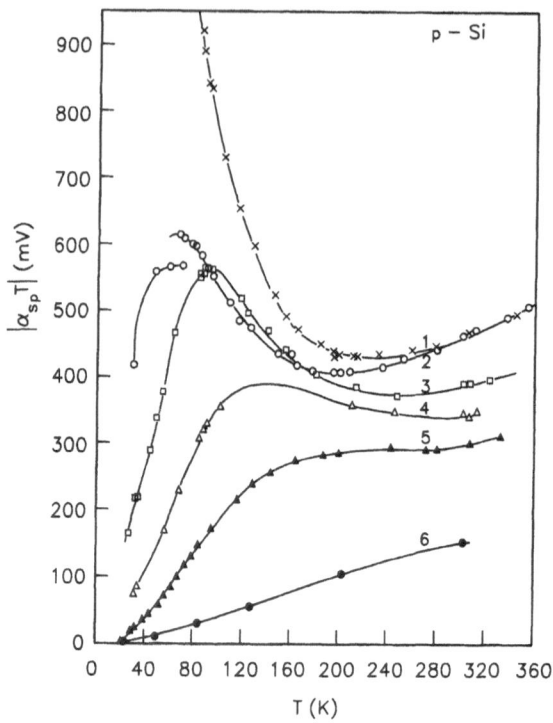

Fig. 5.4 Measured values of the Seebeck coefficient-temperature product $|\alpha_{sp}T|$ as a function of temperature for p-type Si at various doping levels. Here, **1**: $N_D = 1.9 \times 10^{14}\,cm^{-3}$, $N_A = 10^{15}\,cm^{-3}$; **2**: $N_D = 2.2 \times 10^{14}\,cm^{-3}$, $N_A = 9.2 \times 10^{14}\,cm^{-3}$; **3**: $N_D = 2.3 \times 10^{15}\,cm^{-3}$, $N_A = 2.6 \times 10^{16}\,cm^{-3}$; **4**: $N_D = 4.9 \times 10^{15}\,cm^{-3}$, $N_A = 2 \times 10^{17}\,cm^{-3}$; **5**: $N_A = 10^{18}\,cm^{-3}$; **6**: $N_A = 1.5 \times 10^{19}\,cm^{-3}$. Adapted from [67]

material. Hence, as with material resistivity data seen in Sect. 2.4, the Seebeck coefficient is sensitive to process parameters as well as thin film geometry. A number of theoretical models have been proposed to predict the Seebeck coefficient in polycrystalline materials taking into account film thickness, the average grain size, and the various scattering mechanisms (see [68]). However, the models are complex and only capable of predicting trends associated with doping and temperature dependence, rather than providing quantitative agreement with measurements. Because of the involved nature of carrier transport models for these materials, it is preferable to employ values retrieved experimentally, through use of dedicated test structures [69].

Table 5.1 lists measured values of the Seebeck coefficient compiled for various bulk and thin film materials. With the latter, process conditions are given where available; alternatively, we provide the name of the commercial IC foundry. As expected, there is a general trend of increase

Table 5.1. *Values of Seebeck coefficient for various thin film and bulk semiconductor materials.* Unless otherwise indicated, values shown are for 300 K. AMS, EM, and ES2 denote CMOS foundries

Material	Seebeck Coefficient (μV/K)	Conditions
n-Polysilicon	-525, 1.11×10^{-3}/K [70]	P: $N_D = 10^{19}$ cm^{-3}, $\rho \approx 300$ mΩ-cm
	-425, 1.47×10^{-3}/K [70]	P: $N_D = 2 \times 10^{19}$ cm^{-3}, $\rho \approx 50$ mΩ-cm
	-120 [69]	$n+$poly1, $\rho = 0.85$ mΩ-cm, AMS
	-108 [69]	$n+$poly1, $\rho = 1.03$ mΩ-cm, EM
	-520 [69]	$n-$poly2, $\rho = 96$ mΩ-cm, EM
	-111 [69]	$n+$poly3, $\rho = 1.22$ mΩ-cm, EM
	-108 [69]	$n+$poly1, $\rho = 1.58$ mΩ-cm, ES2
	-128 [69]	$n+$poly2, $\rho = 2.25$ mΩ-cm, ES2
	-240, -176 [71]	P: $R_s = 60$ Ω/sq., 100 Ω/sq.
p-Polysilicon	380, 0.99×10^{-3}/K [70]	B: $N_A = 10^{19}$ cm^{-3}, $\rho \approx 60$ mΩ-cm
	315, 1.75×10^{-3}/K [70]	B: $N_A = 2 \times 10^{19}$ cm^{-3}, $\rho \approx 20$ mΩ-cm
	190 [69]	$p+$poly2, $\rho = 5.8$ mΩ-cm, AMS
	330 [69]	$p+$poly4, $\rho = 16.2$ mΩ-cm, EM
	117 [71]	B: $R_s = 60 \Omega$ /sq.
n-Type a-Si:H	-360 to -780 [72]	$\rho \approx 100$ Ω-cm to 1 MΩ-cm
	-120 to -230, 8×10^{-3}/K [73]	$\rho \leq 0.05$ Ω-cm
	-42 to -267 [74]	33 K $\leq T \leq$ 146 K
p-Type a-Si:H	180 to 230, 7×10^{-3}/K [73]	$\rho \leq 0.05$ Ω-cm
μc-SiGe:H [75]	-100	n-type: $\rho \approx 0.1$ Ω-cm to 25 mΩ-cm
	100 to 180	p-type: $\rho \approx 11$ mΩ-cm to 2.5 mΩ-cm
μc-Ge:H [75]	-420 to -125	n-type: $\rho \approx 0.2$ Ω-cm to 1 mΩ-cm
	62	p-type $\rho \approx 1.4$ mΩ-cm
In$_x$Ga$_{1-x}$As [76]	-450 to -50, 790 to 390	$x = 0.53$, N_D, N_A : 10^{16} to 10^{18} cm^{-3}
GaAs [76, 77]	-500 to -100, 750 to 360	N_D, N_A : 10^{16} to 10^{18} cm^{-3}
Al$_x$Ga$_{1-x}$As [76]	-720 to -320, 770 to 380	$x = 0.3$, N_D, N_A : 10^{16} to 10^{18} cm^{-3}
Bi$_{0.5}$Sb$_{1.5}$Te$_3$ [78]	230	p-type: $\rho = 17.2$ mΩ-cm
Bi$_{0.87}$Sb$_{0.13}$ [78]	-100	n-type: $\rho = 0.7$ mΩ-cm

in Seebeck coefficient with increasing resistivity, within the limits of measurement error ($\pm 5\%$) in the values associated with commerical foundries [69]. Unlike crystalline Si, the Seebeck coefficient in heavily doped $n+$ polysilicon is smaller than the equivalent heavily doped $p+$ material ($\alpha_{sp} > \alpha_{sn}$). This behavior can possibly be attributed to the type of dopant species, its concentration, and/or associated trap density at grain boundaries. For example, phosphorus (P)-doped polysilicon gives rise to a higher trapping state density at grain boundaries resulting in a larger barrier voltage, thus lowering the concentration of mobile carriers. In contrast, boron (B) leads to a smaller barrier voltage and hence, a higher concentration of mobile carriers. Other factors include the nature of material crystallinity (microcrystalline or polycrystalline), which is influenced by deposition temperature, and the possible differences in phonon contributions associated with thickness variations, although the role of the latter is believed to be minimal [68]. Values given for poly-silicon are from low pressure chemical vapor deposited (LPCVD) films and those for hydrogenated amorphous Si (a-Si:H) and related microcrystalline (μc) materials are based on plasma-enhanced chemical vapor deposition (PECVD) processes.

Modeling of the thermoelectric behavior in metals can be even more complicated than that of polycrystalline materials. For qualitative insight, we can express the Seebeck coefficient as [79]

$$\alpha_s = \frac{\pi^2 k^2 T}{3q} \left[\frac{\partial \log_e \sigma(E)}{\partial E} \right]_{E=E_F}, \tag{5.58}$$

where E denotes the electron energy relative to the Fermi energy, E_F. As seen in Sect. 2.5, the conductivity (σ) is a function of the relaxation time and the electron concentration at the Fermi energy. The latter is proportional to the density of states, which is largest near the Fermi energy, and weakly dependent on temperature. The relaxation time is determined by the scattering due to phonons, other electrons, impurities,

Table 5.2. *Values of the Seebeck coefficient in selected metals.* Source [3]

Metal	$\alpha_s(\mu V/K)$ 273 K	$\alpha_s(\mu V/K)$ 300 K
Ag	1.38	1.51
Al		− 1.7
Au	1.79	1.94
Cr	18.8	17.3
Cu	1.7	1.83
Mo	4.71	5.57
Ni	− 18.0	
Pb	− 0.995	− 1.047
Pd	− 9.00	− 9.99
Pt	− 4.45	− 5.28
Rh	0.48	0.4
V	0.13	1.0
W	0.13	1.07

and defects. Here, the first two scattering mechanisms are dependent on temperature. Thus accurate prediction of the Seebeck coefficient requires models for the energy- and temperature-dependence of the various scattering parameters. In addition, we need to include phonon drag effects. Again, as with polysilicon, we cannot rely on theoretical or empirical models for reliable simulations; instead, we resort to empirical values given in handbooks. Values of the Seebeck coefficient for selected metals are compiled in Table 5.2.

5.3.2 Thermal Conductivity in Semiconductors, Metals, and Dielectrics

The thermal properties of conducting and insulating materials are determined by the mobile carriers (electrons and holes) and the quantized vibration of atoms in the lattice (phonons). For example, heat conduction in metals is dominated by electron transport in a thermal gradient while the contribution to heat capacity comes from phonons. In insulators, there are no free carriers and the thermal properties are determined by phonons, which also dominate the thermal properties of semiconductors. In what follows, we briefly review the thermal properties and underlying mechanisms, along with models and/or material data, for semiconductors, metals, and insulators.

A. Crystalline Semiconductors
The thermal conductivity κ in crystalline semiconductors comprises an electronic component, $\kappa_e = \kappa_n + \kappa_p$, due to mobile electrons (κ_n) and holes (κ_p), and a lattice (phonon) component (κ_L). In intrinsic and low doped semiconductors (carrier concentrations $< 10^{18}\,\mathrm{cm}^{-3}$) and at not too high temperatures $(T < 800\,\mathrm{K})$, heat conduction by phonons is predominant, with a growing significance of the electronic component with increasing temperatures [80, 81]. Experimentally it has been observed that κ in a pure crystal increases approximately exponentially from low temperatures to 20 K and subsequently decays with increasing temperature; first rapidly and then as $1/T$ until it reaches the melting point. The decay in κ stems from increased scattering of phonons with other phonons, impurities, and defects. The significance of the electronic contribution varies with different crystals. With Si and Ge near room temperature, the electronic contribution is negligible; it becomes significant at temperatures above 800 K. For example, at their melting point, the relative contributions are $\kappa_e \approx 40\%$ and $\kappa_L \approx 60\%$.

The behavior of thermal conductivity for low doped crystalline materials (e.g., Si or Ge) at temperatures in the range, $T < 300\,\mathrm{K}$, can be

Table 5.3. *Values of model parameters for thermal conductivity in intrinsic Si and Ge*

Parameter	Si	Ge
Debye temperature θ_D	674 K	395 K
Melting point	1681 K	1210 K
Average sound velocity v	6.4×10^5 cm/s	3.94×10^5 cm/s
Volume/atom V_0	1.99×10^{-23} cm^3	2.26×10^{-23} cm^3
Impurity scattering parameter Γ	1.65×10^{-5}	4.90×10^{-5}
$\kappa_L^{-1}(T) = a + bT + cT^2$ cm-K/W	$a = 0.03$ cm-K/W	$a = 0.17$ cm-K/W
	$b = 1.56 \times 10^{-3}$ cm/W	$b = 3.95 \times 10^{-3}$ cm/W
	$c = 1.65 \times 10^{-6}$ cm/W-K	$c = 3.38 \times 10^{-6}$ cm/W-K
	$337\,\mathrm{K} \leq T \leq 1000\,\mathrm{K}$	$198\,\mathrm{K} \leq T \leq 700\,\mathrm{K}$

modeled by [80]

$$\kappa = \kappa_L = \frac{4k\pi}{v}\left(\frac{kT}{h}\right)^3 \int_0^{\theta_D/T} \tau \frac{x^4 e^x}{(e^x - 1)^2}\,dx, \tag{5.59}$$

where k denotes Boltzmann's constant, v the average sound (phonon) velocity in the lattice, T the absolute temperature, h the Planck's constant, θ_D the Debye temperature, and τ the total scattering relaxation time. In Eq. (5.59), the integration variable comes from the substitution, $x = h\omega/(2\pi kT)$, where ω is the phonon frequency. Values of selected parameters drawn from [80] are given in Table 5.3. The relaxation time accounts for the relevant scattering processes, *viz.*,

$$\tau^{-1} = \tau_b^{-1} + \tau_i^{-1} + \tau_u^{-1}, \tag{5.60}$$

where τ_b, τ_i, and τ_u denote the respective relaxation times associated with scattering at device boundaries, scattering due to imperfections, and Umklapp scattering. The relaxation time for diffuse scattering of phonons at device boundaries is given as

$$\tau_b^{-1} = v/L, \tag{5.61}$$

where L denotes the average diameter of sample considered (4400 μm in this case, see [80]) and the value of the average sound velocity v is given in Table 5.3. The scatttering due to imperfections in the material can be described by

$$\tau_i^{-1} = \frac{3V_0 \Gamma \omega^4}{\pi v^3} \tag{5.62}$$

where V_0 denotes the average volume per atom in the crystal and Γ is an impurity scattering parameter. With Umklapp scattering, which is a three-phonon scattering process, the associated relaxation time is modeled empirically [80]

$$\tau_u^{-1} = B_u \omega^2 T. \tag{5.63}$$

Here, B_u is a scattering coefficient which can be approximated as

$$B_u \approx \frac{h\gamma^2}{2\pi m \theta_D v^2} \exp\left(-\theta_D/3T\right), \tag{5.64}$$

where m is the average mass of a single atom and γ is Grüneisen's constant, which is usually assumed to have a value of 2.

The model Eq. (5.59), along with the relaxation time accounting for various scattering mechanisms described by (5.60) through (5.63), yields good agreement with measurement data particularly at low temperatures where the phonon contribution to heat transport is dominant. At higher temperatures ($T > 300$ K), we need to consider the added contribution of electrons and holes to heat transport. Assuming only weak interactions between the various carriers of thermal energy, we add the respective contributions

$$\kappa = \kappa_L + \kappa_e. \tag{5.65}$$

At these temperatures, we can assume that effects of phonon scattering at crystal boundaries are negligible. This assumption must be treated with caution in thin film materials, where film thicknesses can be of the same order as the phonon scattering length. While the other (imperfection and Umklapp) scattering processes encountered at low temperature remain, an additional higher order four-phonon scattering process needs to be included at higher temperature. The associated relaxation time reads

$$\tau_h^{-1} = B_h \omega^2 T^2 \tag{5.66}$$

and the total relaxation time now becomes:

$$\tau^{-1} = \tau_i^{-1} + \tau_u^{-1} + \tau_h^{-1}. \tag{5.67}$$

In (5.66), B_h is a scattering coefficient and following measurement data (see [80]), it can be approximated using the relation, $B_h = B_u/[(\tau_h/\tau_u)T]$, where the value of τ_h/τ_u is 1 for Si and 3 for Ge, at $T = \theta_D$. The values for this ratio become smaller as $T > \theta_D$. For Si, we can summarize the relative importance of the four-phonon scattering process as follows: for $T < \theta_D$, $\tau_h > \tau_u$; at $T = \theta_D$, $\tau_h \approx \tau_u$; and for $T > \theta_D$, $\tau_h < \tau_u$ (which causes a faster decrease in thermal conductivity with increasing temperature).

At higher temperatures ($T > 300$ K), $h\omega \ll 2\pi kT$ and the integrand in Eq. (5.59) can be simplified to yield a simpler expression for the phonon contribution (κ_L) to the thermal conductivity

$$\kappa_L = \frac{4k\pi}{v} \left(\frac{kT}{h}\right)^3 \int_0^{\theta_D/T} \tau x^2 dx. \tag{5.68}$$

The values predicted by the above model vary linearly with temperature. However, agreement with measurements is only satisfactory for $T \geq \theta_D/2$, where the electronic component κ_e becomes significant. In the linear regime, the thermal conductivity behavior can be described as:

$$\frac{1}{\kappa_L(T)} = a + bT + cT^2 \qquad (5.69)$$

valid in the temperature range: $337\,\text{K} \leq T \leq 1000\,\text{K}$ and $198\,\text{K} \leq T \leq 700\,\text{K}$ for Si and Ge, respectively. Values for the coefficients are given in Table 5.3 [80].

At very high temperatures, the thermal generation of mobile carriers is large and the electronic contribution to thermal conductivity, due to ambipolar diffusion of electrons and holes, can become comparable to the phonon contribution. Here, we observe a deviation from the linear behavior with temperature, and the electronic contribution to the total thermal conductivity can be modeled as

$$\kappa_e = 2\left(\frac{k}{q}\right)^2 \sigma T \left[1 + \frac{b}{(1+b)^2}\left(\frac{E_g}{2kT} + 2\right)\right], \qquad (5.70)$$

where σ denotes the electrical conductivity, E_g the band gap energy, and b the mobility ratio (μ_n/μ_p). All of these parameters are temperature-dependent, with the exception of $b/(1+b)^2$ which turns out to have a value of 0.21 in Si for a broad range of temperatures. Relation (5.70) is obtained under highly idealized conditions; it does not account for effects of doping and non-parabolicity in valence and conduction band structures. Also, only lattice scattering of mobile carriers is taken into account; inter-

Table 5.4. *Values of thermal conductivity in selected semiconductors.* Unless otherwise indicated, values for thermal conductivity shown are for 300 K. Sources include [80, 81]

Material	Thermal Conductivity (W/cm-K)
GaAs	0.44
GaP	0.77
Ge	200 K: 0.95, 250 K: 0.73, 300 K: 0.60
	400 K: 0.44, 500 K: 0.33, 600 K: 0.26
InP	0.68
InSb	0.17
Si	200 K: 2.66, 250 K: 1.95, 300 K: 1.56
	400 K: 1.05, 500 K: 0.80, 600 K: 0.64
SiC	5
Diamond	20

valley scattering associated with multiple conduction and valence bands are neglected. Despite the various assumptions, the model provides reasonable agreement with measurement data [80]. Values of the thermal conductivity in selected crystalline semiconductors is given in Table 5.4. The values given are for room temperature; for convenience, values for Si and Ge, consistent with values predicted by model Eqs. (5.59) and (5.68), are also given for different temperatures.

With increasing doping, despite the larger concentration of free carriers, the phonon contribution to thermal conductivity is still dominant at not too high temperatures. However, the thermal conductivity degrades due to scattering of phonons with electrons and impurities [81]. For example, κ at room temperature can decrease from 2 W/cm-K for low doped Si ($\sigma \approx 0.5/\Omega$-cm) to 1 W/cm-K for high doped Si ($\sigma \approx 103/\Omega$-cm), along with a decrease in temperature dependence with increasing doping [82].

B. Polycrystalline Semiconductors

Unlike the bulk material, values of thermal conductivity of thin films are not readily available. Thin film values are quite different from bulk and vary widely in view of the process-dependence of their microstructure. For example, the process-dependent grain size and scattering at grain boundaries in polycrystalline Si (poly-Si) lead to a large reduction in the thermal conductivity from the bulk crystalline material, as well as to a large spread in values with different process technologies. The modeling of thermal conductivity in poly Si can turn out to be even more daunting than that of the electrical conductivity (see Sect. 2.4) despite the simplifying assumptions related to the microstructure and underlying transport mechanisms. Thus for simulations, we resort to measured values of thermal conductivity retrieved from dedicated micromachined test structures [69, 83].

The thermal conductivity in thin film poly-Si has been experimentally characterized by various research groups. With IC interconnect poly-Si, which is generally n-type with a sheet resistance of around 18 Ω/sq., the thermal conductivity at room temperature ranges from about 0.25 W/cm-K to about 0.34 W/cm-K. This is a reduction from the corresponding (doped) crystalline value by a factor of 3 to 4. Measurement data for poly-Si for various process technologies are given in Table 5.5. The temperature dependence of the thermal conductivity in the temperature range, 130 K $\leq T \leq 400$ K, is shown in Fig. 5.5 for the (n-doped) gate and (p-doped) capacitor polysilicon layers of a double poly commercial CMOS process [83]. The respective layers have a sheet resistance of 21.4 Ω/sq. and 213 Ω/sq. at 300 K. The corresponding thermal conductivities at 300 K are 0.235 ± 0.012 W/cm-K and 0.172 ± 0.080 W/cm-K, respectively. Compared to crystalline Si, the temperature dependence of thermal conductivity is very weak.

Table 5.5. *Values of thermal conductivity for n- and p-doped polysilicon.* Unless otherwise indicated, values shown are for 300 K

Material	κ (W/cm-K)	Conditions
n-Polysilicon	0.24 [69]	$n +$ poly1, $\rho = 0.85$ mΩ-cm, AMS
	0.19 [69]	$n +$ poly1, $\rho = 1.03$ mΩ-cm, EM
	0.22 [69]	$n -$ poly2, $\rho = 96$ mΩ-cm, EM
	0.17 [69]	$n +$ poly3, $\rho = 1.22$ mΩ-cm, EM
	0.16 [69]	$n +$ poly1, $\rho = 1.58$ mΩ-cm, ES2
	0.24 [69]	$n +$ poly2, $\rho = 2.25$ mΩ-cm, ES2
	0.3 [84]	P: $\rho = 1$ mΩ-cm, density: 2.3 g/cm^3
	0.29 [85]	1.2 μm CMOS double poly process, AMS
	0.41 [86]	bulk poly-Si, $\rho \approx 0.26$ mΩ-cm
	0.34 [46]	n-doped 2.0 μm CMOS process, MITEL
p-Polysilicon	0.17 [69]	$p +$ poly2, $\rho = 5.8$ mΩ-cm, AMS
	0.20 [69]	$p +$ poly4, $\rho = 16.2$ mΩ-cm, EM

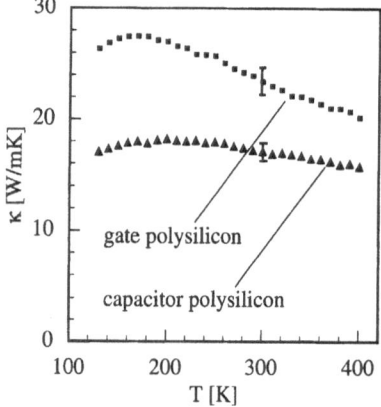

Fig. 5.5 Thermal conductivities of integrated circuit (IC) gate and capacitor polysilicon as a function of temperature. Source [83]

Although the values given in Table 5.5 do not indicate any thickness dependence and directional property of the thermal conductivity, it is important to recognize that since film thicknesses are of the order of 1 μm or less, phonon scattering at film boundaries may turn out to be significant enough to yield a κ that is dependent on film thickness and even anisotropic. In particular, the thermal conductivity in the direction of heat flow tranverse to film boundaries (κ_t) can be substantially lower than the corresponding bulk or longitudinal (κ_l) value parallel to the film boundaries, viz., $\kappa_l > \kappa_t$. Thus analysis and interpretation of simulation results need to be treated with caution.

C. Metals

The thermal conductivity in metals is made up of electron and phonon contributions, as given by (5.65). The relative significance of each contribution is dependent on material purity, electrical conductivity, and temperature. The electronic heat conduction component is consistent with the electrical conductivity of the material and obeys the Wiedemann-Franz law. In highly conducting pure metals, the electronic contribution (κ_e) is dominant for a broad range of temperatures. In high-resistivity metals and alloys, the phonon contribution (κ_L) becomes significant only at low temperatures; otherwise electrons remain the prevalent thermal carrier.

As with the electrical conductivity in metals seen in Sect. 2.5, we adopt quantum mechanical considerations whereby only electrons with an energy close to the Fermi level E_F participate in the conduction process [87]. This yields

$$\kappa \equiv \kappa_e = \frac{1}{9}\pi^2 k^2 n(E_F)v_F^2 \tau T, \tag{5.71}$$

where k denotes Boltzmann's constant, $n(E_F)$ the electron concentration at the Fermi level, v_F the velocity of electrons at the Fermi level, and τ the scattering relaxation time. Thus similar to the electrical conductivity, the thermal conductivity is high in metals which have a large concentration of electrons at the Fermi level and large electron velocity. The former is proportional to the density of states and the latter is a function of the scattering length, and hence, the relaxation time. The ratio of the thermal conductivity, (5.71), to the electrical conductivity (see Sect. 2.5), following the Wiedemann-Franz law, is reasonably well obeyed by most bulk (see Table 5.6) and thin film metals, including CMOS IC metallization (see Fig. 5.6) [69]. Values of the thermal conductivity, and associated electrical parameters, of metallization layers in commercial IC processes are given in Table 5.7 [83], along with their temperature dependence in Fig. 5.7. The results clearly show that the materials employed for IC metallization are Al-based alloys whose thermal conductivity values are quite different from that of the pure material [88, 89]. For example, at 300 K, the first and second metallization layers have thermal conductivities of 1.94 W/cm-K and 1.73 W/cm-K [69], respectively, compared to 2.35 W/cm-K for pure Al.

Similar to poly-Si, the thermal conductivity of thin film metal can turn out to be sensitive to layer thickness [90] as a result of electron scattering at film boundaries [91], along with possible directional behavior transverse or longitudinal to film boundaries. Again, $\kappa_l > \kappa_t$, with a substantial degradation in κ_t as the metal film thickness becomes comparable to the electron mean free path length. However, practical thicknesses of

Table 5.6. *Comparison of the Lorenz number L for selected metals.* Theoretical value: $L = 2.44 \times 10^{-8} \, W\Omega/K^2$

Metal	σ $10^7/\Omega$-m	κ W/m-K	$\kappa/\sigma T$ $10^{-8} \, W\Omega/K^2$
Ag	6.15	422.58	2.29
Al	3.55	209.20	1.96
Au	4.09	297.06	2.42
Cd	1.30	100.42	2.57
Cu	5.82	384.92	2.20
Fe	1.00	66.94	2.23
Ni	1.28	58.58	1.53
Pb	0.45	33.47	2.48
W	1.80	200.83	3.72

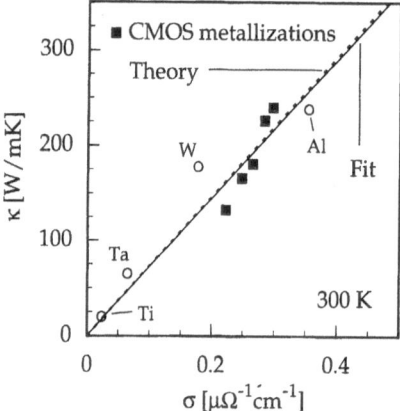

Fig. 5.6 Thermal versus electrical conductivities for CMOS metallization. The solid line corresponds to the Wiedemann-Franz law and a fit through measurement data. Source [69]

metallization layers in IC processes tend to be much larger than the mean free path length of electrons.

D. Dielectrics

Thermal conduction in dielectrics, due to the absence of mobile carriers, is governed solely by phonons drifting down a thermal gradient from the hot to the cold region. The thermal conductivity is governed by the (constant volume) heat capacity of the lattice, c_v (see Sect. 5.3.3 for discussion on heat capacity), the average sound (phonon) velocity, v, and the phonon mean free scattering length, l. Following the kinetic theory of gases, the thermal conductivity can be stated as [87]

$$\kappa = \frac{1}{3} c_v v l \tag{5.72}$$

Table 5.7. *Values of thermal conductivity for IC metallization.* Unless otherwise indicated, values shown are for 300 K.

IC metallization	κ (W/cm-K)	Conditions/Process
Metal 1 [69]	1.81 ± 0.08	AMS: CMOS, R_s: 65 ± 0.7 mΩ/sq.
	2.26 ± 0.10	EM: CMOS, R_s: 44 ± 0.2 mΩ/sq.
Metal 2 [69]	1.66 ± 0.08	AMS: CMOS, R_s: 36 ± 0.4 mΩ/sq.
	2.40 ± 0.10	EM: CMOS, R_s: 29 ± 0.2 mΩ/sq.
	1.33 ± 0.06	ES2: CMOS, R_s: 30 ± 0.3 mΩ/sq.
Al [90]	0.73, 0.84, 0.95	Thickness: 20, 24, 30 nm
		ρ: 0.10 μΩ-cm, 0.12 μΩ-cm, 0.13μΩ-cm

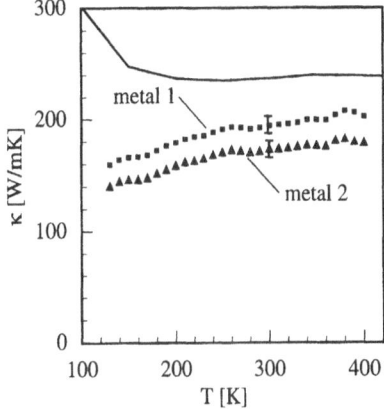

Fig. 5.7 Thermal conductivities of CMOS metallization as a function of temperature. Solid line corresponds to recommended values for aluminium. Source [83]

where the factor 1/3 appears to be somewhat arbitrary; in [92], it is stated as 1/4.

In crystalline or polycrystalline insulators, e.g., quartz, the temperature dependence of thermal conductivity $\kappa(T)$ appears to be fairly well understood to yield consistent results. The phonon velocity ($v \approx 5 \times 10^5$ cm/s) is only weakly dependent on temperature but not the mean free scattering length, which is determined by scattering processes associated with other phonons, lattice imperfections, and material boundaries. The last of the scattering mechanisms is insignificant in the bulk material. At low temperatures, the phonon density, governed by the Bose-Einstein statistics, decreases and so do phonon-phonon interactions. The resulting scattering length is large ($l \approx 10$ μm at 20 K) and virtually constant. Here, the temperature dependence of thermal conductivity stems from that of the specific heat which is proportional to T^3. At higher temperatures, phonon-phonon interactions, including Umklapp and higher order scattering processes prevail, thus decreasing the mean free path by several orders of magnitude ($l \approx 0.01\,\mu m$ at $300\,K$). Here, the thermal conductivity varies approximately as $1/T$ in a manner similar to that of crystalline semiconductors.

In non-crystalline insulators, e.g., fused quartz or integrated circuit SiO_2, the mechanisms underlying heat conduction, and in particular, the associated scattering processes, can be rather involved. For example, measurement results of κ for bulk and thin film SiO_2 appear to be widely different not only with respect to room temperature values but also in terms of temperature dependence [69, 83, 92–95] (see Table 5.8). Unlike the crystalline counterpart, the thermal conductivity in bulk amorphous glasses increases with increasing temperature. Here, it is believed that the mean free path is limited by the microstructural disorder of the material; the

Table 5.8. *Values of thermal conductivity for oxides and IC dielectrics.* Unless otherwise indicated, values shown are for $300\,K$

Material	Thermal Conductivity (W/cm-K)	Temperature (K)	Conditions
Bulk SiO_2	0.0140	300	
(see [93])	0.0152	378	
	0.0160	453	
	0.0165	528	
Phosphorus-doped	0.00677, 0.00968	300	Thicknesses: 1.74 and 3.04 µm
oxide [93]	0.00599, 0.00820	378	LPCVD
	0.00518, 0.00717	453	
	0.00468, 0.00614	528	
IC inter-metal	0.00175 – 0.00075	100 – 550	Thicknesses: 1.4 µm
oxide [94]		.	MITEL 2.0 µm double-poly
			double-metal CMOS
IC inter-poly	0.004 – 0.001	200 – 300	Thicknesses: 0.05 µm
oxide [94]	0.0005 – 0.0015	300 – 450	MITEL 2.0 µm double-poly
	0.0010 – 0.0025	450 – 600	double-metal CMOS
Thermal oxide [69]	0.0128 ± 0.0011	300	AMS 1.2 µm double-poly
Contact oxide [69]	0.0132 ± 0.0018		double-metal CMOS
Intermetal oxide [69]	0.016 ± 0.0024		
Passivation [69, 83]	0.015 ± 0.0025		Oxide/nitride sandwich
	0.0165 ± 0.0025		or oxynitride layer
Contact oxide [69]	0.015 ± 0.002	300	ES2 1.0 µm single-poly
Intermetal oxide [69]	0.0125 ± 0.002		double-metal CMOS
Intermetal oxide [95]	0.0107 – 0.012	280 – 400	Thickness: 0.57 to 2.28 µm APCVD and PECVD, measurement uncertainty: – 11.3% to + 7%
Si_3N_4 [96]	0.032 ± 0.005		LPCVD
a-SiN:H [97]	0.045 ± 0.007	300	PECVD

scattering length is assumed to be of the same magnitude as the scale of the disorder (≈ 7 Å in amorphous glasses) [92]. Since the specific heat increases with increasing temperature, so too does the thermal conductivity, on account of a constant scattering length.

The inconsistency in measured data, particularly in IC thin film dielectrics, may be qualitatively interpreted as follows. The microstructure of the dielectric is highly process-dependent, not to mention the presence of dopants or impurities and defects, which constitute additional scattering centers. All of these factors have a strong bearing on the mean free scattering length of the phonons. In addition, the thickness of the thin film can also play a crucial role since phonons can scatter from film boundaries. The relative importance of the various scattering mechanisms can be viewed in terms of the associated relaxation times. To first order, they can be assumed proportional to the respective scattering length, $\tau_j \sim l_j$, where subscript j signifies a scattering mechanism. Assuming that scattering rates are additive, the total relaxation time, following (5.60) but with slight change of notation, can be stated as

$$\tau^{-1} = \tau_{ph-ph}^{-1} + \tau_d^{-1} + \tau_i^{-1} + \tau_b^{-1} \tag{5.73}$$

where the subscripts $ph-ph$, d, i, and b denote scattering mechanisms associated with phonon-phonon interactions, microstructure disorder, impurities (and defects), and film boundaries. Thus the thermal conductivity, which is proportional to the relaxation time, $\kappa \propto \tau$, is limited by the dominant scattering process with the smallest relaxation time. Its temperature-dependence can be interpreted qualitatively as being due to an interplay between the temperature-dependence of the specific heat and that of the relaxation time, viz.,

$$\frac{\partial \kappa}{\partial T} \sim c_v \frac{\partial \tau}{\partial T} + \tau \frac{\partial c_v}{\partial T}. \tag{5.74}$$

Values of the thermal conductivity for various IC materials are given in Table 5.8. The thin film dielectric layers have values consistently lower than the bulk counterpart. The values for passivation layers, which are often either a silicon oxide/silicon nitride composite layer or a silicon oxynitride layer, lie between those known for low pressure chemical vapor deposited (LPCVD) oxide and nitride materials.

Not all of the published works report a consistent behavior of $\kappa(T)$ for the oxides; in some cases the dependence follows that of bulk SiO_2, i.e., $\kappa(T)$ increases with increasing temperature (see Fig. 5.8 [83]), and in other cases, one observes the opposite behavior. Similarly, the observations on the dependence of κ on film thickness are also not consistent. We believe that these discrepancies can be qualitatively explained in terms of relations (5.73) and (5.74); a quantitative interpretation requires further materials

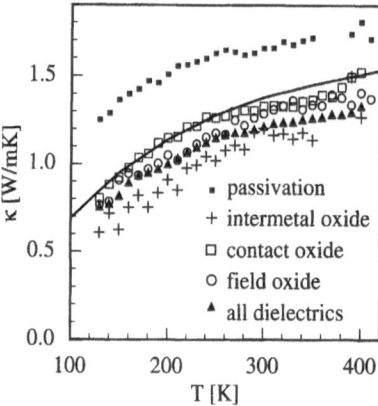

Fig. 5.8 Thermal conductivities of CMOS dielectrics as a function of temperature. Solid
line corresponds to recommended values for fused silica. Source [83]

characterization and theoretical modeling. Nevertheless, it is clear that
these values have to be used with caution, as they can strongly affect
simulation results.

5.3.3 Specific Heat in Semiconductors, Metals, and Dielectrics

The specific heat capacity is basic to thermal capacitance needed in
simulation of transient heat transfer. The heat capacity defines the amount
of thermal energy required for a material to reach a certain temperature. It
can be viewed as being analogous to electrical capacity; the larger the
capacitance, the more electrical energy is required to reach a certain
voltage. However, unlike the electrical case determined by induced
electrical charge, the contribution to specific heat comes mainly from
phonons. At room temperature, the electronic contribution is less than 1%;
it is appreciable only at very low or very high temperatures. Among the
various definitions for the heat capacity, the two most commonly used
definitions are the specific heat capacity (*viz.*, heat capacity per unit mass)
at constant volume (c_v) and at constant pressure (c_p) (see [87]). The latter
is more useful from an experimental standpoint since measurement at
constant pressure is much easier. The difference between c_p and c_v is about
5% for solids at room temperature and negligible at low temperatures.
Another useful definition to compare materials with the same number of
atoms or molecules is the molar heat capacity ($C_{v,p}$). This is defined as the
product of specific heat with the molar mass (m), $C_{v,p} = c_{v,p} m$, and has
units of J/mol-K.

Most metals at room temperature have a molar heat capacity of
$C_v = 25$ J/mol-K, which is known as the Dulong-Petit value. However, the

heat capacity is a function of temperature and vanishes at absolute zero. Experimental observations show an increase in specific heat as T^3 to reach 96% of its final value at the Debye temperature, θ_D. At high temperatures, $T > \theta_D$, the specific heat can be explained in terms of the harmonic oscillator model and the classical equipartition law [87]. At low temperatures, $T < \theta_D$, the quantum mechanical Debye model approximates the molar heat capacity as

$$C_v = 9\,kN_0 \left(\frac{T}{\theta_D}\right)^3 \int_0^{\theta_D/T} \frac{x^4 e^x}{(e^x - 1)^2}\, dx \tag{5.75}$$

which, when $T < \theta_D$ reduces to

$$C_v = \frac{12\pi^4}{5} kN_0 \left(\frac{T}{\theta_D}\right)^3. \tag{5.76}$$

Here, k is Boltzmann's constant, $x = h\omega/(2\pi kT)$ as in (5.59) with ω as the phonon frequency, and N_0 the Avogadro's number (6.022×10^{23}/mole).

Table 5.9. *Values of specific heat for selected materials.* Unless otherwise indicated, values shown are for 300 K. Sources include [87]

Material	Specific heat capacity (J/kg-K)	Volumetric heat capacity (J/cm^3-K)	Molar heat capacity (J/mole-K)	Conditions
Al	899		24.3	
Cu	385		24.4	
Ni	456		26.8	
IC metallization [98]		2.23 ± 1.87		Metal 1 EM 2.0 μm double-metal CMOS
Bulk Si	700			
Poly-Si [99]	770			LPCVD
Thermal, contact, and intermetal oxides [69]		1.05 ± 0.1		AMS 1.2 μm double-poly double-metal CMOS
Passivation [69]		2.7 ± 0.25		Oxide/nitride sandwich or oxynitride layer
Composite IC dielectrics [98]		1.74 ± 0.13		EM 2.0 μm double-metal CMOS
Si$_3$N$_4$ [96]	700 ± 100			LPCVD
a-SiN:H [97]	1500 ± 230			PECVD
Carbon (graphite)	904		10.9	
Water	4184		75.3	

Relation (5.75) is only an approximation; it does not account for the material structure and the associated density of vibrational modes, which in the case of thin film microfabrication materials would be hard to predict. Again, for reliable simulations, we resort to measurement data. Table 5.9 lists measured values of the specific heat capacity for selected materials including those in commerical IC technologies. The values show clear differences between thin film and bulk data.

5.4 Electro-Thermo-Magnetic Interactions

Interaction between electric and magnetic fields, and associated galvano-magnetic effects and coefficients under isothermal conditions were reviewed in Chapt. 4. In Sect. 5.1, we described the interaction between electrical and non-uniform thermal fields along with the family of electrothermal effects and underlying coefficients. We now extend the treatment of coupled fields to include interactions of galvanomagnetic, electrothermal, and thermomagnetic effects. Fig. 5.9 illustrates the various effects stemming from these interactions. The basis for the coupling is the action of the Lorentz force on mobile carriers (electrons and holes) which drift or diffuse in the presence of potential and temperature gradients. In

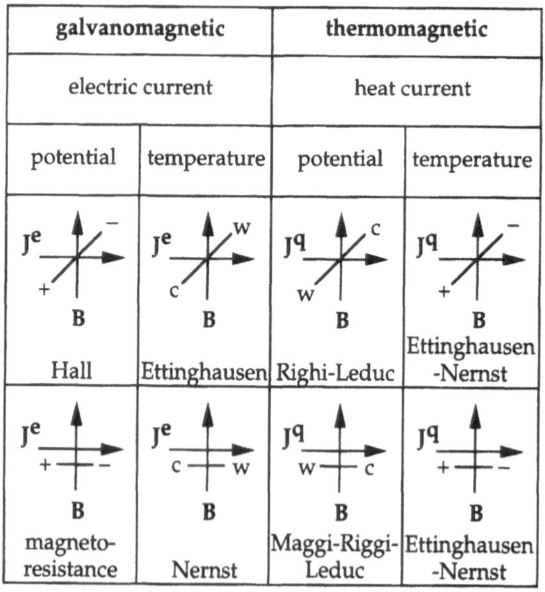

Fig. 5.9 Illustration of galvanomagnetic and thermomagnetic effects. Transverse effects: 4th row, longitudinal effects: 5th row, $(+, -)$: potential gradient, (w, c): temperature gradient. Adapted from [103]

Sect. 5.3, however, we learned that the electronic contribution to heat transport is very small in semiconductors; hence we anticipate the galvano-thermo-magnetic interaction to be weak also. Indeed, the Ettinghausen-Nernst coefficients in n- and p-type polycrystalline material are merely $77 \pm 3 \times 10^{-8}$ V/T-K and $-6 \pm 3 \times 10^{-8}$ V/T-K, respectively [100]. Although small, the effects could turn out to be of significance in magnetic sensors from the standpoint of device offset and possible thermal modulation of the output (voltage or current) response.

Non-uniform electric fields, and in particular, singularities in the magnetic field dependent electric current density arise at the union of Dirichlet and Neumann boundaries (e.g., contact edges) and at device corners. These singularities give rise to Joule heating and consequently, to a non-uniform temperature distribution in the device. Additionally, temperature gradients can arise from heating or cooling at material junctions (e.g., contacts) stemming from Joule, Peltier, and Thomson effects. Gradients in temperature, as seen in Sect. 5.1.1, result in an additional current component which can turn out to be large enough to contribute to device offset. Additionally, this may lead to an undesirable modulation of the sensor's output (voltage or current) response due to the Ettinghausen, Righi-Leduc, and Nernst effects. In particular, the Ettinghausen effect could be important in practical Hall devices; a temperature gradient transverse to current flow induces a voltage at the Hall probes by virtue of the Seebeck effect [101].

Modeling of galvano-thermo-magnetic interactions can be approached using either microscopic transport theory or by Onsager's linear irreversible thermodynamics; both arrive at similar results (see [102, 103]). We follow Onsager's formalism, where the relevant fluxes and associated driving forces for a bipolar system (i.e., electrons and holes) can be cast into the following form [103]:

$$
\begin{bmatrix} \mathbf{J}_n^e \\ \mathbf{J}_p^e \\ \mathbf{J}_{n,p}^q \end{bmatrix} = \begin{bmatrix} L_{nn} & 0 & L_{nT} \\ 0 & L_{pp} & L_{pT} \\ L_{Tn} & L_{Tp} & L_{TT} \end{bmatrix} \begin{bmatrix} \operatorname{grad} \phi_n \\ \operatorname{grad} \phi_p \\ (1/T)\operatorname{grad} T \end{bmatrix}. \tag{5.77}
$$

Here, $\mathbf{J}_{n,p}^e$ and $\mathbf{J}_{n,p}^q$ denote the respective electrical and heat current densities associated with electrons and holes, L_{jk} the transport coefficients, $\phi_{n,p}$ the Fermi potentials, and T the temperature. Because of the magnetic field, the coefficients L_{jk} are tensors, whose components are magnetic field dependent. By virtue of Onsager's theorem, the coefficients satisfy [60, 61]

$$
L_{jk}(\mathbf{B}) = L_{jk}^t(-\mathbf{B}), \tag{5.78}
$$

where L^t is the transpose tensor of L. This gives rise to a maximum of fifteen independent coefficients in relation (5.77); for a unipolar system, the number of independent coefficients reduces to nine. Since the Lorentz force involves the vector product of the driving force and magnetic field, the coefficients can be assumed to also contain vector product terms (see [103]), which leads to the following form for the electric and heat current densities:

$$\mathbf{J}_{nB}^{e} - (\sigma_n R_{Hn}\mathbf{J}_{nB}^{e} + \sigma_n\eta_n\,\mathrm{grad}\,T) \times \mathbf{B} = -\sigma_n\,\mathrm{grad}\,\phi_n \\ + \sigma_n\alpha_{sn}\,\mathrm{grad}\,T, \tag{5.79}$$

$$\mathbf{J}_{pB}^{e} - (\sigma_p R_{Hp}\mathbf{J}_{pB}^{e} + \sigma_p\eta_p\,\mathrm{grad}\,T) \times \mathbf{B} = -\sigma_p\,\mathrm{grad}\,\phi_p \\ - \sigma_p\alpha_{sp}\,\mathrm{grad}\,T, \tag{5.80}$$

$$\mathbf{J}_{nB}^{q} + (\eta_n T\mathbf{J}_{nB}^{e} + \kappa_n L_n\,\mathrm{grad}\,T) \times \mathbf{B} = -\alpha_{sn}T\mathbf{J}_{nB}^{e} - \kappa_n\,\mathrm{grad}\,T, \tag{5.81}$$

$$\mathbf{J}_{pB}^{q} + (\eta_p T\mathbf{J}_{pB}^{e} + \kappa_p L_p\,\mathrm{grad}\,T) \times \mathbf{B} = \alpha_{sp}T\mathbf{J}_{pB}^{e} - \kappa_p\,\mathrm{grad}\,T. \tag{5.82}$$

Here, we have adopted a slight change in notation for the current densities to reflect their dependence on magnetic field, $\mathbf{B}$. Omitting subscripts, σ denotes the electrical conductivity, R_H the Hall coefficient, η the Nernst coefficient, α the Seebeck coefficient, κ the thermal conductivity, and L the Righi-Leduc coefficient. Although not explicitly stated in the given form, these coefficients are not only functions of temperature but also of magnetic field. Equations (5.79) through (5.82) constitute a three-dimensional generalization [103] of Callen's two-dimensional model [60].

Equations (5.79) and (5.80) have to be supplemented with the carrier continuity equations for electrons and holes and Poisson's equation for the electric potential, as described in Sects. 2.1 and 2.2. For the heat transport, we must account for the lattice (phonon) contribution (given by: $-\kappa_L\,\mathrm{grad}\,T$), since the total heat current comprises the electronic (electron and hole) and the lattice contributions. The generalized heat conduction equation is then constructed following the principle of total energy conservation, using the procedure given in Sect. 5.2.1. Expressions for the various terms in the equation are similar in form to (5.22), except that the net thermal conductivity and heat generation now contain cross product terms in magnetic field (see [103]). The boundary conditions for the galvano-thermo-magnetic system are formulated along the same lines discussed in the electrothermal case. In certain, if not all, cases, the system of model equations, boundary conditions, and interface conditions can be simplified taking into consideration that electrical conduction is unipolar and heat transport by phonons is dominant.

5.5 Heat Transfer in Thermal Microstructures

Heat transfer is an integral part of the transduction mechanism in thermal microtransducers. It determines the transduction efficiency of the microstructure for the application in question. For example, in vacuum sensing applications, we maximize the component of heat transfer to the surrounding fluid. But in infrared (IR) sensing applications, we maximize the in-membrane heat conduction component. In all cases, we minimize the non-useful (or parasitic) components by suitable structural design. To illustrate the various heat transfer components, we revisit the simplified generalization of the thermal microstructure considered in Sect. 1.2. For convenience, Fig. 5.10 replicates Fig. 1.5 for illustration of the thermal balance

$$Q_{\text{Joule}} + Q_{\text{radin}} = Q_{\text{cond}} + Q_{\text{fluid}} + Q_{\text{conv}} + Q_{\text{radout}}. \tag{5.83}$$

The terms on the left-hand-side constitute heat sources; Q_{Joule} denotes the Joule heat associated with the input electrical power to the resistive heating element and Q_{radin} the heat generation associated with absorption of incident radiation. The terms on the right-hand side of (5.83) constitute heat sinks. The component Q_{cond} describes the heat loss through the solid (membrane to Si substrate) by conduction. The component Q_{fluid} describes heat loss to the surrounding fluid (above and below the membrane) by conduction. If the temperature distribution of the membrane microstructure is highly non-uniform, then part of Q_{fluid} may return to a cooler part of the membrane. Thus Q_{fluid} is bi-directional. The component Q_{conv} describes heat loss by forced and free convection. Forced convection occurs in the presence of a fluid flow stream or when the microtransducer moves through a stagnant fluid. Free convection occurs when fluid motion is caused by internal buoyancy forces stemming from temperature gradients within the fluid, which give rise to density variations. Free convection can be

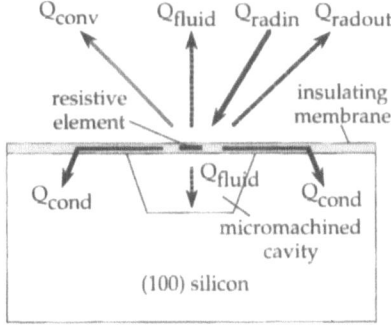

Fig. 5.10 Heat transfer components associated with thermal microstructures

neglected in most microstructures operating at not too large temperatures [46]. The component Q_{radout} denotes the radiative heat loss. Although this component is relatively small at standard temperature and pressure, and for not too large membrane temperatures, it can be appreciable in operation at vacuum.

Our brief description of heat transfer will focus on the latter three mechanisms, Q_{fluid}, Q_{conv}, and Q_{radout}, which appear on the right-hand side of Eq. (5.83). The other components, Q_{Joule} and Q_{cond}, were discussed in Sect. 5.2 and selected aspects of Q_{radin} were treated in Chapt. 3 when dealing with the wavelength-dependent absorption of electromagnetic radiation. We begin by stating, in Sect. 5.5.1, the model equations governing fluid motion and energy transport. These equations stem from the conservation laws of mass, momentum, and energy. Further discussion of the governing physical laws and constitutive properties is given in Sect. 7.3. The conservation equations are essential to modeling heat transfer coefficients associated with conduction (Q_{fluid}) and convection (Q_{conv}). An extended treatment of the former is given in Sect. 5.5.2. Sect. 5.5.3 follows with the modeling of (pressure-dependent) fluid thermal conductivity in the continuum, transitional, and free molecular regimes. Following a short discussion of radiative heat loss in Sect. 5.5.4, we present in Sect. 5.5.5, simplification of model equations and boundary conditions, along with associated assumptions pertinent to quasi three-dimensional analysis of thermal microsensors.

5.5.1 Governing Equations for Convective Heat Transfer

The mathematical description for convective heat transfer must account for motion of the fluid in addition to energy transport. Three key equations are needed to model heat transport from the surface of a microstructure in the presence of a flow stream [104]: the continuity equation, the Navier-Stokes equations of motion, and the energy equation.

The continuity equation is derived from the law of conservation of mass. For an incompressible (constant property) fluid, the density is independent of space and time, and the continuity equation, in terms of velocity components, reads

$$\frac{\partial u}{\partial x} + \frac{\partial v}{\partial y} + \frac{\partial w}{\partial z} = 0. \tag{5.84}$$

Here, u, v, w denote the x, y, and z components of the velocity field, respectively.

The equations of fluid motion are derived from Newton's second law which states that the sum of forces acting on the fluid must equal the mass-acceleration product. For an isotropic Newtonian fluid, the constitutive relations are linear (see Sect. 7.3.1). This leads to the Navier-Stokes equations of motion, which for an incompressible fluid read:

$$\rho\left(\frac{\partial u}{\partial t}+u\frac{\partial u}{\partial x}+v\frac{\partial u}{\partial y}+w\frac{\partial u}{\partial z}\right)=-\frac{\partial p}{\partial x}+\mu\left(\frac{\partial^2 u}{\partial x^2}+\frac{\partial^2 u}{\partial y^2}+\frac{\partial^2 u}{\partial z^2}\right)+\rho g_x,$$

$$(5.85)$$

$$\rho\left(\frac{\partial v}{\partial t}+u\frac{\partial v}{\partial x}+v\frac{\partial v}{\partial y}+w\frac{\partial v}{\partial z}\right)=-\frac{\partial p}{\partial y}+\mu\left(\frac{\partial^2 v}{\partial x^2}+\frac{\partial^2 v}{\partial y^2}+\frac{\partial^2 v}{\partial z^2}\right)+\rho g_y,$$

$$(5.86)$$

$$\rho\left(\frac{\partial w}{\partial t}+u\frac{\partial w}{\partial x}+v\frac{\partial w}{\partial y}+w\frac{\partial w}{\partial z}\right)=-\frac{\partial p}{\partial z}+\mu\left(\frac{\partial^2 w}{\partial x^2}+\frac{\partial^2 w}{\partial y^2}+\frac{\partial^2 w}{\partial z^2}\right)+\rho g_z,$$

$$(5.87)$$

where, ρ denotes the fluid density, p the pressure, μ the dynamic viscosity (assumed a constant and position independent), and g the gravity vector. Here, it is assumed that the only body force acting on the fluid is gravity.

The law of conservation of energy forms the basis of energy transport within the fluid. For not too large flow velocities, viscous dissipation (see Sect. 7.3.2) becomes insignificant and the resulting energy equation for an incompressible fluid reads

$$\rho c_p\left(\frac{\partial T}{\partial t}+u\frac{\partial T}{\partial x}+v\frac{\partial T}{\partial y}+w\frac{\partial T}{\partial z}\right)=\frac{\partial}{\partial x}\left(\kappa\frac{\partial T}{\partial x}\right)+\frac{\partial}{\partial y}\left(\kappa\frac{\partial T}{\partial y}\right)$$
$$+\frac{\partial}{\partial z}\left(\kappa\frac{\partial T}{\partial z}\right). \qquad (5.88)$$

Here, c_p denotes the specific heat (at constant pressure) and κ the thermal conductivity, both of which are dependent on temperature, T.

Equations (5.84) through (5.88) describe the complete system needed for three-dimensional simulation of heat transport in a fluid that is in motion relative to the microstructure. The system yields five equations for the five unknowns; the three velocity components (u, v, and w), the pressure (p), and the temperature (T), which are solved subject to appropriate boundary conditions. For a viscous fluid, there is no relative velocity between the surface of bounding walls (including the micromachined cavity walls) and the fluid. This yields vanishing normal and tangential components of fluid velocity. A non-viscous fluid slips over the surface and here, the normal component of velocity vanishes. At regions far from bounding surfaces, the velocity components can assume free stream values ($\mathbf{v}_\infty$). A more detailed discussion of the velocity boundary conditions is given in Sect. 7.3.2. The value of pressure is set to that of the ambient pressure at non-bounding surfaces. At bounding surfaces, we assume a vanishing normal component of the pressure gradient. The conditions for the temperature at the fluid-sensor interface must be consistent with the total energy balance. Here, the total energy current density in the microstructure, (5.30), must satisfy the continuity condition, (5.31), at the interface. At regions far from the thermal boundary layer (usually a distance of at least seven times the thickness of the boundary layer), the temperature assumes the fluid's free-steam value (T_∞) or that of the cold wall, where appropriate.

The system of equations can be simplified following standard boundary layer flow assumptions (see [104]) and under certain conditions related to the nature of the fluid flow and its profile. This enables treatment of heat loss from the microstructure as a heat sink term expressed in terms of a heat transfer coefficient. The special one-dimensional case is discussed in Sect. 5.5.5.

5.5.2 Zero Flow Two-Dimensional Heat Transfer Coefficient

Accurate modeling of heat loss to the ambient under zero forced flow conditions is crucial, since microstructures, with their high degree of thermal isolation, can lose a significant percentage of the input power to the surroundings. For example, the pressure sensor illustrated in Sect. 5.7 loses over 99% of its input power to the surrounding fluid at standard temperature and pressure. This heat loss, computed using analytical and semi-analytical approaches, has limited accuracy. The accuracy is limited by the geometry and temperature distribution of the microstructure; for example, the discrepancy can be as large as 50% [46]. In what follows, we describe a technique based on the boundary element (panel) method (see [105]) to compute accurately the two-dimensional heat transfer coefficient for arbitrary geometries and temperature distribution.

In the absence of flow, the fluid can be considered stagnant since buoyancy forces associated with microstructures, even for temperatures as high as 400 °C, are insufficient to produce significant natural convective currents. The Grashof number (see, e.g., [106]), which is the ratio of buoyancy to viscous fluid forces, is typically less than 0.1 for most microstructures, implying the device is operating in the stagnant (diffusive) regime. In fact, measurements of the electrical terminal behavior in various sensor designs do not show a dependence on the sensor orientation with respect to the gravity vector [46]. Thus for all practical purposes, we can neglect natural convection in a stagnant fluid, and the temperature distribution can be computed using a slightly modified form of (5.88). Under steady-state conditions and for spatially uniform thermal conductivity, this reads

$$\frac{\partial^2 T}{\partial x^2} + \frac{\partial^2 T}{\partial y^2} + \frac{\partial^2 T}{\partial z^2} = 0. \tag{5.89}$$

Boundary conditions for the solution of Eq. (5.89) include the surface temperature distribution of the microstructure, $T_s(\mathbf{r})$; the temperature of the micromachined cavity walls, $T_w(\mathbf{r})$; and the free stream temperature, T_∞. Here, $\mathbf{r}$ denotes the position vector. The three-dimensional solution of Eq. (5.89), using the panel method [46], yields the temperature gradient, $\partial T_s(\mathbf{r})/\partial n$, normal to the surface of the microstructure at position $\mathbf{r}$. The gradient in temperature is proportional to the local heat density distribution. Its integral over the surface area (Γ) of the microstructure yields the net surface heat loss to the fluid,

$$Q_{\text{fluid}} = \int_\Gamma (\partial q_s(\mathbf{r})/\partial n)\, d\Gamma, \tag{5.90}$$

where $d\Gamma$ denotes an elemental area and $q_s(\mathbf{r})$ is the local heat density, which can be represented in the following form:

$$q_s(\mathbf{r}) = G(\mathbf{r}) \kappa_{\text{fluid}} [T_s(\mathbf{r}) - T_\infty]. \tag{5.91}$$

Here, $G(\mathbf{r})$ is a normalized local heat density function defined as the heat flux from the surface per unit temperature per unit thermal conductivity (κ_{fluid}). This function needs to be computed only once for a given microstructure geometry. Substitution of (5.91) into (5.90) yields the heat loss to the fluid. The value of the computed heat loss is accurate within numerical precision for any arbitrary excess temperature distribution $(T_s(\mathbf{r}) - T_\infty)$ and for a spatially uniform thermal conductivity. Eq. (5.91) is treated as a heat source/sink term in the solution of the heat conduction equation in the electrothermal system of equations. Here, the discrete form is constructed by evaluating Q_{fluid} over a control area to reflect the local heat loss from that region.

In arriving at the heat transfer function $G(\mathbf{r})$, we assumed a fluid thermal conductivity that is independent of position. In practice, the thermal conductivity is a function of temperature, which in turn is a function of position. As an approximation, we can employ a value of thermal conductivity based on an average temperature, viz., $\kappa_{\text{fluid}}(T_{av})$, where $T_{av} = (T_s(\mathbf{r}) + T_\infty)/2$. However, a more accurate way of treating the temperature dependence is to employ Kirchhoff's transformation (see [45]), since the temperature dependence of κ_{fluid} is generally differentiable. In this case, the governing equation is no longer Laplace's equation but

$$\text{div}\,(\kappa_{\text{fluid}}\,(T)\,\text{grad}\,T) = 0. \tag{5.92}$$

We define a new variable ω such that

$$\omega = \int_{T_0}^{T} \kappa_{\text{fluid}}(\zeta)d\zeta, \tag{5.93}$$

where T_0 denotes some arbitrary reference temperature and ζ an integration variable. From (5.93), we obtain

$$\text{grad}\,\omega = \kappa_{\text{fluid}}(T)\,\text{grad}\,T, \tag{5.94}$$

the divergence of which yields Laplace's equation

$$\text{div}\,\text{grad}\,\omega = 0. \tag{5.95}$$

From here on, we employ the procedure described earlier to construct the position-dependent heat transfer function, $G(\mathbf{r})$. But this time the $G(\mathbf{r})$ accounts for the thermal properties of the fluid. Often, the Laplace's equation, (5.89) or (5.95), and associated boundary conditions are expressed in terms of a dimensionless variable for convenience of solution. For example, considering (5.95), if we set $\varphi = (\omega_s - \omega_\infty)$, where ω_s and ω_∞ (as with T_s and T_∞) denote the surface and free stream values, then we obtain the following system:

$$\text{div}\,\text{grad}\,\varphi = 0 \quad \text{in the fluid,}$$

$$\varphi = 1 \quad \text{on the surface of microstructure,}$$

$$\varphi = 0 \quad \text{at the free-stream,} \tag{5.96}$$

where we have assumed all wall temperatures to be at the free-stream value, $T_w(\mathbf{r}) = T_\infty$.

5.5.3 Thermal Conductivity of Gases

Under steady state conditions, the heat loss from a thermal microstructure is largely governed by the thermal conductivity of the surrounding gas. This in turn is a function of the gas species, its temperature, and its pressure. In fact, the variation of thermal conductivity (κ_{fluid}) with different gases and its dependence on pressure is widely used for detection/discrimination of gas species and vacuum sensors based on the Pirani principle, respectively.

The dependence of thermal conductivity on temperature can be modeled using a second order polynomial of the form

$$\kappa_{\text{fluid}} = \kappa_0 + \alpha T + \beta T^2, \tag{5.97}$$

where κ_0 is the thermal conductivity at 273 K and α and β are fitting coefficients. Generally, a linear fit is sufficient with most gases for deviations not too large from room temperature. For example, values for air at standard pressure are: $\kappa_0 = 0.0241$ W/m-K at 273 K, $\alpha = -7.7862 \times 10^{-6}$ W/m-K^2, and $\beta = 5.1065 \times 10^{-8}$ W/m-K^3.

However, modeling of the dependence of thermal conductivity on pressure is more complex. Close to atmospheric pressure, a gas can be treated as a continuum whereby properties such as density, velocity, and temperature can be assumed to vary continuously in space and time. At lower pressures, the gas loses its continuum properties. The critical parameter which determines the applicability of continuum methods is the mean free path of gas molecules, λ. For air at standard temperature and pressure, λ has a value less than 0.1 µm. As long as λ is much smaller than the characteristic length (L_c) of the system being examined, the normal continuum equations of motion and heat transfer apply. This implies that velocity and thermal boundary layers will be formed next to stationary surfaces. If λ increases, or L_c decreases, to reach approximately the same order of magnitude, then the interaction of molecules with each other is reduced and they are unable to reach an equilibrium state with the stationary surface. Further increases in λ, or decreases in L_c, to the point where $\lambda \gg L_c$, results in a free molecular regime. In this state, intermolecular interactions are assumed to be negligible. Since the molecules incident on the surface do not interact with those reflected (or re-emitted) from the surface, the boundary layers disappear.

The effects of the continuum, transitional, and free molecular regimes on the velocity and thermal boundary layers over a heated horizontal plate are shown in Fig. 5.11 [107]. In the continuum regime, the free-stream velocity v_∞ decays to zero at the surface and the temperature varies smoothly from the free-stream value of T_∞ to T_w at the surface. In the transitional regime labeled "b", the impinging molecules do not reach complete equilibrium with the surface, resulting in a free-stream velocity that does not decay all the way to zero. This leads to a temperature discontinuity at the surface. In the free molecular regime (curves labeled

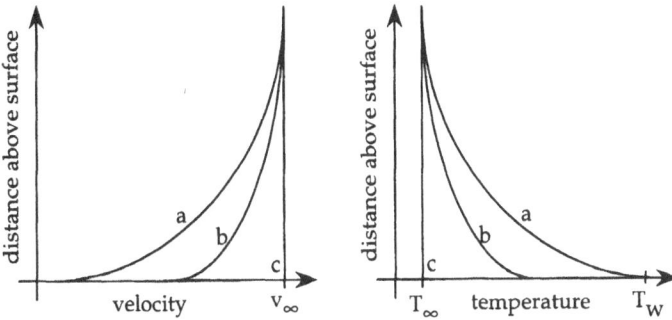

Fig. 5.11 Velocity and thermal boundary layers associated with (a) continuum, (b) transitional, and (c) free molecular pressure regimes

"c"), there is no boundary layer formed, and the free-stream velocity and temperature are maintained even immediately adjacent to the stationary surface.

The transitions between the different conduction regimes are characterized by the Knudsen number (Kn) defined by

$$Kn = \frac{\lambda}{L_c}.$$
(5.98)

The continuum regime exists for Kn < 0.01, the transitional regime for 0.01 < Kn < 3, and the free molecular regime for Kn > 3 [108]. Following Knudsen's theory of energy interchanges in non-equilibrium interactions of gas molecules with a surface, the associated heat transfer density (q_f) can be empirically modeled as [109]:

$$q_f = \alpha_T G_{fm} p \sqrt{\frac{273.16}{T_i}} (T_s - T_i).$$
(5.99)

Here, α_T denotes the thermal accommodation coefficient, G_{fm} the free molecular coefficient of thermal conductivity, and p the gas pressure. The above relation shows that the energy transfer is linearly dependent on pressure and the temperature difference between the hot surface (T_s) and that of the gas molecules (T_i). The accommodation coefficient is defined as the ratio of the actual energy transferred (to the surface) to the energy transfer under equilibrium conditions, viz., $\alpha_T = (T_r - T_i)/(T_s - T_i)$, where T_r and T_i denote the temperature of reflected and incident molecules, respectively. The free molecular coefficient G_{fm} is given as [109]

$$G_{fm} = \frac{0.1486}{\sqrt{M}} \left(\frac{\gamma + 1}{\gamma - 1} \right),$$
(5.100)

with units W/m^2-K-mTorr. Here, M denotes the molecular mass and γ the ratio of specific heats at constant pressure (c_p) and constant volume (c_v), $\gamma = c_p/c_v$. Eqs. (5.99) and (5.100) can be applied to monoatomic, diatomic, and polyatomic molecules. In the case of monoatomic gases, $\gamma = \frac{5}{3}$. Values of the free molecular coefficient of thermal conductivity and associated parameters are given in Table 5.10 for selected gases [109].

From relation (5.100), it is apparent from dimensional analysis that the free molecular coefficient of thermal conductivity and the thermal conductivity do not represent the same physical properties. In fact, G_{fm} has a character that is similar to a heat transfer coefficient. The latter is generally specified as the ratio of the thermal conductivity to the characteristic length (L_c) of the system. Following this argument, the fluid thermal conductivity at very low pressures can be approximated

Table 5.10. *Free molecular coefficients of thermal conductivity for selected gases*

Gas	M	γ	G_{fm}
He	4.003	1.67	0.2935
N_2	28.020	1.40	0.1663
O_2	32.000	1.40	0.1557
Ar	39.940	1.67	0.0929

as [110]:

$$\kappa_{\text{fluid}} = L_c p\, G_{fm}. \tag{5.101}$$

The accuracy of modeled values relies on the accuracy of L_c. In simple geometries, e.g., parallel plate structures, L_c can be chosen to be the plate separation. For more complex geometries, L_c may need to be determined experimentally. However, even with the best possible value for L_c, the simulated temperature distribution will not be entirely meaningful because the temperature, by its usual defintion, is itself a continuum property. In the free molecular regime, each molecule could have its own temperature since it does not interact with other molecules. Simplifications such as incorporating a temperature jump at surfaces may be used. However, this is difficult to incorporate in a general purpose simulator since the models employed are required to cover a wide range of pressures. However, despite the breakdown of assumptions associated with the continuum hypothesis, good agreement between simulated and measured results have been reported [110].

In the transitional regime, i.e., between the free molecular and continuum regimes, an effective thermal conductivity may be approximated using a combination of the free molecular (κ_{fm}) and standard pressure (κ_a) thermal conductivities, *viz.*,

$$\kappa_{\text{fluid}} = \left(\frac{1}{\kappa_{fm}} + \frac{1}{\kappa_a} \right)^{-1} = \frac{\kappa_{fm}\kappa_a}{\kappa_{fm} + \kappa_a}. \tag{5.102}$$

The thermal conductivity at standard pressure can be readily found in the literature (see, e.g., [109]); values for selected gases are given in Table 5.11.

5.5.4 Radiative Heat Transfer

Heat loss by radiation in a large majority of microstructures is generally insignificant relative to the other components at standard (atmospheric) pressure. However, it could become important at decreasing pressures as

Table 5.11. *Thermal conductiv-*
ities at standard temperature and
pressure for selected gases

Gas	Thermal conductivity (W/m-K)
Ar	0.01728
N_2	0.02625
Air	0.02636
He	0.14990

conductive heat loss to the fluid diminishes. Even at low pressures and relatively high operating temperatures, unless the structure is extremely well thermally isolated, heat loss by conduction via the supporting structure can be significantly larger than heat loss by radiation. Nevertheless, radiative heat loss is part of the thermal balance and the associated flux density is modeled following the Stefan-Boltzmann law.

For the simple case of a grey body with a surface temperature distribution $T_s(\mathbf{r})$, surrounded by a medium in equilibrium at a temperature T_∞, the emitted local heat density is [2]

$$q_{rad}(\mathbf{r}) = \varepsilon \sigma_s \left[T_s^4(\mathbf{r}) - T_\infty^4 \right], \qquad (5.103)$$

where ε denotes the emissivity of the surface and σ_s the Stefan-Boltzmann constant, *viz.*, $\sigma = 5.6696 \times 10^{-8}$ W/m^2-K^4. The term grey body implies a structure that is neither 100% absorbing or reflecting. A black body is a perfect absorber (emissivity, $\varepsilon = 1$). According to Kirchhoff, the emissivity is equal to the absorptivity (i.e., the fraction of radiation absorbed) for each wavelength. A perfect reflector has emissivity zero. A grey body is somewhere in between; $0 < \varepsilon < 1$.

Again as with the other heat loss components, (5.103) represents a heat density and thus needs to be integrated over a control region to evaluate the local heat loss for subsequent use as a heat sink term in the solution of the heat conduction equation.

5.5.5 Model Simplification for Quasi Three-Dimensional Analysis

In view of the planar nature of thermally isolated microstructures and because of the small thicknesses of thin film layers relative to the other linear dimensions, the model equations describing the electrothermal behavior within the microstructure can be reduced to a two-dimensional system [46, 111]. The approximations underlying this simplification, although reasonable in terms of electrical behavior, may not be intuitively obvious from a thermal standpoint and hence require justification. With most structures surrounded by a fluid that is stagnant (see Section 5.5.2), the Biot number (Bi) is relatively small. This dimensionless number, defined as $Bi = h\tau/\kappa_{eff}$, describes the ratio of surface heat conductance to the internal heat conductance across the thickness of microstructure. Here, h denotes a surface heat transfer coefficient, τ the planar thickness, and κ_{eff} the effective thermal conductivity, which is a function of the various layers comprising the microstructure. The smaller the Bi, the smaller is the temperature gradient across the thickness. Indeed, in most applications, Bi turns out to be very small. For example, if the microstructure is 1 μm thick with an effective conductivity of $\kappa_{eff} = 0.1$ W/m-K (which is small from a practical standpoint), this yields $Bi = 10^{-5} h$. For a 100 μm square area subject to laminar air flow of 100 m/s, assuming that the velocity and thermal boundary layers both begin at the leading edge of the microstructure, h is approximately 10^4. This implies a Bi of the order of 0.1, which is encouraging. But in a strict sense, the Biot number is only meaningful for homogeneous materials and for structures with no heat sources and sinks. Thus experimental verification is necessary to justify the model simplification.

Fig. 5.12 illustrates a measurement test structure with three independent polysilicon and metal resistor coils (for temperature measurement) stacked above each other and separated by insulating dielectric layers found in a typical CMOS process. Measurements were performed at standard temperature and pressure using the bottom resistor coil as a heating element. The average temperature, retrieved from a priori characterization of the temperature coefficient of resistance, of the various coils is shown in Fig. 5.13. The results show that the temperature drop across the thickness of the microstructure is very small. Indeed the temperature gradient in the thickness direction is negligible compared to the lateral temperature gradients, thus permitting a two-dimensional solution of the heat conduction equation without significant loss of accuracy.

Given these considerations, electrical transport within the microstructure, assuming unipolar conduction, negligible thermoelectric effects, and more importantly, thin film thicknesses much smaller than other linear dimensions, can be described by

$$\text{div} [\sigma(T) \, \text{grad} \, \psi(x, y)] = 0 \tag{5.104}$$

where the active region of the microstructure lies in the $(x-y)$ plane. Relation (5.104) constitutes a reduced form of Eq. (5.16) or (5.17) with a slight change in solution variable from the Fermi potential (ϕ) to the electrostatic potential (ψ) to reflect unipolar analysis. Here, σ denotes the electrical conductivity of the electrically active thin film region(s). To arrive at an equivalent description for two-dimensional heat transport, we

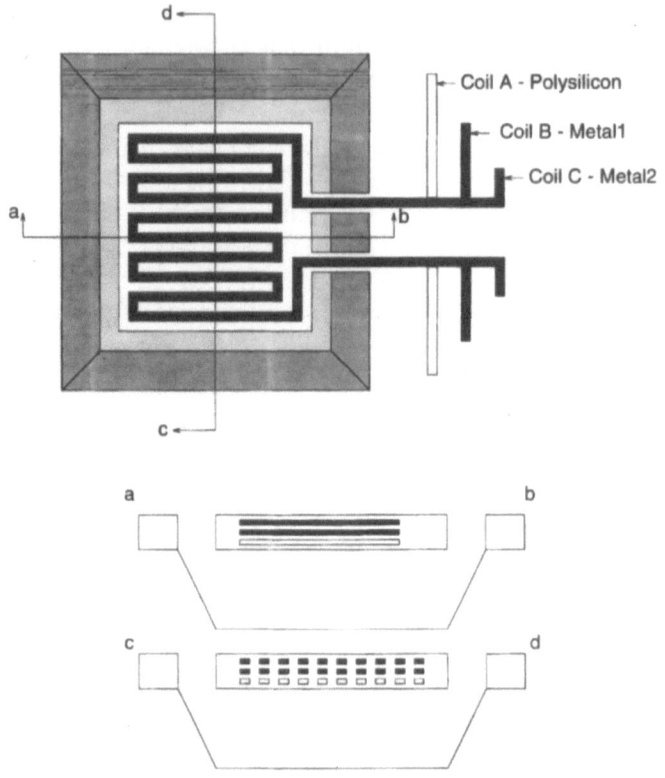

Fig. 5.12 Schematic of test structure for measurement of temperature gradient across thickness of microstructure. Source [46]

integrate the steady-state form of the heat conduction Eq. (5.22) over the thickness τ of the microstructure in the z-direction [112]. Following the usual assumptions,

$$\int_\tau \text{div}\,[\kappa(T)\,\text{grad}\,T]dz = \int_\tau \sigma(T)[\,\text{grad}\,\psi]^2\,dz \qquad (5.105)$$

yields the following reduced form:

$$\text{div}\,[\kappa(T)\,\text{grad}\,T(x,\,y)] = \sigma(T)[\,\text{grad}\,\psi(x,\,y)]^2$$
$$+\frac{1}{\tau}\kappa_{\text{fluid}}(T)\left(\frac{\partial T}{\partial n}\right)_{\text{surfaces}}. \qquad (5.106)$$

Here, the heat loss from the edges of the microstructure is assumed negligibly small in comparison to heat losses from the top and bottom

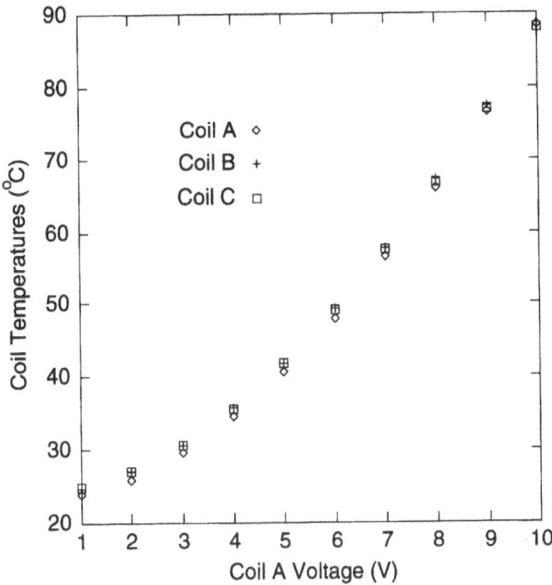

Fig. 5.13 Average temperature of measurement coils A, B, and C as a function of applied voltage on coil C. Source [46]

surfaces in the $(x-y)$ plane. In (5.106), $\kappa(T)$ and $\sigma(T)$ denote the integral averages of the temperature-dependent thermal and electrical conductivities, respectively, and $\kappa_{fluid}(T)$ the thermal conductivity of the surrounding fluid. The first term on the right-hand-side of (5.106) denotes the usual Joule heat and the second term describes the heat loss from the top and bottom surfaces. Specifically, $\partial/\partial n$ denotes the outward normal derivative to the $(x-y)$ plane, viz., $(\partial/\partial n)_{top} = \partial/\partial z$ and $(\partial/\partial n)_{bottom} = -\partial/\partial z$. The heat loss from these surfaces accounts for conduction through the stagnant fluid. In fact, we can add to (5.106) heat loss by forced convection (in the presence of a flow stream) and radiation, and maintain consistency with energy conservation. With these heat loss components incorporated, Eq. (5.106) can no longer be termed as a heat conduction equation; we refer to it as the reduced energy equation.

The heat loss by forced convection, described by the system of equations, (5.84) through (5.88), can be simplified under special cases. Assuming a laminar flow profile with comparatively small in-fluid conduction, the heat transfer from the surface of the microstructure to the moving fluid is predominantly one-dimensional. Under the usual boundary layer approximations and assuming constant gas properties, the local surface heat flux (q_s) from a microstructure with non-uniform temperature distribution can be approximated using linear superposition (see, e.g., [104]), *viz.*,

$$q_s(x) = \int_0^x h(x, \xi) \frac{d\theta(\xi)}{d\xi} d\xi,$$ (5.107)

where h denotes the heat transfer coefficient, defined as

$$h(x, \xi) = 0.332 \frac{\kappa \mathrm{Pr}^{1/3}}{x} \mathrm{Re}_x^{1/2} \frac{1}{[1 - (\xi/x)^{3/4}]^{1/3}}.$$ (5.108)

In (5.107) and (5.108), x denotes the position from the leading edge, ξ the integration variable, θ the difference in temperature between the microstructure surface and that of the free-stream, Pr the fluid Prandtl number, and Re the Reynolds number. The Prandtl number serves to compare the momentum and thermal diffusivities, which constitute the two dissipative forces that act perpendicular to the flow,

$$\mathrm{Pr} = \left(\frac{\mu}{\rho}\right) \bigg/ \left(\frac{\kappa}{\rho c_p}\right) = \frac{\mu c_p}{\kappa}.$$ (5.109)

Here, μ denotes the dynamic viscosity, ρ the density, κ the thermal conductivity, and c_p the specific heat at constant pressure.

The Reynolds number Re_x compares inertial forces (momentum convection) with viscous forces in the fluid,

$$\mathrm{Re}_x = \left(\frac{\rho v_\infty^2}{x}\right) \bigg/ \left(\frac{\mu v_\infty}{x^2}\right) = \frac{\rho v_\infty x}{\mu},$$ (5.110)

and classifies the flow regime as being either laminar or turbulent. For example, in a circular tube, laminar flow is present for $\mathrm{Re} < 2300$ while the flow is fully turbulent for $\mathrm{Re} > 10^4$ [113]. Between these limits, the flow is assumed to be in a transitionary state. Eq. (5.107) provides a suitable approximation to the true laminar flow heat loss only when the temperature gradients in the stream direction are comparatively small and when the fluid properties are constant. Despite these simplifying assumptions, the model provides good agreement with measurements performed for flow streams in rectangular ducts (see [13] and also Sect. 1.2). Expressions similar to (5.107) and (5.108) exist for other flow configurations such as tubes or parallel plates [114].

Equations (5.104) and (5.106) constitute the reduced system of model equations for quasi three-dimensional steady-state analysis of thermal microsensors. Depending on device structure, operating conditions, and intended application, the system must be supplemented with additional relations to account for the relevant heat transfer components, i.e., fluidic conduction and/or convection, and radiation. The boundary and interface conditions for solution of the two-dimensional system are a straightforward adaptation of the conditions discussed in Sect. 5.2.2. A summary of governing equations, physical models, boundary and interface conditions for the thermal microsensor family is given in Sect. 5.6 along with the computational process for self consistent solutions of electrostatic potential and temperature.

5.6 Summary of Equations and Computational Procedure

The system of coupled equations together with appropriate boundary conditions can be numerically solved using a variety of discretization schemes such as finite element, finite difference, or control volume approximations, in conjunction with pertinent upwinding methods to deal with possible instabilities stemming from convective-dominated heat transport. A review of the full family of discretization schemes is beyond the present scope; detailed treatment of the underlying approximations can be found in several texts (see, e.g., [115]). However, we review one class of discretization schemes based on the control volume approxi-

Table 5.12. *Summary of model equations pertinent to thermal microtransducers.* Physical models/data for solids: electrical parameters – Chapt. 2, thermophysical parameters – Sect. 5.3; for gases: thermal conductivity – Sect. 5.5.3, other parameters – Sect. 7.3

Analysis Type	Governing Equation	Boundary Conditions	Solution/Coupling Variable	Application
3D electrical (bipolar)	(5.16)–(5.19) Sections 2.1 & 2.2 (5.22)	(5.25)–(5.35)	Electrostatic potential ψ Carrier concentrations n, p Temperature T	Thermal microtransducer family
3D heat transfer	(5.84)–(5.88)	Section 5.5.1 Section 7.3	Velocity components u, v, w Pressure p Temperature T	
Steady-state quasi 3D (Unipolar)	(5.104), (5.106) (5.107), (5.108)	Section 5.2.2	Electrostatic potential ψ Temperature T	Flow sensor
	(5.104), (5.106) (5.91), (5.95)	Section 5.2.2 (5.96)	Electrostatic potential ψ Temperature T	Vacuum sensor IR sensor Thermal converter Bimorph actuator
	(5.104), (5.106) (5.91), (5.95) (5.103)	Section 5.2.2 (5.96)	Electrostatic potential ψ Temperature T	Thermal imager

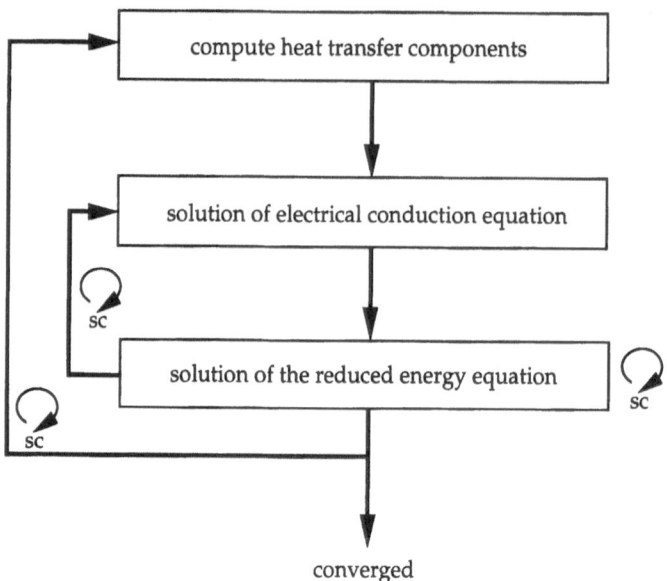

Fig. 5.14 Computational procedure for quasi three-dimensional simulation of thermal microsensors; *sc* denotes iterations for self-consistent solutions. Corresponding equation systems are given in Table 5.12

mation, coupled with upwinding, in Sect. 6.7.3. Although this has been applied to analysis of electrical transport, it can be easily extended to thermal analysis. A summary of relevant model equations is given in Table 5.12.

In quasi three-dimensional analysis, the numerical procedure for self-consistent solutions of the electrostatic potential (ψ) and temperature (T) begins with the computation of heat transfer components based on a suitable initial guess of the surface temperature of the microstructure. If the conductive heat transfer to the fluid is significant, then Laplace's equation needs to be solved in three-dimensions in order to compute the zero flow two-dimensional heat transfer coefficient (see Sect. 5.5.2). This follows with the two-dimensional solution of the coupled system of electrothermal equations comprising the electrical conduction equation and the reduced energy equation. Here, the solution of the former is restricted to within the electrically active regions while the latter is solved over the entire simulation domain. Since this coupled system can be potentially non-linear, an inner iterative loop may be required. The system of three equations are iterated until self-consistent solutions of ψ and T are obtained (see Fig. 5.14). Upon convergence, solution verification is necessary, which can be based on a straightforward check on conservation of the electrical and heat fluxes.

5.7 Illustrative Simulation Example–Micro Pirani Gauge

As seen in Sect. 5.5.3, the heat transfer from a resistive heating element to the ambient can be modulated by the pressure of the surrounding gas since the mean free scattering length of molecules, and hence, the thermal conductance is dependent on gas pressure. This effect, first utilized by Pirani, has found widespread use for pressure measurement in vacuum systems. Miniaturized versions fabricated using IC technologies yield good performance due to the favorable thermal scaling laws coupled with high thermal isolation attainable with silicon. In this example, we illustrate a CMOS-based coil structure [20], which unlike other integrated devices reported hitherto, relies extensively on the lateral component of heat transfer from an active current-heated polysilicon coil to a passive neighboring coil. A photomicrograph of the device along with its cross section are shown in Figs. 5.15 and 5.16, respectively.

For insight into device operation, including underlying heat transport mechanisms, and for optimization of device design to meet key requirements of fast thermal response time, low operating power, and low operating temperature, numerical solutions to the coupled system of equations governing electrothermal behavior is mandatory. For the application intended, we would like to maximize the heat transfer from

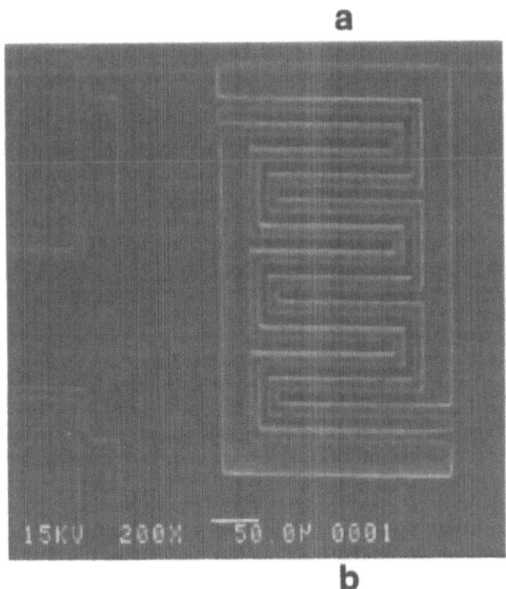

Fig. 5.15 Photomicrograph of CMOS micro Pirani gauge

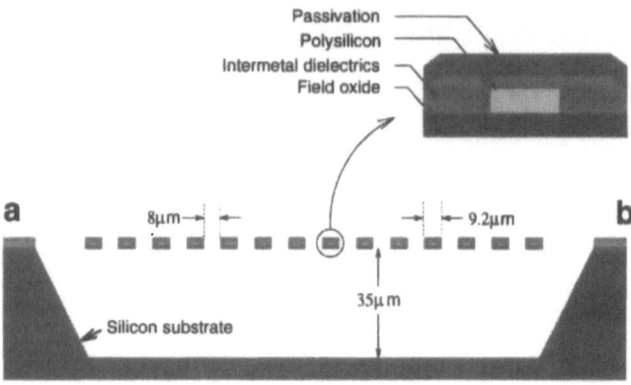

Fig. 5.16 Cross section of device viewed at a–b

active to passive coils (through the ambient fluid) to maximize transduction efficiency. In particular, with pressure measurement, a large Knudsen number, (5.98), is desirable. The device considered here has a Knudsen number Kn ∼ 0.01 at standard temperature and pressure. The Knudsen number improves at low pressures but degrades at high pressures. However, despite the small Kn, high transduction efficiency near atmospheric pressure can be achieved through optimization of the interaction length.

The simulated temperature distribution of the device in air is illustrated in Fig. 5.17. The peaks and valleys correspond to the temperature of the active (current carrying) and passive (temperature sensing) coils, respectively. The numerical simulations were performed using the two-dimensional electrothermal model equations (see Sect. 5.5.5) coupled with the three-dimensional solution of the heat conduction equation to model heat

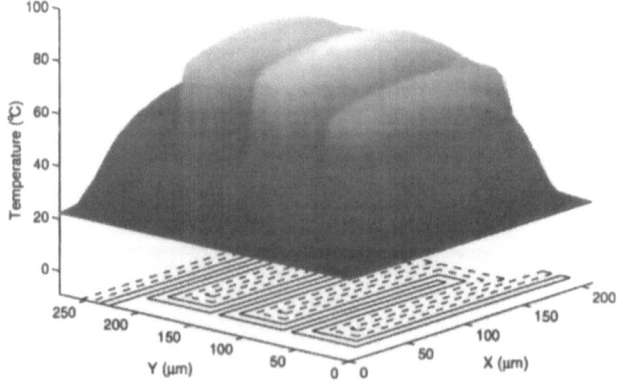

Fig. 5.17 Simulated temperature distribution at standard temperature and pressure. Dashed line: active coil, solid line: passive coil

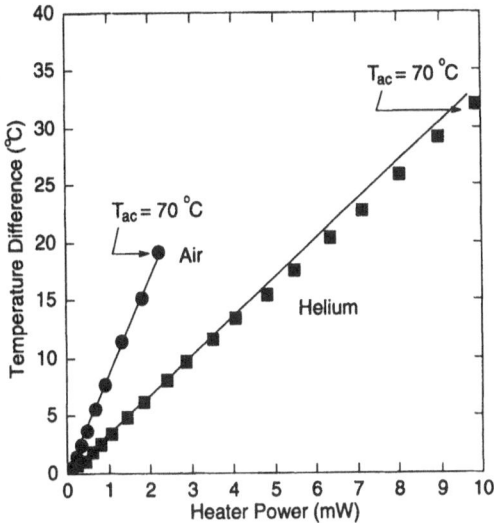

Fig. 5.18 Temperature difference between coils as a function of input power for air and helium at standard temperature and pressure. Solid line: simulations, dots and squares: measured values. T_{ac} denotes temperature of active coil

loss to the ambient (see Sect. 5.5.2). The simulated and measured thermal behaviors of active and passive coils in both air and helium are illustrated in Fig. 5.18.

The values of temperature shown are based on measurements of coil resistance from which an average coil temperature is retrieved following

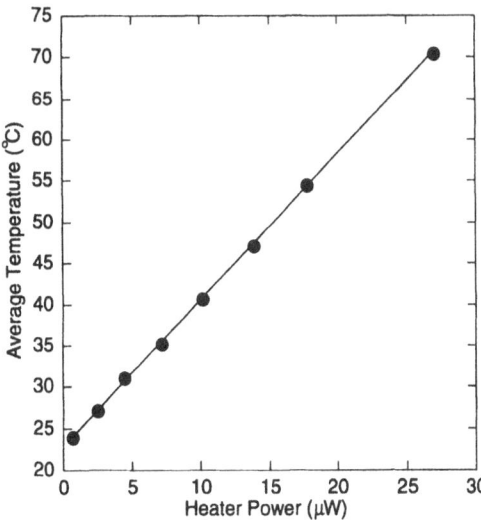

Fig. 5.19 Average temperature of active coil as a function of input power in vacuum. Solid line: simulation, dots: measured values

prior temperature coefficient characterization of the polysilicon layer. While maintaining the active coil average temperature at 70 °C, we clearly see that the input power (i.e., heat loss) is consistent with the thermal conductivities of the two gases. This implies an almost 100% transduction efficiency. The larger temperature difference in the case of helium is due to the conductive heat loss to the surrounding which dominates over the inter-coil lateral heat transfer. The high sensitivity to thermal conductivity makes the device attractive for applications in gas sensing and subsequent discrimination in binary mixtures [116]. The device has a power consumption of 2 mV in air (see Fig. 5.18). However, in vacuum (30 mTorr), the input power is a mere 27 μW at an active coil temperature of 70 °C (see Fig. 5.19). In fact, simulations indicate that the combined heat loss through conduction along coils and radiation is less than 1%. Although not shown, the device has a peak pressure sensitivity of 400 mV/Torr. The performance of the device is very encouraging at atmospheric pressure and higher. Here, further optimization in relation to the intercoil spacing and hence, interaction length, is needed to cope with the reduced Knudsen number (Kn < 0.01) stemming from the small mean free scattering length.

5.8 References

[1] Meijer, G. C. M., van Herwaarden, A. W., *Thermal Sensors*, Bristol: Adam Hilger, 1994.
[2] van Herwaarden, A. W., Meijer, G. C. M., Thermal Sensors, in: *Semiconductor Sensors*, Sze, S. M. (Ed.), Chapt. 7, New York: Wiley, 1994, pp. 331–382.
[3] van Herwaarden, A. W., Sarro, P. M., Thermal Sensors Based on the Seebeck Effect, *Sensors and Actuators*, 10 (1986), 321–346.
[4] Baltes, H., Paul, O., Jaeggi, D., Thermal CMOS Sensors – An Overview, in: *Sensors Update*, Baltes, H., Göpel, W., Hesse, J. (Eds.), Weinheim: VCH, 1996, pp. 121–142.
[5] Tabata, O., Fast-Response Silicon Flow Sensor with an On-Chip Fluid Temperature Sensing Element, *IEEE Trans. Electron Devices*, 33 (1986), 361–365.
[6] Johnson, R. G., Higashi, R. E., A Highly Sensitive Silicon Chip Microtransducer for Air Flow and Differential Pressure Sensing Applications, *Sensors and Actuators*, 11 (1987), 63–72.
[7] Tai, Y., Muller, R. S., Lightly Doped Polysilicon Bridge as an Anemometer, *Digest of Technical Papers*, Transducers '87, Tokyo, 1987, pp. 360–363.
[8] Stemme, G., A CMOS Integrated Silicon Gas-Flow Sensor with Pulse-Modulated Output, *Sensors and Actuators*, 14 (1988), 293–303.
[9] Qin-Yi, T., Jin-Biao, H., Integrated Multi-Function Sensor for Flow Velocity, Temperature and Vacuum Measurements, *Sensors and Actuators*, 19 (1989), 3–11.
[10] van Oudheusden, B. W., Huijsing, J. H., Integrated Silicon Flow-Direction Sensor, *Sensors and Actuators*, 16 (1989), 109–119.
[11] Moser, D., Lenggenhager, R., Baltes, H., Silicon Gas Flow Sensors Using Industrial CMOS and Bipolar IC Technology, *Sensors and Actuators A*, A25–A27 (1991), 577–581.
[12] van Oudheusden, B. W., Silicon Thermal Flow Sensors, *Sensors and Actuators A*, 30 (1992), 5–26.

[13] Nagata, M., Swart, N. R., Stevens, M., Nathan, A., Thermal Based Micro Flow Sensor Optimization Using Coupled Electrothermal Numerical Simulations, *Digest of Technical Papers*, Vol. 2, Transducer's '95, Stockholm, 1995, pp. 447–450.

[14] van Herwaarden, A. W., Sarro, P. M., Floating-Membrane Thermal Vacuum Sensor, *Sensors and Actuators*, 14 (1988), 259–268.

[15] Mastrangelo, C. H., Muller, R. S., µ-Pirani Pressure Gauge with Digital Readout, *Digest of Technical Papers*, Transducers '91, San Francisco, 1991, pp. 245–248.

[16] Völklein, F., Schnelle, W., A Vacuum Microsensor for the Low-Vacuum Range, *Sensors and Materials*, 3 (1991), 41–48.

[17] Robinson, A. M., Haswell, P., Lawson, R. P. W., Parameswaran, M., A Thermal Conductivity Microstructural Pressure Sensor Fabricated in Standard Complementary Metal-Oxide Semiconductor, *Rev. Sci. Instrum.*, 63 (1992), 2026–2029.

[18] Baltes, H., Moser, D., CMOS Vacuum Sensors and Other Applications of CMOS Thermopiles, *Digest of Technical Papers*, Transducers '93, Yokohama, Japan, 1993, pp. 736–741.

[19] Bonne, U., Kubisiak, D., Burstproof, Thermal Pressure Sensor for Gases, *Technical Digest*, IEEE Solid-State Sensors and Actuators Workshop, Hilton Head Is., 1994, pp. 78–81.

[20] Swart, N. R., Nathan, A., An Integrated CMOS Polysilicon Coil-Based Micro-Pirani Gauge with High Heat Transfer Efficiency, *Technical Digest*, IEEE IEDM, San Francisco, 1994, pp. 135–138.

[21] Paul, O., Brand, O., Lenggenhager, R., Baltes, H., Vacuum Gauging with Complementary Metal-Oxide-Semiconductor Microsensors, *J. Vac. Sci. Technology*, A13 (1995), 503–508.

[22] Paul, O., Baltes, H., Novel Fully CMOS-Compatible Vacuum Sensor, *Sensors and Actuators A*, 46–47 (1995), 143–146.

[23] Gajda, M. A., Ahmed, H., Applications of Thermal Silicon Sensors on Membranes, *Sensors and Actuators A*, 49 (1995), 1–9.

[24] Suehle, J. S., Cavicchi, R. E., Gaitan, M., Semancik, S., Tin Oxide Gas Sensor Fabricated Using CMOS Micro-Hotplates and in-situ Processing, *IEEE Electron Device Letts.*, 14 (1993), 118–120.

[25] Cavicchi, R. E., Suehle, J. S., Chaparala, P., Kreider, K. G., Micro-Hotplate Gas Sensor, *Technical Digest*, IEEE Solid-State Sensors and Actuators Workshop, Hilton Head Is., 1994, pp. 53–56.

[26] Swart, N. R., Nathan, A., Design Optimization of Integrated Microhotplates, *Sensors and Actuators A*, 43 (1994), 3–10.

[27] Baltes, H., Moser, D., Lenggenhager, R., Brand, O., Jaeggi, D., Thermomechanical Microtransducers by CMOS and Micromachining, *Micromechanical Sensors, Actuators, and Systems*, 32 (1991), 61–75.

[28] Jaeggi, D., Leme, C. A., O'Leary, P., Baltes, H., Improved CMOS AC Power Sensor, *Digest of Technical Papers*, Transducers '93, Yokohama, 1993, pp. 462–465.

[29] Yoon, E., Wise, K. D., A Wideband Monolithic RMS-DC Converter Using Micromachined Diaphragm Structures, *IEEE Trans. Electron Devices*, 41 (1994), 1666–1668.

[30] Wood, R. A., Han, C.-J., Kruse, P. W., Integrated Uncooled Infrared Detector Imaging Array, *Technical Digest*, IEEE Solid-State Sensors and Actuators Workshop, Hilton Head Is., 1992, pp. 132–135.

[31] Lenggenhager, R., *CMOS Thermoelectric Infrared Sensors*, Ph.D. Dissertation, ETH Zürich, No. 10744, Switzerland, 1994.

[32] Lenggenhager, R., Jaeggi, D., Malcovati P., Duran, H., Baltes, H., Doering, E., CMOS Membrane Infrared Sensors and Improved TMAHW Etchant, *Technical Digest*, IEEE IEDM, San Francisco, 1994, pp. 531–534.

[33] Srinivas, T. A. S., Timans, P. J., Ahmed, H., A High-Performance Infrared Detector Using MOS Technology, *Sensors and Materials*, 8 (1996), 317–326.

[34] Parameswaran, M., Robinson, A. M., Blackburn, D. L., Gaitan, M., Geist, J., Micromachined Thermal Radiation Emitter from a Commercial CMOS Process, *IEEE Electron Device Letts.*, 12 (1991), 57–59.

[35] Mastrangelo, C. H., Yeh, J. H.-J., Muller, R. S., Electrical and Optical Characteristics of Vacuum-Sealed Polysilicon Microlamps, *IEEE Transactions on Electron Devices*, 39 (1992), 1363–1375.

[36] Cole, B. E., Han, C.-J., Higashi, R. E., Ridley, J., Holmen, J., Anderson, J., Nielsen, D., Marsh, H., Newstrom, K., Zins, C., Wilson, P., Beaudoin, K., 512 × 512 Infrared Scene Projector Array for Low-Background Simulations, *Technical Digest*, IEEE Solid-State Sensors and Actuators Workshop, Hilton Head Is., 1994, pp. 7–12.

[37] Bouwstra, S., Kemna, P., Legtenberg, R., Thermally Excited Resonating Membrane Mass Flow Sensor, *Sensors and Actuators*, 20 (1989), 213–223.

[38] Brand, O., Baltes, H., Baldenweg, U., Thermally Excited Silicon Oxide Beam and Bridge Resonators in CMOS Technology, *IEEE Trans. Electron Devices*, 40 (1993), 1745–1753.

[39] Brand, O., *Micromachined Resonators for Ultrasound Based Proximity Sensing*, Ph.D. Dissertation, ETH Zürich, No. 10896, Switzerland, 1994.

[40] Mastrangelo, C. H., *Thermal Applications of Microbridges*, Ph.D. Dissertation, Department of Electrical Engineering and Computer Sciences, University of California, Berkeley, USA, 1991.

[41] van Oudheusden, B. W., The Thermal Modelling of a Flow Sensor Based on Differential Convective Heat Transfer, *Sensors and Actuators A*, 29 (1991), 93–106.

[42] Wachutka, G., Lenggenhager, R., Moser, D., Baltes, H., Analytical 2D-Model of CMOS Micromachined Gas Flow Sensors, *Digest of Technical Papers*, Transducers '91, San Francisco, 1991, pp. 22–25.

[43] Dillner, U., Thermal Modeling of Multilayer Membranes for Sensor Applications, *Sensors and Actuators A*, 41–42 (1994), 260–267.

[44] van Duyn, D. C., Munter, P. J. A., Finite-Element Modelling of Thermoelectric Materials and Devices, *Sensors and Actuators A*, 32 (1992), 413–418.

[45] Allegretto, W., Shen, B., Lai, Z., Robinson, A. M., Numerical Modelling of Time Response of CMOS Micromachined Thermistor Sensor, *Sensors and Materials*, 6 (1994), 71–83.

[46] Swart, N. R., *Heat Transport in Thermal-Based Microsensors*, Ph.D. Thesis, University of Waterloo, Waterloo, Ontario N2L 3G1, Canada, 1994.

[47] Jaeggi, D., Funk, J., Häberli, A., Baltes, H., Overall System Analysis of a CMOS Thermal Converter, *Digest of Technical Papers*, Vol. 2, Transducers '95, Stockholm, 1995, pp. 112–115.

[48] Kriegl, W., Steiner, P., Folkmer, B., Lang, W., MICROTHERM: A Program for Thermal Modelling of Microstructures, *Sensors and Actuators A*, 46–47 (1995), 637–639.

[49] Nathan, A., Microtransducer CAD, *Proc. ESSDERC '96*, Baccarani, G., Rudan, M. (Eds.), Bologna, 1996, pp. 707–715.

[50] Mayer, F., Paul, O., Baltes, H., Influence of Design Geometry and Packaging on the Response of Thermal CMOS Flow Sensors, *Digest of Technical Papers*, Vol. 1, Transducers '95, Stockholm, 1995, pp. 528–531.

[51] Park, S., Kim, H., Kang, Y., Study of Flow Sensor Using Finite Difference Method, *Sensors and Materials*, 7 (1995), 43–51.

[52] Mayer, F., Salis, G., Funk, J., Paul, O., Baltes, H., Scaling of Thermal CMOS Gas Flow Microsensors: Experiment and Simulation, *Proc. IEEE MEMS*, 1996, pp. 116–121.

[53] Akimori, H., Owada, N., Taneoka, T., Uda, H., Reliability Study of Polycrystalline Silicon Thin Film Resistors Used in LSI Under Thermal and Electrical Stress, *Proc. IEEE Int. Reliability Phys. Symp.*, 1990, pp. 276–280.

[54] Swart, N. R., Nathan, A., Reliability Study of Polysilicon for Microhotplates, *Technical Digest*, IEEE Solid-State Sensor and Actuator Workshop, Hilton Head Is., 1994, pp. 119–122.

[55] Swart, N. R., Parameswaran, M., Nathan, A., Optimization of the Dynamic Response of an Integrated Silicon Thermal Scene Simulator, *Digest of Technical Papers*, Transducers '93, Yokohama, 1993, pp. 750–753.

[56] Paul, O. M., Korvink, J., Baltes, H., Determination of the Thermal Conductivity of CMOS IC Polysilicon, *Sensors and Actuators A*, 41–42 (1994), 161–164.

[57] Kells, K., Müller, S., Fichtner, W., Simulating Temperature Effects in Multi-Dimensional Silicon Devices with Generalized Boundary Conditions, *Simulation of Semiconductor Devices and Processes*, Vol. 4, Fichtner, W., Aemmer, D. (Eds.), Zürich: Hartung-Gorre, 1991, pp. 141–148.

[58] Ciampolini, P., Pierantoni, A., Baccarani, G., Three-Dimensional Self-Consistent Modeling of MOS Devices Under Non-Isothermal Regime, *Simulation of Semiconductor Devices and Processes*, 4, Fichtner, W., Aemmer, D. (Eds)., Zürich: Hartung-Gorre, 1991, pp. 149–156.

[59] Middlehoek, S., Audet, S. A., *Silicon Sensors*, New York: Academic Press, 1989.

[60] Callen, H. B., *Thermodynamics*, 2nd Ed., New York: Wiley, 1961.

[61] Madelung, O., *Introduction to Solid-State Theory*, Berlin: Springer-Verlag, 1981.

[62] Wachutka, G. K., Rigorous Thermodynamic Treatment of Heat Generation and Conduction in Semiconductor Device Modeling, *IEEE Trans. CAD of ICAS*, 9 (1990), 1141–1149.

[63] Selberherr, S., *Analysis and Simulation of Semiconductor Devices*, Vienna: Springer-Verlag, 1984.

[64] van Herwaarden, A. W., The Seebeck Effect in Silicon ICs, *Sensors and Actuators*, 6 (1984), 245–254.

[65] Blatt, F. J., *Physics of Electronic Conduction in Solids*, New York: McGraw-Hill, 1968.

[66] Seeger, K., *Semiconductor Physics*, 3rd Ed., Berlin: Springer-Verlag, 1985.

[67] Geballe, T. H., Hull, G. W., Seebeck Effect in Silicon, *Phys. Rev.*, 98 (1955), 940–947.

[68] Völklein, F., Review of the Thermoelectric Efficiency of Bulk and Thin-Film Materials, *Sensors and Materials*, 8 (1996), 389–408.

[69] Paul, O., von Arx, M., Baltes, H., Process-Dependent Thermophysical Properties of CMOS IC Thin Films, *Digest of Technical Papers*, Vol. 1, Transducers '95, Stockholm, 1995, pp. 178–181.

[70] Völklein, F., Baltes, H., Thermoelectric Properties of Polysilicon Films Doped with Phosphorus and Boron, *Sensors and Materials*, 3 (1992), 325–334.

[71] Lahiji, G. R., Wise K. D., A Batch-Fabricated Silicon Thermopile Infrared Detector, *IEEE Trans. Electron Devices*, ED-29 (1982), 14–22.

[72] Jones, D. I., Le Comber, P. G., Spear, W. E., Thermoelectric Power in Phosphorous Doped Amorphous Silicon, *Philosophical Magazine*, 36 (1977), 541–551.

[73] Kodato, S., Nishida, S., Konagai, M., Takahashi, K., High-conductivity a-Si:H:F Film and Its Performance for a Power Sensor and a Strain Gauge, *J. Non-Crystalline Solids*, 59/60 (1983), 1207–1210.

[74] Bhatnagar, Y. K., Nathan, A., A Thermal Sensor Based on The Seebeck-Effect in a-Si:H, International Conf. on Amorphous Semiconductors, Cambridge, 1993.

[75] Kodato, S., Naitoh, Y., Kuroda, K., The Seebeck Effect in Highly Conductive µc-Ge:H Films and Its Application to Sensors, *Sensors and Actuators A*, 34 (1992), 161–166.

[76] Dehé, A., Pavlidis, D., Hong, K., Hartnagel, H. L., InGaAs/InP Thermoelectric Infrared Sensor Utilizing Surface Bulk Micromachining Technology, *IEEE Trans. Electron Devices*, 44 (1997), 1052–1059.

[77] Chong, N., Srinivas, T. A. S., Ahmed, H., Performance of GaAs Microbridge Thermocouple Infrared Detectors, *J. of Microelectromechanical Systems*, 6 (1997), 136–141.

[78] Völklein, F., Wiegand, A., Baier, V., High-Sensitivity Radiation Thermopiles Made of Bi-Sb-Te Films, *Sensors and Actuators A*, 29 (1991), 87–91.

[79] Ziman, J. M., *Principles of the Theory of Solids*, 2nd Ed., Cambridge: Cambridge University Press, 1972.

[80] Glassbrenner, C. J., Slack, G. A., Thermal Conductivity of Silicon and Germanium from 3 °K to the Melting Point, *Phys. Rev.*, 134 (1964), A1058–A1069.

[81] Maycock, P. D., Thermal Conductivity of Silicon, Germanium, III-V Compounds and III-V Alloys, *Solid-State Electron.*, 10 (1967), 161–168.

[82] Brinson, M. E., Dunstan, W., Thermal Conductivity and Thermoelectric Power of Heavily Doped n-Type Silicon, *J. Phys. C: Solid State Physics*, 3 (1970), 483–491.

[83] Paul, O., von Arx, M., Baltes, H., CMOS IC Layers: Complete Set of Thermal Conductivities, *Semiconductor Characterization Workshop*, NIST, Jan. 1995, AIP, 1996, pp. 197–201.

[84] Tai, Y. C., Mastrangelo, C. H., Müller, R. S., Thermal Conductivity of Heavily Doped LPCVD Polycrystalline Silicon Films, *J. Appl. Phys.*, 63 (1988), 1442–1447.

[85] Völklein, F., Baltes, H., A Microstructure for Measurement of Thermal Conductivity of Polysilicon Thin Films, *J. of Microelectromechanical Systems*, 1 (1992), 193–196.

[86] Slack, G. A., Thermal Conductivity of Pure and Impure Silicon, Silicon Carbide, and Diamond, *J. Appl. Phys.*, 35 (1964), 3460–3466.

[87] Hummel, R. E., *Electronic Properties of Materials*, 2nd Ed., Berlin: Springer-Verlag, 1993.

[88] Liley, P.E., *Heat Exchanger Handbook*, Vol. 5, New York: Hemisphere, 1983.

[89] Touloukian, S., Powell, R. W., Ho, C. Y., Klemens, P. G., *Thermophysical Properties of Matter*, Vols. 1-2, New York: Plenum Press, 1970.

[90] Stärz, T., Schmidt, U., Völklein, F., Microsensor for in situ Thermal Conductivity Measurements of Thin Films, *Sensors and Materials*, 7 (1995), 395–403.

[91] Kumar, S., Vradis, G. C., Thermal Conduction by Electrons Along Thin Films: Effects of Thickness According to Boltzmann Transport Theory, *Micromechanical Sensors, Actuators, and Systems*, 32 (1991), 89–101.

[92] Kittel, C., Interpretation of the Thermal Conductivity of Glasses, *Phys. Rev.*, 75 (1949), 972–974.

[93] Schafft, H. A., Suehle, J. S., Mirel, P. G. A., Thermal Conductivity Measurements of Thin-Film Silicon Dioxide, *Proc. IEEE Int. Conf. on Microelectronic Test Structures*, Vol. 2, 1989, pp. 121–125.

[94] Orchard-Webb, J. H., A New Structure for Measuring the Thermal Conductivity of Integrated Circuit Dielectrics, *Proc. IEEE Int. Conf. on Microelectronic Test Structures*, Vol. 4, 1991, pp. 41–45.

[95] Kleiner, M. B., Kühn, S. A., Weber, W., Thermal Conductivity Measurements of Thin Silicon Dioxide Films in Integrated Circuits, *IEEE Trans. Electron Devices*, 43 (1996), 1602–1609.

[96] Mastrangelo, C. H., Tai, Y. C., Muller, R. S., Thermophysical Properties of Low-Residual Stress, Silicon-Rich, LPCVD Silicon Nitride Films, *Sensors and Actuators A*, 21–23 (1990), 856–860.

[97] Eriksson, P., Andersson, J. Y., Stemme, G., Thermal Characterization of Surface-Micromachined Silicon Nitride Membranes for Thermal Infrared Detectors, *J. of Microelectromechanical Systems*, 6 (1997), 55–61.
[98] von Arx, M., Paul, O., Baltes, H., Test Structures to Measure the Heat Capacity of CMOS IC Layer Sandwiches, *Proc. IEEE Int. Conf. on Microelectronic Test Structures*, Monterey, 1997, pp. 203–208.
[99] Mastrangelo, C. H., Muller, R. S., Thermal Diffusivity of Heavily Doped Low Pressure Chemical Vapor Deposited Polycrystalline Silicon Films, *Sensors and Materials*, 3 (1988), 133–142.
[100] Paul, O., Baltes, H., Measuring Thermogalvanomagnetic Properties of Polysilicon for the Optimization of CMOS Sensors, *Digest of Technical Papers*, Transducers '93, 1993, pp. 606–609.
[101] Popovic, R. S., *Hall Effect Devices*, Bristol: Adam Hilger, 1991.
[102] Nathan, A., Manku, T., The Thermomagnetic Carrier Transport Equation, *Sensors and Actuators A*, 36 (1993), 193–197.
[103] Rudin, S., Wachutka, G., Baltes, H., Thermal Effects in Magnetic Microsensor Modeling, *Sensors and Actuators A*, 25–27 (1991), 731–735.
[104] Kays, W. M., Crawford, M. E., *Convective Heat and Mass Transfer*, New York: McGraw-Hill, 1993.
[105] Harrington, R. F., *Field Computation by Moment Methods*, New Jersey: IEEE Press, 1993.
[106] Özisik, M. N., *Basic Heat Transfer*, New York: McGraw-Hill, 1977.
[107] Rohsenow, W. M., Choi, H., *Heat, Mass, and Momentum Transfer*, New Jersey: Prentice-Hall, Englewood Cliffs, 1961.
[108] Berman, A., *Total Pressure Measurements in Vacuum Technology*, Orlando: Academic Press, 1985.
[109] Dushman, S., *Scientific Foundations of Vacuum Technique*, New York: Wiley, 1962.
[110] Stevens, M. E., *CMOS Electrothermal Microsensors for Flow and Pressure Measurement*, M.A.Sc. Thesis, University of Waterloo, Waterloo, Ontario N2L 3G1, Canada, 1996.
[111] Nathan, A., Swart, N. R., Quasi Three-Dimensional Simulation of Heat Transport in Thermal-Based Microsensors, in: *Simulation of Semiconductor Devices and Processes*, Vol. 6, Ryssel, H., Pichler, P. (Eds.), 1995, pp. 30–33.
[112] Allegretto, W., Shen, B., Haswell, P., Lai, Z., Robinson, A. M., Numerical Modeling of a Micromachined Thermal Conductivity Gas Pressure Sensor, *IEEE Trans. CAD of ICAS*, 13 (1994), 1247–1256.
[113] Rohsenow, W. M., Hartnett, J. P., *Handbook of Heat Transfer*, New York: McGraw-Hill, 1973.
[114] Sellars, J. R., Tribus, M., Klein, J. S., Heat Transfer to Laminar Flow in a Round Tube or Flat Conduit – the Gratez Problem Extended, *Trans ASME*, 78 (1956), 441–448.
[115] Hirsch, C., *Numercial Computation of Internal and External Flow, Vol. 1, Fundamentals of Numerical Discretization*, New York: Wiley, 1989.
[116] Swart, N. R., Stevens, M., Nathan, A., Karanassios, V., A Flow-Insensitive Thermal Conductivity Microsensor and Its Application to Binary Gas Mixtures, *Sensors and Materials*, 9 (1997), 387–394.

6 Mechanical Effects on Carrier Transport

Effects of mechanical stress on electrical properties of semiconductor materials and devices have been known since the invention of the transistor. The pioneering work of Bardeen and co-workers [1], on effects of pressure on the *p-n* junction, and subsequently, by Smith on stress modulation of resistivity [2], have led to what we presently term as the piezojunction and piezoresistance effects, respectively. Today, these effects are routinely exploited for sensing of mechanical signals. Integrated circuit (IC) mechanical microsensors and microactuators, including micro-electro-mechanical systems in closed loop operation, rely on modulation of the inherent electrical carrier transport induced by an external mechanical signal such as pressure, force, or acceleration. Here, the transduction mechanism from the mechanical to the electrical domain is by virtue of piezoresistance. A large number of applications and associated devices have been developed, along with a corresponding amount of literature. An extensive early review of the interaction of the mechanical signal with electrical transport in semiconductors is given in [3]. Specific references to recent developments are given in the sections that follow. In addition to piezoresistive and piezojunction effects, there is the piezoelectric effect, which however, is present in ferroelectric materials. Here, mechanical stress gives rise to electric polarization. The piezoelectric effect and associated applications are described in Sect. 8.6. Recent progress in mechanical sensors and micro-electro-mechanical systems is reviewed in [4]. The latter along with other mechanical microactuators is discussed in Chapt. 8.

Mechanical strain has also become an important design variable in newly emerging (molecular beam epitaxy and chemical vapor depostion) technologies for fabrication of high performance devices such as strained-layer double heterojunction bipolar transistors and strained-layer metal-oxide-semiconductor (MOS) field effect transistors. Here, in contrast to microtransducers, the strain intrinsic to the material is tailored to yield enhanced electrical performance; the strain stems from lattice mismatched semiconductor materials and associated alloys (e.g., Si, Ge, and SiGe alloys). A review of the physics of strained materials, and their fabrication technologies and associated devices can be found in [5–7].

While mechanical stress or strain can be highly beneficial, it can, on the other hand have adverse effects on the operation of most microsensors, in material characterization experiments, as well as in very large scale integrated (VLSI) bipolar and MOS devices. Here, mechanical stress stems from encapsulation (see, e.g., [8, 9]) of the device or IC chip as well as from thin film conducting and dielectric microfabrication materials (see, e.g., [10] and Sect. 7.2.3) overlying the electrical active region. For example, in magnetic microsensors, mechanical stress gives rise to device offset, which disturbs the detection of static and low frequency magnetic fields [11–13]. Additionally, the offset undermines the accuracy and long term stability of the magnetic output response. In Hall-based material characterization experiments, the values retrieved for the Hall mobility may not be meaningful due to stress induced anisotropy in the galvanomagnetic transport coefficients such as conductivity and Hall coefficient [14, 15]. Here, measured values strongly depend on the orientation of the test structure (i.e., current flow) with respect to the stress components. In VLSI technologies, the stress induced change in resistivity, particularly at chip edges, can be as large as 35% [8]. Consideration of these effects is now becoming an integral part of the device and circuit design process [16–19]. In particular, in trench isolation VLSI technologies, stress components in regions at the trench vicinity can be of the order of hundreds of MPa. Depending on processing conditions (oxide topology and fill material), the change in collector current of a bipolar transistor can be close to 100% [16], which clearly undermines the highly needed device matching in circuit design.

Thus modeling of the stress dependence of transport coefficients is crucial for reliable prediction of device and circuit behavior. Here, models are needed not only for design of mechanical sensors but also for other microsensors and VLSI circuits. The key issues are related to the optimization of shape, size, placement, and orientation of device or circuit to enhance or limit the effects of mechanical stress. For example, in mechanical sensors we can reduce the non-linearity in output response, for a given pressure range, to avoid compensation by resistor trimming or by electronic circuitry. In design of readout circuits in microsystems, optimal placement and orientation of devices is crucial for device matching in, for example, differential amplifiers. These design issues can only be addressed by numerical modeling, particularly when we are dealing with complex device geometries, tensorial transport coefficients, and arbitrary stress distributions [20–22]. Analytical solutions, if at all possible, only apply to very special cases of device geometry and stress distribution (see, e.g., [15]).

The presence of mechanical stress alters the band structure of the semiconductor; the bands either shift in energy, get distorted, become non-degenerate, or undergo any combination of the three [23]. Consequently, the band gap energy gets affected and so do the effective masses as well as the distributions of carrier concentrations. More importantly, the transport coefficients (e.g., carrier mobility) become strongly direction-dependent or anisotropic, thus requiring tensorial descriptions. Section 6.1 reviews piezoresistivity and the generalized Ohm's law for an anisotropic material, along with values of associated coefficients for crystalline and polycrystalline semiconductors. Sections 6.2 and 6.3 give a detailed treatment of the strained conduction and valence band structures, providing the physics underlying piezoresistivity as well as the piezojunction effect; the latter is described in Sect. 6.4. The band structure analysis leads to a simple description of the carrier mobilities in terms of the piezoresistance coefficients for subsequent formulation of the piezo drift-diffusion model. So far, we deal with a stress distribution that is uniform. Presence of a stress gradient gives rise to second order effects in both electrical and heat transport. This is discussed in Sect. 6.5. Interactions of strained electrical transport with the magnetic field, relevant for simulation of magnetic microsensors, is reviewed in Sect. 6.6. Section 6.7 summarizes the anisotropic drift diffusion model, along with a description of numerical discretization and solution procedures. These procedures are also applicable to stress effects in other microsensors associated with earlier chapters on radiant, magnetic, and thermal signals. Sect. 6.8 provides an illustrative example, namely the modeling of offset in strained Hall effect microsensors.

Throughout the subsequent discussion on model equations and associated coefficients, the question of coordinate transformation will surface. The associated formulas and rules will be given in Chapt. 7, which also describes the governing equations and constitutive relations needed to compute the mechanical variables such as stress and strain.

6.1 Piezoresistive Effect

Piezoresistivity refers to the alteration of resistivity in a material when subject to a uniform mechanical stress. The effect is most pronounced in low doped semiconductors; in the degenerately doped material the effect is much smaller, and in metals it is virtually negligible (see [24]).

Mechanical stress alters the energy band structure in the semiconductor, and consequently, its electrical properties, which now become anisotropic. In n-type semiconductors, the interpretation of piezoresistivity is relatively straightforward. For not too large stress, the conduction band can be assumed to only shift in energy with minimal distortion. In multi-valley semiconductors such as Si and Ge, this leads to redistribution of electrons in the conduction band valleys giving rise to an anisotropic conductivity. However, the effects of stress on p-type semiconductors are more involved due to the degenerate nature of the valence band structure. Here, in addition to energy shifts, the shape of the valence band is altered thus reducing its symmetry, leading to anisotropic effective mass and mobility. Because of the multitude of effects associated with the valence band, piezoresistivity, particularly that stemming from shear stresses, is larger in p-type materials. The physics underlying piezoresistivity is treated in more detail in Sects. 6.2 and 6.3. In this section, our discussion is restricted to piezoresistance coefficients in mono- and polycrystalline semiconductors.

6.1.1 Piezoresistance Coefficients in Monocrystalline Semiconductors

When an isotropic monocrystalline semiconductor is subject to mechanical stress, its resistivity (ρ), as described in Ohm's law, $\mathbf{E} = \rho \mathbf{J}$, becomes a symmetric second rank tensor. Here, $\mathbf{E}$ and $\mathbf{J}$ denote the electric field and current density vectors, respectively. Ohm's law can be generalized for the anisotropic material to read

$$E_i = \sum_j \rho_{ij} J_j \quad \text{with} \quad i, j = 1, 2, 3 \tag{6.1}$$

or, as follows, in expanded form:

$$E_1 = \rho_{11} J_1 + \rho_{12} J_2 + \rho_{13} J_3,$$
$$E_2 = \rho_{21} J_1 + \rho_{22} J_2 + \rho_{23} J_3,$$
$$E_3 = \rho_{31} J_1 + \rho_{32} J_2 + \rho_{33} J_3. \tag{6.2}$$

Each component of the electric field is linearly related to all three components of the current density. The nine components, ρ_{11} through ρ_{33}, form the resistivity tensor, which is symmetric; thus the number of independent components is reduced to six. For not too large stress, the fractional change in resistivity can be approximated as being linearly related to stress via the fourth rank tensor (π) of piezoresistance coefficients [2]:

$$\frac{(\Delta\rho)_{ij}}{\rho_0} = -\frac{(\Delta\sigma)_{ij}}{\sigma_0} = \sum_{k,l} \pi_{ijkl} T_{kl} \quad \text{with} \quad i, j, k, l = 1, 2, 3.$$

$$\tag{6.3}$$

Here, ρ_0 and σ_0 denote the zero-stress resistivity and conductivity, respectively, and T_{kl} the stress tensor (which we denote by σ later in Chapt. 7). Following (6.3), we re-express the generalized Ohm's law in terms of the piezoresistance coefficients, *viz.*,

$$E_i = \rho_0 \left(1 + \sum_{k,l} \pi_{ijkl} T_{kl} \right) J_j \quad \text{with} \quad i,j,k,l = 1,2,3. \quad (6.4)$$

The number of coefficients in the piezoresistance tensor can be reduced considerably by symmetry considerations. The symmetry in stress reduces the number of independent coefficients to thirty-six, which can be further reduced by invoking material symmetry [25]. For example, with cubic semiconductors (e.g., Si or Ge) there are three independent coefficients [2], which can be stated in the following form using engineering (matrix) notation

$$\pi = \begin{bmatrix} \pi_{11} & \pi_{12} & \pi_{12} & 0 & 0 & 0 \\ \pi_{12} & \pi_{11} & \pi_{12} & 0 & 0 & 0 \\ \pi_{12} & \pi_{12} & \pi_{11} & 0 & 0 & 0 \\ 0 & 0 & 0 & \pi_{44} & 0 & 0 \\ 0 & 0 & 0 & 0 & \pi_{44} & 0 \\ 0 & 0 & 0 & 0 & 0 & \pi_{44} \end{bmatrix}. \quad (6.5)$$

Here, use of the matrix notation allows to adopt a single suffix, *viz.*, $11 \to 1$, $22 \to 2$, $33 \to 3$, $(23, 32) \to 4$, $(31, 13) \to 5$, $(12, 21) \to 6$, for the first two and last two suffixes in these coefficients. Values for the piezoresistance coefficients π_{11}, π_{12}, π_{44}, as measured by Smith for Si and Ge [2], are given in Table 6.1. Following (6.5), Eq. (6.4) can now be expanded to read

$$E_1 = \rho_0 J_1 + \rho_0 \pi_{11} T_1 J_1 + \rho_0 \pi_{12} (T_2 + T_3) J_1 + \rho_0 \pi_{44} (J_2 T_6 + J_3 T_5),$$
$$E_2 = \rho_0 J_2 + \rho_0 \pi_{11} T_2 J_2 + \rho_0 \pi_{12} (T_1 + T_3) J_2 + \rho_0 \pi_{44} (J_1 T_6 + J_3 T_4),$$
$$E_3 = \rho_0 J_3 + \rho_0 \pi_{11} T_3 J_3 + \rho_0 \pi_{12} (T_1 + T_2) J_3 + \rho_0 \pi_{44} (J_1 T_5 + J_2 T_4).$$

$$(6.6)$$

Here, the first term on each right-hand side reproduce the usual Ohm's law for an isotropic (unstressed) material, the second term describes the effect of stress parallel to current flow, the third describes the effect of stress transverse to current flow, and the last term of each right-hand side describes the effect associated with shear stress.

Equation (6.3) and subsequent relations are valid for not too large uniform stress. Smith retrieved the piezoresistance coefficients under this condition. When stresses are large, the linear expansion leading to (6.3) does not hold but requires the inclusion of higher order terms (see Sect. 6.1.3). Any non-uniformities in the stress distribution lead to second order effects associated with non-uniform band structure. Effects of stress gradients are dealt with in Sect. 6.5. Also, all of the above transport

Table 6.1. *Measured values of piezoresistance coefficients for n- and p-type Si and Ge at 300 K [2]*

Material	Piezoresistance coefficients (10^{-11}) Pa^{-1}			Resistivity (Ω-cm)
	π_{11}	π_{12}	π_{44}	
n-Type Si	− 102.2	+ 53.4	− 13.6	11.7
p-Type Si	+ 6.6	− 1.1	+ 138.1	7.8
n-Type Ge	− 2.3	− 3.2	− 138.1	1.5
	− 2.7	− 3.9	− 136.8	5.7
	− 4.7	− 5.0	− 137.9	9.9
	− 5.2	− 5.5	− 138.7	16.6
p-Type Ge	− 3.7	+ 3.2	+ 96.7	1.1
	− 10.6	+ 5.0	+ 98.6	15.0

relations describe purely ohmic situations. The relations do not account for effects of diffusion associated with spatially varying doping profiles, which are typical for diffused or implanted piezoresistors. Incorporation of diffusion effects will be described in Sects. 6.2 and 6.3.

The various coefficients in relations (6.1) to (6.4) are specified with respect to the crystal coordinate system. The relations for an arbitrarily oriented Cartesian coordinate system can be obtained following standard transformation rules [25] (discussed later in Sect. 7.1.1). In special configurations involving uniaxial stress parallel or transverse to current flow, the transformations lead to simple descriptions for the piezoresistance coefficients in certain crystal orientations (see [4]). The associated directional behavior for selected orientations is depicted in Figs. 6.1 and 6.2 [26]. Here, π_l describes the fractional change in resistivity when the uniaxial stress, current and electric field are all in the same direction. Correspondingly, π_t is the piezoresistance coefficient when the current flow and electric field are co-linear and the uniaxial stress is perpendicular to both. The results given in Figs. 6.1 and 6.2 stem from the following generalized descriptions for the longitudinal and transverse coefficients (see [4, 26, 27])

$$\pi_l = \pi_{11} - 2(\pi_{11} - \pi_{12} - \pi_{44})(l_{11}^2 l_{12}^2 + l_{11}^2 l_{13}^2 + l_{12}^2 l_{13}^2), \qquad (6.7)$$

$$\pi_t = \pi_{12} - 2(\pi_{11} - \pi_{12} - \pi_{44})(l_{11}^2 l_{21}^2 + l_{12}^2 l_{22}^2 + l_{13}^2 l_{23}^2). \qquad (6.8)$$

Here, the l_{ij}'s are the direction cosines; they relate the "new" axes x_i to the "old" axes x_j (see Sect. 7.1.1). The above expressions along with values for the piezoresistance coefficients given in Table 6.1, allow us to generate simple illustrations (Figs. 6.1 and 6.2) for any orientation (see [27]). They constitute useful heuristic design aids.

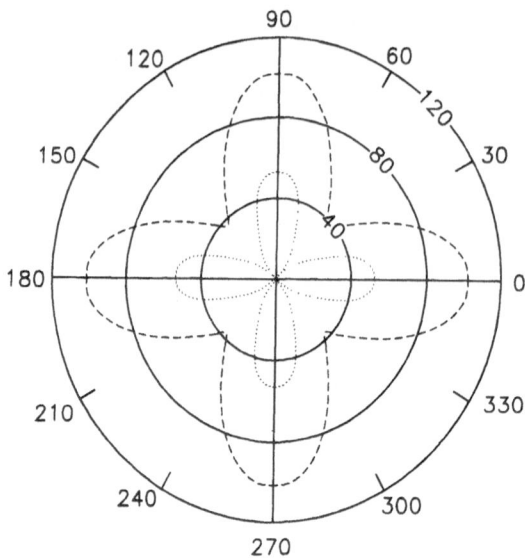

Fig. 6.1 Longitudinal (dashed curve) and transverse (dotted curve) piezoresistance coefficients in *n*-type (100) silicon. Values indicated are in units of 10^{-11} Pa^{-1}

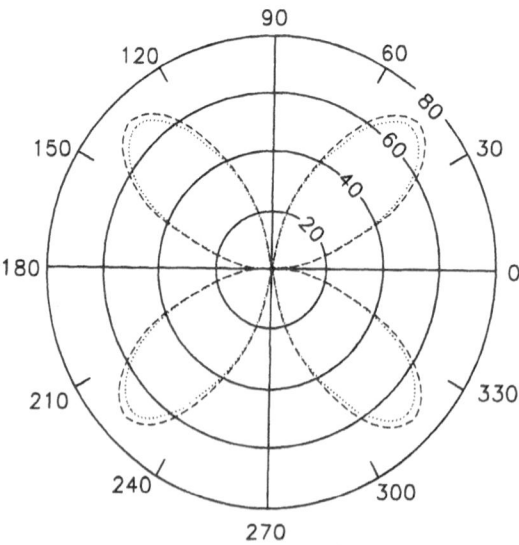

Fig. 6.2 As in Fig. 6.1, but for *p*-type (100) silicon

6.1.2 Doping- and Temperature-Dependence of Piezoresistance Coefficients

It is clear from the measurement results of piezoresistance in Ge (see Table 6.1) that the magnitude of the piezoresistance coefficient decreases with increasing doping for both n- and p-type material. Low doped samples yield large piezoresistance, but the associated coefficients have a strong temperature-dependence. In the highly doped material, the piezoresistance coefficients are smaller and their dependence on temperature is much weaker.

The behavior of the piezoresistance coefficient with doping and temperature is very similar to that of drift mobility. At low doping levels, lattice (phonon) scattering prevails resulting in a strong dependence on temperature. At high doping levels, scattering due to impurities is dominant. This degrades the piezoresistance coefficient but reduces its dependence on temperature. Thus in practical devices, the doping concentration in piezoresistors is around $10^{18} \, \mathrm{cm}^{-3}$ where the piezoresistance effect is still significant but the associated temperature coefficient is small (see [4]).

Figures 6.3 and 6.4 illustrate the behavior of piezoresistance as a function of temperature for various impurity concentrations (N) in n- and

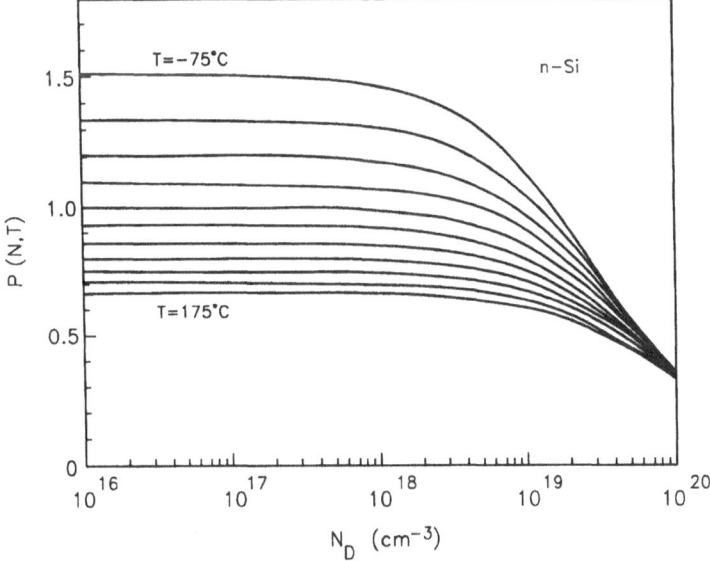

Fig. 6.3 Doping-dependence of piezoresistivity in n-type Si at various temperatures ($-75\,^{\circ}\mathrm{C} \leq T \leq 175\,^{\circ}\mathrm{C}$, in steps of $25\,^{\circ}\mathrm{C}$), expressed in terms of a dimensionless factor, $P(N, T) = \pi(N, T)/\pi(300\,K)$; see Eqn. (6.9). Adapted from [27]

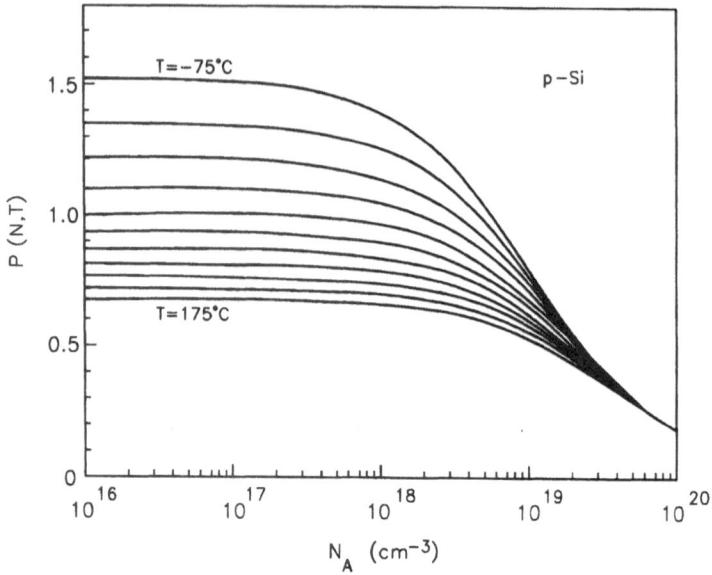

Fig. 6.4 As in Fig. 6.3, but for p-type Si. Adapted from [27]

p-type Si, respectively [27]. Here, the behavior is expressed in terms of a dimensionless factor, $P(N,T) = \pi(N,\,T)/\pi(300\,\mathrm{K})$, computed using the relaxation time approximation

$$P(N,T) = \frac{300}{T}\,\frac{F'_{r+(1/2)}(E_F/kT)}{F_{r+(1/2)}(E_F/kT)}, \tag{6.9}$$

where T is the temperature [K], F is the Fermi integral, the prime indicates its derivative with respect to (E_F/kT), and r is the scattering factor, which takes into account only lattice scattering, *viz.*, $r = -1/2$. Despite the assumption on scattering mechanisms, the agreement between computed and measured values is good for the n-doped material at all concentrations. In p-doped Si, the agreement is reasonable only at low doping concentrations ($N < 10^{17}\,\mathrm{cm}^{-3}$). At higher doping concentrations, the discrepancies are 12% for $N = 5 \times 10^{18}\,\mathrm{cm}^{-3}$ and 20% for $N = 3 \times 10^{19}\,\mathrm{cm}^{-3}$ [27]. In diffused or implanted piezoresistors, the doping concentration decreases from the surface to the junction depth, x_j. However, the piezoresistance behaves in exactly the opposite manner because of doping-dependence. Assuming that stress remains uniform over x_j, we can determine an average piezoresistance coefficient π_{av}, using [28, 29]:

$$\pi_{av} = \frac{\int_0^{x_j} \pi[N(x)]\sigma[N(x)]dx}{\int_0^{x_j} \sigma[N(x)]dx} \tag{6.10}$$

to account for the weighted distribution of current flow due to the doping induced conductivity variation, $\sigma[N(x)]$. Here, $N(x)$ is the doping profile. The value of π_{av} is only slightly larger than the corresponding value obtained for a doping level equal to the surface concentration in the diffused layer [28, 30].

6.1.3 Non-Linear Piezoresistance Coefficients

Relation (6.3) is based on a linear expansion valid for not too large stress. It does not account for possible non-linearities arising from higher-order piezoresistance and stress terms. The degree of non-linearity depends on the magnitude, type (compressive or tensile), and orientation of stress, and the temperature [31, 32]. If second order terms are included in the expansion, the fractional change in resistivity, in matrix notation, reads [28]:

$$\frac{(\Delta\rho)_i}{\rho_0} = \sum_{j=1}^{6} \pi_{ij}T_j + \sum_{j,k=1}^{6} \pi_{ijk}T_jT_k. \tag{6.11}$$

Here, π_{ijk} are the second-order piezoresistance coefficients. In Si, because of cubic symmetry, there are nine independent coefficients. Their measured values are given in Table 6.2 [33, 34] for diffused n- and p-type Si layers with surface doping concentration of 10^{18} cm^{-3}.

Table 6.2. *Measured values of some second-order piezoresistance coefficients for n- and p-type Si at 300 K [33, 34]*

Coefficient $(10^{-8}\,\text{MPa}^{-2})$	n-Type Si	p-Type Si
π_{111}	≈ 0	$+71$
π_{122}	≈ 0	-36
π_{661}	-22	-5
π_{112}	≈ 0	-35
π_{166}	$+98$	≈ 0
π_{123}	≈ 0	
π_{144}	-57	
π_{441}	$+44$	
π_{456}	-51	

6.1.4 Piezoresistance Coefficients in Polycrystalline Semiconductors

Polycrystalline silicon (poly-Si) has been found to be a viable alternative to the monocrystalline counterpart for realization of piezoresistors. For example, poly-Si exhibits a piezoresistance as high as 70% of monocrystalline Si for the same doping concentration [35]. The notable feature with poly-Si stems from the increased electrical isolation of the piezoresistor since the material can be deposited on standard IC dielectrics, thus extending sensor operation to temperatures in excess of 200 °C [36, 37].

Modeling of piezoresistance in polysilicon, as with its other physical properties, is complex due to presence of grain boundaries, varying grain orientations, and irregularity in mechanical properties in the grains (see [36]). The model must reflect the dependence of magnitude and orientation of the piezoresistance effect on the material's microstructure and its mechanical properties, both of which are highly process-dependent [37]. Furthermore, because of the random orientation of grains in the material, an averaging of the components in the associated tensor elements is needed. Various averaging procedures have been proposed (see [37]) based on constant local strain or constant local stress approximations. The choice in averaging procedure depends on the grain size relative to film thickness, which also determines the nature of mechanical behavior in the grains.

Various models of piezoresistance in poly-Si have been reported [35, 37–40]. The underlying assumptions in most cases are that the mechanical properties are invariant with respect to rotation around the axis perpendicular to film surface, and that the strain is continuously varying from the substrate to the film. Additionally, we assume that the grain size is smaller than the layer thickness and that mechanical strain does not influence electrical transport across grain boundaries. Following these assumptions, the average fractional change in resistivity is given as [37]

$$\left\langle \frac{\delta\rho_i'}{\rho_0} \right\rangle = \left\langle \sum_{j=1}^{6} m_{ij}'\varepsilon_j' \right\rangle = \sum_{j=1}^{6} \left\langle m_{ij}' \right\rangle \varepsilon_j' \qquad \text{with} \quad i = 1, 2.$$

(6.12)

Here, the $i = 1, 2$ refer to the respective longitudinal and transverse directions in the plane of the thin film and the primed notation is used to indicate the associated system of Cartesian coordinates. In (6.12), the m_{ij}' denote the elastoresistivity components (the elastoresistivity tensor being the product of piezoresistance and stiffness tensors) and ε_j' the strain components. To determine the average value of the elastoresistivity elements, $\langle m_{ij}' \rangle$, we need to compute the average value of the direction cosines associated with the transformation over all orientations. In the case of isotropically oriented grains, the averaging function reads [35, 37]

$$\langle f_{ij} \rangle = \frac{1}{8\pi^2} \int_0^{2\pi} \int_0^{\pi} \int_0^{2\pi} f_{ij}(\phi, \theta, \psi) \sin\theta \, d\phi \, d\theta \, d\psi \tag{6.13}$$

$$\text{with} \quad i, j = 1, 2, 3,$$

where f_{ij} is the function of direction cosines and ϕ, θ, and ψ denote Euler's angles. For an anisotropically oriented grain distribution, which is generally what is observed in X-ray diffraction measurements, the averaging is performed as [35, 37]

$$\langle f_{ij} \rangle = \frac{1}{2\pi} \int_0^{2\pi} f_{ij}(\phi_0, \theta_0, \psi) d\psi \qquad \text{with} \quad i, j = 1, 2, 3, \tag{6.14}$$

where ϕ_0 and θ_0 are defined by grain orientations. Following the averaging procedure on the m'_{ij} in (6.12), only three independent coefficients remain [35]

$$\frac{\delta\rho'_1}{\rho_0} = \langle m'_{11} \rangle \varepsilon'_1 + \langle m'_{12} \rangle \varepsilon'_2 + \langle m'_{13} \rangle \varepsilon'_3, \tag{6.15}$$

$$\frac{\delta\rho'_2}{\rho_0} = \langle m'_{12} \rangle \varepsilon'_1 + \langle m'_{11} \rangle \varepsilon'_2 + \langle m'_{13} \rangle \varepsilon'_3 \tag{6.16}$$

which can be approximated as

$$\langle m'_{12} \rangle \approx -2 \, m_{44} \langle f_{ik} \rangle \qquad \text{with} \quad k = 1, 2, 3 \tag{6.17}$$

for p-type poly-Si. The value of the elastoresistance coefficient m_{44} is given in Table 6.3 (see [41]) and the averaging function $\langle f_{ik} \rangle$ is given by (6.13) or (6.14).

We assume that the thin film has only two independent strain components; one longitudinal (ε_l) and one transverse (ε_t) with respect to the oblong film boundaries in the plane of the thin film. The fractional change in resistance can then be stated as [37]:

$$\frac{\delta R}{R} = G_l \varepsilon_l + G_t \varepsilon_t, \tag{6.18}$$

where $G_{l,t}$ are the longitudinal and transverse gauge factors, respectively. Here, the gauge factor is defined as $G = \delta R / (R\varepsilon)$. The gauge factors can be approximated as

$$G_{l,t} \approx \left(\frac{\rho_{\text{mono}}}{\rho_{\text{poly}}} \right) m_{44} \langle F_{l,t} \rangle \tag{6.19}$$

where $\langle F_{l,t} \rangle$ are functions of the averaging functions, (6.13) or (6.14), and

Table 6.3. *Values of longitudinal (G_l) and transverse (G_t) gauge factors. Unless otherwise indicated, values shown are for 300 K, films are low pressure chemical vapor deposited (LPCVD), p-type films are boron-doped, and gauge factor decreases with increasing doping. Abbreviations – TCG: temperature coefficient of gauge factor, TCR: temperature coefficient of resistance/resistivity, and ZMR: zone melt recrystallization*

Material	Gauge factor, G	Doping concentration/ temperature coefficient
p-Type poly-Si	$G_l \approx 50$, $G_t \approx 17$ [42, 43]	Approximately 10^{17} to 10^{19} cm^{-3}
	$G_{shear} \approx 9$ to 14 [44]	10^{19} to 10^{20} cm^{-3}, TCG $\approx (-0.25$ to $-0.15)\%/K$
	$G_l \approx 37$, $G_t \approx -9$ [39]	5×10^{18} to 10^{20} cm^{-3}, TCG $\approx (-1.7$ to $-2.3)\%/K$
	$G_l \approx 3$ to 40 [45]	10^{19} to 10^{20} cm^{-3}, TCG: $-0.1\%/K$, TCR: $0.1\%/K$
	$G_l \approx 30$ to 45 [35]	8×10^{18} to 10^{20} cm^{-3}, TCR $\approx (-0.03$ to $0.1)\%/K$
	$G_l = 55$ [41]	10^{17} cm^{-3}, $m_{44} = 27.5$
	$G_l = 42$ [41]	10^{18} cm^{-3}, $m_{44} = 21$, TCG: $-0.16\%/K$, TCR: $0.023\%/K$
	$G_l = 26$ [41]	10^{20} cm^{-3}, $m_{44} = 13$, TCG: $-0.15\%/K$, TCR: $0.11\%/K$
	$G_l \approx 22$ to 30 [46]	10^{18} to 10^{20} cm^{-3}, TCG: $\approx (-0.2$ to $-0.1)\%/K$
	$G_l \approx 22$, $G_t \approx -10$ [36, 40]	10^{20} cm^{-3}, TCG$_l$: $-0.06\%/K$, TCG$_t$: $-0.11\%/K$, TCR: $0.09\%/K$
	$G_l \approx 94$, $G_t \approx -75$ [47]	ZMR-LPCVD, 1.7×10^{18} cm^{-3}, TCR: $\approx 0\%/K$
	$G_l \approx 20$ to 32 [49]	2 to 4×10^{19} cm^{-3}
	$G_l \approx 20$ to 40 [49]	10^{19} to 10^{20} cm^{-3}
n-Type poly-Si	$G_l = -22$ [41]	10^{21} cm^{-3}, $m_{11} = -55$, TCG: $-0.011\%/K$, TCR: $0.11\%/K$
p-Type µC-Si : H	$G_l = 15-17$ [50]	Resistivity: 4Ω-cm, TCG: $< -0.03\%/K$, TCR: $0.01\%/K$
a-Si : H : F	$G_l = 33$, $G_t = -19$ [51]	
p-Type poly β-SiC	$G_l = 7$ to 20 [52]	Plasma-enhanced CVD at $600\,^\circ$C
n-Type β-SiC	$G_l = -31.8$ [53]	Resistivity: $0.7\,\Omega$-cm
n+β-SiC	$G_l = -26.6$ [53]	Resistivity: $0.02\,\Omega$-cm
p-Type poly diamond	$G_l = 6$ to 25 [54]	Hot filament CVD, resistivity: 5 to $30\,\Omega$-cm

the average Poisson's ratio (ν_a),

$$\langle F_l \rangle = 2[\langle f_{11} \rangle - \nu_a \langle f_{13} \rangle], \tag{6.20}$$

$$\langle F_t \rangle = 2[\langle f_{12} \rangle - \nu_a \langle f_{13} \rangle]. \tag{6.21}$$

The average Poisson's ratio can be computed as

$$\nu_a = -\left[\frac{C_{12} + (C_{11} - C_{12} - 2C_{44})\langle f_{13}\rangle}{C_{11} + (C_{11} - C_{12} - 2C_{44})\langle f_{33}\rangle}\right], \qquad (6.22)$$

where the C's denote components of the stiffness tensor for monocrystalline Si. In (6.19), the premultiplicative factor, ρ_{mono}/ρ_{poly}, denotes the ratio of resistivities in monocrystalline and polycrystalline material, which can be determined from resistance measurements of identical doping concentrations. The doping-dependence of m_{44} can be assumed similar to that of the piezoresistance coefficient. For isotropic grain orientation, the averaging factors $\langle F_l\rangle$ and $\langle F_t\rangle$ are 0.91 and -0.29, respectively [37]. Generally, the isotropic grain orientation approximation underestimates G_l and overestimates G_t. In the case of textured films with $\langle 110\rangle$ and $\langle 330\rangle$ as dominant orientations, the values are 1.1 and -0.27, respectively [35]. In all cases, as confirmed by measurement [35, 37], the longitudinal gauge factor is larger than the transverse component, $|G_l| > |G_t|$. The agreement of the model (6.19) with measurement is acceptable; the discrepancy is significant only for non-highly doped poly-Si films. This could be attributed to effects associated with strain-induced modulation of grain boundary transport [35, 37]. Measured values of the gauge factor for various polycrystalline materials are given in Table 6.3, along with temperature coefficients.

6.2 Strain and Electron Transport

In the presence of mechanical stress, the conduction band of most semiconductors (e.g., Si, Ge, or GaAs), to first order, only shifts in energy [7, 55]. Since shifts do not alter the shape of the conduction band, the effective masses (associated with the density of states and the carrier concentration) are unaltered. However, the concentration of electrons is redistributed if the conduction band is multi-valleyed (e.g., Si or Ge). Figs. 6.5 and 6.6 illustrate the constant energy surfaces of the conduction band in Si under conditions of zero-strain and when subject to an in-plane compressive strain (in the [100] and [010] directions), respectively.

6.2.1 Conduction Band

The conduction band edge in Si as well as Ge consists of a set of spheroidal energy shells located at equivalent positions in k-space. In the case of Si, six equivalent energy shells are located in the $\langle 100\rangle$ directions (with Ge, there are eight equivalent energy shells located in the $\langle 111\rangle$ directions). If

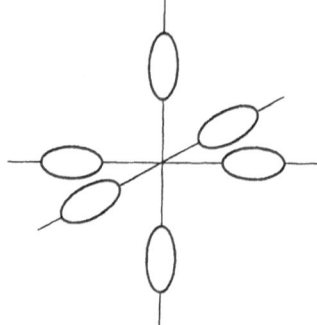

Fig. 6.5 Constant energy surfaces of the conduction band in unstrained Si

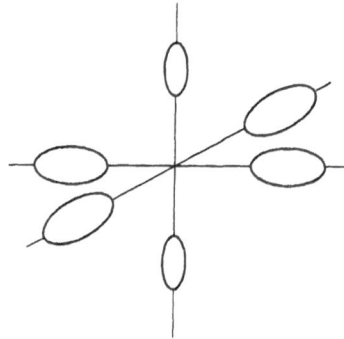

Fig. 6.6 Constant energy surfaces of the conduction band in Si subject to compressive strain in the [100] and [010] directions. These directions are associated with the larger ellipsoids

we choose the z-axis to be parallel to the major axis of the spheroid (see Fig. 6.7), and measure the wave vector from the center, the energy of each band can be described by [56]

$$E_c(\mathbf{k}) = \frac{h^2 k_z^2}{8\pi^2 m_l^*} + \frac{h^2(k_x^2 + k_y^2)}{8\pi^2 m_t^*}. \tag{6.23}$$

Here, m_l^* and m_t^* denote the longitudinal and transverse effective masses. In Si, $m_l^*/m_0 = 0.98$ and $m_t^*/m_0 = 0.19$, where m_0 is the free electron rest mass. Eq. (6.23) is valid for a wide range of $\mathbf{k}$, as verified by Dresselhaus, Kip, and Kittel [57]. Under strain, the bands shift in energy by an amount that is linearly related to the strain in the semiconductor. The shape of the bands, however, can be assumed unaltered. Using deformation potential theory, Herring and Vogt [55] obtained the shifts of each of the conduction

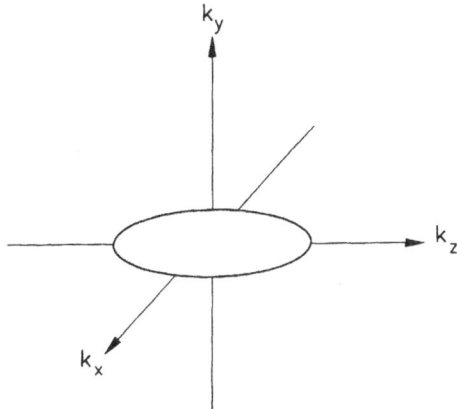

Fig. 6.7 Illustration of one conduction band on translated axis

band edges. In Si, these shifts can be described by:

$$\Delta E_c^{[100],[\bar{1}00]} = -\Xi_d(\varepsilon_{xx} + \varepsilon_{yy} + \varepsilon_{zz}) - \Xi_u \varepsilon_{xx}, \tag{6.24}$$

$$\Delta E_c^{[010],[0\bar{1}0]} = -\Xi_d(\varepsilon_{xx} + \varepsilon_{yy} + \varepsilon_{zz}) - \Xi_u \varepsilon_{yy}, \tag{6.25}$$

$$\Delta E_c^{[001],[00\bar{1}]} = -\Xi_d(\varepsilon_{xx} + \varepsilon_{yy} + \varepsilon_{zz}) - \Xi_u \varepsilon_{zz}. \tag{6.26}$$

Here, Ξ_d and Ξ_u denote the conduction band deformation potentials, and we have used the sign convention that the shifts are positive when the band edges move towards the gap. The former is a dilation potential involving only changes in volume of the strained material and the latter is a shear potential involving a uniaxial stretch along the major axis, which results in a symmetrical compression along the minor axis. In (6.24) through (6.26), the ε's denote strain components expressed in terms of Cartesian notation. The relation between strain and stress, following the generalized Hooke's law (see Sect. 7.2.1), reads for a cubic crystal

$$\varepsilon_{xx} = S_{xxxx}T_{xx} + S_{xxyy}(T_{yy} + T_{zz}), \tag{6.27}$$

$$\varepsilon_{yy} = S_{xxxx}T_{yy} + S_{xxyy}(T_{xx} + T_{zz}), \tag{6.28}$$

$$\varepsilon_{zz} = S_{xxxx}T_{zz} + S_{xxyy}(T_{xx} + T_{yy}), \tag{6.29}$$

$$\varepsilon_{xy} = S_{xyxy}T_{xy}, \tag{6.30}$$

$$\varepsilon_{yz} = S_{xyxy}T_{yz}, \tag{6.31}$$

$$\varepsilon_{xz} = S_{xyxy}T_{xz}, \tag{6.32}$$

here the S's denote compliance coefficients and the T's the stress components. Substituting (6.27) through (6.32) into (6.24) through (6.26), we obtain the shifts in energy of the conduction band in terms of the stress tensor,

$$\Delta E_c^{[100],[\bar{1}00]} = \alpha T_{xx} + \beta(T_{yy} + T_{zz}), \tag{6.33}$$

$$\Delta E_c^{[010],[0\bar{1}0]} = \alpha T_{yy} + \beta(T_{xx} + T_{zz}), \tag{6.34}$$

$$\Delta E_c^{[001],[00\bar{1}]} = \alpha T_{zz} + \beta(T_{xx} + T_{yy}). \tag{6.35}$$

Here,

$$\alpha = -\Xi_d(S_{xxxx} + 2S_{xxyy}) - \Xi_u S_{xxxx}, \tag{6.36}$$

$$\beta = -\Xi_d(S_{xxxx} + 2S_{xxyy}) - \Xi_u S_{xxyy}. \tag{6.37}$$

The values of α and β, with $\Xi_d = 0.412\,\text{eV}$, $\Xi_u = 9.2\,\text{eV}$, $S_{xxxx} = 7.69 \times 10^{-13}\,\text{cm}^2/\text{dyne}$, $S_{xxyy} = -2.24 \times 10^{-13}\,\text{cm}^2/\text{dyne}$, and $S_{xyxy} = 1.25 \times 10^{-12}\,\text{cm}^2/\text{dyne}$, are $\alpha = -7.21 \times 10^{-12}\,\text{eV}\,\text{cm}^2/\text{dyne}$ and $\beta = 1.83 \times 10^{-12}\,\text{eV}\,\text{cm}^2/\text{dyne}$.

Thus for a given stress distribution, we can compute the position of the conduction band edge, relative to the unstrained position, through relations (6.33) to (6.37). Similar relations for the band shifts can also be obtained for Ge (see [7]). But here, the eight equivalent valleys are located in the $\langle 111 \rangle$ directions.

In the above relations, we have assumed only first order deformation effects, whereby the bands only shift but do not distort. A second order perturbation analysis [7] shows that the contribution to band distortion is negligible and can be ignored for all practical purposes. For example, at a stress of $1 \times 10^9\,\text{dyne}/\text{cm}^2$, the change in m_t^* and m_l^* is a mere 0.1%. However, the shifts in energy can be significant. For example, with a torsional strain of $\varepsilon_{yz} = 0.01$, the resulting energy shift in the [100] direction is 32 meV.

Using the energy spectrum given in Eq. (6.23), the total density of states for $E > E_c$, can be determined as [58]:

$$g_c(E) = \alpha \frac{(2m_n^*)^{3/2}}{h^3} \sum_{i=1}^{6} \left[E - E_c^{(i)} \right]^{1/2}. \tag{6.38}$$

Here, α is a parameter related to the shape of the band structure and it is approximately $3\pi/4$. For a spherical parabolic band structure, $\alpha = 2\pi/3$, which yields the ideal (classical) density of states. The superscript (i) in (6.38) denotes the ith conduction band valley and $E_c^{(i)}$ the associated band edge. In the unstrained case, all the conduction band valleys are in line (see Fig. 6.5), hence reducing the density of states to the form (see (2.21)):

$$g_c(E) = 6\alpha \frac{(2m_n^*)^{3/2}}{h^3} (E - E_c)^{1/2}. \tag{6.39}$$

6.2.2 Electron Mobility and Piezoresistance

Although there is no distortion in the conduction band valleys, the redistribution of electrons associated with band shifts gives rise to anisotropic conductivity, which can be viewed as an anisotropy in mobility. To evaluate the electron mobility components, we take into account the shifts in band edges through the electron concentration. Here, the total mobility is given by the weighted average of the unstrained electron mobility (since there is no distortion) of the ith band with the corresponding strained electron concentration [58], $viz.$,

$$\mu_{n,\text{str}} = \frac{\sum_{i=1}^{6} n_{\text{str}}^{(i)} \mu_{n,\text{unstr}}^{(i)}}{\sum_{i=1}^{6} n_{\text{str}}^{(i)}}. \tag{6.40}$$

Here, μ_n denotes the electron mobility, n the electron concentration, the subscripts (str, unstr) refer to the associated (strained, unstrained) condition, and the superscript (i) refers to the different valleys. The unstrained mobility in the various valleys take the following form:

$$\mu_{n,\text{unstr}}^{[100],[\bar{1}00]} = \begin{pmatrix} \mu_l & 0 & 0 \\ 0 & \mu_t & 0 \\ 0 & 0 & \mu_t \end{pmatrix}, \tag{6.41}$$

$$\mu_{n,\text{unstr}}^{[010],[0\bar{1}0]} = \begin{pmatrix} \mu_t & 0 & 0 \\ 0 & \mu_l & 0 \\ 0 & 0 & \mu_t \end{pmatrix}, \tag{6.42}$$

$$\mu_{n,\text{unstr}}^{[001],[00\bar{1}]} = \begin{pmatrix} \mu_t & 0 & 0 \\ 0 & \mu_t & 0 \\ 0 & 0 & \mu_l \end{pmatrix}, \tag{6.43}$$

where μ_l and μ_t denote the respective mobility components along the major and minor axes in each ellipsoid.

The strained electron concentration, for non-degenerate doping concentrations, can be computed using Maxwell-Boltzmann statistics

$$n_{\text{str}}^{(i)} = \frac{N_c}{6} \exp\left[\frac{E_F - E_c^{(i)}}{kT}\right]. \tag{6.44}$$

Here, N_c is the effective density of states for the conduction band, see (2.27). The ith valley measured with respect to the conduction band edge yields

$$n_{\text{str}}^{(i)} = \frac{n_{\text{unstr}}}{6} \exp\left[\Delta E_c^{(i)}/kT\right], \tag{6.45}$$

where n_{unstr} denotes the unstrained electron concentration. Equations (6.44) and (6.45) lead to the following form for the strained mobility

$$\mu_{n,\text{str}} = \frac{\sum_{i=1}^{6} \mu_{n,\text{unstr}}^{(i)} \exp\left[\Delta E_c^{(i)}/kT\right]}{\sum_{i=1}^{6} \exp\left[\Delta E_c^{(i)}/kT\right]}. \tag{6.46}$$

In the case of compressive strain in the [100] and [010] directions (see Fig. 6.6), the electron population in the [001] valley decreases (from the unstrained value of $n_{\text{unstr}}/3$) as $\exp(-\Delta E/kT)/[\exp(-\Delta E/kT) + 2]$ with increasing strain. Here, ΔE (which is a positive number) denotes the strain-induced energy shift between the [001] and the [100] or [010] valleys. Since the [001] valleys move further away from the Fermi level, the effective density of states can be expressed as

$$N_{c,\text{str}} = \frac{N_{c,\text{unstr}}}{3} \left[\frac{3\exp[-\Delta E/kT]}{2 + \exp[-\Delta E/kT]} + 2 \right]. \tag{6.47}$$

Denoting the [100], [010], and [001] directions as x, y, and z respectively, Eq. (6.46) along with (6.33) to (6.35), yield the following expressions for the mobility components [58]

$$\mu_{n,xx} = \frac{(\mu_t + \mu_l) \exp\left[\Delta E^x/kT\right] + \mu_t \exp\left[\Delta E^z/kT\right]}{2\exp\left[\Delta E^x/kT\right] + \exp\left[\Delta E^z/kT\right]} = \mu_{n,yy}, \tag{6.48}$$

$$\mu_{n,zz} = \frac{2\mu_t \exp\left[\Delta E^x/kT\right] + \mu_l \exp\left[\Delta E^z/kT\right]}{2\exp\left[\Delta E^x/kT\right] + \exp\left[\Delta E^z/kT\right]}, \tag{6.49}$$

where the in-plane components are equal. In the absence of strain, $\Delta E = 0$ and we recover the well-known descriptions for the mobility (see, e.g., [24]), viz.,

$$\mu_{n,xx} = \frac{(\mu_t + \mu_l) + \mu_t}{3} = \frac{2\mu_t + \mu_l}{3} = \mu_{n,yy}, \tag{6.50}$$

$$\mu_{n,zz} = \frac{2\mu_t + \mu_l}{3} \tag{6.51}$$

to yield identical values for the components, $\mu_{n,xx} = \mu_{n,yy} = \mu_{n,zz}$, thus recovering its scalar form.

The variation of mobility with stress can be related to, or also obtained from, the piezoresistance coefficients measured by Smith [2]. Following intuitive reasoning, the unstrained conductivity,

$$\sigma_n = \sum_{i=1}^{6} \sigma_n^{(i)} = \sum_{i=1}^{6} qn^{(i)} \mu_n^{(i)} \tag{6.52}$$

gets modified in the presence of strain as [56]

$$\Delta\sigma_n = \sum_{i=1}^{6} q\Delta n^{(i)}\mu_n^{(i)} + \sum_{i=1}^{6} qn^{(i)}\Delta\mu_n^{(i)}. \tag{6.53}$$

Assuming strain does not induce generation-recombination centers, then all electrons in the material participate in the conduction process. Thus there is no change in total concentration, $\Delta n = 0$, and relation (6.3), by virtue of (6.53), can be stated as

$$\frac{\Delta\sigma_{nij}}{\sigma_{n0}} = \frac{\Delta\mu_{nij}}{\mu_{n0}} = -\sum_{k,l=1}^{3} \pi_{ijkl}T_{kl}. \tag{6.54}$$

Thus relation (6.54) now allows incorporation of strain effects on electron transport in the drift-diffusion model whereby the tensor description for the mobility components is in terms of the piezoresistance coefficients. The resulting model is given in Sect. 6.7.

6.3 Strain and Hole Transport

Unlike the conduction band, the effects of strain on the valence band are very different due to its degenerate nature. Here, not only does the strain shift the bands, but it also reduces the symmetry of the band structure; the distortion in bands leads to a change in effective mass and mobility. In view of the multitude of different effects, it is clear why the piezoresistance effect is more significant in p-type materials than in n-type materials, as evident from the maximum value of the associated piezoresistance coefficients.

The valence band structure can be computed using various approaches [23]: the pseudo-potential method [59], the linear coupled atomical orbital method [60], the free electron approximation method, and the $\mathbf{k} \cdot \mathbf{p}$ perturbation method [61]. Each of the methods varies in terms of simplicity, accuracy, and computational resources. The most powerful approach is the $\mathbf{k} \cdot \mathbf{p}$ perturbation method, also referred to as the effective mass method, since it allows incorporation of spin-orbit coupling effects without much difficulty [62]. In what follows, we present an analysis of the strained valence band structure, applicable to group-IV semiconductors, using $\mathbf{k} \cdot \mathbf{p}$ perturbation coupled with deformation potential theory [23].

6.3.1 Valence Band

The valence band structure of an unstrained group-IV semiconductor was first discussed by Shockley [63], whose solution ignored spin-orbit coupling. Subsequently, Dresselhaus, Kip, and Kittel [57] included spin orbit interactions and obtained a 6×6 Hamiltonian whose eigenvalues revealed the energy spectrum $E(\mathbf{k})$ of the valence band. However, the eigenvalues of the Hamiltonian had to be computed numerically, which could turn out to be very tedious when computing the transport parameters. Kane [62] obtained a secular equation describing the eigenvalues of the 6×6 Hamiltonian, thus making it easier to compute the valence band spectrum.

The effects of strain on the valence band were first taken into account by Pikus and Bir [61], who obtained a closed-form expression for the valence band spectrum under strain. Their expression, however, neglected the effects of spin-orbit coupling. Spin-orbit coupling is particularly important for Si and Ge since the spin-orbit band is relatively close ($\sim 50\,\text{meV}$) to the heavy and light hole bands at $\mathbf{k} = 0$, and hence, can affect the valence band considerably in the presence of strain. Only recently, the effects of spin-orbit coupling were incorporated to yield an analytical expression for the strained valence band structure [23]. Here, the formalism is based along the lines adopted in [57, 62] with spin-orbit effects included as a perturbation. The resulting expression that governs the strained valence band energy spectrum takes the following form:

$$\sum_{i=0}^{3} \sum_{j=0}^{3-i} a_{ij} E^{j}(\mathbf{k}) k^{2i} = 0, \tag{6.55}$$

where the components a_{ij} are functions of deformation potentials, strain elements (ε_{ij}), and band parameters (see [23]). Eq. (6.55) yields three distinct values of $E(\mathbf{k})$ for a particular value of $\mathbf{k}$, each of which corresponds to the heavy hole (HH), light hole (LH), or the spin-orbit (SO) bands. The computed band structure is shown in Figs. 6.8 and 6.9 [64]. Fig. 6.8 illustrates the energy spectra (E-k) in the [100] and [111] directions for the HH band at zero stress, uniaxial stress (10^{9} dynes/cm^{2}), and shear stress (10^{9} dynes/cm^{2}). As expected, the stress introduces significant asymmetry and the bands become non-degenerate at $\mathbf{k} = 0$; the HH band moves upwards and the LH band, although not shown, moves downwards. Here, the shift associated with the latter is less pronounced implying a larger proportion of the hole concentration in the HH band. The constant energy surfaces, taken at $-15\,\text{meV}$ from the top of the HH band, at zero stress and shear stress of 10^{9} dynes/cm^{2} are shown in Fig. 6.9. The band distortion with shear stress is the largest. As we will see in Sect. 6.3.2, this leads to the largest change in mobility, which in turn yields the largest piezoresistance coefficient (π_{44}).

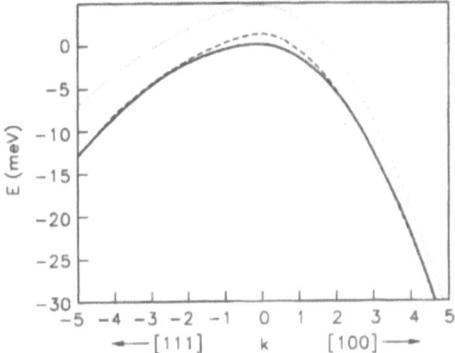

Fig. 6.8 Energy spectra for the heavy hole (HH) band at zero stress (solid line), uniaxial stress (dashed line) and shear stress (dotted line). The energy (E) is in meV and the wave vector ($\mathbf{k}$) in 10^8/m

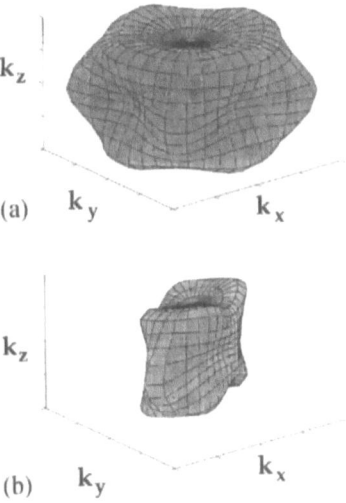

Fig. 6.9 Constant energy surface of HH band in $\mathbf{k}$-space (a.u.): (a) at zero stress (b) under shear stress

The splitting of the bands can be computed by setting $\mathbf{k} = 0$ in (6.55), to yield the reduced form

$$\sum_{i=0}^{3} a_i E^i = 0,$$ (6.56)

where the stress-dependent coefficients a_i read as follows,

$$a_0 = \frac{\Delta}{3}(pq + pr + qr - N^2 T_c^2)$$
$$+ N^2(2N T_{xy} T_{xz} T_{yz} - p T_{yz}^2 - q T_{xz}^2 - r T_{xy}^2) + pqr, \tag{6.57}$$

$$a_1 = \frac{2\Delta}{3}(p + q + r) - (pq + pr + qr - n^2 T_c^2), \tag{6.58}$$

$$a_2 = p + q + r - \Delta, \tag{6.59}$$

$$a_3 = -1 \tag{6.60}$$

with

$$T_c^2 = T_{xy}^2 + T_{xz}^2 + T_{yz}^2, \tag{6.61}$$

$$p = L T_{xx} + M T_{yy} + M T_{zz}, \tag{6.62}$$

$$q = M T_{xx} + L T_{yy} + M T_{zz}, \tag{6.63}$$

$$r = M T_{xx} + M T_{yy} + L T_{zz}. \tag{6.64}$$

Here, the T's denote stress components, Δ, L, M, N are band parameters, which take on values as follows: $\Delta = 44\,\text{meV}$, $L = 1.25 \times 10^{-12}\,\text{eV cm}^2/$ dyne, $M = 3.76 \times 10^{-13}\,\text{eV cm}^2/\text{dyne}$, and $N = -7.36 \times 10^{-12}\,\text{eV cm}^2/$ dyne. In (6.56), the three independent solutions of E correspond to the HH, LH, and SO band shifts. For any configuration of stress, the HH band moves upwards (i.e., towards the conduction band) and the LH, downwards. Hence, the valence band edge is defined by that of the HH band whose contribution to conduction is also the highest. The band shifts at $\mathbf{k} = 0$ are illustrated in Fig. 6.10.

6.3.2 Hole Mobility and Piezoresistance

Because of the strain induced distortion in band structure, the mobility is a tensor. Unlike the conduction band, the reduction in symmetry of the valence band does not provide any straightforward or analytical means of determining the hole mobility. The strained mobility components have to be calculated using a model derived from the Boltzmann transport equation. For not too high electric fields, the mobility components are computed as [7]

$$\mu_{pij} = \frac{\dfrac{4\pi^2 q}{h^2}\displaystyle\int d^3 k \tau(E) \frac{\partial E}{\partial k_i} \frac{\partial E}{\partial k_j} \frac{\partial f}{\partial E}}{\displaystyle\int d^3 k f(E, E_F, T)}, \tag{6.65}$$

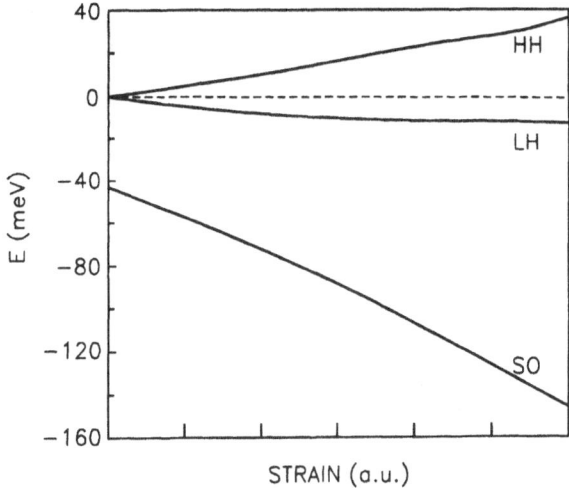

Fig. 6.10 Strain induced energy shifts of the HH, LH and SO bands at $\mathbf{k} = 0$

where f is the Boltzmann distribution, E_F denotes the Fermi energy, T the temperature, and E the energy spectrum of the valence band as described by Eq. (6.55). In (6.65), $\tau(E)$ denotes the energy dependent relaxation time which stems from the sum of the various scattering rates associated with acoustic, optical, and impurity scattering mechanisms. In computing the relaxation time, we assume (see Sect. 2.3.5): (i) optical scattering occurs inelastically, (ii) acoustic scattering is perfectly elastic, (iii) the holes scatter isotropically, (iv) the scattering mechanisms only change due to the distortion in the band structure, and (v) hole degeneracy effects are negligible to yield insignificant hole-hole scattering. The mobilities are computed for each individual band (i.e., HH, LH, and SO bands) and a weighted-average is performed to obtain the total (i.e., measured) mobility [7]. Here, since the heavy hole band shifts upwards in energy, we should expect that its contribution to the total mobility will be the greatest.

The denominator in (6.65) is directly related to the hole concentration, *viz.,*

$$p = \frac{1}{4\pi^3} \int d^3k f(E, E_F, T) \tag{6.66}$$

which upon evaluation is reduced to its usual form,

$$p = N_v \exp\left[\frac{E_v - E_F}{kT}\right]. \tag{6.67}$$

Since the effective density of states N_v decreases with increasing strain, we expect the mobility to increase by virtue of (6.65).

Table 6.4. *Comparison between measured [2] and calcu-lated values of piezoresistance coefficients in p-type Si. Values are in units of* $10^{-11}\,Pa^{-1}$

Coefficient	Measured	Calculated
π_{11}	$+6.6$	$+8$
π_{12}	-1.1	-2
π_{44}	$+138.1$	$+140$

Table 6.5. *Comparison between measured [2] and calcu-lated values of piezoresistance coefficients in p-type Ge. Values are in units of* $10^{-11}\,Pa^{-1}$

Coefficient	Measured	Calculated
π_{11}	-10.6	-11
π_{12}	$+5$	$+6$
π_{44}	$+96.7$	$+100$

The mobility components computed using (6.65) can be related to the piezoresistance coefficients for the p-type material. Following the same arguments as before, i.e., that the global carrier concentration does not change with stress, $\Delta p = 0$, the strained mobilities can be expressed in terms of the piezoresistance coefficients [56, 64], *viz.*,

$$\frac{\Delta\mu_{pij}}{\mu_{p0}} = -\sum_{k,l=1}^{3} \pi_{ijkl} T_{kl}. \tag{6.68}$$

Here, $\Delta\mu_{pij}/\mu_{p0}$ is the fractional change in hole mobility and the rest in the usual notation (see Sect. 6.2.2). Relation (6.68), as with Eq. (6.3), is valid for not too large stress whose distribution is uniform over the region of interest. Relation (6.68) now allows us to incorporate the effects of strain on hole transport in the drift-diffusion model, whereby the tensorial mobility components can now be described in terms of the piezoresistance coefficients. A comparison of calculated piezoresistance coefficients [7], using relations (6.65) and (6.68), with measured values [2] is given in Tables 6.4 and 6.5, respectively. The calculated values are consistently larger in magnitude, which can be attributed to numerical errors arising from evaluation of the integrals over the entire (whole) energy surface required in order to compute the difference in mobility, $\Delta\mu_p$. In the absence of stress, the computations are simplified. Here, the integrals need to be evaluated over just $1/8$ of the total energy surface due to symmetry in the band structure [64].

We now provide a qualitative analysis of the stress induced variation in mobility components, governed by relation (6.68), by inspection of

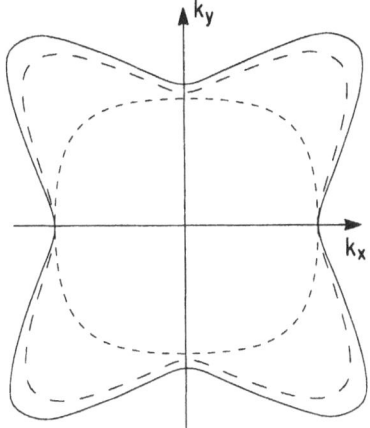

Fig. 6.11 The constant energy surfaces in a normalized $k_x - k_y$ plane of the HH band for compressively strained Si. The solid line is for unstrained Si, the large dashed line is for $T_{xx} = 1 \times 10^9$ dynes/cm^2, and the small dashed line is for $T_{xx} = 1 \times 10^{10}$ dynes/cm^2. The value of the constant energy surface is taken to be -10 meV from the top of the corresponding band

constant energy surfaces associated with the HH and LH valence bands in Si [23]. The constant energy surfaces, illustrated in Figs. 6.11 to 6.13, follow from the solution of Eq. (6.55). Here, the values of the needed strain components are computed using the generalized Hooke's law in accordance to relations (6.27) through (6.32). We begin with stating relations (6.68) in a more intuitive form:

$$\frac{\Delta\mu_{pxx}}{\mu_{p0}} = -\pi_{xxxx}T_{xx} - \pi_{xxyy}(T_{yy} + T_{zz}), \qquad (6.69)$$

$$\frac{\Delta\mu_{pyy}}{\mu_{p0}} = -\pi_{xxxx}T_{yy} - \pi_{xxyy}(T_{xx} + T_{zz}), \qquad (6.70)$$

$$\frac{\Delta\mu_{pzz}}{\mu_{p0}} = -\pi_{xxxx}T_{zz} - \pi_{xxyy}(T_{xx} + T_{yy}), \qquad (6.71)$$

$$\frac{\Delta\mu_{pxy}}{\mu_{p0}} = -\pi_{xyxy}T_{xy}, \qquad (6.72)$$

$$\frac{\Delta\mu_{pyz}}{\mu_{p0}} = -\pi_{xyxy}T_{yz}, \qquad (6.73)$$

$$\frac{\Delta\mu_{pxz}}{\mu_{p0}} = -\pi_{xyxy}T_{xz}. \qquad (6.74)$$

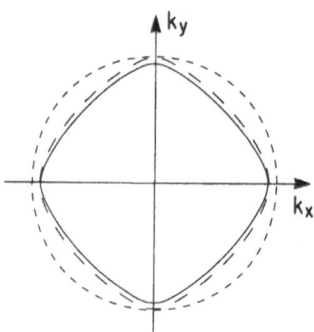

Fig. 6.12 As Fig. 6.11 but for the LH band

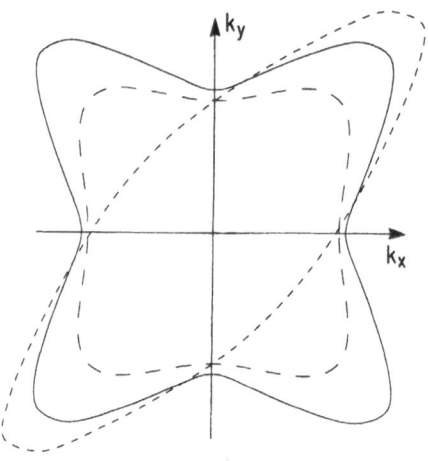

Fig. 6.13 The constant energy surfaces in the $k_x - k_y$ plane of the HH band under shear stress. The solid line is for unstrained Si, the small dashed line is for $T_{xy} = 1 \times 10^9$ dynes/cm^2, and the large dashed line is for $T_{yz} = 1 \times 10^9$ dynes/cm^2

We consider the case in which $T_{xx} \neq 0$ and $T_j = 0$ for $j \neq xx$. Equations (6.69) and (6.70) yield

$$\frac{\Delta \mu_{pxx}}{\mu_{p0}} = -\pi_{xxxx} T_{xx}, \tag{6.75}$$

$$\frac{\Delta \mu_{pyy}}{\mu_{p0}} = -\pi_{xxyy} T_{xx} = \frac{\Delta \mu_{pzz}}{\mu_{p0}}, \tag{6.76}$$

where for convenience, we restate the values of the piezoresistance coefficients in Si: $\pi_{xxxx} = 6.6 \times 10^{-12}$ cm^2/dyne and $\pi_{xxyy} = -1.1 \times 10^{-12}$ cm^2/dyne. Relations (6.75) and (6.76) suggest that the mobility component in the x-direction decreases when uniaxially stressed whereas the components in the y- and z-directions increase. Indeed, we can predict the same result intuitively by inspection of the HH and LH energy contours given in Figs. 6.11 and 6.12. These energy contours are taken at $-10\,\text{meV}$ from the top of the corresponding band for a uniaxial stress of $T_{xx} = 10^9$ and $10^{10}\,\text{dynes/cm}^2$. Since the mobility is inversely proportional to the tensorial effective mass, $1/m_{ij} = (4\pi^2/h^2)(\partial^2 E/\partial k_i \partial k_j)$, we see that the mobility for $T_{xx} = 10^9\,\text{dynes/cm}^2$ remains approximately the same for the HH band but becomes smaller for the LH band. Consequently, the overall mobility in the x-direction decreases; see $\Delta\mu_{pxx}$ in (6.75) and note that π_{xxxx} is positive. The mobility in the y- (or z-) direction appears to have increased for the HH band, but a decrease is noted for the LH band. Therefore, the overall mobility in the y- (or z-) direction increases since the HH band contributes much more to the conduction process; see $\Delta\mu_{pyy}$ or $\Delta\mu_{pzz}$ in (6.76) noting that π_{xxyy} is negative.

We now look at the case in which we apply a shear stress component, $T_{xy} \neq 0$ and $T_j = 0$ for $j \neq xy$. In this case, we shall only consider the HH band (see Fig. 6.13) since its contribution to the conduction process is most significant. Equation (6.72), which has a shear piezoresistance coefficient, $\pi_{xyxy} = 138.1 \times 10^{-12}\,\text{cm}^2/\text{dyne}$, implies that if we apply an electric field E_x to the piezoresistor, there would also be a current component in the y-direction, but in the negative $(-y)$ direction since π_{xyxy} is positive. This can also be observed in the energy contours under shear stress of 10^9 and $10^{10}\,\text{dynes/cm}^2$ illustrated in Fig. 6.13. Here, an electric field E_x will result in a hole velocity that points approximately $-45°$ from the k_x axis (see Fig. 6.14), recalling that $\mathbf{v} = 2\pi/h\ \text{grad}\ E(\mathbf{k})$. Resolving the velocity into its x and y components, it is clear that hole movement takes place in the positive x- and negative y- directions. Note also that the energy contours are much more distorted compared to the uniaxial stress configuration

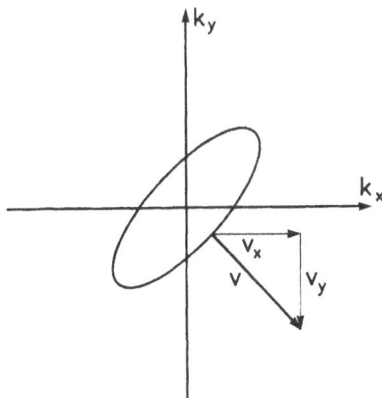

Fig. 6.14 An illustration of the velocity components of carriers with a shear stress component T_{xy} under an electric field in the x-direction

shown in Fig. 6.11. This is also reflected in the magnitude of the piezoresistance coefficients in the two cases: $|\pi_{xyxy}| \gg |\pi_{xxxx}|$, $|\pi_{xxyy}|$. Also in Fig. 6.13, we have superimposed an energy contour for shear stress in the x-z plane. According to (6.72), which follows from a linear expansion, there is no predicted change in the mobility, although a change is clearly evident in the energy contour. Although not quantitatively discussed here, this change only manifests itself when $T_{xz} > 10^9$ dynes/cm^2. This can be attributed to higher order effects not included in the linear expansion, (6.68) [23].

It is clear from the above analysis, and in particular, with the comparisons given in Tables 6.4 and 6.5, that although mechanical strain gives rise to energy shifts of the HH and LH bands, their associated contribution to piezoresistivity in the p-type material is very small. The main contribution to piezoresistivity stems from strain induced distortions in the valence band structure.

An important parameter that is strongly related to the band structure is the carrier concentration (CC) effective mass m_{cc}^*. The CC effective mass, which is of particular importance in analysis and simulation of strained semiconductor devices, provides a convenient description of the mass of the carrier (holes in this case) needed to produce the correct effective density of states N_v, by virtue of the ideal free-carrier relationship. Hence, it is a function of temperature as well as Fermi level. Its dependence on the latter becomes insignificant in the case of a non-degenerate semiconductor. The CC effective mass can be defined as follows [23]:

$$\left(\frac{m_{cc}^*}{m}\right)^{3/2} = \frac{\int g_v(E)f(E)dE}{\int g_{v0}(E)f(E)dE},$$

(6.77)

where $f(E)$ is the Boltzmann distribution at energy E (assuming non-degenerately doped semiconductor), $g_{v0}(E)$ the ideal density of states which follows a $E^{1/2}$ dependence, and $g_v(E)$ the density of states obtained from the band structure equation, (6.55). Since the band structure is a function of the applied stress (T), so is the effective mass, i.e., $m_{cc}^* = m_{cc}^*(T)$. Following a linear expansion in stress terms, the stress dependent CC effective mass can be written as

$$\left(\frac{m_{cc}^*}{m_{cc,0}^*}\right) = \sum_{i,j=0}^{3} \Lambda_{ij} T_{ij},$$

(6.78)

where

$$\Lambda_{ij} = \frac{1}{m_{cc,0}^*} \frac{\partial m_{cc}^*}{\partial T_{ij}}.$$

(6.79)

In materials with cubic symmetry (e.g., Si), Eq. (6.78) can be stated in the

Table 6.6. *Values of the effective mass coefficient Λ_{xx} and Λ_{xy} (units: $10^{-12}\,cm^2/dyne$) for various magnitudes of uniaxial (T_{xx}) and shear (T_{xy}) stress (units: $dynes/cm^2$), respectively. N_v denotes the density of states in the valence band (units: $10^{19}\,cm^{-3}$)*

Stress orientation, $T_{\alpha\alpha}$	N_v	$\dfrac{\Delta m_{cc}^*}{m_{cc,0}^*}$	$\Lambda_{\alpha\alpha} = \dfrac{\Delta m_{cc}^*}{m_{cc,0}^* T_{\alpha\alpha}}$
$T_{xx} = 0.00$	3.100 48		
$T_{xx} = 10^7$	3.099 34	-2.44×10^{-4}	-24.4
$T_{xx} = 10^8$	3.089 08	-2.45×10^{-3}	-24.5
$T_{xx} = 10^9$	2.985 17	-2.49×10^{-2}	-24.9
$T_{xy} = 0.00$	3.100 48		
$T_{xy} = 10^7$	3.095 51	-1.07×10^{-3}	-107
$T_{xy} = 10^8$	3.046 93	-1.14×10^{-2}	-114
$T_{xy} = 10^9$	2.553 76	-1.22×10^{-1}	-122

following reduced form

$$\left(\frac{m_{cc}^*}{m_{cc,0}^*}\right) = \Lambda_{xx}(T_{xx} + T_{yy} + T_{zz}) + \Lambda_{xy}(T_{xy} + T_{yz} + T_{xz}), \quad (6.80)$$

where we have only two independent components, *viz.*, Λ_{xx} and Λ_{xy}. Values for these coefficients are approximately -24.7×10^{-12} and $-114 \times 10^{-12}\,cm^2/dyne$, respectively, at $300\,K$ (see Table 6.6). The coefficients take into account the strain induced distortion in the band structure and associated energy shifts. As expected, the value of the shear component is larger, in fact, Λ_{xy} is approximately four times larger in magnitude than the component Λ_{xx}.

6.4 Piezojunction Effect

Strain induced energy shifts associated with the conduction and valence bands, discussed in Sects. 6.2 and 6.3, respectively, lead to variations in the band gap energy, E_g, which can be exploited in *p-n* junction based devices. Here, the dependence of E_g on stress gives rise to a modulation of the intrinsic carrier concentration and hence, in the minority carrier injection, leading to a modification of the output characteristics of diodes or transistors (see [1, 65–68]). However, the change in output characteristics is only significant when the stress almost reaches the fracture limit of the semiconductor material. The intrinsic carrier concentration n_{ie} depends exponentially on band gap energy. Any strain induced variations in the latter can lead to an appreciable change in n_{ie}. The strain dependence of n_{ie} is modeled along the same lines discussed in Sect. 2.3.3, in association with treatment of heavy doping effects on the band gap energy. Assuming that the strain does not affect the ionization energy of dopants, the majority carrier concentrations remain independent of strain, and the strained intrinsic carrier concentration $n_{ie,\text{str}}$ can be modeled as

$$n_{ie,\text{str}} = n_{ie,0} \exp\left[\Delta E_g / 2kT\right]. \tag{6.81}$$

Here, $n_{ie,0}$ denotes the effective intrinsic carrier concentration taking into account effects of temperature and band gap narrowing. Following our discussion on the strained band structure in Sects. 6.2 and 6.3, the change in gap energy ΔE_g can, to first order, be assumed proportional to strain. Here, ΔE_g comprises the conduction (ΔE_c) and valence (ΔE_v) band components, viz., $\Delta E_g = \Delta E_c + \Delta E_v$. Both components are positive quantities; here, positive $\Delta E_{c,v}$ imply movement of bands toward each other.

The conduction band shifts can be computed using relations (6.24) to (6.26) and the shifts associated with the valence band can be obtained through solution of (6.56). With the latter, we need to consider only the shift of the heavy hole band due to its relative dominance in hole conduction.

Following relation (6.81) for the intrinsic carrier concentration, we can state the relations for the strain-dependent minority carrier concentrations, $n_{p,\text{str}}$, $p_{n,\text{str}}$. For a non-degenerately doped p-n junction, Boltzmann's injection law yields (see [69]),

$$\frac{n_{p,\text{str}}}{n_{p,0}} = \frac{p_{n,\text{str}}}{p_{n,0}} = \exp\left(\Delta E_g / kT\right), \tag{6.82}$$

where $n_{p,0}$ and $p_{n,0}$ denote the minority carrier concentrations in the p- and n-type material, respectively, under conditions of zero strain.

So far we have assumed a strain (or stress) distribution that is spatially invariant over the region of interest. Spatial variations in strain give rise to a position-dependent band structure, and hence, gap energy $E_g(r)$, due to $\Delta E_c(r)$ and $\Delta E_v(r)$. If $\Delta E_c(r) \neq \Delta E_v(r)$, we need to employ a modified form of the current density relations to account for the additional electric field stemming from $E_g(r)$. Variations in gap energy also give rise to additional second order effects. These considerations, along with model equations and effects, will be treated in what follows.

6.5 Effects of Stress Gradients

Non-uniformities in the stress distribution are typical in most micro-mechanical structures. For example, in the micromechanical (cantilever) structure shown in Fig. 6.15, there are stress gradients over most of the active sensing region. For maximum sensitivity, the piezoresistor is

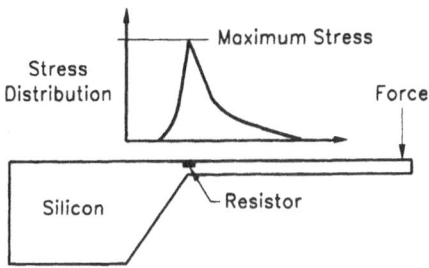

Fig. 6.15 Stress distribution in a micromechanical cantilever structure

generally located at the region where the stress is largest, which is also where the gradient is the largest.

When a device is subject to a stress gradient, a non-uniform band structure results, which gives rise to two components of the electrical current density [56]: one due to a bending of the band edges and the other arising from a non-uniform distribution of the density of states. These effects and coefficients will be described in what follows. We begin with the generalized transport equation (see [70]), which accounts for the position dependent band structure derived using a first-order perturbation solution of the Boltzmann transport equation. The resulting generalized descriptions for the electron and hole transport equations can be stated as

$$\mathbf{J_n} = -q\mu_n n \operatorname{grad}\left[\psi - \Delta E_c/q\right] + q D_n \operatorname{grad} n - q D_n \operatorname{grad}^* n, \quad (6.83)$$

$$\mathbf{J_p} = -q\mu_p p \operatorname{grad}\left[\psi + \Delta E_v/q\right] - q D_p \operatorname{grad} p + q D_p \operatorname{grad}^* p. \quad (6.84)$$

Here, $\mathbf{J_{n,p}}$ denote the respective electron and hole current densities, $\mu_{n,p}$ the drift mobilities, n and p the electron and hole concentrations, $\Delta E_{c,v}$ the position dependent conduction and valence band shifts, and $D_{n,p}$ the diffusion coefficients. The drift mobility and diffusion coefficient adopt tensor descriptions in the presence of strain. The last term in (6.83) and (6.84) can be defined as [56]

$$\operatorname{grad}^* n \equiv \int_{E_c}^{\infty} dE f(E - E_c, E_F, T) \left[\operatorname{grad} g_c\right]_{E - E_c}, \quad (6.85)$$

$$\operatorname{grad}^* p \equiv \int_{-\infty}^{E_v} dE \left[1 - f(E_v - E, E_F, T)\right] \left[\operatorname{grad} g_v\right]_{E_v - E}, \quad (6.86)$$

where $f(E, E_F, T)$ is the Fermi distribution, and $g_{c,v}$ are the density of states in the conduction and valence bands, respectively. In (6.85) and (6.86), the gradient of the density of states (DOS) is evaluated at constant energy, $|E - E_{c,v}|$. Note that for a uniform DOS distribution, $\operatorname{grad} g_{c,v}$

vanishes and we retrieve the usual form for the carrier concentration relations. The electrical current associated with a non-uniform DOS can be viewed in the following physical terms. If there is a localized region in the material which has a smaller DOS compared to its surrounding, there will be carrier flow from the region with smaller DOS to regions of higher DOS, to occupy lower energy states, despite identical doping conditions. When viewed in terms of effective masses, carriers flow from regions with lighter effective mass to regions where the heavier carriers are located. The first term in (6.83) and (6.84) arises from a gradient in the conduction and valence bands, which comprises of two components; one due to the applied potential, and the other due to the position dependence of the conduction or valence band edges. Here, we have used the sign convention such that when the band edges move towards the forbidden gap, both ΔE_c and ΔE_v are taken to be positive.

6.5.1 Electron Transport

Extending the results (6.33) to (6.35) for the strain induced conduction band shifts, the grad ΔE_c term in (6.83) can be written as

$$\text{grad } \Delta E_c = - \sum_{i,j=1}^{3} \chi_{ij}^c \text{ grad } T_{ij}. \tag{6.87}$$

Here, a new coefficient χ_{ij}^c is introduced, which has the following properties:

- Since shear stress does not alter the conduction band edge, $\chi_{ij}^c = 0$ for $i \neq j$.
- If the maximum shift turns out to be the [100], [$\bar{1}$00] bands, then $\chi_{xx}^c = \alpha$ and $\chi_{yy}^c = \chi_{zz}^c = \beta$, where α and β are given as (6.36) and (6.37).
- If the maximum shift is associated with the [010], [0$\bar{1}$0] bands then $\chi_{yy}^c = \alpha$ and $\chi_{xx}^c = \chi_{zz}^c = \beta$.
- If the maximum shift is associated with the [001], [00$\bar{1}$] bands, then $\chi_{zz}^c = \alpha$ and $\chi_{yy}^c = \chi_{xx}^c = \beta$.

Values for α and β, reproduced from Sect. 6.2.1, are $\alpha = -7.21 \times 10^{-12}$ eV cm^2/dyne and $\beta = 1.83 \times 10^{-12}$ eV cm^2/dyne.

We now turn to the last term in (6.83) to obtain a simplified expression for grad$^*n$ described by (6.85). Assuming non-degenerately doped material, we replace the Fermi distribution with the Boltzmann distribution, and upon evaluation of the integral in (6.85), we obtain [56]

$$\text{grad}^*n = (n/N_c) \text{ grad } N_c, \tag{6.88}$$

where N_c denotes the effective density of states in the conduction band. For convenience, we assume that the maximum shift is associated with the [100], [$\bar{1}$00] bands. Substituting (6.24) through (6.26) into (6.38) yields

$$\text{grad}^* n = \frac{n}{kT} \sum_{i,j=1}^{3} \Lambda_{ij}^c \, \text{grad} \, T_{ij}, \tag{6.89}$$

where the coefficients Λ_{ij}^c are defined as follows:

$$\Lambda_{xx}^c = -\frac{(\alpha+\beta)\left(e^{(\alpha+\beta)(T_{yy}-T_{xx})/kT} + e^{(\alpha+\beta)(T_{zz}-T_{xx})/kT}\right)}{1 + e^{(\alpha+\beta)(T_{yy}-T_{xx})/kT} + e^{(\alpha+\beta)(T_{zz}-T_{xx})/kT}}, \tag{6.90}$$

$$\Lambda_{yy}^c = \frac{(\alpha+\beta)e^{(\alpha+\beta)(T_{yy}-T_{xx})/kT}}{1 + e^{(\alpha+\beta)(T_{yy}-T_{xx})/kT} + e^{(\alpha+\beta)(T_{zz}-T_{xx})/kT}}, \tag{6.91}$$

$$\Lambda_{zz}^c = \frac{(\alpha+\beta)e^{(\alpha+\beta)(T_{zz}-T_{xx})/kT}}{1 + e^{(\alpha+\beta)(T_{yy}-T_{xx})/kT} + e^{(\alpha+\beta)(T_{zz}-T_{xx})/kT}}. \tag{6.92}$$

Here, $\Lambda_{ij}^c = 0$ for $i \neq j$. Substituting relations (6.87) and (6.89) into (6.83), and utilizing Einstein's relation, we obtain the following equation for the current density in the presence of a stress gradient, grad $T \neq 0$, etc.

$$\mathbf{J_n} = -q\mu_n n[\text{grad}\,\psi + \frac{1}{q}\sum_{i,j=1}^{3}(\chi_{ij}^c + \Lambda_{ij}^c)\,\text{grad}\,T_{ij}] + qD_n\,\text{grad}\,n. \tag{6.93}$$

Relation (6.93) clearly identifies the stress gradient as an additional driving force of the electron current density. Equation (6.93) is valid only within the cubic cell coordinate system. It can be transformed following standard transformation rules to the Cartesian coordinate system of arbitrary orientation.

The current component induced by the stress gradient can be viewed as being equivalent to a generation-recombination current. Recall that the carrier continuity equation under negligible generation-recombination reads: div $\mathbf{J_n} = 0$. For a given stress distribution, the band edge, density of states, mobility, and diffusivity are spatially fixed. Hence, the resulting electron current continuity equation becomes [56]

$$\text{div}\,\mathbf{J_{n0}} = \text{div}\left[\mu_n n \sum_{i,j=1}^{3}(\chi_{ij}^c + \Lambda_{ij}^c)\,\text{grad}\,T_{ij}\right], \tag{6.94}$$

where

$$\mathbf{J_{n0}} = -q\mu_n n\,\text{grad}\,\psi + qD_n\,\text{grad}\,n \tag{6.95}$$

and the right-hand side of (6.94) is equivalent to a net generation-recombination rate. In (6.94) and (6.95), μ_n and D_n are tensors whose components are related to the piezoresistance coefficients; see relation (6.54).

6.5.2 Hole Transport

Adopting the same procedure described earlier for the electron transport, we first compute the gradient of the strained induced valence (heavy hole) band shift following solution of Eq. (6.56). As before, the gradient term in (6.84) can be put into the form:

$$\text{grad } \Delta E_v = \sum_{i,j=1}^{3} \chi_{ij}^{v} \text{ grad } T_{ij}, \tag{6.96}$$

where the coefficients χ_{ij}^{v} have the same interpretation as χ_{ij}^{c} except that the off-diagonal terms of χ_{ij}^{v} are non-zero; $\chi_{xx,yy,xx}^{v} = 2.19 \times 10^{-12}\,\text{eV cm}^2/$ dyne, and $\chi_{xy,yz,xz}^{v} = 6.31 \times 10^{-12}\,\text{eVcm}^2/\text{dyne}$. The treatment of the last term in (6.84) is analogous to that of electrons,

$$\text{grad}^{*}p = (p/N_v) \text{ grad } N_v, \tag{6.97}$$

where N_v denotes the effective density of states in the valence band. To obtain an expression for grad N_v, we employ expression (6.78), which relates the change in carrier concentration effective mass with stress. Relation (6.78) along with $N_v \propto (m_p^*)^{3/2}$ reduces (6.97) to

$$\text{grad}^{*}\,p = \frac{3p}{2} \sum_{i,j=1}^{3} \Lambda_{ij} \text{ grad } T_{ij}. \tag{6.98}$$

In arriving at (6.98), we made the approximation $m_{cc}^{*}/m_{cc,0}^{*} \approx 1$, which is reasonable considering the relative magnitude of $\Lambda_{ij}T_{ij}$ for a piezoresistor. Values of Λ_{ij}, reproduced from Sect. 6.3.2 (see Table 6.6), are $\Lambda_{xx} = -24.7 \times 10^{-12}\,\text{cm}^2/\text{dyne}$ and $\Lambda_{xy} = -114 \times 10^{-12}\,\text{cm}^2/\text{dyne}$ at 300 K. Substituting relations (6.96) and (6.98) into (6.84), and utilizing Einstein's relation, the hole current density in the presence of a stress gradient, grad $T \neq 0$, reads

$$\mathbf{J_p} = -q\mu_p p[\text{grad } \psi - \frac{1}{q} \sum_{i,j=1}^{3} (\chi_{ij}^{v} + \Lambda_{ij}^{v}) \text{ grad } T_{ij}] - qD_p \text{ grad } p, \tag{6.99}$$

where $\Lambda_{ij}^{v} = \frac{3}{2}kT\Lambda_{ij}$. Again, we identify the stress gradient as an additional driving force for the hole current density. Following the same procedures leading to Eq. (6.94), the divergence of the hole current continuity equation reads

$$\operatorname{div} \mathbf{J_{p0}} = -\operatorname{div} \left[\mu_p p \sum_{i,j=1}^{3} (\chi_{ij}^v + \Lambda_{ij}^v) \operatorname{grad} T_{ij} \right] \qquad (6.100)$$

with

$$\mathbf{J_{p0}} = -q\mu_p p \operatorname{grad} \psi - qD_p \operatorname{grad} p. \qquad (6.101)$$

Again we identify the right-hand side of (6.100) as an equivalent net generation-recombination rate. The μ_p and D_p are tensors, which can be related to the piezoresistance coefficients; see relation (6.68).

The transport equations (6.93) and (6.99) have significant implications. Consider, for example, a p-type diffused piezoresistor on a cantilever micromechanical structure that is located at a region where the stress gradient is large (see Fig. 6.16). The piezoresistor is oriented orthogonal to the longitudinal axis of the cantilever. Following earlier discussion, in relation to (6.85) and (6.86), the DOS of holes decreases with increasing stress; holes move from high stress to low stress regions. But at the same time, the valence moves upwards causing holes to move to a region of higher stress. These two effects do not cancel each other (compare the relative magnitudes of χ_{ij}^v and Λ_{ij}^v). In cases where the valence band shifts dominate, there is hole accumulation at the region of maximum stress. In contrast, if the change in DOS is more dominant, there is depletion of holes from the region of maximum stress. Thus the resistance of the p-diffused piezoresistor changes not only due to change in mobility but also due to change in hole concentration as well. The latter being induced by the stress gradient. These arguments equally apply to electrons.

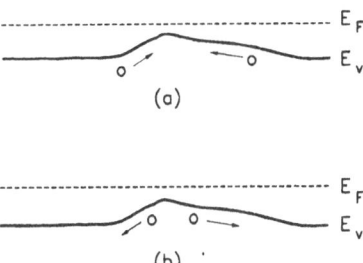

Fig. 6.16 An illustration of how the stress distribution alters the valence band edge. The band edge in (a) illustrates when the "band bending" effect is more dominant, and in (b) when the "non-uniform density of states effect" is the more dominant. The peak in band edges corresponds to the point of maximum stress in Fig. 6.15

6.5.3   Phonon Transport and Heat Flux

A non-uniform stress distribution alters the lattice spacing in the semiconductor by virtue of Hooke's law. Subsequent distortion in the lattice induces a change in phonon frequency, and hence, in the phonon distribution. This leads to a diffusion of phonons, which can be associated to heat flux at the microscopic level [71].

The dependence of phonon frequency on the lattice structure can be described in terms of the mode-dependent Grüneisen constant (γ_{kp}) [71–73],

$$\gamma_{kp} = -\frac{1}{3}\frac{a}{da}\frac{d\omega_{kp}}{\omega_{kp}}. \tag{6.102}$$

Here, a denotes the lattice constant of the crystal and ω_{kp} the phonon frequency for a given wave number (k) and polarization (p). We note from (6.102) that stress-induced spatial variations in lattice constant lead to spatial variations in phonon frequency.

In equilibrium, the phonon distribution function obeys the Bose-Einstein statistics whereby it is a function of phonon energy and temperature. In non-equilibrium, the distribution function is space and time dependent. Assuming that the rate of change of the distribution function is diffusion dominated, the spatial variation in ω_{kp} induced by a stress gradient, grad $T \neq 0$, leads to (see [71])

$$\left.\frac{\partial n_{np}}{\partial t}\right|_{\text{diff}} = -c_{kp}\frac{\partial n_{kp}}{\partial T}\,\text{grad}\,T. \tag{6.103}$$

which constitutes a heat flux. Here, c_{kp} denotes the phonon group velocity and n_{kp} defines the deviation of the phonon distribution from equilibrium for given wave vector and polarization.

6.5.4 Thermodynamic Consideration of Electro-Thermo-Mechanical Interactions

The effects associated with interactions of electrical, thermal, and mechanical signals, although traditionally considered to be second order in macroscopic systems, may no longer be ignored in microsystems. In the latter, owing to their small physical dimensions, the electrical, thermal, and mechanical conditions may become relatively extreme leading to significant mutual coupling [74].

Earlier in Sects. 6.5.1 and 6.5.2, we employed microscopic transport theory to show how stress gradients lead to flow of electrical current due to the combined effects of variation in band energy and the electronic density

of states. Subsequently, in Sect. 6.5.3, we glimpsed phonon diffusion
stemming from the change in phonon distribution induced by a stress
gradient. We now develop, following principles of irreversible thermodynamics, the generalized system of phenomenological equations, which
describe the interaction of electrical and heat fluxes with the mechanical
stress gradient. Here, we present an additional equation, which constitutes a
so-called strain flux that stems from the various driving forces, i.e.,
gradients in potential, temperature, and stress.

To arrive at the generalized coupled system of equations, we adopt the
same formalisms employed for the electrothermal system (Sect. 5.1.1) and
electro-thermo-magnetic interactions in Sect. 5.4. To include stress
gradients, we add, to the entropy relation, an extra term to account for
the work done on the system by stress. This work is the product of the
applied force and the associated distance, hence, the work per unit volume
is $\chi\varepsilon$, where χ denotes stress and ε the strain. Here, the notation used for
stress is not to be confused with the band structure parameter χ in (6.87)
and (6.96); the change in notation from T to χ was adopted to avoid
confusion with temperature. For a unipolar system (i.e., single carrier) that
is homogeneous (no-carrier diffusion), the presence of a stress gradient
leads to the following system of dynamic equations [74]

$$-\mathbf{J}^{\mathrm{e}} = L_{11}\,\mathrm{grad}\,(\psi/T) + L_{12}\,\mathrm{grad}\,(1/T) + L_{13}\,\mathrm{grad}\,(\chi/T),$$
$$(6.104)$$

$$\mathbf{J}^{\mathrm{q}} = L_{21}\,\mathrm{grad}\,(\psi/T) + L_{22}\,\mathrm{grad}\,(1/T) + L_{23}\,\mathrm{grad}\,(\chi/T), \quad (6.105)$$

$$-\mathbf{J}^{\varepsilon} = L_{31}\,\mathrm{grad}\,(\psi/T) + L_{32}\,\mathrm{grad}\,(1/T) + L_{33}\,\mathrm{grad}\,(\chi/T),$$
$$(6.106)$$

where $\mathbf{J}^{\mathrm{e}}$, $\mathbf{J}^{\mathrm{q}}$, $\mathbf{J}^{\varepsilon}$ denote the electrical current, heat current, and strain flux
densities, respectively, ψ the electric potential, T the temperature, and χ the
tensor of stress components (*viz.*, grad $\chi = \sum_{ij}\,\mathrm{grad}\,\chi_{ij}$). In (6.104)
through (6.106), the L's denote kinetic coefficients, which are dependent on
the intensive parameters (T, ψ, and χ) and are tensors due to stress-induced
material anisotropy. The kinetic coefficients are related by virtue of
Onsager's reciprocity theorem. The first two terms on the right-hand sides
of Eqs. (6.104) and (6.105) represent the familiar electrothermal effects
(see Sect. 5.1.1). The corresponding third term represents charge and
phonon flow induced by a stress gradient. The former was treated in Sects.
6.5.1 and 6.5.2 for electrons and holes, respectively, and the latter in Sect.
6.5.3. Here, the absence of a stress gradient, grad $\chi = 0$, will alter the
material isotropically so that there is no stress induced electrical or heat
flux components. The remaining equation in the system describes a strain
flux, which can be interpreted in terms of an anisotropic expansion or
contraction leading to a net flow (or migration) of the material in some

preferred direction [74]. Here, a vanishing last term on the right-hand side of (6.106) implies isotropic material deformation with no net flow. The corresponding first and second terms can be associated with the piezoelectric and thermoelastic effects, respectively, whereby the presence of a potential or temperature gradient leads to anisotropic deformation. In all of these cases, deformation ceases when equilibrium between the externally induced force and the internal restoring force in the material is reached. These considerations are not accounted for in the present equation system.

Equation (6.106) reduces to the steady state linear thermoelasticity equation in the absence of potential or stress gradients (see [74]).

6.6 Galvano-Piezo-Magnetic Effects

In the presence of stress, the Hall coefficient R_H, discussed in Chapt. 4, is no longer a scalar quantity. It is described by a tensor which takes the same form as the stress-dependent electrical resistivity [14]. Anisotropies in conductivity and Hall coefficient have a strong bearing on Hall-based material characterization experiments and integrated magnetic field sensors. It is well-known from classical Hall effect experiments that measurement of both Hall coefficient and material conductivity allows retrieval of the Hall mobility, *viz.*, $\mu_H = |\sigma R_H|$. This procedure leads to meaningful results only when the material coefficients are isotropic. In integrated Hall sensors, besides stress induced device offset (by virtue of piezoresistance effect), there may be a stress dependence of the magnetic sensitivity. This is due to stress induced modulation of the Hall coefficient by virtue of the so-called piezo-Hall effect. The resulting anisotropies lead to an orientation dependence of measured quantities from the test structure in question, i.e., the direction of the applied current density with respect to the stress components.

6.6.1 Piezo-Hall Coefficients

In *n*-type Si, the conduction band valleys in the unstrained material, as illustrated in Fig. 6.5, have the same number of electrons in the different crystal directions, leading to a scalar conductivity and Hall coefficient. If, for example, we apply a stress in the [100] direction, the two sets of valleys perpendicular to the [100] direction are lowered. Electrons from the higher [100], [$\bar{1}$00] valleys then empty into the lower valleys (in directions parallel to applied stress). Thus there are more electrons with a larger mobility, μ_t. Hence, in the presence of a magnetic field, the larger electron mobility leads to a larger Lorentz force, which in turn yields a larger Hall voltage. For not too large stress levels, the change in Hall coefficient can be linearly related to stress and carries the same form as piezoresistance

$$\frac{(\Delta R_H)_{ij}}{R_{H0}} = \sum_{k,l} P_{ijkl} T_{kl}, \tag{6.107}$$

where P_{ijkl} is the tensor of piezo-Hall coefficients, R_{H0} is the scalar (isotropic) Hall coefficient, and T_{kl} the stress tensor. In the cubic material system (e.g., Si), the coefficients can be represented as

$$P = \begin{bmatrix} P_{11} & P_{12} & P_{12} & 0 & 0 & 0 \\ P_{12} & P_{11} & P_{12} & 0 & 0 & 0 \\ P_{12} & P_{12} & P_{11} & 0 & 0 & 0 \\ 0 & 0 & 0 & P_{44} & 0 & 0 \\ 0 & 0 & 0 & 0 & P_{44} & 0 \\ 0 & 0 & 0 & 0 & 0 & P_{44} \end{bmatrix}. \tag{6.108}$$

In n-type Si, the three independent coefficients have values $P_{11} \approx -88 \times 10^{-12}\,\mathrm{cm}^2/\mathrm{dyne}$, $P_{12} = (40 \text{ to } 45) \times 10^{-12}\,\mathrm{cm}^2/\mathrm{dyne}$ and $P_{44} \approx 10 \times 10^{-12}\,\mathrm{cm}^2/\mathrm{dyne}$. The measured value of P_{12} (for electron concentrations in the range: 10^{14} to $10^{16}\,\mathrm{cm}^{-3}$) corroborates well with the theoretically estimated value of $+44 \times 10^{-12}\,\mathrm{cm}^2/\mathrm{dyne}$ [14] obtained using the standard conduction band transport theory of Herring and Vogt [55] (see Sect. 6.2). The coefficient P_{12} decreases almost linearly with

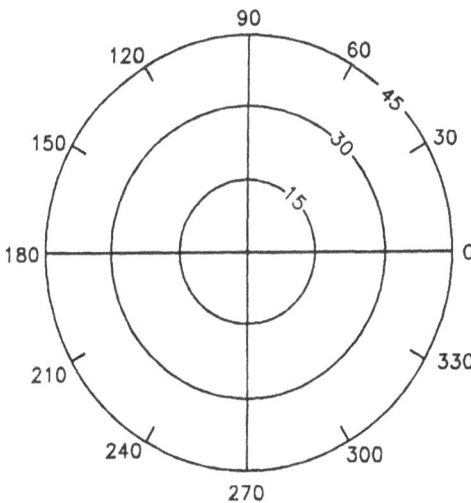

Fig. 6.17 Variation of longitudinal and transverse piezo-Hall coefficients (P_{12}) in (100) n-type Si. Here, both the coefficients are $45 \times 10^{-11}\,\mathrm{Pa}^{-1}$

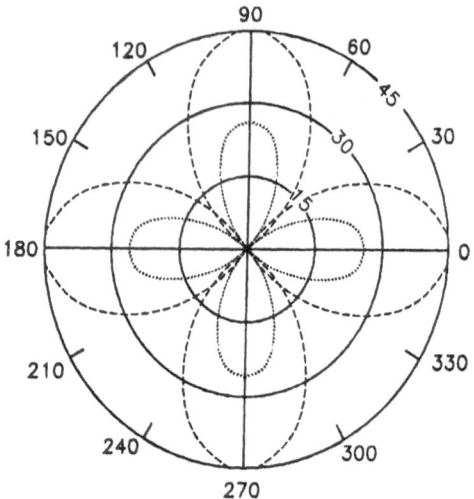

Fig. 6.18 Variation of longitudinal (dashed curve) and transverse (dotted curve) piezo-Hall coefficients (P_{12}) in (110) n-type Si. Values indicated are in units of 10^{-11} Pa^{-1}

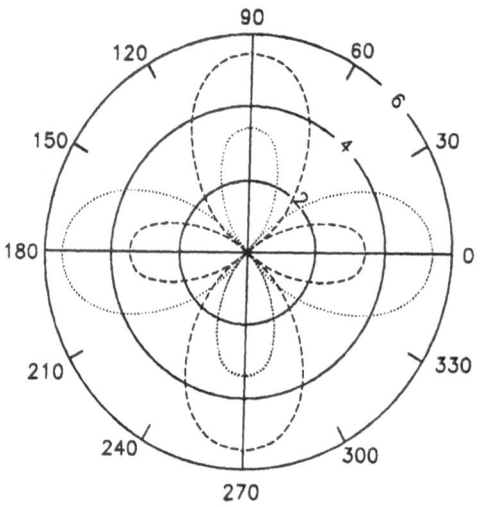

Fig. 6.19 As in Fig. 6.18, but in (111) n-type Si

temperature over the range $-80\,°C$ to $100\,°C$. Its variation is approximately -0.08×10^{-12} cm^2/dyne-°C [14].

Although the values of the piezo-Hall for n-type Si are of the same order of magnitude as the corresponding piezoresistance coefficients, their variation with respect to orientation is completely different. This is due to the presence of the cross product term in the transport equation, which for an n-type material, reads

$$\mathbf{J_{nB}} + \sigma_n R_{Hn}\mathbf{B} \times \mathbf{J_{nB}} = \sigma_n \mathbf{E}. \tag{6.109}$$

Here, σ_n is the conductivity tensor, which is related to the piezoresistance coefficients as given by (6.3), and R_{Hn} is the Hall coefficient tensor as described by (6.107). For example, for a magnetic field along the $\langle 100 \rangle$ direction and current density along the $\langle 010 \rangle$ direction, the coefficient P_{12} is invariant on rotation within the (100) plane, $P_l = P_t = P_{12}$ (see Fig. 6.17), However, this is not quite so in the (110) and (111) planes. Here, P_l and P_t are highly directional (see Figs. 6.18 and 6.19).

The piezo-Hall coefficients for p-type Si have not yet been measured nor have there been any reliable theoretical predictions. Drawing from our experience with piezoresistance in the p-type material, the computation of the piezo-Hall coefficients would be an equally challenging task.

6.7 The Piezo Drift-Diffusion Transport Model

Following earlier results in Sects. 6.2 and 6.3, which relate the drift mobilities of electrons and holes to the piezoresistance coefficients, we can now state the drift-diffusion model in the presence of mechanical stress.

6.7.1 Transport Relations

The diffusion coefficient, in the presence of stress, becomes a tensor and using Einstein's relation, $D_{ij} = (kT/q)\mu_{ij}$, the standard isothermal drift-diffusion relations governing electron and hole transport can be expressed as [64]:

$$\mathbf{J_n} = qD_n\left[\operatorname{grad} n - n\operatorname{grad}\left(\frac{q\psi}{kT}\right)\right], \tag{6.110}$$

$$\mathbf{J_p} = -qD_p\left[\operatorname{grad} p + p\operatorname{grad}\left(\frac{q\psi}{kT}\right)\right]. \tag{6.111}$$

The relations are valid only when the stress distribution is uniform; in the presence of a stress gradient, the transport relations take the form given in (6.93) and (6.99). In the presence of magnetic field, relations (6.110) and (6.111) become modified to read [64]:

$$\mathbf{J_{nB}} + q\mu_n n[R_{Hn}\mathbf{B} \times \mathbf{J_{nB}}] = qD_n\left[\operatorname{grad} n - n\operatorname{grad}\left(\frac{q\psi}{kT}\right)\right], \tag{6.112}$$

$$\mathbf{J_{pB}} - q\mu_p p[R_{Hp}\mathbf{B} \times \mathbf{J_{pB}}] = -qD_p\left[\operatorname{grad} p + p\operatorname{grad}\left(\frac{q\psi}{kT}\right)\right]. \tag{6.113}$$

In (6.110) through (6.113), $\mu_{n,p}$, $D_{n,p}$, and $R_{Hn,p}$ are tensors, and $\mathbf{B}$ denotes the magnetic field vector. In the absence of stress, the mobility, diffusion coefficient, and Hall coefficient become scalars. In particular, (6.112) and (6.113) reduce to the well-known form in terms of the Hall mobility, since $\mu_{Hn,p} = |q\mu_{n,p}(n,p)R_{Hn,p}|$.

The above transport relations together with Poisson's equation and the carrier continuity equations constitute the complete system of model equations required for simulation of carrier transport in the presence of mechanical stress and/or magnetic field.

6.7.2 Complete System and Summary of Model Equations

For convenience, we restate the system of equations given in Chapt. 2,

$$\mathrm{div}\,(\varepsilon\,\mathrm{grad}\,\psi) = -q(p - n + N_D - N_A), \tag{6.114}$$

$$\mathrm{div}\,\mathbf{J}_{nf} = qR + q\frac{\partial n}{\partial t}, \tag{6.115}$$

$$\mathrm{div}\,\mathbf{J}_{pf} = -qR - q\frac{\partial p}{\partial t}, \tag{6.116}$$

where $\mathbf{J}_{nf}$ and $\mathbf{J}_{pf}$ are transport relations as described by relations (6.110) through (6.113) and R, following (6.94) and (6.100), is modified to include an additional term in the presence of a stress gradient. For the sake of generality and for compatibility with standard drift-diffusion simulation code, the transport relations can be stated in a reduced form regardless of the presence of a stress field and/or magnetic field. For convenience, considering only two-dimensional electron transport, Eq. (6.112) can be reduced to the following form

$$\mathbf{J}_{nf} = \begin{bmatrix} f_{11} & f_{12} \\ f_{21} & f_{22} \end{bmatrix} \mathbf{J}_{n0}, \tag{6.117}$$

where $\mathbf{J}_{n0}$ denotes the zero (stress or magnetic) field current density

$$\mathbf{J}_{n0} = qD_n\left[\mathrm{grad}\,n - n\,\mathrm{grad}\left(\frac{q\psi}{kT}\right)\right]. \tag{6.118}$$

In (6.117), the elements f_{ij} are a function of the stress and/or magnetic field, the relevant physical parameters (e.g., Hall coefficient and mobility) as well as parameters pertinent to coordinate transformation. The elements are evaluated using relations (6.54) and (6.107) for the piezoresistance and piezo-Hall effects, respectively. However, they are formulated in a manner such that in the absence of the external field, $[f]$ reduces to an identity

Table 6.7. *Models for simulation of mechanical effects on electrical transport applicable to mechanical and other (e.g., Hall) microsensors as well as semiconductor devices. Remaining models for physical parameters and boundary/interface conditions are given in Chapt. 2*

Models	Reference	Description/Validity
Governing electrical transport equations	(6.114)–(6.116)	
Transport relations		
Drift	(6.3) & (6.4)	Generalized Ohm's law
Drift-diffusion	(6.110), (6.111)	Uniform stress
	(6.112), (6.113)	Uniform stress with magnetic field
	(6.93), (6.99)	Stress gradient
Electro-thermo-mechanical coupling	(6.104)–(6.106)	Electrical, thermal, and strain flux relations with stress gradient
Physical Parameters		
Piezoresistance coefficients: π	See Sects. 6.1.1 to 6.1.3	n- and p-type mono-Si and -Ge Doping- and temperature-dependence, and non-linearity effects
	See Section 6.1.4	n- and p-Type poly Si
Carrier mobillity: $\mu_{n,p}$	(6.54), (6.68)	Relation to piezoresistance
Effective mass	(6.78), Table 6.6	Carrier concentration effective mass
Intrinsic concentration: n_{ie}	(6.81)	Piezojunction effect
Band edge shifts	(6.87), (6.96)	Band gap variation due to stress gradient
"Generation-recombination"	(6.94), (6.100)	Equivalent generation-recombination current due to stress gradient
Piezo-Hall coefficients: P	(6.107) See Sect. 6.61	n-Type mono-Si
Boundary conditions		
Outer boundaries	(4.73) in Chapt. 4	See Sects. 2.6.4 and 4.3.2 for physical and mathematical plausibility

matrix. If $T \neq 0$ but $B = 0$, then $[f]$ is symmetric. If $T \neq 0$, $B \neq 0$, then $[f]$ is asymmetric, and for $T \neq 0$ and $B \neq 0$, the matrix comprises of a symmetric part and an asymmetric part; the latter fulfilling Onsager's reciprocity relations.

A summary of governing equations, physical models, and boundary and interface conditions for simulation of stress effects on electrical carrier transport is given in Table 6.7. Due to the stress-induced anisotropy in transport, the standard boundary conditions on the electric potential ψ may not apply. Again, as in analysis of bipolar magnetic sensors, conditions based on physical or mathematical plausibility have to be imposed. Here, the boundary conditions for ψ can be modeled along the lines described in Sect. 4.3.2, and in particular, Eq. (4.73).

In the following, we present a numerical solution scheme [75], for the system of Eqs. (6.114) through (6.118), which can be employed for simulation of the different microsensors discussed in relation to Chapts. 2 to 6.

6.7.3 Discretization Scheme

The discretization scheme considered is based on the control region approximation [76] or more commonly known as the box integration method (see [77]). This method when applied to any of the equations, (6.114) to (6.116) transforms it by means of Gauss' divergence theorem into a system of discrete algebraic equations expressed as a function of the respective unknowns, ψ, n, and p. The procedures for mesh generation and refinement can be found in [78].

Following standard procedure, the continuous form of Poisson's equation in (6.114) is first linearised by a Newtonian scheme around a known potential, ψ^0. For small deviations in potential, it reads

$$
\begin{aligned}
&- \operatorname{div} (\varepsilon \operatorname{grad} \psi) + (q/V_t)(p^0 + n^0)\psi \\
&\quad - (q/V_t)(p^0 + n^0)\psi^0 - \rho^0 = 0,
\end{aligned}
\tag{6.119}
$$

where

$$
\begin{aligned}
p^0 &= n_{ie} \exp\left[(\phi_p^0 - \psi^0)/V_t\right], \\
n^0 &= n_{ie} \exp\left[(\psi^0 - \phi_n^0)/V_t\right], \\
\rho^0 &= p^0 - n^0 + N.
\end{aligned}
\tag{6.120}
$$

Here, V_t denotes the thermal voltage, kT/q, $\phi_{n,p}$ the electron and hole Fermi potentials, and N the net doping, $N_D - N_A$. The superscripted quantities denote known values. Applying the divergence theorem to Eq. (6.119) transforms it into a system of algebraic equations found by solving for any node i, the equation

$$
\begin{aligned}
&- \int_{\partial\Omega_i} \varepsilon \operatorname{grad} \psi \cdot \mathbf{n}\, d(\partial\Omega_i) + \int_{\Omega_i} (q/V_t)(p^0 + n^0)\psi\, d\Omega_i \\
&\quad - \int_{\Omega_i} [(q/V_t)(p^0 + n^0)\psi^0 + \rho^0]\, d\Omega_i = 0,
\end{aligned}
\tag{6.121}
$$

where Ω_i is the cell (or subdomain) formed by the union of perpendicular bisectors of element edges that converge to node i (see Fig. 6.20), $\partial\Omega_i$ denotes the boundary of Ω_i, and $\mathbf{n}$ the outward normal vector to $\partial\Omega_i$. The choice of perpendicular bisectors of element edges keeps calculations of flux emanating from node i, relatively simple. The electrostatic potential ψ is assumed to be spatially linear within an element (i.e., the electric fields are piecewise constant in the domain). By assuming a piecewise constant material permittivity ε, the discretized form of Eq. (6.121) can easily be evaluated. For a chosen element (see Fig. 6.21), the discretized equation for

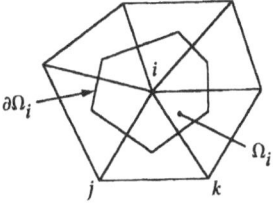

Fig. 6.20 The cell or subdomain, Ω_i (bounded by $\partial\Omega_i$) formed by the union of perpendicular bisectors of elemental edges converging to node i

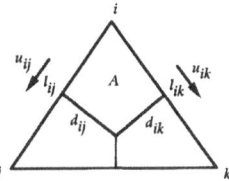

Fig. 6.21 An element with vertices i, j, and k, constituting part of the cell shown in Fig. 6.20

node i reads

$$V_t \varepsilon [(d_{ij}/l_{ij})(\psi_i - \psi_j) + (d_{ik}/l_{ik})(\psi_i - \psi_k)]$$
$$+ \int_A (q/V_t)(p^0 + n^0)\psi dA \qquad (6.122)$$
$$- \int_A [(q/V_t)(p^0 + n^0)\psi^0 + \rho^0]dA = 0,$$

where the subscripts associated with d and l denote scalar distances of the perpendicular bisectors and elemental edges respectively, and A denotes a portion of the subdomain Ω_i that is encompassed by d_{ij} and d_{ik} (see Fig. 6.21). The charge density terms ($p^0 + n^0$) and ρ^0 in Eq. (6.122) can be assumed to be either constant or spatially linear functions within the element. In the case of the former, they assume the values at node i. The integrals over A are evaluated by simple numerical integration using triangular area coordinates [79]. Alternatively, the nonlinear charge density integrals can also be evaluated using an n-point Gauss quadrature. A sufficiently high accuracy could be achieved with a six point formula [80]. The procedure employed in obtaining Eq. (6.122) is then repeated on an element by element basis covering the entire simulation domain. The entries for the stiffness and the right-hand side matrices are updated in the process.

The discretization of the electron and hole continuity equations is based on the two-dimensional extension (to triangular grids [81]) of the Scharfetter-Gummel approach [82] but with the anisotropies introduced by the external (stress and/or magnetic) field suitably incorporated. Due to these anisotropies, variations in the standard procedure need to be employed in determining the projected current density along the edge of an element under consideration. This involves taking into account all three nodal values of ψ, n, and p pertaining to the element.

We now present the steps leading to the discretized form for the electron continuity equation. An analogous approach holds for holes. We recall the field dependent electron current density (6.117) and (6.118). Expressing $\mathbf{J}_{n0}$ in divergence form and projecting it onto an element side connected by nodes a and b yields

$$\mathbf{J}_{n0} \cdot \mathbf{u}_{ab} \exp\left(-\psi/V_t\right) = qD_n \operatorname{grad}\left[n \exp\left(-\psi/V_t\right)\right] \cdot \mathbf{u}_{ab}, \quad (6.123)$$

where $\mathbf{u}_{ab}$ denotes the unit vector parallel to the side. For the element under consideration (say element with vertices i, j, k shown in Fig. 6.21), the integrals of (6.123) along paths i to j and i to k yield

$$\begin{aligned}
\mathbf{J}_{n0} \cdot \mathbf{u}_{ij} &= (qD_n/l_{ij})[n_j B(\psi_{ji}/V_t) - n_i B(\psi_{ij}/V_t)], \\
\mathbf{J}_{n0} \cdot \mathbf{u}_{ik} &= (qD_n/l_{ik})[n_k B(\psi_{ki}/V_t) - n_i B(\psi_{ik}/V_t)],
\end{aligned} \quad (6.124)$$

respectively. The left-hand side of Eqs. (6.124) denote scalar current densities and $n_{i,j,k}$ denote the respective nodal electron concentrations. To arrive at (6.124), the scalar current density, $\mathbf{J}_{n0} \cdot \mathbf{u}_{ab}$, is assumed constant along the elemental edge, the scalar mobility μ_n (and consequently D_n) is assumed constant in the element and is replaced by an average elemental value. The electric potential is as before assumed to be spatially linear. As in zero field cases, B denotes the Bernoulli function which is defined as $B(\Delta) = \Delta/[\exp(\Delta) - 1]$ where Δ denotes the potential difference between nodes, $\psi_{ab} = \psi_a - \psi_b$. We can extract the zero field current density $\mathbf{J}_{n0}$ in the element;

$$\mathbf{J}_{n0} = U^{-1}F_n, \quad (6.125)$$

where

$$\mathbf{J}_{n0} = \begin{bmatrix} J_{n0}^x \\ J_{n0}^y \end{bmatrix}, \qquad U = \begin{bmatrix} U_{ij}^x & U_{ij}^y \\ U_{ik}^x & U_{ik}^y \end{bmatrix}, \qquad F_n = \begin{bmatrix} F_{n1} \\ F_{n2} \end{bmatrix},$$

$$\begin{aligned}
F_{n1} &= (qD_n/l_{ik})[n_j B(\psi_{ji}/V_t) - n_i B(\psi_{ij}/V_t)] \\
F_{n2} &= (qD_n/l_{ik})[n_k B(\psi_{ki}/V_t) - n_i B(\psi_{ik}/V_t)].
\end{aligned} \quad (6.126)$$

Substituting the zero field current (6.125) into (6.117), the field dependent current density can now be fully determined, $viz.$,

$$\mathbf{J}_{nf} = fU^{-1}F_n. \quad (6.127)$$

Having constructed the field current density $\mathbf{J}_{nf}$ in the element, the standard procedure is now employed in obtaining the discretized form of the electron continuity equation. By assuming $\mathbf{J}_{nf}$ is constant in the element, the flux or the current emanating from node i is evaluated in the usual manner. The divergence theorem applied to the steady state continuity equation (6.115) yields for a subdomain Ω_i

$$\int_{\Omega_i} \text{div} \, \mathbf{J_{nf}} \, d\Omega_i = \int_{\partial\Omega_i} \mathbf{J_{nf}} \cdot \mathbf{n} \, d(\partial\Omega_i) = \int_{\Omega_i} qR d\Omega_i, \qquad (6.128)$$

where the integral over $\partial\Omega_i$ denotes the flux related to node i. In the element under consideration, the contribution of flux to node i from its neighbors, j and k, can be expressed in the form

$$\mathbf{J_{nf}}^T U^T \mathbf{D} = \mathbf{F_n}^T [U^{-1}]^T f \, U^T \mathbf{D}, \qquad (6.129)$$

where $\mathbf{D} = [d_{ij} \quad d_{ik}]^T$. The discretised form of the electron continuity equation based on SRH recombination (see Sect. 2.3.6) reads

$$\mathbf{F_n}^T [U^{-1}]^T f \, U^T \mathbf{D} = qA(n_i p_i - n_{iei}^2)/[\tau_p(n_i + n_0) + \tau_n(p_i + p_0)], \qquad (6.130)$$

where n_i, p_i, and n_{iei} denote the electron (n), hole (p), and intrinsic concentrations (n_{ie}), respectively, at node i and $\tau_{n,p}$ denote the electron and hole lifetimes. The right-hand side of Eq. (6.128) is evaluated by assuming a constant net recombination in A. Again a more accurate treatment of the recombination integral can be performed by employing an n-point Gauss quadrature formula [80].

An analogous treatment yields the discretized form for the hole continuity equation. The entries to the stiffness matrix are assembled on an element basis, skipping those elements in the oxide region. This way, the solution to the continuity equations is restricted to within the device domain with the no-flow current density condition at the device/oxide interface naturally recovered.

The above numerical scheme, which is based on the extension of the one-dimensional Gummel-Scharfetter approximation, may not locally preserve the true character of the field induced anisotropy; there may be issues related to local flux conservation. More recently, a family of flux conserving box discretization schemes has been developed [83, 90]. Here, there are three free nodal parameters, in the case of a rectangular grid, which can be employed to identify pertinent nodes, consistent with the local physics, to construct the discrete form of the anisotropic current continuity equation.

6.7.4 Solution Scheme

We solve the discretized equations discussed in the previous section using a non-linear procedure which employs a successive scheme (Gummel's algorithm, see [84]). The solution procedure is initiated by solving the non-linear Poisson's equation (inner loop), with the quasi-Fermi potentials kept known from the previous iteration. The non-linear procedure may require an acceleration algorithm where the update for the potential is damped,

$$t\delta u = \psi^{k+1} - \psi^k. \tag{6.131}$$

Here, t is a damping parameter (see [85]). When the inner loop iterations are satisfactorily terminated, the concentrations n and p need to be updated with the new values of ψ, which is in turn kept as known during the subsequent solution of the continuity equations.

An iterative process is employed in the solution of continuity equations to yield self-consistent values of n and p for the given ψ. At each iteration step, the recombination terms on the right-hand side of the continuity equation are evaluated with the most recent updates of n and p. This procedure appears to reduce the total number of outer loop iterations necessary for the complete convergence of ψ, n, and p.

To obtain solutions of the variables for a desired device operating point, the applied bias is incremented in steps starting from the zero bias (or equilibrium) solution. Convergence problems encountered at higher injection levels, particularly with the solution to the continuity equations primarily due to the strong coupling of ψ, n, and p through the recombination terms, could be alleviated with the partial linearisation of the recombination terms in the continuity equations as shown in [86].

The discretized equations can be solved by a family of direct methods or iterative schemes, although the former demands large storage requirements. Although the variables in the system are ψ, n, and p, convergence checks are applied to the potentials: ψ, ϕ_n, and ϕ_p. The quasi-Fermi potentials are updated at the end of each continuity equation solution. The convergence criteria for each node are as follows

$$\left(u^{k+1} - u^k\right)/u^{k+1} < \gamma_1 \qquad \text{for} \quad u^{k+1} \geq 1.0$$

$$\text{and} \quad \left(u^{k+1} - u^k\right) < \gamma_2 \qquad \text{for} \quad u^{k+1} < 1.0, \tag{6.132}$$

where u represents the electric or quasi-Fermi potentials. The precise values of γ_1 and γ_2 are chosen depending on the problem at hand.

6.7.5 Evaluation of Terminal Currents

Terminal currents are evaluated once the desired solution to the variables has been achieved, i.e., as a post processing step. In most cases, apart from solution verification, it is desirable to perform an accurate computation of the terminal currents to determine true estimates of the device sensitivity to the field.

A well-known technique of calculating a terminal current is by integrating the flux crossing the border of the contact and the device. Since the total current density involves numerical derivatives of the solution variables, large errors may result particularly at the union of Dirichlet and Neumann boundary conditions. Although the solutions to the variables are smooth at these regions, there may be singularities in the derivatives which yield singularities in the current density.

The technique outlined in what follows employs an alternative approach to terminal current evaluation [75]. This procedure appears to provide excellent conservation of terminal currents and its implementation is straightforward in terms of computational complexity. Considering just the electrons for simplicity, the continuity equation can be represented in the weak form, *viz.*,

$$\int_\Gamma \zeta \, \mathrm{div} \, \mathbf{J}_{\mathbf{nf}} \, d\Gamma = \int_\Gamma \zeta q R \, d\Gamma, \tag{6.133}$$

where ζ is a smooth test function and Γ represents the device domain. Eq. (6.133) can be expressed as

$$\int_{\partial \Gamma} (\zeta \, \mathbf{J}_{\mathbf{nf}}) \cdot \mathbf{n} \, d(\partial \Gamma) = \int_\Gamma \mathbf{J}_{\mathbf{nf}} \cdot \mathrm{grad} \, \zeta \, d\Gamma + \int_\Gamma \zeta q R \, d\Gamma, \tag{6.134}$$

where $\mathbf{n}$ is an outward normal vector and $\partial \Gamma$ represents the device boundary. If ζ is chosen to be 1 over an electrode α and zero at all other electrodes, the outward electron current at electrode α is then given by

$$I_{n\alpha} = \int_\Gamma \mathbf{J}_{\mathbf{nf}} \cdot \mathrm{grad} \, \zeta \, d\Gamma + \int_\Gamma \zeta q R \, d\Gamma. \tag{6.135}$$

Hence, the total current $I_{n\alpha}$ through electrode α is obtained by

$$I_{T\alpha} = I_{n\alpha} + I_{p\alpha} = \int_\Gamma (\mathbf{J}_{\mathbf{nf}} + \mathbf{J}_{\mathbf{pf}}) \cdot \mathrm{grad} \, \zeta \, d\Gamma. \tag{6.136}$$

An elegant test for the conservation of total electron device current can be performed by setting the test function ζ to be 1 over the entire device domain. This reduces (6.134) to

$$\int_{\partial \Gamma} \mathbf{J}_{\mathbf{nf}} \cdot \mathbf{n} \, d(\partial \Gamma) = I_{nL} = \int_\Gamma q R \, d\Gamma, \tag{6.137}$$

where I_{nL} denotes the total electron leakage current. The right-hand side integral in (6.137) can be conveniently evaluated since it only requires summing the various recombination integrals which have been performed earlier for each cell as seen in Eq. (6.130). The procedure employed in the test for conservation of hole current is analogous. The above weak form current calculation can also be adopted for computation of heat fluxes in thermal microsensors.

6.8 Illustrative Simulation Example – Stress Effects on Hall Sensors

The encapsulation of a silicon die induces significant mechanical stress on active devices on the chip surface. In Hall sensors, this leads to device offset which is undesirable when detecting low frequency magnetic fields [11, 12, 87]. In this example, we illustrate the significance of device orientation and device location, in terms of control/elimination of the stress-induced offset, using simple n-type Hall sensor structures [20]; the classical Hall and split electrode devices (see Fig. 6.22). The results follow from the numerical solution of the piezo-Hall transport relation based on stress distributions obtained using the finite-element simulation package, ANSYS. The corresponding offset voltages or currents are obtained for various device orientations, substrate orientations, and micromachined packaging configurations.

Mechanical stress created during encapsulation (die bonding to chip carrier) stems from differences in thermal expansion coefficients of the tri-material system comprising the die, the epoxy, and the substrate (chip carrier) [88]. The magnitude of the stress induced at active device regions is determined by the thermoelastic properties of these different materials, although the dominant factor is the expansion coefficient of the substrate (see [26]). Accurate prediction of the stress distribution requires numerical simulations, which can be performed using a variety of commercial finite-element packages, e.g., ANSYS (see Sect. 8.2).

The induced stress modulates the resistivity of the device by virtue of the piezoresistance effect (see Sect. 6.1). In addition, the effects of stress coupled with magnetic field can potentially affect the output magnetic response due to the piezo-Hall effect (see Sect. 6.6). Restating Eq. (6.109), which governs electrical transport in the n-type Hall sensor in the presence

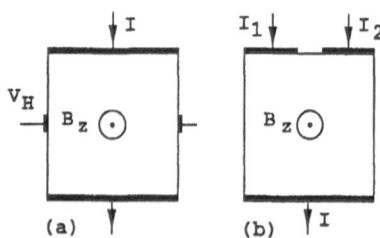

Fig. 6.22 (a) Hall and (b) split-electrode device structures

of homogeneous stress and magnetic field,

$$\mathbf{J}_{nf} = \sigma_n \mathbf{E} - \sigma_n [(R_{Hn}\mathbf{B}) \times \mathbf{J}_{nf}]. \tag{6.138}$$

Here, $\mathbf{J}_{nf}$ denotes the electron current density, $\mathbf{E}$ the applied electric field, σ_n the conductivity, and R_{Hn} the Hall coefficient. The σ_n and R_{Hn} are both symmetric second-rank tensors and are related to stress via the respective fourth-rank tensor of piezo-coefficients; *viz.*, see relations (6.3) and (6.107).

The piezoresistance and piezo-Hall coefficients are transformed from the cubic crystallographic axes to the working plane of interest. The transformation, which is based on the Euler angles (see Sect. 7.1), is constructed by rotating the cubic coordinate system into the plane of the Hall device. The current density equation (6.138) incorporating effects of both stress and magnetic field can be written as

$$\mathbf{J}_{nf} = \sigma'(\mathbf{B}, T)\mathbf{E}. \tag{6.139}$$

Here, the effective conductivity σ', which is an asymmetric second-rank tensor, is a function of both magnetic field $\mathbf{B}$ and stress T; these quantities are now defined in the device working plane. The conductivity matrix consists of two parts: a symmetric part comprising the conductivity and an asymmetric part which is given by the cross product term in (6.138). The entries in σ' associated with the latter satisfy Onsager's reciprocity relations. We consider a two-dimensional system with the orientations of current, magnetic field, and uniaxial stress as depicted in Fig. 6.23. Assuming negligible generation or recombination in the device, the divergence of the current density (6.139) under steady-state conditions can be cast into the form [89]

$$\sum_{i,j=1,2} D_i(a_{ij}D_j\psi) = 0, \tag{6.140}$$

where $D_1 = \partial/\partial x, D_2 = \partial/\partial y$ and ψ denotes the electrostatic potential. The boundary conditions for (6.140) consist of Dirichlet conditions (prescribed by the applied potential) at the current electrodes and the Neumann condition at insulating boundaries, $\mathbf{J}_{nf} \cdot \mathbf{n} = 0$, where $\mathbf{n}$ denotes the outward normal vector. Eq. (6.140) together with the boundary conditions are numerically solved based on a standard Galerkin finite-

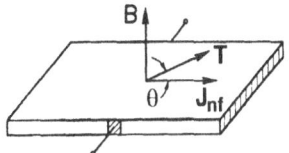

Fig. 6.23 Illustration of the orientations of current density ($\mathbf{J}_{nf}$), magnetic field ($\mathbf{B}$), and uniaxial stress ($\mathbf{T}$)

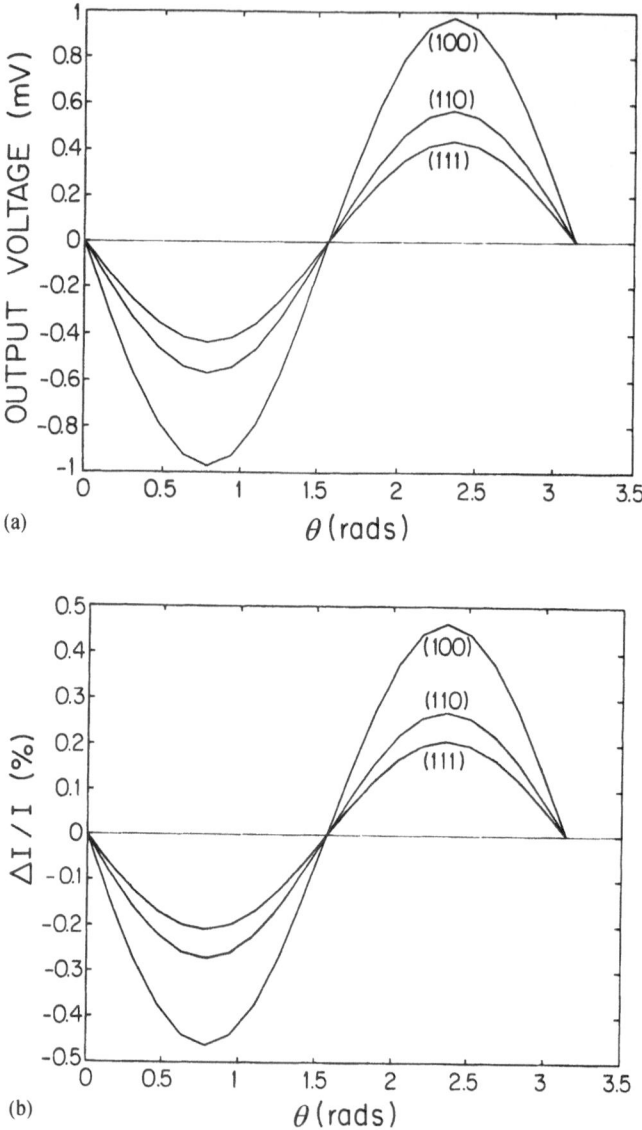

Fig. 6.24 Output response of $\langle 100 \rangle$, $\langle 110 \rangle$, and $\langle 111 \rangle$ Si (a) Hall voltage and (b) Hall current, sensors as a function of orientation θ under uniaxial stress of 10^8 dynes/cm^2 at $\mathbf{B} = 0$ T

element discretization procedure, using an adaptively generated grid (see [89] and references therein).

Simulation results of output response are illustrated in Figs. 6.24 to 6.26. The device dimensions are $100\,\mu m \times 100\,\mu m$ and the devices are biased at 1.25 V. For the classical Hall device geometry (Fig. 6.22(a)), the output voltage is taken across the two Hall probes, which produce a Hall

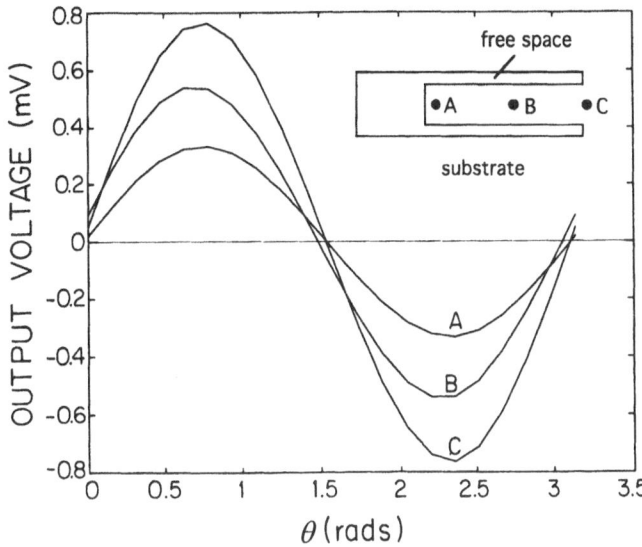

Fig. 6.25 The offset voltage at different locations on micromachined cantilever beam as a function of device orientation

voltage in the presence of a magnetic field perpendicular to the chip surface. For the split-electrode geometry, the current imbalance, $\Delta I = I_1 - I_2$, is taken to be a measure of the perpendicular field strength. Note that the output can be non-zero (device offset) in the absence of a magnetic field even in geometrically symmetric structures, due to stress-induced anisotropy.

The output response in the absence of magnetic field is shown in Fig. 6.24, under a uniaxial stress of 10^8 dynes/cm^2, for various substrate orientations. Here, the uniaxial stress acting on the sample has a fixed direction and the device (current flow) is rotated with the angle between stress and current directions, θ, varying from 0 to π radians. The offset voltage and normalized offset current $(\Delta I/I)$ vary sinusoidally with the rotation angle. As expected, the $\langle 100 \rangle$ substrate has the largest offset (compared to $\langle 110 \rangle$ and $\langle 111 \rangle$ counterparts) due to its relatively larger piezoresistance coefficient. The offsets are largest when the angle between the current density and uniaxial stress is 45°. For the $\langle 100 \rangle$ sample, this corresponds to current flow parallel to the $\langle 011 \rangle$ direction. The output is the lowest when the uniaxial stress is either parallel ($\theta = 0, \pi$) or perpendicular ($\theta = \pi/2$) to the direction of current flow. Although not shown, the magnetic response for both device geometries is simply shifted by the corresponding Hall voltage or current imbalance [15, 20]. This is intuitively reasonable since the principal piezo-Hall and piezoresistance coefficients are approximately of the same value. Hence, the conductivity-

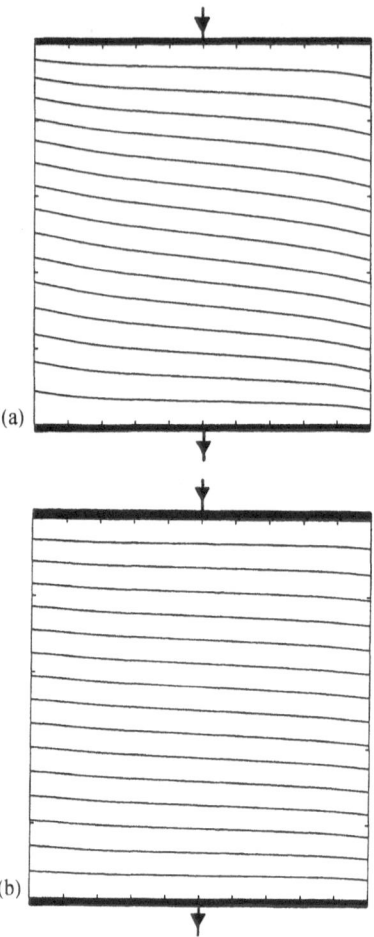

Fig. 6.26 Equipotential lines for a Hall device with (a) $\mathbf{B} = 2$ T, $T_{yy} = 0$ dyne/cm^2 and (b) $\mathbf{B} = 0$ T, $T_{yy} = 10^9$ dynes/cm^2

Hall coefficient product is approximately unity and the magnetic field response $(V_H(\mathbf{B})$ or $\Delta I(\mathbf{B}))$ is superimposed on the response at the zero measurand (offset).

The stresses can be reduced by suitably mechanically isolating the location of the Hall device. Various micromachined configurations along the lines reported in ref. [13] were investigated. For illustrative purposes, we consider the location of a (100)-Hall device on a cantilever beam and investigate the corresponding device offset. The simulated offset voltages, as a function of device orientation, are illustrated in Fig. 6.25 for three different locations on the device (see inset shown). As one would intuitively expect, the offset voltages increase as the location approaches the clamped end.

No significant non-linear coupling between the stress and the magnetic field was noted for magnetic field strengths as high as 2 Tesla [20]. This can be noted from the equipotential lines shown in Fig. 6.26, which appear to be virtually identical.

Drawing from various simulations, we find that although the encapsulation stress can be reduced by micromachining, thus mechanically isolating the location of the Hall device, the stress reduction stems rather from reduction of the pertinent (die bonding) contact area than from mechanical isolation. The encapsulation stress generally decreases in proportion to the decrease of contact area. However, there are practical limitations associated with minimizing contact area in view of heat sinking and mechanical integrity considerations. Thus some degree of compromise is necessary. Alternatively, a more feasible approach to offset reduction technique is to etch grooves on the chip surface, using maskless micromachining, to prevent stress creep into the device active region (see [26]).

6.9 References

[1] Hall, H. H., Bardeen, J., Pearson, G. L., The Effects of Pressure and Temperature on the Resistance of $p-n$ Junctions in Germanium, *Phys. Rev.*, 34, No. 1 (1951), 129–132.
[2] Smith, C. S., Piezoresistance Effect in Germanium and Silicon, *Phys. Rev.*, 94 (1954), 42–49.
[3] Keyes, R. W., The Effects of Elastic Deformation on the Electrical Conductivity of Semiconductors, *Solid State Physics*, Vol. 11, Seitz, F., Turnbull, D. (Eds.), New York: Academic Press, 1960, pp. 149–221.
[4] Kloeck, B., De Rooij, N. F., Mechanical Sensors, in: *Semiconductor Sensors*, Sze, S. M. (Ed.), Chapt. 4, New York: Wiley, 1994, pp. 153–204.
[5] Pearsall, T. P., Silicon Germanium Alloys and Heterostructures: Optical and Electronic Properties, *CRC Critical Reviews in Solid State and Material Sciences*, 15 (1989), 551–600.
[6] Jain, S. C., *Germanium-Silicon Strained Layers and Heterostructures*, New York: Academic Press, 1994.
[7] Manku, T., *Electronic Transport Properties of Strained and Relaxed* $Si_{1-x}Ge_x$ *Alloys*, Ph. D. Dissertation, Department of Electrical and Computer Engineering, University of Waterloo, Waterloo, Ontario N2L 3G1, Canada.
[8] Tummula, R. R., Rymaszewski, E. J., *Microelectronics Packaging Handbook*, New York: Van Nostrand-Reinhold, 1989.
[9] Chin, S.-W., Rajan, S. D., Nagaraj, B. K., Mahalingam, M., Automated Design Tool for Examining Microelectronic Packaging Design Alternatives, *IEEE Transactions on Components, Packaging, and Manufacturing Technology-Part B: Advanced Packaging*, 17 (1994), 76–82.
[10] Hu, S. M., Stress-Related Problems in Silicon Technology, *J. Appl. Phys.*, 70 (1991), R53–R67.
[11] Baltes, H. P., Popovic, R. S., Integrated Semiconductor Magnetic Field Sensors, *Proc. IEEE*, 74 (1986), 1107–1132.
[12] Popovic, R. S., *Hall Effect Devices*, New York: Adam Hilger, 1991.

[13] Hälg, B., Popovic, R. S., How to Liberate Integrated Sensors from Encapsulation Stress, *Sensors and Actuators*, A21–A23 (1990), 908–910.

[14] Hälg, B., Piezo-Hall Coefficients of *n*-Type Silicon, *J. Appl. Phys.*, 64 (1988), 276–282.

[15] Nathan, A., Manku, T., Modeling of Piezo-Hall Effects in *n*-Doped Silicon Devices, *Appl. Phys. Lett.*, 62 (1993), 2947–2949.

[16] Egley, J. L., Chidambarrao, D., Strain Effects on Device Characteristics: Implementation in Drift-Diffusion Simulators, *Solid-State Electronics*, 36 (1993), 1653–1664.

[17] Miura, H., Tanizaki, Y., Effect of Process-Induced Mechanical Stress on Circuit Layout, *Simulation of Semiconductor Devices and Processes*, Vol. 6, Ryssel, H., Pichler, P. (Eds.), Wien-New York: Springer-Verlag, 1995, pp. 147–150.

[18] Wang, Z. Z., Suski, J., Collard, D., Silicon Piezoresistivity Modelling: Application to the Simulation of MOSFETs, *Sensors and Actuators A*, 46–47 (1995), 628–631.

[19] Ferreira, P., Senez, V., Baccus, B., Mechanical Stress Analysis of an LDD MOSFET Structure, *IEEE Transactions on Electron Devices*, 43 (1996), 1525–1532.

[20] Manku, T., Nathan, A., O, N., Aflatooni, K., Allegretto, W., Modeling of Encapsulation Stress Effects on Output Response of Hall Sensors, *Sensors and Materials*, 6 (1994), 225–234.

[21] Lades, M., Frank, J., Funk, J., Wachutka, G., Analysis of Piezoresistive Effects in Silicon Structures Using Multidimensional Process and Device Simulation, *Simulation of Semiconductor Devices and Processes*, Vol. 6, Ryssel, H., Pichler, P., (Eds.), Wien-New York: Springer-Verlag, 1995, pp. 22–25.

[22] Ciampolini, P., Pierantoni, A., Rudan, M., A CAD Environment for the Numerical Simulation of Integrated Piezoresistive Transducers, *Sensors and Actuators A*, 46–47 (1995), 618–622.

[23] Manku, T., Nathan, A., Valence Energy-Band Structure for Strained Group-IV Semiconductors, *J. Appl. Phys.*, 73 (1993), 1205–1213. (Erratum: J. Appl. Phys., 74 (1993), 4803).

[24] Middelhoek, S., Audet, S. A., *Silicon Sensors*, New York: Academic Press, 1989.

[25] Nye, J. F., *Physical Properties of Crystals*, Oxford: Oxford University Press, 1957.

[26] Aflatooni, K., *Strained Silicon Hall Effect Devices*, M.A.Sc. Thesis, University of Waterloo, Waterloo, Ontario N2L 3G1, Canada, 1994.

[27] Kanda, Y., A Graphical Representation of the Piezoresistance Coefficients in Silicon, *IEEE Trans. Electron Devices*, ED-29 (1982), 64–70.

[28] Kanda, Y., Piezoresistance Effect in Silicon, *Sensors and Actuators A*, 28 (1991), 83–91.

[29] Tufte, O. N., Steizer, E. L., Piezoresistive Properties of Silicon Diffused Layers, *J. Appl. Phys.*, 34 (1963), 313–318.

[30] Kerr, D. R., Milnes, A. G., Piezoresistance of Diffused Layers in Cubic Semiconductor, *J. Appl. Phys.*, 34 (1963), 727–731.

[31] Yamada, K., Nishihara, M., Shimada, S., Tanabe, M., Shimazoe, M., Matsuoka, Y., Nonlinearity of the Piezoresistance Effect of *p*-Type Silicon Diffused Layers, *IEEE Trans. Electron Devices*, ED-29 (1982), 71–77.

[32] Matsuda, K., Kanda, Y., Yamamura, K., Suzuki, K., Nonlinearity of Piezoresistance Effect in *p*- and *n*-Type Silicon, *Sensors and Actuators*, A21-A23 (1990), 45–48.

[33] Matsuda, K., Kanda, Y., Suzuki, K., Second-Order Piezoresistance Coefficients of *n*-Type Silicon, *Jpn. J. Appl. Phys.*, 28 (1989), 1676–1677.

[34] Matsuda, K., Kanda, Y., Yamamura, K., Suzuki, K., Second-Order Piezoresistance of *p*-Si, *Jpn. J. Appl. Phys.*, 29 (1990), 1941–1942.

[35] Suski, J., Mosser, V., Goss, J., Polysilicon SOI Pressure Sensor, *Sensors and Actuators*, 17 (1989), 405–414.

[36] Gridchin, V. A., Lubimsky, V. M., Sarina, M. P., Polysilicon Strain Gauge Transducers, *Sensors and Actuators A*, 30 (1992), 219–223.

[37] Mosser, V., Suski, J., Goss, J., Obermeier, E., Piezoresistive Pressure Sensors Based on Polycrystalline Silicon, *Sensors and Actuators A*, 28 (1991), 113–132.

[38] French, P. J., Evans, A. G. R., Polysilicon Strain Sensors, *Sensors and Actuators*, 7 (1985), 135–142.

[39] Schubert, D., Jenschke, W., Uhlig, T., Schmidt, F. M., Piezoresistive Properties of Polycrystalline and Crystalline Silicon Films, *Sensors and Actuators*, 11 (1987), 145–155.

[40] Gridchin, V. A., Lubimsky, V. M., Sarina, M. P., Piezoresistive Properties of Polysilicon Films, *Sensors and Actuators A*, 49 (1995), 67–72.

[41] Voronin, V. A., Druzhinin, A. A., Marjamova, I. I., Kostur, V. G., Pankov, Ju. M., Laser-Recrystallized Polysilicon Layers in Sensors, *Sensors and Actuators A*, 30 (1992), 143–147.

[42] Binder, J., Henning, W., Obermeier, E., Schaber, H., Cutter, D., Laser-Recrystallized Polysilicon Resistors for Sensing and Integrated Circuits Applications, *Sensors and Actuators*, 4 (1983), 527.

[43] Dössel, O., Longitudinal and Transverse Gauge Factors of Polycrystalline Strain Gauges, *Sensors and Actuators*, 6 (1994), 169–179.

[44] French, P. J., Evans, A. G. R., Polysilicon Strain Sensors Using Shear Piezoresistance, *Sensors and Actuators*, 15 (1988), 257–272.

[45] Schäfer, H., Graeger, V., Kobs, R., Temperature-Independent Pressure Sensors Using Polycrystalline Silicon Strain Gauges, *Sensors and Actuators*, 17 (1989), 521–527.

[46] Obermeier, E., Kopystynski, P., Polysilicon as a Material for Microsensor Applications, *Sensors and Actuators A*, 30 (1992), 149–155.

[47] Jeanjean, P., Sicart, J., Robert, J. L., Dutartre, D., Conedera, V., Electrical and Piezoresistive Properties of Boron-Implanted ZMR-SOI Films, *Sensors and Actuators A*, 36 (1993), 241–245.

[48] Semmache, B., Kleimann, P., Le Berre, M., Lemiti, M., Barbier, D., Pinard, P., Rapid Thermal Processing of Piezoresistive Polycrystalline Silicon Films: An Innovative Technology for Low Cost Pressure Sensor Fabrication, *Sensors and Actuators A*, 46–47 (1995), 76–81.

[49] Le Berre, M., Lemiti, M., Barbier, D., Pinard, P., Cali, J., Bustarret, E., Sicart, J., Robert, J. L., Piezoresistance of Boron-Doped PECVD and LPCVD Polycrystalline Silicon Films, *Sensors and Actuators A*, 46–47 (1995), 166–170.

[50] Germer, W., Microcrystalline Silicon Thin Films for Sensor Applications, *Sensors and Actuators*, 7 (1985), 135–142.

[51] Kodato, S., Sugiura, I., Ikeda, A., Otake, S., Nishida, S., Konagai, M., Takahashi, K., Piezoresistance Effects in a-Si:H:F Films and Application to Strain Gauges, *Proc. 2nd Sensor Symposium*, Tsukuba, Japan, 1982, p. 185.

[52] Onuma, Y., Kamimura, K., Nagura, Y., Yi, C. H., Kiuchi, M., Polycrystalline Silicon Carbide Films for Piezoresistive Elements, *Sensors and Materials*, 2 (1991), 207–216.

[53] Shor, J. S., Goldstein, D., Kurtz, A. D., Evaluation of β-SiC for Sensors, *Digest of Technical Papers*, Transducers '91, San Francisco, 1991, pp. 912–915.

[54] Taher, I., Aslam, M., Tamor, M. A., Potter, T. J., Elder, R. C., Piezoresistive Microsensors Using p-Type CVD Diamond Films, *Sensors and Actuators A*, 45 (1994), 35–43.

[55] Herring, C., Vogt, E., Transport and Deformation-Potential Theory for Many-Valley Semiconductors with Anisotropic Scattering, *Phys. Rev.*, 101 (1956), 944–961.

[56] Manku, T., Nathan, A., Electrical Properties of Silicon Under Nonuniform Stress, *J. Appl. Phys.*, 74 (1993), 1832–1837.

[57] Dresselhaus, G., Kip, A. F., Kittel, C., Cyclotron Resonance of Electrons and Holes in Silicon and Germanium, *Phys. Rev.*, 98 (1955), 368–383.

[58] Manku, T., Nathan, A., Electron Drift Mobility for Devices Based on Unstrained and Coherently Strained $Si_{1-x}Ge_x$ Grown on $\langle 001 \rangle$ Silicon Substrate, *IEEE Trans. Electron Devices*, 39 (1992), 2082–2089.

[59] Bassani, F., Brust, D., Effect of Alloying and Pressure on the Band Structure of Germanium and Silicon, *Phys. Rev.*, 131 (1963), 1524–1529.

[60] Moriarty, J. A., Krishnamurthy, S., Theory of Silicon Superlattices: Electronic Structure and Enhanced Mobility, *J. Appl. Phys.*, 54 (1983), 1892–1902.

[61] Pikus, G. E., Bir, G. L. Effects of Deformation on the Hole Energy Spectrum of Germanium and Silicon, *Soviet Physics-Technical Physics* (1958), 2194.

[62] Kane, E. O., Energy Band Structure in *p*-Type Germanium and Silicon, *J. Phys. Chem. Solid*, 1 (1956), 82–99.

[63] Shockley, W., Energy Band Structure of Semiconductors, *Phys. Rev.*, 78 (1950), 173–174.

[64] Nathan, A., Manku, T., Piezoresistance and the Drift-Diffusion Model in Strained Silicon, *Simulation of Semiconductor Devices and Processes,* Vol. 6, Ryssel, H., Pichler, P. (Eds.),Wien-New York: Springer-Verlag, 1995, pp. 94–97.

[65] Wortman, J. J., Hauser, J. R., Burger, R. M., Effect of Mechanical Stress on *p-n* Junction Device Characteristics, *J. Appl. Phys.*, 35 (1964), 2122–2131.

[66] Sansen, W., Vandeloo, P., Puers, B., A Force Transducer Based on Stress Effects in Bopilar Transistors, *Sensors and Actuators*, 3 (1982), 343–354.

[67] Puers, B., Reynaert, L., Snoeys, W., Sansen, M. C., A New Unixial Accelerometer in Silicon Based on the Piezojunction Effect, *IEEE Trans. on Electron Devices*, 35, (1988), 764–770.

[68] Schellin, R., Mohr, R., A Monolithically-Integrated Transistor Microphone: Modelling Theoretical Behaviour, *Sensors and Actuators A*, 37-38 (1993), 666–673.

[69] Nathan, A., Baltes, H. P., Sensor Modelling, in: *Sensors*, Chapt. 3, Grandke, T, Ko, W. H. (Eds.), Vol. 1, Weinheim: VCH, 1989, pp. 45–77.

[70] Marshak, A. H., van Vliet, K. M., Electrical Current in Solids with Position-Dependent Band Structure, *Solid-State Electronics*, 21 (1987), 417–427.

[71] Aflatooni, K., Nathan, A., Heat Transport Properties of Semiconductors Under Non-Uniform Stress, *Appl. Phys. Lett.*, 66 (1995), 1110–1111.

[72] Yin, M. T., Cohen, M. L., Theory of Lattice-Dynamical Properties of Solids: Application to Si and Ge, *Phys. Rev. B*, 26 (1982), 3259–3272.

[73] Soma, T., Kudo, K., Thermal Expansion and Pressure Effect on the Lattice Vibration of Tetrahedrally Covalent Compounds, *J. Phys. Soc. Jpn.*, 48 (1980), 115–122.

[74] Aflatooni, K., Hornsey, R., Nathan, A., Thermodynamic Treatment of Mechanical Stress Gradients in Coupled Electro-Thermo-Mechanical Systems, *Sensors and Materials*, 9 (1997), 449–456.

[75] Allegretto, W., Nathan, A., Baltes, H., Numerical Analysis of Magnetic-Field-Sensitive Bipolar Devices, *IEEE Trans. CAD of ICAS*, 10 (1991), 501–511.

[76] McCartin, B. J., Discretisation of the Semiconductor Device Equations, in: *New Problems and New Solutions for Device and Process Modeling*, Miller, J. J. H., (Ed.), Dublin: Boole Press, 1985, pp. 72–82.

[77] Rudan, M., Guerrieri, R., Ciampolini, P., Baccarani, G., Discretisation Strategies and Software Implementation for a General-Purpose 2D-Device Simulator, in: *New Problems and New Solutions for Device and Process Modeling*, Miller, J. J. H., (Ed.), Dublin: Boole Press, 1985, pp. 110–121.

[78] Selberherr, S., *Analysis and Simulation of Semiconductor Devices*, Wien-New York: Springer-Verlag, 1984.

[79] Zienkiewicz, O. C., *The Finite Element Method*, 3rd Ed., New York: McGraw-Hill, 1977.

[80] Strang, G., Fix, G. F., *An Analysis of the Finite Element Method*, Englewood Cliffs, New Jersey: Prentice-Hall, 1973.

[81] Cottrell, P. E., Buturla, E. M., Two-Dimensional Static and Transient Simulation of Mobile Carrier Transport in a Semiconductor, *Proc. NASECODE 1 Conf.*, Browne, B. T., Miller, J. J. H., (Eds.), Dublin: Boole Press, 1979, pp. 31–64.

[82] Scharfetter, D. L., Gummel, H. K., Large-Signal Analysis of a Silicon Read Diode Oscillator, *IEEE Trans. Electron Devices*, ED-16 (1969), 64–77.

[83] Thangaraj, D., Nathan, A., The Discretization of Anisotropic Drift-Diffusion Equations, *Technical Report*, UW E&CE 97-11, University of Waterloo, Waterloo, Ontario N2L 3G1, Canada, 1997.

[84] Gummel, H. K., A Self-Consistent Iterative Scheme for One-Dimensional Steady State Transistor Calculations, *IEEE Trans. Electron Devices*, ED-11 (1964), 455–465.

[85] Bank, R. E., Rose, D. J., Parameter Selection for Newton-Like Methods Applicable to Nonlinear Partial Differential Equations, *SIAM J. Num. Anal.*, 17 (1980), 806–822.

[86] Seidman, T. I., Choo, S. C., Iterative Scheme for Computer Simulation of Semiconductor Devices, *Solid-State Electronics*, 15 (1972), 1229–1235.

[87] Munter, P. J. A., *Spinning-Current Method for Offset Reduction in Silicon Hall Plates*, Ph. D. Thesis, Department of Electrical Electrical Engineering, Delft University of Technology, Mekelweg 4, 2600 GA Delft, the Netherlands, 1992.

[88] Bar-Cohen, A., Kraus, A. D., *Advances in Thermal Modeling of Electronic Components and Systems*, Bristol, PA: Hemisphere Publishing Corp., 1988.

[89] Allegretto, W., Nathan, A., Manku, T., Numerical Simulation of Piezo-Hall Effects in n-Doped Silicon Magnetic Sensors, in *Simulation of Semiconductor Devices and Processes*, Vol. 5, Selberherr, S., Stipel, H., Strasser, E., (Eds.), Wien-New York: Springer-Verlag, 1993, pp. 377–380.

[90] Thangaraj, D., Nathan, A., A Rotated Monotone Difference Scheme for the Two-Dimensional Anisotropic Drift-Diffusion Equation, *J. of Computational Physics*, 145 (1998), 1–17.

7 Mechanical and Fluidic Signals

Simulation of the static and dynamic behavior of solid structural and fluid mechanical variables, e.g., stress, strain, strain-rate, displacement, force, and velocity, is critical to the design and analysis of microsensors and microactuators in the mechanical domain. For example, in Chapt. 6, we saw how electrical transport is modified by piezoresistance. This, in addition to deflection-induced capacitance change, can be effectively utilized for conversion of signals from the mechanical to the electrical domain. Alternatively, as we will see in Chapt. 8, a micromechanical structure subject to an electrical, thermal, magnetic, or mechanical excitation signal, gives rise to micro-actuation in the mechanical domain. In this chapter, we deal with model equations and constitutive relations relevant to: simulation of mechanical (e.g., pressure) microsensors; computation of velocity profiles relevant to flow microsensors (needed in Chapt. 5) and selected microfluidic systems; computation of mechanical stresses induced by packaging or encapsulation of microtransducers or integrated circuits (needed in Chapt. 6); and simulation of mechanical microactuators, including fluidic damping effects (needed in Chapt. 8).

The model equations presented here stem from classical theory based on the continuum hypothesis. Here, we assume that the physical properties (e.g., density) of the medium have uniquely prescribed values at any position. Hence, the associated mechanical variables at any point in the interior are continuous functions of space and time. The assumptions underlying the continuum hypothesis are justified provided the physical dimensions of the microtransducer are at least hundred times larger than the molecular size. In this case, the statistical fluctuation in number of molecules is small relative to the total number of molecules in the medium, and the physical state of individual molecules is insignificant.

In what follows, we begin with definitions of mechanical variables and their properties, including rules for coordinate transformation to deal with the anisotropies in the medium. Next we discuss the governing equations and constitutive relations, pertinent to solid structural and fluidic mechanical analysis. An excellent treatise of the continuum hypothesis and associated physical laws can be found in, e.g., [1], and its applications to solid and fluid media can be found in [2] and [3], respectively.

7.1 Definitions

In this Section, we describe the various mechanical variables pertinent to the analysis of solid and fluid media. For convenience, we switch from Cartesian to index notation, as appropriate, either to accommodate our reference to particular coordinate systems or for compact representation. We begin with an introduction to transformations pertinent to coordinate systems, vectors, and tensors.

7.1.1 Transformations

The orientation of orthogonal axes systems to describe the mechanical field variables (e.g., displacements) and material properties (e.g., stiffness coefficients) can be chosen arbitrarily. Thus, if values of these quantities are prescribed for a given axes orientation, then in moving to a different orientation, the values can change and can be determined if the relation between the two orientations is known. This does not imply that the properties of the medium are dependent on the choice of the coordinate system; the properties remain the same, but their values can depend on orientation.

The relations between two sets of mutually orthogonal axes $(x_1, x_2, x_3$ and $x_1', x_2', x_3')$ with the same origin can be expressed in terms of directional cosines, see e.g., [4],

$$
\begin{array}{c|ccc}
 & x_1 & x_2 & x_3 \\
\hline
x_1' & l_{11} & l_{12} & l_{13} \\
x_2' & l_{21} & l_{22} & l_{23} \\
x_3' & l_{31} & l_{32} & l_{33}
\end{array}
\tag{7.1}
$$

where the primed axes represent the new orientation. The entries, l_{ij} represent the cosine of the angle between the new (i) and the old (j) axes. For example, the direction cosines of x_1' from x_1, x_2, and x_3, are l_{11}, l_{12}, and l_{13}, respectively (see Fig. 7.1). Thus their mutual orientation can be specified by three angles, which are referred to as the Euler angles, i.e., $\cos^{-1} l_{ij}$. The entries l_{ij} are not independent, viz.,

$$
l_{ik} l_{jk} = \delta_{ij} \qquad \text{with} \quad i, j, k = 1, 2, 3,
\tag{7.2}
$$

where δ_{ij} denotes the Kronecker delta; $\delta_{ij} = 1$ when $i = j$ and zero

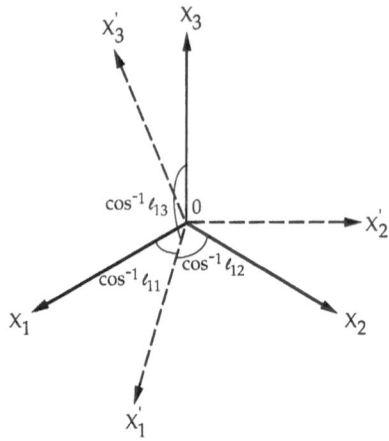

Fig. 7.1 Transformation of mutually orthogonal coordinate systems from x_1, x_2, x_3 (old) to x_1', x_2', x_3' (new) axes

otherwise. In Eq. (7.2) and hereafter, we adopt the summation convention, i.e., a repeated index implies summation over its range. Condition (7.2) serves as a convenient check for valid transformation since the squares of the entries in matrix (7.1) should add to unity, both row-wise and column-wise.

Vectors are transformed in a similar manner. A vector **p** in the old orientation when transformed reads,

$$p_i' = l_{ij} p_j \qquad \text{with} \quad i, j = 1, 2, 3, \tag{7.3}$$

where p_i' are the components of the transformed vector and the l_{ij} are direction cosines, (7.1). The coordinates of a point (position vector) can be transformed as

$$x_i' = l_{ij} x_j \qquad \text{with} \quad i, j = 1, 2, 3, \tag{7.4}$$

where x_i' corresponds to the coordinates of the point in the transformed axes.

Tensors specified relative to one set of axes can be transformed to a different set of axes, again using direction cosines. For example, a second rank tensor T_{kl} transforms as

$$T_{ij}' = l_{ik} l_{jl} T_{kl} \qquad \text{with} \quad i, j, k, l = 1, 2, 3 \tag{7.5}$$

where T_{ij}' and T_{kl} are the respective tensors in the new and old orientations. Transformation (7.5) holds for the tensors of conductivity (see Chapt. 6), permittivity and permeability (see Chapt. 8), and mechanical stress and strain (this chapter). Similarly, we obtain for a third rank tensor.

$$T_{ijk}' = l_{il} l_{jm} l_{kn} T_{lmn} \qquad \text{with} \quad i, j, k, l, m, n = 1, 2, 3. \tag{7.6}$$

Relation (7.6) holds for the tensor of piezoelectric coefficients (see Chapt. 8). A fourth rank tensor transforms according to

$$T'_{ijkl} = l_{im} l_{jn} l_{ko} l_{lp} T_{mnop} \quad \text{with} \quad i,j,k,l,m,n,o,p = 1,2,3$$

(7.7)

which holds for the tensor of piezoresistance coefficients (see Chapt. 6) and stiffness coefficients (see Sect. 7.2).

7.1.2 Forces

There are two kinds of forces that act on a solid medium; body forces and surface forces. The body force

$$\int_V f_i^b \, dV \quad \text{with} \quad i = 1,2,3,$$

(7.8)

acts throughout the volume of a solid. Here, the f_i^b denote components of the body force density (unit: N/m^3), V the volume of the solid, and dV the volume element. Examples of body forces include gravitational forces, electrodynamic forces, magnetostatic forces, and inertial forces (see Chapt. 8). The surface force f_i^s (unit: N) acts on the surface of a body and following Cauchy's stress principle [5], it can be expressed in terms of stress components (to be introduced in Sect. 7.1.3 below):

$$f_i^s = \int_\Gamma \sigma_{ij} n_j \, d\Gamma \quad \text{with} \quad i,j = 1,2,3,$$

(7.9)

where σ_{ij} denote stress components (unit: N/m^2), n_j the jth component of the unit vector normal to the surface, Γ the integration surface, and $d\Gamma$ the surface element. Eq. (7.9) describes the stress vector for a given orientation of the surface at a point. The integrand in (7.9) is also referred to as surface traction. Examples of surface forces include electrostatic forces (Sect. 8.3), hydrostatic pressure, and fluidic damping forces (Sect. 7.3).

7.1.3 Stress

Regardless of their origin, forces acting on a solid give rise to a mechanical stress which exists on the surface of every internal element in the body. With reference to the internal elemental unit cube shown in Fig. 7.2, the force acting on each face can be resolved into three components. For a face perpendicular to the x_1-axis, there is a normal component, σ_{11}, and two tangential components, σ_{12} and σ_{13}. For the face perpendicular to the x_2-axis, the normal component is σ_{22} and the tangential components are σ_{21}

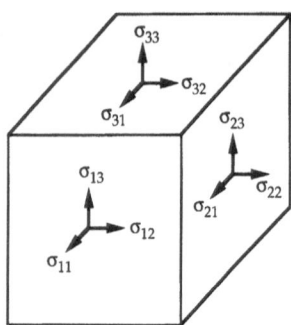

Fig. 7.2 Stress components on the faces of an internal elemental cube

and σ_{23}. Similarly, the face perpendicular to the x_3-axis has a normal component, σ_{33}, and tangential components, σ_{31} and σ_{32}. Collectively, these nine components form a second rank stress tensor,

$$\sigma = \begin{bmatrix} \sigma_{11} & \sigma_{12} & \sigma_{13} \\ \sigma_{21} & \sigma_{22} & \sigma_{23} \\ \sigma_{31} & \sigma_{32} & \sigma_{33} \end{bmatrix}. \tag{7.10}$$

Each component is characterized by two indices; the first describes the plane on which the stress is acting while the second represents the direction of stress. The diagonal terms of the tensor σ_{dd} denote normal stress components while the off-diagonal terms signify shear stresses. In general, the σ_{ij} are functions of space and time. Following equilibrium conditions, we can write

$$\sigma_{ij} = \sigma_{ji}, \tag{7.11}$$

which holds regardless of the nature of the stress distribution (i.e., homogeneous or inhomogeneous) including body forces and acceleration of internal elements in the solid. Relation (7.11) is transformation invariant. Following (7.11), the number of stress components is reduced from nine to six independent components. By transformation of the tensor to the principal axes, the shear components vanish to yield a diagonalized tensor of principal stresses [4]. The three associated directions are referred to as principal directions of stress and the planes perpendicular to the principal directions are referred to as the principal planes of stress.

7.1.4 Strain and Thermal Expansion

The strain in a solid medium describes the state of deformation of the solid. The deformation refers to the change in distance between two points in the solid and is represented by a displacement vector **u**. The spatial derivatives of the displacement components yield the Lagrangian strain components, $\varepsilon_{ij}(x_1, x_2, x_3)$, viz.,

$$\varepsilon_{ij} = \frac{1}{2} \left[\frac{\partial u_i}{\partial x_j} + \frac{\partial u_j}{\partial x_i} + \frac{\partial u_k}{\partial x_i} \frac{\partial u_k}{\partial x_j} \right] \quad \text{with} \quad i, j, k = 1, 2, 3, \quad (7.12)$$

where the position of the internal points x_i in the solid in the undeformed state are the independent variables. The values of strain components, and in particular, the product term can be assumed small even for large deformation, although it could turn out to be important in thin film membrane or plate structures. In the case of infinitesimal strain, (7.12) yields a linear strain-displacement relation,

$$\varepsilon_{ij} = \frac{1}{2} \left[\frac{\partial u_i}{\partial x_j} + \frac{\partial u_j}{\partial x_i} \right] \quad \text{with} \quad i, j = 1, 2, 3. \quad (7.13)$$

The resulting strain tensor has the form

$$\varepsilon = \begin{bmatrix} \varepsilon_{11} & \varepsilon_{12} & \varepsilon_{13} \\ \varepsilon_{21} & \varepsilon_{22} & \varepsilon_{23} \\ \varepsilon_{31} & \varepsilon_{32} & \varepsilon_{33} \end{bmatrix}. \quad (7.14)$$

Here, the diagonal terms $(i = j)$ signify normal strain components (parallel to the coordinate axes) and the off-diagonal $(i \neq j)$ terms signify components of shear strain (rotations about the coordinate axes). Again, symmetry considerations lead to $\varepsilon_{ij} = \varepsilon_{ji}$ and hence reduce the strain tensor to six independent components. Analogous to principal stresses, transformation of the strain tensor to the principal axes yields vanishing shear components which results in a diagonalized tensor of principal strain components.

Temperature variations in the material give rise to mechanical deformation, even in the absence of external forces, by virtue of thermal expansion or contraction. If the change in temperature is homogeneous, so too is the associated deformation, and the strain components can be related to the change in temperature (ΔT) via the coefficients of thermal expansion. For a small change in temperature, the resulting thermal strain components ε_{ij}^T, neglecting higher order terms, read [6]

$$\varepsilon_{ij}^T = \alpha_{ij} \Delta T \quad \text{with} \quad i, j = 1, 2, 3, \quad (7.15)$$

where α_{ij} is the second rank tensor of linear thermal expansion coefficients. Since the strain tensor is symmetric, so is α_{ij}. Relation (7.15) is only valid in the elastic regime. Referring the tensors to the principal axes, we get

$$\varepsilon_i^T = \alpha_i \Delta T \quad \text{with} \quad i = 1, 2, 3, \quad (7.16)$$

where the α_i are the principal thermal expansion coefficients, which for most materials are positive with values dependent on crystal symmetry [4]. In crystalline materials of cubic symmetry (e.g., Si), there is only one independent coefficient and the thermal expansion behavior is similar to that of an isotropic material. Polycrystalline materials (e.g., poly Si) can also be considered isotropic since their grain size is generally small compared to the deformation [7]. Values of expansion coefficients for selected materials relevant to microtransducers are given in Sect. 7.2.3.

7.1.5 Strain-Rate

Unlike solids, fluids have a limited ability to sustain even minute applied forces, and therefore need to be characterized in terms of the deformation- or strain-rate. The instantaneous locations of interior points in the fluid are described by a velocity field. In analogy to strain in a solid material defined by displacement gradients (7.13), fluids are characterized by the velocity gradient tensor, whose symmetric part is the strain-rate tensor (see, e.g., [5]):

$$\dot{\varepsilon}_{ij} = \frac{1}{2}\left(\frac{\partial v_i}{\partial x_j} + \frac{\partial v_j}{\partial x_i}\right) \qquad \text{with} \quad i,j = 1,2,3 \tag{7.17}$$

and the anti-symmetric part is the spin tensor

$$\Omega_{ij} = \frac{1}{2}\left(\frac{\partial v_j}{\partial x_i} - \frac{\partial v_i}{\partial x_j}\right) \qquad \text{with} \quad i,j = 1,2,3. \tag{7.18}$$

Here, $v_i = \partial u_i / \partial t$ denotes the velocity field. Eq. (7.18) is not considered in analysis of solid structures.

7.2 Model Equations for Mechanical Analysis

Computation of the mechanical field variables (e.g., stress, displacement) is basic to the design of micromechanical devices ranging from mechanical sensors for measurement of pressure, force, acceleration, angular rate (gyroscopes), acoustic fields (microphones), to micromechanical actuation systems. Micromechanical devices are inherently three-dimensional structures realized using integrated silicon microfabrication processes coupled with application-specific thin film deposition, bulk and/or surface micromachining technologies. In particular, mechanical microsensors rely on stress, deflection, or change in resonant frequency induced by, e.g., pressure, force, or acceleration, of a cantilever, bridge, or diaphragm [8, 9]. The conversion of the mechanical signal to an electrical signal is based on the measurement of capacitance or resistance modulation. This section presents differential equations, constitutive relations, and boundary conditions, along with relevant mechanical material models and/or data including extraction procedures, for simulation of mechanical behavior in these devices. Equation systems for analyses of mixed signals pertinent to microactuation are reviewed in Chapt. 8.

Static and dynamic mechanical analysis using analytical or quasi-analytical techniques (based on the Ritz method or Fourier expansions) although simple, compact, fast, and intuitive, entail several simplifying assumptions which limit the range of structures that can be considered for optimization (see, e.g., [10–13]). Here, solutions are only possible with idealized device geometries and boundary conditions. Even for simple geometries, it is difficult, if not impossible, to formulate appropriate boundary conditions that accurately reflect the clamping conditions of the supporting rim and associated degrees of freedom. These have a strong bearing on the stiffness or rigidity of the micromechanical structure [14, 15], and hence, on its static and dynamic behavior. For treatment of more complex structures involving material inhomogeneities and anisotropies, arbitrary device geometries, boundary conditions, and loading conditions, numerical simulation is mandatory [16, 17]. In particular, there are several critical design and manufacturing issues related to micromechanical sensors which need to be addressed:

- Assessment of mechanical performance of structural variants such as corrugated [18], bossed [19], and perforated diaphragms [20].
- Effect of internal stresses and their distribution, on output response (e.g., offset and non-linearity). Internal stresses vary with different materials and deposition conditions; in particular, they can be significant with boron-doped silicon micromachined diaphragms [18]. Also there can be residual stresses arising from thermal mismatch of different materials including packaging and assembly. In particular, residual stresses can lead to temperature-dependent static and dynamic performance [21, 22].
- Detailed insight into distributions of mechanical field variables (stress and deflection) for optimization of device structure/geometry, including piezoresistor placement, with respect to static and dynamic performance such as sensitivity, offset, non-linearity, and resonant frequency [23–33].
- Assessment of integrity of device packaging/assembly with respect to die attach materials and mounting configurations [34–37].
- Assessment of performance sensitivity, including offsets, to process imperfections associated with mask misalignment and over-etching, which give rise to geometrical non-idealities.

7.2.1 Governing Equations and Constitutive Relations

The differential equation that governs static and dynamic mechanical behavior in solids stems from the principle of balance of momentum (see, e.g., [2, 5]). Specifically, Newton's second law of motion states that the rate of change of momentum is equal to the resultant of forces acting on the solid. The components of linear momentum of the solid are

$$p_i = \int_V \rho v_i \, dV \qquad \text{with} \quad i = 1, 2, 3, \tag{7.19}$$

where V denotes the volume of the solid, dV the volume of an associated element, ρ the mass density, and v_i the time-dependent components of the velocity vector field. If the solid body is subject to surface and body forces, the resultant force, following (7.8) and (7.9), is

$$F_i = \int_\Gamma \sigma_{ij} n_j \, d\Gamma + \int_V f_i^b \, dV \qquad \text{with} \quad i, j = 1, 2, 3. \tag{7.20}$$

The surface integral in (7.20) can be transformed to a volume integral by virtue of Gauss' theorem, to yield the following form for the resultant force:

$$F_i = \int_V \left(\frac{\partial \sigma_{ij}}{\partial x_j} + f_i^b \right) dV \qquad \text{with} \quad i, j = 1, 2, 3. \tag{7.21}$$

Following Newton's law, Eqs. (7.19) and (7.21) lead to

$$\frac{\partial \sigma_{ij}}{\partial x_j} + f_i^b = \frac{\partial}{\partial t}(\rho v_i) + \frac{\partial}{\partial x_j}(\rho v_i v_j) \qquad \text{with} \quad i, j = 1, 2, 3, \tag{7.22}$$

where the second term on the right-hand side describes the net change in momentum flux through a control (discrete) surface. Eq. (7.22) can be expanded to yield

$$\frac{\partial \sigma_{ij}}{\partial x_j} + f_i^b = v_i \left[\frac{\partial \rho}{\partial t} + \frac{\partial (\rho v_j)}{\partial x_j} \right] + \rho \left[\frac{\partial v_i}{\partial t} + v_j \frac{\partial v_i}{\partial x_j} \right] \tag{7.23}$$

$$\text{with} \quad i, j = 1, 2, 3.$$

The first term on the right-hand side of (7.23) vanishes by virtue of the principle of mass conservation which postulates that in an infinitesimal volume element, in the absence of sources or sinks, the rate of mass flow per unit volume must equal the time rate of change of mass, viz.,

$$\frac{\partial \rho}{\partial t} + \frac{\partial}{\partial x_i}(\rho v_i) = 0 \qquad \text{with} \quad i = 1, 2, 3. \tag{7.24}$$

Equation (7.24) is also referred to as the continuity equation for mass. Following (7.24), we can reduce (7.23) to

$$\frac{\partial \sigma_{ij}}{\partial x_j} + f_i^b = \rho \frac{\partial v_i}{\partial t} + \rho v_j \frac{\partial v_i}{\partial x_j} \qquad \text{with} \quad i,j = 1,2,3. \qquad (7.25)$$

The first term on the right-hand side describes the rate of increase of momentum per unit volume and the second term is a non-linear term which describes the rate of momentum gained by convection per unit volume. Collectively, these two terms describe the acceleration field a_i associated with an interior point in the solid. Thus Eq. (7.25) can also be stated as

$$\frac{\partial \sigma_{ij}}{\partial x_j} + f_i^b = \rho a_i \qquad \text{with} \quad i,j = 1,2,3. \qquad (7.26)$$

The velocity and acceleration of an interior point in the solid can be described in terms of the associated displacement. Using substantial derivatives, D/Dt (see, e.g., [5]), we can write

$$v_i = \frac{Du_i}{Dt} \equiv \frac{\partial u_i}{\partial t} + v_j \frac{\partial u_i}{\partial x_j} \qquad \text{with} \quad i,j = 1,2,3, \qquad (7.27)$$

$$a_i = \frac{Dv_i}{Dt} \equiv \frac{\partial v_i}{\partial t} + v_j \frac{\partial v_i}{\partial x_j} \qquad \text{with} \quad i,j = 1,2,3. \qquad (7.28)$$

If the displacements are assumed infinitesimal, we can reduce the mathematical complexity through use of a linearized theory. Thus neglecting higher order terms, as we did earlier with (7.12), we can reduce (7.25) to the following well-known form, also referred to as Navier's equation [2, 5], which governs dynamic mechanical behavior in a solid:

$$\frac{\partial \sigma_{ij}}{\partial x_j} = -f_i + \rho \frac{\partial^2 u_i}{\partial t^2} \qquad \text{with} \quad i,j = 1,2,3. \qquad (7.29)$$

Here, we have adopted a slight change in notation for the body force density, f_i (instead of f_i^b).

Equation (7.29) can be extended to the analysis of free vibrations ($f_i = 0$) of micromechanical structures in the linear elastic regime for the computation of the resonant or natural frequencies. The structure is allowed to be mechanically free on part of its surface boundary (Γ_1) and fixed on the remaining surface boundary (Γ_2), viz.,

$$\sigma_{ij} n_j = 0 \quad \text{on} \quad \Gamma_1 \qquad \text{with} \quad i = 1,2,3, \qquad (7.30)$$

$$u_i = 0 \quad \text{on} \quad \Gamma_2 \qquad \text{with} \quad i = 1,2,3 \qquad (7.31)$$

in the usual notation. Assuming that the mechanical field variables (e.g.,

$\sigma, \varepsilon, u)$ are sinusoidal functions of time, the equations of motion become [38]

$$\frac{\partial \sigma_{ij}}{\partial x_j} + \lambda \rho u_i = 0 \qquad \text{with} \quad i,j = 1,2,3, \tag{7.32}$$

where λ is the eigenvalue equal to the square of the resonant or natural frequency, $\lambda = \omega^2$, and ρ denotes the material density. The eigenvalues are non-zero and positive by virtue of a non-vanishing Γ_2; otherwise the structure assumes arbitrary solutions whose uniqueness cannot be guaranteed.

System (7.29) gives rise to three equations. However, the total number of independent unknowns is fifteen and comprise six components of stress, six components of strain, and three components of displacement. Thus the governing system needs to be supplemented with additional equations, some of which can be drawn from the relations between unknown field variables associated with the physical (material) properties of the solid. These are referred to as the constitutive relations. There are six equations stemming from the stress-strain constitutive relations, as we will see below, and there are six additional equations from the strain-displacement relations, (7.13). This completes the specification of the problem as there are as many equations as unknowns.

The stress-strain constitutive behavior can have different character depending on the nature of the solid (i.e., elastic, plastic, viscoelastic, or other). Most materials pertinent to micromechanical sensors and actuators can be idealized as elastic. For not too large stresses, the stress and strain tensors in homogeneous elastic materials are linearly related by the generalized form of Hooke's law

$$\sigma_{ij} = C_{ijkl}\, \varepsilon_{kl} \qquad \text{with} \quad i,j,k,l = 1,2,3. \tag{7.33}$$

Here, C_{ijkl} is the fourth rank tensor of eighty-one elastic stiffness coefficients. Alternatively, relation (7.33) can be written as

$$\varepsilon_{ij} = S_{ijkl}\, \sigma_{kl} \qquad \text{with} \quad i,j,k,l = 1,2,3, \tag{7.34}$$

where S_{ijkl} is the tensor of compliance coefficients. Relations (7.33) and (7.34) describe the stress-strain relationship for a material whose initial state is free of stress or strain. The symmetry in stress ($\sigma_{ij} = \sigma_{ji}$) and strain ($\varepsilon_{kl} = \varepsilon_{lk}$) reduces the number of independent elastic coefficients to thirty-six ($C_{ijkl} = C_{jikl} = C_{ijlk}$). The symmetry in the first two and last two suffixes in these coefficients allows a more convenient representation of (7.33) or (7.34) in terms of engineering (matrix) notation. Adopting a single suffix, viz., $11 \to 1$, $22 \to 2$, $33 \to 3$, $23, 32 \to 4$, $31, 13 \to 5$, $12, 21 \to 6$, we obtain the following compact form for (7.33),

$$\sigma_i = C_{ij}\, \varepsilon_j \qquad \text{with} \quad i,j = 1,2,\ldots,6, \tag{7.35}$$

which can be expanded to read as

$$\begin{bmatrix} \sigma_1 \\ \sigma_2 \\ \sigma_3 \\ \sigma_4 \\ \sigma_5 \\ \sigma_6 \end{bmatrix} = \begin{bmatrix} C_{11} & C_{12} & C_{13} & C_{14} & C_{15} & C_{16} \\ C_{21} & C_{22} & C_{23} & C_{24} & C_{25} & C_{26} \\ C_{31} & C_{32} & C_{33} & C_{34} & C_{35} & C_{36} \\ C_{41} & C_{42} & C_{43} & C_{44} & C_{45} & C_{46} \\ C_{51} & C_{52} & C_{53} & C_{54} & C_{55} & C_{56} \\ C_{61} & C_{62} & C_{63} & C_{64} & C_{65} & C_{66} \end{bmatrix} \begin{bmatrix} \varepsilon_1 \\ \varepsilon_2 \\ \varepsilon_3 \\ \varepsilon_4 \\ \varepsilon_5 \\ \varepsilon_6 \end{bmatrix}. \tag{7.36}$$

The constitutive relations in the form given by (7.35) or (7.36) have lost their tensor properties and can no longer be subject to the usual transformation rules. To carry out a transformation, they have to be reversed back to the tensor notation, (7.33). In (7.36), $\sigma_{1,2,3}$ and $\varepsilon_{1,2,3}$ denote normal components of stress and strain, respectively, and $\varepsilon_4 = 2\varepsilon_{23,32}, \varepsilon_5 = 2\varepsilon_{13,31}$, and $\varepsilon_6 = 2\varepsilon_{12,21}$. The number of independent stiffness coefficients can be further reduced by invoking material symmetry [4]. The form of the elasticity matrices for selected classes of materials, such as class 32 trigonal crystals (e.g., quartz) and crystals with cubic symmetry (e.g., Si) are given in (7.37) below:

$$\begin{bmatrix} C_{11} & C_{12} & C_{13} & C_{14} & & \\ C_{12} & C_{11} & C_{13} & -C_{14} & & \\ C_{13} & C_{13} & C_{33} & & & \\ C_{14} & -C_{14} & & C_{44} & & \\ & & & & C_{44} & \\ & & & & & C_{66} \end{bmatrix},$$

$$\begin{bmatrix} C_{11} & C_{12} & C_{12} & & & \\ C_{12} & C_{11} & C_{12} & & & \\ C_{12} & C_{12} & C_{11} & & & \\ & & & C_{44} & & \\ & & & & C_{44} & \\ & & & & & C_{44} \end{bmatrix}. \tag{7.37}$$

The first matrix represents a class 32 trigonal crystal whereby there are seven independent coefficients. The second matrix is representative of cubic crystalline materials. Here, there are three independent coefficients. Non-crystalline materials including thin films (e.g., poly Si, SiO_2) can be modeled as isotropic materials. The form of the stiffness matrix is similar to that of the cubic system except that there are only two independent coefficients [4]:

$$C_{11} = \frac{E(1-\nu)}{(1+\nu)(1-2\nu)}, \qquad C_{12} = \frac{E\nu}{(1+\nu)(1-2\nu)},$$

$$C_{44} = \frac{1}{2}(C_{11} - C_{12}) = \frac{E}{2+2\nu} \tag{7.38}$$

characterized in terms of Young's modulus (E) and Poisson's ratio (ν).

Usually the strain components are given in terms of the stress components and the compliance coefficients. For convenience, we provide the expanded form of (7.34) applicable to isotropic materials:

$$\varepsilon_1 = S_{11}\sigma_1 + S_{12}\sigma_2 + S_{12}\sigma_3 = \frac{1}{E}[\sigma_1 - \nu(\sigma_2 + \sigma_3)],$$

$$\varepsilon_2 = S_{12}\sigma_1 + S_{11}\sigma_2 + S_{12}\sigma_3 = \frac{1}{E}[\sigma_2 - \nu(\sigma_3 + \sigma_1)],$$

$$\varepsilon_3 = S_{12}\sigma_1 + S_{12}\sigma_2 + S_{11}\sigma_3 = \frac{1}{E}[\sigma_3 - \nu(\sigma_1 + \sigma_2)],$$

$$\varepsilon_4 = 2(S_{11} - S_{12})\sigma_4 = \frac{1}{G}\sigma_4,$$

$$\varepsilon_5 = 2(S_{11} - S_{12})\sigma_5 = \frac{1}{G}\sigma_5,$$

$$\varepsilon_6 = 2(S_{11} - S_{12})\sigma_6 = \frac{1}{G}\sigma_6, \tag{7.39}$$

where

$$S_{11} = \frac{1}{E}, \tag{7.40}$$

$$S_{12} = -\frac{\nu}{E}, \tag{7.41}$$

and G denotes the shear modulus,

$$G = \frac{1}{2(S_{11} - S_{12})} = \frac{E}{2+2\nu}, \tag{7.42}$$

which is identical to C_{44}.

Values of various mechanical coefficients for cubic and non-crystalline thin film materials are given in Sect. 7.2.3 along with a review of pertinent measurement procedures for mechanical parameter extraction. From a simulation standpoint, the use of measured values, retrieved using dedicated characterization microstructures [39], is preferred in view of the sensitivity of material parameters to fabrication process conditions. In particular, with thin films, we need to account for stresses intrinsic to the material which vary with different materials and deposition conditions. In addition, there can be thermal strains induced by different processing temperatures or by temperature variations in the micromechanical structure due to internal heating (see Chapt. 5 and Sect. 8.4). To account for internal (residual) stresses (σ_0) or strains (ε_0), a slightly modified form of (7.35) needs to be employed, *viz.*,

$$(\sigma_i - \sigma_{0i}) = C_{ij}(\varepsilon_j - \varepsilon_{0j}) \qquad \text{with} \quad i,j = 1, 2, \dots, 6. \qquad (7.43)$$

The above stress-strain constitutive relations are valid only in the linear elastic regime; deviation from linearity occurs at the yield stress (σ_y) or correspondingly, at the limiting strain ($\varepsilon_{\lim}$). For small deviations, the material may remain in the elastic regime whereby mechanical behavior is reversible but the stress-strain constitutive relations are non-linear.

At very large deviations, the material becomes plastic and is unable to regain its original shape. Simulation of mechanical behavior in the plastic regime is a complex issue beyond the scope of this chapter. A comprehensive treatment can be found in [40]. In the plastic regime, the displacements may no longer be single valued functions of stress due to hysteresis effects, and the stress-strain relations may become time-dependent. Thus the description of constitutive behavior needs to be not only mathematically simple, but must provide agreement with measurements. Often, idealizations of the constitutive behavior are employed. For example, metals such as Al obey the elastic-plastic idealized behavior, whereby the material behaves linearly elastic until the limiting strain is reached, following which it undergoes plastic deformation with a gradual transition from elastic to plastic regimes. On the other hand, brittle materials (e.g., SiO_2) experience very little or no plastic deformation before fracture.

Modeling the constitutive behavior at large stresses requires a yield criterion to indicate the transition from elastic to plastic regimes along with a description of the stress-strain behavior associated with the latter. Here, the deformation can be modeled along the lines of a proportionality between the stress- and strain-deviators based on the theories of Saint Venant and Reuss (see, e.g., Ref. [1]). As for the yield criterion, it can be retrieved accurately for relatively simple configurations using uniaxial stress test measurements. However, for complex three-dimensional configurations, we need to rely on estimated values of the yield stress using criteria based on theories of Rankine, Tresca, and Von Mises (see, e.g., Ref. [1]).

Equation (7.29) governing dynamic mechanical behavior, the constitutive relations (7.35) or (7.43), the strain-displacement relations (7.13), and the relevant force contributions stemming from external fields (discussed in Chapt. 8) are solved for the displacement as a function of space and time, subject to appropriate boundary and initial conditions. Dirichlet boundary conditions are specified at regions where the mechanical displacement is prescribed on the surface by a fixed value (u_0),

$$u_i = u_0 \qquad \text{with} \quad i = 1, 2, 3 \tag{7.44}$$

or at symmetry planes where the normal derivative of displacement vanishes,

$$(\partial u_i / \partial x_j) n_j = 0 \qquad \text{with} \quad i, j = 1, 2, 3. \tag{7.45}$$

Here, the n_j denote the components of the unit vector normal to the symmetry plane. Neumann boundary conditions are specified at free surfaces where the point-wise surface traction (or surface force) is zero,

$$\sigma_{ij} n_j = 0 \qquad \text{with} \quad i, j = 1, 2, 3 \tag{7.46}$$

or has a prescribed value,

$$\sigma_{ij} n_j = f_{0i} \qquad \text{with} \quad i, j = 1, 2, 3. \tag{7.47}$$

Here, the f_{0i} denote the components of the prescribed surface force density vector, which constitutes the coupling term for the various actuation forces whether they are acting on the surface or in the body. The latter can be converted to a surface force by virtue of the divergence theorem (see Sect. 8.5.3). In the static case, (7.47) should be consistent with the force balance (7.29),

$$\frac{\partial \sigma_{ij}}{\partial x_j} = -f_i \qquad \text{with} \quad i, j = 1, 2, 3. \tag{7.48}$$

Mixed boundary conditions are specified when forces are prescribed on some parts of the surface and displacements prescribed on other parts. These conditions are described by (7.44), (7.46), and/or (7.47). Prescribing a net force (F_0) on a boundary surface constitutes a floating boundary condition whereby the displacement is constant

$$u_i = \text{constant} \qquad \text{with} \quad i = 1, 2, 3 \tag{7.49}$$

but whose value is not known, and the net stresses are governed by

$$\int_\Gamma \sigma_{ij} n_j \, d\Gamma = F_{0i} \qquad \text{with} \quad i, j = 1, 2, 3. \tag{7.50}$$

Here, Γ denotes the integration surface and $d\Gamma$ a surface element. Relation (7.50), in comparison to the electrical domain, is analogous to a floating equipotential with a prescribed net charge [41]. Such floating boundaries are characteristic of infinitely stiff materials where the boundary surface is rigid.

The system of Eqs. (7.13), (7.29), (7.35), solved under boundary conditions, (7.44) to (7.50), along with initial conditions yield unique solutions for the displacement components, u_i (where $i = 1, 2, 3$). As initial conditions, the displacement and its time derivative (i.e., velocity) need to be specified.

If only the strain components, ε_i (where $i = 1, 2, \ldots, 6$), were given, we have six equations for the three unknown displacement components and there is no guarantee that the solutions obtained for u_i are unique. To guarantee uniqueness of solutions, compatibility conditions are needed for integrability of relations (7.13). Differentiation of (7.13) and subsequent elimination of displacement terms yields

$$\frac{\partial^2 \varepsilon_{ik}}{\partial x_i \partial x_m} + \frac{\partial^2 \varepsilon_{lm}}{\partial x_i \partial x_k} - \frac{\partial^2 \varepsilon_{il}}{\partial x_k \partial x_m} - \frac{\partial^2 \varepsilon_{km}}{\partial x_i \partial x_l} = 0 \qquad (7.51)$$

with $i, k, l, m = 1, 2, 3$.

System (7.51), also known as the Saint Venant's compatibility relations, yields eighty-one equations. However, symmetry considerations reduce the number of independent equations to six. These compatibility relations guarantee that the displacements (7.13) can be integrated in a consistent manner.

An alternative approach for solving the governing field equation employs the variational method based on the principle of minimum energy (see [42]). Under static conditions, the differential equation (7.29) and associated boundary conditions comprising prescribed surface force densities (on boundary Γ_1) and displacement (on boundary Γ_2), can be stated as:

$$\frac{\partial \sigma_{ij}}{\partial x_j} + f_i = 0 \qquad \text{with} \quad i, j = 1, 2, 3,$$

$$\sigma_{ij} n_j = f_{0i} \text{ on } \Gamma_1 \qquad \text{with} \quad i, j = 1, 2, 3,$$

$$u_i = u_0 \text{ on } \Gamma_2 \qquad \text{with} \quad i = 1, 2, 3. \qquad (7.52)$$

If the displacements from the equilibrium state (7.52) are denoted by δu_i (where $i = 1, 2, 3$), Eq. (7.52), in terms of its weak formulation, reads

$$\int_V \left[\frac{\partial \sigma_{ij}}{\partial x_j} + f_i \right] \delta u_i \, dV + \int_{\Gamma_1} \left[\sigma_{ij} n_j - f_i^0 \right] \delta u_i \, d\Gamma_1 = 0$$

with $i, j = 1, 2, 3$, $\qquad (7.53)$

where dV and $d\Gamma$ denote the respective volume and surface area of an element, and $\delta u_i = 0$ on Γ_2. Eq. (7.53) can be transformed using Gauss' theorem to read

$$\int_V \sigma_{ij} \, \delta\varepsilon_{ij} \, dV - \int_V f_i \, \delta u_i \, dV - \int_{\Gamma_1} f_{0i} \, \delta u_i \, d\Gamma_1 = 0$$

$$\text{with} \quad i,j = 1,2,3, \tag{7.54}$$

where

$$\delta\varepsilon_{ij} = \frac{1}{2} \left[\frac{\partial(\delta u_i)}{\partial x_j} + \frac{\partial(\delta u_j)}{\partial x_i} \right] \qquad \text{with} \quad i,j = 1,2,3. \tag{7.55}$$

The integrand in the first term of (7.54) is associated with the strain energy function W whereby

$$\delta W = \sigma_{ij} \, \delta\varepsilon_{ij} \qquad \text{with} \quad i,j = 1,2,3. \tag{7.56}$$

In terms of displacement components, the strain energy function becomes

$$W = \frac{1}{8} C_{ijkl} \left[\frac{\partial u_k}{\partial x_l} + \frac{\partial u_l}{\partial x_k} \right] \left[\frac{\partial u_i}{\partial x_j} + \frac{\partial u_j}{\partial x_i} \right] \qquad \text{with} \quad i,j,k,l = 1,2,3. \tag{7.57}$$

Following (7.56), we can express (7.54) as

$$\delta \left[\int_V W \, dV \right] - \int_V f_i \, \delta u_i \, dV - \int_{\Gamma_1} f_{0i} \, \delta u_i \, d\Gamma_1 = 0$$

$$\text{with} \quad i = 1,2,3. \tag{7.58}$$

Eq. (7.58) can also be stated as

$$\delta \left[\int_V W \, dV - \int_V f_i \, u_i \, dV - \int_{\Gamma_1} f_{0i} \, u_i \, d\Gamma_1 \right] = 0$$

$$\text{with} \quad i = 1,2,3. \tag{7.59}$$

Since there is no variation in the prescribed body forces and surface forces, we obtain for the total energy

$$E = \int_V [W - f_i u_i] \, dV - \int_{\Gamma_1} f_{0i} \, u_i \, d\Gamma_1 \qquad \text{with} \quad i = 1,2,3 \tag{7.60}$$

which must be a minimum, *viz.*,

$$\delta E = 0. \tag{7.61}$$

The above system is solved for the displacement components for the given force and displacement boundary conditions. In particular, the boundary conditions for the forces are naturally recovered from (7.60) (see, e.g., [43]). Once the distribution u_i is known, the distributions of relevant stress and strain components can be computed using the strain-displacement relations (7.13) and the stress-strain constitutive relations (7.33).

7.2.2 Simplified Analysis for Single- and Multi-Layer Diaphragms

In special cases of device structure comprising a thin homogeneous plate (e.g., diaphragm) whose thickness is much smaller than the other geometrical dimensions, mechanical analysis can be simplified greatly by reducing the model equations to two dimensions using the plane stress approximation. The underlying assumptions employed in problem simplification stem from the classical thermoelastic thin-plate theory [16, 44]:

- The plate is isothermal with a low rate of change of temperature.
- Body forces and inertial effects are negligible.
- The plate material is elastic and obeys linear elasticity theory.
- Plate deflection is small compared to other geometrical dimensions and the plate has uniform cross section.

For convenience, we will adopt Cartesian notation. Invoking the plane stress approximation, which assumes that the components of stress in the direction of thickness are negligible, the equilibrium equation relating the bending moments in a diaphragm (in the $y - z$ plane) when subject to an applied pressure p is given by [44]:

$$\frac{\partial^2 M_{yy}}{\partial y^2} + 2\frac{\partial^2 M_{yz}}{\partial y \, \partial z} + \frac{\partial^2 M_{zz}}{\partial z^2} = -p. \tag{7.62}$$

Here, the origin is chosen at the center of the diaphragm (see Fig. 7.3). The bending moments M_{yy}, M_{zz}, M_{yz} are given in terms of the stress

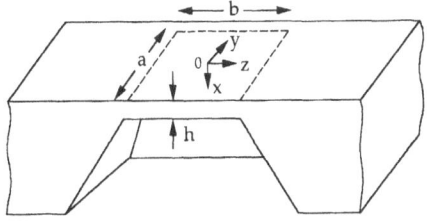

Fig. 7.3 Coordinate system used in simplified two-dimensional mechanical analysis

components σ_{yy}, σ_{zz}, and σ_{yz},

$$M_{yy} = \int_{-h/2}^{h/2} \sigma_{yy} x \, dx, \tag{7.63}$$

$$M_{zz} = \int_{-h/2}^{h/2} \sigma_{zz} x \, dx, \tag{7.64}$$

$$M_{yz} = \int_{-h/2}^{h/2} \sigma_{yz} x \, dx, \tag{7.65}$$

where h denotes diaphragm thickness. In case of a silicon diaphragm parallel to the $\langle 110 \rangle$ orientation, the stress components are given by [45]

$$\begin{bmatrix} \sigma_{yy} \\ \sigma_{zz} \\ \sigma_{yz} \end{bmatrix} = -x \begin{bmatrix} \dfrac{E}{1-\nu^2} & \dfrac{\nu E}{1-\nu^2} & 0 \\ \dfrac{\nu E}{1-\nu^2} & \dfrac{E}{1-\nu^2} & 0 \\ 0 & 0 & G \end{bmatrix} \begin{bmatrix} \dfrac{\partial^2 w}{\partial y^2} \\ \dfrac{\partial^2 w}{\partial z^2} \\ 2\dfrac{\partial^2 w}{\partial y \, \partial z} \end{bmatrix}, \tag{7.66}$$

where ν denotes Poisson's ratio, E the Young's modulus, G the shear modulus, and w the diaphragm displacement. Here, ν, E, and G are not isotropic and assume values pertinent to the chosen orientation. For a *thin rectangular single-layer silicon diaphragm* that is fully clamped at the edges and subject to an applied pressure p, the governing equation and boundary conditions can be expressed as [10, 44]

$$D\frac{\partial^4 w}{\partial y^4} + 2H\frac{\partial^4 w}{\partial y^2 \partial z^2} + D\frac{\partial^4 w}{\partial z^4} = p, \tag{7.67}$$

$$w = 0, \frac{\partial w}{\partial y} = 0 \quad \text{at} \quad y = \pm\frac{a}{2},$$

$$w = 0, \frac{\partial w}{\partial z} = 0 \quad \text{at} \quad z = \pm\frac{b}{2}, \tag{7.68}$$

where

$$D = \frac{E h^3}{12(1-\nu^2)}, \qquad H = \nu D + \frac{G h^3}{6}. \tag{7.69}$$

The displacement of the diaphragm is obtained by first solving the problem with the assumption that all the edges are simply supported and then applying appropriate magnitudes of bending moments along the edges so as to eliminate the rotations produced along these edges by the action of the lateral load [10, 44]. Hence, w is a superposition of w_0, the displacement of a simply supported diaphragm, and w_y and w_z, which are the displacements caused by the moments distributed along the edges for the clamped-edge boundary conditions, $viz., w = w_0 + w_y + w_z$. The displacement components take the following form [46, 47]:

$$
\begin{aligned}
w_0(y, z) = \frac{4 p a^4}{\pi^5 D} \sum_{m>0}^{\infty} & \frac{\sin\left(\frac{m\pi}{a}y\right)}{m^5} \\
& \times \left[1 - R(m) \cos\left(D_-\frac{m\pi}{a}z\right) \cosh\left(D_+\frac{m\pi}{a}z\right) \right. \\
& \left. + S(m) \sin\left(D_-\frac{m\pi}{a}z\right) \sinh\left(D_+\frac{m\pi}{a}z\right) \right],
\end{aligned}
\tag{7.70}
$$

$$
\begin{aligned}
w_z(y, z) = \frac{-a^2}{\pi^2 D} \sum_{m>0}^{\infty} & \frac{(-1)^{m/2-1/2} E_m \cos\left(\frac{m\pi}{a}y\right)}{m^2 A(m)} \\
& \times \left[\sin\left(D_-\frac{m\pi}{a}z\right) \sinh\left(D_+\frac{m\pi}{a}z\right) \right. \\
& \left. - \tan D_-\alpha_m \tanh D_+\alpha_m \cos\left(D_-\frac{m\pi}{a}z\right) \cosh\left(D_+\frac{m\pi}{a}z\right) \right],
\end{aligned}
\tag{7.71}
$$

$$
\begin{aligned}
w_y(y, z) = \frac{-b^2}{\pi^2 D} \sum_{m>0}^{\infty} & \frac{(-1)^{m/2-1/2} F_m \cos\left(\frac{m\pi}{a}z\right)}{m^2 B(m)} \\
& \times \left[\sin\left(D_-\frac{m\pi}{a}y\right) \sinh\left(D_+\frac{m\pi}{a}y\right) \right. \\
& \left. - \tan D_-\beta_m \tanh D_+\beta_m \cos\left(D_-\frac{m\pi}{a}y\right) \cosh\left(D_+\frac{m\pi}{a}y\right) \right].
\end{aligned}
\tag{7.72}
$$

Here,

$$
D_- = \sqrt{\frac{D - H}{2D}}, \qquad D_+ = \sqrt{\frac{D + H}{2D}},
\tag{7.73}
$$

$$
\alpha_m = \frac{m\pi b}{2a}, \qquad \beta_m = \frac{m\pi a}{2b},
\tag{7.74}
$$

$$A(m) = \left(\frac{H}{D} + \frac{\sqrt{D^2 - H^2}}{D} \tan D_{-\alpha_m} \tanh D_{+\alpha_m}\right)$$

$$\times \sin D_{-\alpha_m} \sinh D_{+\alpha_m}$$

$$+ \left(\frac{\sqrt{D^2 - H^2}}{D} - \frac{H}{D} \tan D_{-\alpha_m} \tanh D_{+\alpha_m}\right)$$

$$\times \cos D_{-\alpha_m} \cosh D_{+\alpha_m}, \tag{7.75}$$

$$B(m) = \left(\frac{H}{D} + \frac{\sqrt{D^2 - H^2}}{D} \tan D_{-\beta_m} \tanh D_{+\beta_m}\right)$$

$$\times \sin D_{-\beta_m} \sinh D_{+\beta_m}$$

$$+ \left(\frac{\sqrt{D^2 - H^2}}{D} - \frac{H}{D} \tan D_{-\beta_m} \tanh D_{+\beta_m}\right)$$

$$\times \cos D_{-\beta_m} \cosh D_{+\beta_m}, \tag{7.76}$$

$$R(m) = \frac{\cos D_{-\alpha_m} \cosh D_{+\alpha_m} + \dfrac{H}{\sqrt{D^2 - H^2}} \sin D_{-\alpha_m} \sinh D_{+\alpha_m}}{\cos^2 D_{-\alpha_m} \cosh^2 D_{+\alpha_m} + \sin^2 D_{-\alpha_m} \sinh^2 D_{+\alpha_m}}, \tag{7.77}$$

$$S(m) = \frac{\dfrac{H}{\sqrt{D^2 - H^2}} \cos D_{-\alpha_m} D_{+\alpha_m} - \sin D_{-\alpha_m} \sinh D_{+\alpha_m}}{\cos^2 D_{-\alpha_m} \cosh^2 D_{+\alpha_m} + \sin^2 D_{-\alpha_m} \sinh^2 D_{+\alpha_m}}, \tag{7.78}$$

and E_m and F_m are found by solving the system of linear equations resulting from the following boundary conditions:

$$\frac{\partial w}{\partial y} = \frac{\partial w_0}{\partial y} + \frac{\partial w_y}{\partial y} + \frac{\partial w_z}{\partial y} = 0 \qquad \text{at} \quad y = \pm\frac{a}{2}, \tag{7.79}$$

$$\frac{\partial w}{\partial z} = \frac{\partial w_0}{\partial z} + \frac{\partial w_y}{\partial z} + \frac{\partial w_z}{\partial z} = 0 \qquad \text{at} \quad z = \pm\frac{b}{2}. \tag{7.80}$$

Equation (7.67), which governs static behavior, can be modified slightly to perform free vibration analysis for computation of the plate's resonant or natural frequencies. For thin plates undergoing small amplitude motion in the absence of damping forces, the loading term p is replaced with an inertial body force, stemming from the oscillatory plate, that acts in opposition to the acceleration [48]. The differential equation that governs free vibration then takes the form

$$D\frac{\partial^4 w(y,z,t)}{\partial y^4} + 2H\frac{\partial^4 w(y,z,t)}{\partial y^2 \partial z^2} + D\frac{\partial^4 w(y,z,t)}{\partial z^4} + \rho\frac{\partial^2 w(y,z,t)}{\partial t^2} = 0,$$

$$(7.81)$$

whereby the displacement w is now a function of time, in addition to spatial coordinates, and ρ denotes the material density of the plate.

The displacement function $w(y,z,t)$ can be expressed as a product of two functions, one as a function of space and the other as a function of time. Employing separation of variables, we obtain the following homogeneous partial differential equation

$$D\frac{\partial^4 w(y,z)}{\partial y^4} + 2H\frac{\partial^4 w(y,z)}{\partial y^2 \partial z^2} + D\frac{\partial^4 w(y,z)}{\partial z^4} - \omega^2 \rho w(y,z) = 0,$$

$$(7.82)$$

which involves the mode shape $w(y,z)$ and the oscillation frequency, ω. The latter implicitly accounts for the time dependence of (7.82). Again, Fourier series-type solutions for Eq. (7.82) can be obtained to satisfy the prescribed boundary conditions which include plate edges that are simply supported, clamped, or free (see [48]).

The model Eqs. (7.62) to (7.80), for the static analysis of single-layer structures, can be extended to *composite multi-layer structures* with the aid of additional simplifying assumptions [49, 50]. Here, the diaphragm is still assumed to be of thickness h but composed of n layers. Simplifying assumptions include a constant Poisson's ratio throughout the thickness and an identical x-dependence of the stiffness, E_{ij},

$$E_{ij}^{(k)} = r^{(k)} E_{ij}^0.$$

$$(7.83)$$

Here, $E_{ij}^{(k)}$ and $r^{(k)}$ vary from layer to layer (Fig. 7.4) and the E_{ij}^0 are normalization constants. With a translation in the origin of the coordinate system, the stresses in each layer, expressed in terms of diaphragm displacement, take the same form as before, *viz.*,

$$\begin{bmatrix} \sigma_{yy} \\ \sigma_{zz} \\ \sigma_{yz} \end{bmatrix}^{(k)} = -x \begin{bmatrix} \dfrac{E}{1-\nu^2} & \dfrac{\nu E}{1-\nu^2} & 0 \\ \dfrac{\nu E}{1-\nu^2} & \dfrac{E}{1-\nu^2} & 0 \\ 0 & 0 & G \end{bmatrix}^{(k)} \begin{bmatrix} \dfrac{\partial^2 w}{\partial y^2} \\ \dfrac{\partial^2 w}{\partial z^2} \\ 2\dfrac{\partial^2 w}{\partial y \partial z} \end{bmatrix}.$$

$$(7.84)$$

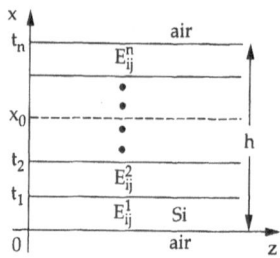

Fig. 7.4 Schematic of multilayer diaphragm structure

The governing equation and boundary conditions for the multi-layer diaphragm are similar to that of the single layer, described by (7.67) and (7.68), but with D and H given as:

$$D = \frac{1}{1 - v^2} \int_{-x_0}^{h - x_0} E^{(k)} x^2 \, dx, \qquad (7.85)$$

$$H = vD + \int_{-x_0}^{h - x_0} G^{(k)} x^2 \, dx. \qquad (7.86)$$

Here,

$$x_0 = \frac{\frac{1}{2} \sum_{k=1}^{n} r^{(k)} \left(t_k^2 - t_{k-1}^2 \right)}{\sum_{k=1}^{n} r^{(k)} \left(t_k - t_{k-1} \right)}, \qquad (7.87)$$

denotes the distance of a neutral plane from an arbitrary chosen reference at the lower face of the plate [50], and t_k is the interface coordinate measured from the lower face, with $t_n = h$. The origin of the coordinate system is translated to this neutral plane in evaluating (7.84).

The boundary conditions can be rewritten as a system of equations which carries the following form:

$$E_i u_1(i) + \sum_{m=1,3,\ldots,\infty}^{\infty} F_m v_1(i, m) = c_1(i),$$

$$F_i u_2(i) + \sum_{m=1,3,\ldots,\infty}^{\infty} E_m v_2(i, m) = c_2(i), \qquad (7.88)$$

which can be solved numerically for the coefficients E_m and F_m. Here, the first $(n + 1)/2$ equations from each set are used with m allowed to range from 1 to n. System (7.88) can be cast into the following form

$$\begin{bmatrix} [U_1] & [V_2] \\ [V_1] & [U_2] \end{bmatrix} \begin{bmatrix} [E] \\ [F] \end{bmatrix} = \begin{bmatrix} [C_1] \\ [C_2] \end{bmatrix}, \tag{7.89}$$

where the matrix of coefficients results from a combination of 4 sub-matrices with $[U_1]$ and $[U_2]$ denoting diagonal matrices. The system of equations is then solved using standard techniques based on direct or iterative schemes.

The choice of single or multi-layer models depends on the relative thickness of layers involved. When the thickness of the silicon layer is approximately that of the other layers, the use of a single-layer model leads to a discrepancy in the computed displacement that can be greater than 30%. In contrast, if the thickness of the silicon layer is much greater than that of other layers, the corresponding discrepancy is less than 10% [47]. Thus the use of the single-layer model for analysis of multi-layer structures is acceptable only as long as the thickness of the silicon layer is much larger than that of the other layers.

The boundary conditions discussed in relation to the above system assume a diaphragm that is rigidly clamped at the edges. In practice, the clamping conditions at the supporting edges are non-ideal. Non-idealities can be accounted for by introducing an edge factor g [49] into the boundary conditions, thereby modifying (7.68). In the absence of thermally-induced bending moments, the boundary conditions (7.68) become

$$w = 0 \quad \text{at} \quad y = \pm \frac{a}{2}, z = \pm \frac{b}{2},$$

$$gD\frac{\partial^2 w}{\partial y^2} + \frac{\partial w}{\partial y} = 0 \quad \text{at} \quad y = \pm \frac{a}{2},$$

$$gD\frac{\partial^2 w}{\partial z^2} + \frac{\partial w}{\partial z} = 0 \quad \text{at} \quad z = \pm \frac{b}{2}. \tag{7.90}$$

The edge factor in effect represents the degree of elastic deformation of the supporting rim and is a function of diaphragm geometry and thickness as well as the geometrical parameters of the supporting rim. The edge factor is zero for rotation free edges, as evident by (7.68), and infinity for simply supported edges allowing full rotation. The value of g in practical device structures is finite. For example, g is of the order of 10^{-6} for a 27 µm thick diaphragm realized with (100) Si with rim slopes oriented in the $\langle 111 \rangle$ direction [15, 49].

The above two-dimensional model based on the plane-stress approximation can be easily extended to account for packaging-induced thermal strains or other external forces, thermal mismatch between Si and other layers integral to the diaphragm, and effects of process-induced intrinsic stresses which vary with the different layers. These effects modify the stress terms, σ_{yy}, σ_{zz}, and σ_{yz}, as well as the loading terms p on the right-hand side of the governing equation. With the computed distributions of stress and displacement in the diaphragm, we can evaluate the change in resistance or capacitance of piezoresistive- and capacitive-based micromechanical sensors.

The relative change in resistance $(\Delta R/R_0)$ can be computed from the stress components,

$$\frac{\Delta R}{R_0} = \pi_l \sigma_l + \pi_t \sigma_t + \pi_s \sigma_s. \tag{7.91}$$

Here, σ_l and σ_t denote the average normal stresses longitudinal (parallel) and transverse (perpendicular) to the current flow, respectively, and π_l and π_t denote the corresponding longitudinal and transverse piezoresistance coefficients. The last term in (7.91) accounts for the effects of shear stress (σ_s) described by the shear piezoresistance coefficient (π_s). The coefficients π_l, π_t, and π_s are functions of the principal components π_{11}, π_{12}, and π_{44}. Since the latter components are specified for the crystallographic axes system, the coefficients π_l, π_t, and π_s have to be evaluated by coordinate transformation (see Sect. 7.1) to the diaphragm axes system.

With capacitive structures, the total effective capacitance (C) between a displaced diaphragm and reference plate (electrode) is computed using the distribution of displacement, $w(y, z)$ [16]

$$C = \varepsilon_0 \int\int \frac{dy\,dz}{[s - w(y, z)]} + C_p. \tag{7.92}$$

Here, C_p accounts for presence of parasitic capacitance, s denotes the zero pressure separation between the diaphragm and reference plate, ε_0 the permittivity of the air gap, and the integral is performed over the plate area.

Although the above analysis technique is able to take into account anisotropies in mechanical properties and is not constrained to non-square (rectangular) diaphragm geometries, the governing equations and boundary conditions are only valid when the geometrical parameters of the structure and device operation satisfy certain well prescribed conditions. Arbitrary device geometries and operating conditions can only be treated with three-dimensional numerical schemes using either in-house or commercially available software tools (see Table 8.2 in Chapt. 8). Again, in such cases, specification of appropriate boundary conditions is crucial for valid solutions. Here, rather than imposing an assumed edge factor, g, the supporting rim and possibly part of the surrounding non-active region should be included in the simulation domain to allow standard (fully clamped) conditions to be imposed at the outer boundaries.

7.2.3 Material Parameters and Extraction

Meaningful simulation of micromechanical devices requires accurate and reliable material property data such as Young's modulus or stiffness coefficients, Poisson's ratio, thermal expansion coefficients, the yield stress, and the internal stress state of materials involved. In most cases, the materials employed are in thin film form (e.g., poly-Si, SiO_2, Si_3N_4, polyimides). The mechanical properties of these films can not only be quite different from their bulk counterparts, but also depend on process conditions [51]. Despite the wealth of experimental characterization data, it is virtually impossible to predict reliably the mechanical properties of microfabricated materials, needed for device simulation, with respect to process parameters. Access to accurate material property data is best achieved through use of material characterization micromachined test structures and associated measurement and retrieval techniques. A number of techniques have been developed along with a correspondingly large volume of published works for the measurement and retrieval of mechanical properties of thin films. The choice in technique depends on the material parameter of interest, desired measurement accuracy, the internal stress state in the structure, the test structure geometry, and a variety of other factors. For example, the load-deflection behavior of beams and membranes is used for structures with tensile stress (see, e.g., [52–55]), the beam buckling technique is used for compressively stressed structures (see, e.g., [56]), the wafer curvature technique is useful for both tensile and compressive films (see, e.g., [57]), and the bulge test for elastic limit properties (see, e.g., [52, 53, 119]). Moreover, there are techniques based on analysis of dynamic behavior such as resonant frequency (see, e.g., [58–60]), and electrostatic pull-in behavior (see [61] and references therein) of beams, including lateral resonators, micro-bridges, and membranes.

A procedure that is simple, effective, and gaining wide usage is the load-deflection technique which enables simultaneous retrieval of the internal stress and elastic coefficients of the thin film. These parameters are retrieved by fitting measurements of center deflection as a function of applied differential pressure (see Fig. 7.5) to a model where coefficient values are constructed from semi-analytical methods, numerical simulations, or both. The relation between the center deflection (d) and applied pressure (p), assuming negligible bending stiffness of the

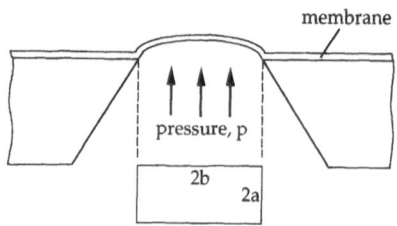

Fig. 7.5 The load-deflection method, applied to a rectangular membrane of edge lengths $2a$ and $2b$, illustrated by membrane's cross section and top view

membrane, takes the form [52, 53]

$$p = \frac{C_1 \sigma_0 t d}{a^2} + \frac{C_2 E t d^3}{a^4},$$
(7.93)

where σ_0 denotes the in-plane residual stress, t the thin film thickness, E is Young's modulus, and C_1 and C_2 are dimensionless coefficients which depend on membrane geometry (e.g., square, rectangular) and aspect ratio, $n = a/b$ (where a and b denote edge lengths), and Poisson's ratio (ν) of the thin film. Other forms of (7.93) can also be employed (see [39, 62]).

The first term in (7.93) describes the stiffness stemming from the internal stress in the film and the second term describes the intrinsic stiffness of the material characterized by the Young's modulus. Eq. (7.93) is valid for very thin membranes and for large deflection. The coefficients C_1 and C_2 can be obtained analytically from variational principles [52, 54]. The form of these coefficients is determined by the choice of shape function. For a product of two cosine functions,

$$C_1 = \frac{\pi^4 (1 + n^2)}{64},$$
(7.94)

$$C_2 = \frac{\pi^6}{32(1 - \nu^2)} \left(\frac{9 + 2n^2 + 9n^4}{256} \right.$$

$$\left. - \frac{\{4 + n + n^2 + 4n^3 - 3n\nu(1 + n)\}^2}{2\{81\pi^2(1 + n^2) + 128n + \nu[128n - 9\pi^2(1 + n^2)]\}} \right).$$
(7.95)

As b/a increases (i.e., n decreases), the coefficient C_1 decreases to approach the constant value of 1.52. The coefficient C_2 decreases with increasing b/a when $b < 2a$. In this regime, C_2 is strongly dependent on the Poisson ratio of the material. However, for large b/a (i.e., rectangular membrane $b \gg a$), the dependence of C_2 on membrane geometry vanishes and also, in this limit, the dependence of C_2 on Poisson's ratio is very weak. Here, the retrieved value of Young's modulus is not as sensitive to

Poisson's ratio for large b/a. The computed values of C_1 and C_2 are 3.04 and 1.83, respectively, for a square $(b = a, n = 1)$ membrane and an assumed value of $\nu = 0.25$ [54].

The above analytical descriptions may have restricted validity; the underlying shape function employed must describe adequately the true deflection behavior. Recently, a more rigorous technique for simultaneous retrieval of Young's modulus, internal stress, and Poisson's ratio has been reported [120], using a single compressively prestressed long rectangular membrane. While Young's modulus and internal stress are extracted from the usual load-deflection response, Poisson's ratio is retrieved from the associated behavior of the ripple (deflection profile) transition.

The load-deflection technique is only applicable to thin films that are in tensile stress. With compressively stressed films, composite layer structures can be employed whereby the thin film to be characterized is deposited on a well-characterized highly tensile membrane to yield a net tensile stress in the composite structure [54, 55]. The net residual stress of the composite membrane can still be retrieved using Eq. (7.93), however, the retrieved values of mechanical properties represent that of the composite layer. Thus additional equations, which relate the composite material property to that of the individual films, are needed [54, 55], *viz.*,

$$\sigma_{0u} = \frac{\sigma_{0c}(t_k + t_u) - \sigma_{0k} t_k}{t_u} \tag{7.96}$$

and, assuming an identical Poisson's ratio for the films,

$$E_u = \frac{E_c(t_k + t_u) - E_k t_k}{t_u}. \tag{7.97}$$

Here, the subscripts u, c, and k are associated with the unknown layer, the composite layer, and the known layer, respectively. The error associated with the assumption on Poisson's ratio can be minimized by using a rectangular membrane with $b \gg a$.

Values of relevant mechanical parameters are given in Tables 7.1 to 7.3 for selected crystalline and thin film materials. No distinction is made on the measurement/retrieval procedure for the latter. This can be obtained from the references cited.

In crystalline semiconductors, metals, and insulators, the coefficients, e.g., Young's moduli and Poisson's ratios, are anisotropic (see Table 7.1). The mechanical properties of thin films are strongly affected by film geometry (thickness), deposition temperature, doping concentration, and subsequent thermal annealing. For example, in poly-Si thin films, both Young's modulus and the residual (compressive) stress increase with increasing film thickness, before decreasing to a steady thickness-independent (bulk) value. This could be attributed to the evolving

Table 7.1. *Orientation-dependent (anisotropic) values of Young's moduli and relevant temperature coefficients at room temperature for selected crystalline semiconductor, metal, and insulator materials*

Material	Stiffness C (GPa), compliance S $(10^{-12}\,\mathrm{Pa}^{-1})$, and other coefficients

Relation between stiffness and compliance coefficients for cubic systems (e.g., Si, diamond, Al, Cu, Ni, W):

$$C_{11} = \frac{S_{11} + S_{12}}{(S_{11} - S_{12})(S_{11} + 2S_{12})}, C_{12} = \frac{-S_{12}}{(S_{11} - S_{12})(S_{11} + 2S_{12})}, C_{44} = \frac{1}{S_{44}}$$

p-Type Si [64] (4 Ω cm)
$C_{11} = -80.5 \pm 3.3 \times 10^{-6}/{}^{\circ}\mathrm{C}, C_{12} = -115.0 \pm 9.0 \times 10^{-6}/{}^{\circ}\mathrm{C};$
$C_{44} = -52.8 \pm 0.5 \times 10^{-6}/{}^{\circ}\mathrm{C}$
$S_{11} = 64.0 \pm 0.6 \times 10^{-6}/{}^{\circ}\mathrm{C}, S_{12} = 38.8 \pm 3.5 \times 10^{-6}/{}^{\circ}\mathrm{C};$
$S_{44} = 58.2 \pm 0.5 \times 10^{-6}/{}^{\circ}\mathrm{C}$

n-Type Si [64] (0.05 Ω cm)
$C_{11} = -97.1 \pm 3.7 \times 10^{-6}/{}^{\circ}\mathrm{C}, C_{12} = -54.8 \pm 0.4 \times 10^{-6}/{}^{\circ}\mathrm{C};$
$C_{44} = -172.0 \pm 10.0 \times 10^{-6}/{}^{\circ}\mathrm{C}$
$S_{11} = 61.5 \pm 0.6 \times 10^{-6}/{}^{\circ}\mathrm{C}, S_{12} = 7.1 \pm 4.9 \times 10^{-6}/{}^{\circ}\mathrm{C};$
$S_{44} = 54.8 \pm 0.4 \times 10^{-6}/{}^{\circ}\mathrm{C}$

Poisson's ratio for $\langle 100 \rangle$ Si : 0.28; $\langle 111 \rangle$ Si : 0.36

The temperature dependent biaxial modulus (in units of GPa) of crystalline Si for $30{}^{\circ}\mathrm{C} \leq T \leq 400{}^{\circ}\mathrm{C}$ [65]:

$$E/(1-\nu) = 176.4 - 5.46 \times 10^{-3}\,T - 2.46 \times 10^{-5}\,T^2 + 1.72 \times 10^{-8}\,T^3$$

Temperature coefficients of thermal expansion for Si [64]:

$$2.84 \pm 0.04 \times 10^{-6}/{}^{\circ}\mathrm{C}, 8.5 \pm 0.5 \times 10^{-9}/{}^{\circ}\mathrm{C}^2, -32.0 \pm 2.0 \times 10^{-12}/{}^{\circ}\mathrm{C}^3$$

Temperature coefficients of mass density for Si [64]:

$$-8.5 \pm 0.1 \times 10^{-6}/{}^{\circ}\mathrm{C}, -25.5 \pm 1.5 \times 10^{-9}/{}^{\circ}\mathrm{C}^2, 95.3 \pm 6.7 \times 10^{-12}/{}^{\circ}\mathrm{C}^3$$

Diamond [66]
$C_{11} = 1079 \pm 5, C_{12} = 124 \pm 5, C_{44} = 578 \pm 2$
Poisson's ratio for (111) planes (values for other planes, see [66]):
0.0791 – longitudinal-induced orthogonal elongation
0.0435 – transverse-induced elongation

Cubic Metals [4]
Al: $S_{11} = 15.9, S_{12} = -5.8; S_{44} = 35.2;$
Cu: $S_{11} = 14.9, S_{12} = -6.3; S_{44} = 13.3$
Ni: $S_{11} = 14.9, S_{12} = -6.3, S_{44} = 13.3;$
W: $S_{11} = 2.57, S_{12} = -0.73; S_{44} = 6.60$

Relation between elasticity and compliance coefficients for trigonal class 32 crystals (e.g., quartz):

$$C_{11} + C_{12} = \frac{S_{33}}{S}, C_{11} - C_{12} = \frac{S_{44}}{S'}, C_{13} = \frac{-S_{13}}{S}, C_{14} = \frac{-S_{14}}{S'},$$
$$C_{33} = \frac{S_{11} + S_{12}}{S}, C_{44} = \frac{S_{11} - S_{12}}{S'}$$

$$S = S_{33}(S_{11} + S_{12}) - 2S_{13}^2, S' = S_{44}(S_{11} - S_{12}) - 2S_{14}^2$$

Quartz   See Table 8.9 for values of stiffness coefficients

Table 7.2. *The Young's and biaxial modulus and residual stress for various thin film semiconductor, metal, and insulator materials. Where available, process conditions are listed along with associated sources; Tdep denotes deposition temperature*

Material	Modulus (GPa)	Process Conditions and Residual Stress
Poly-Si (LPCVD)	150–170 [67]	Undoped, Tdep: 620 °C, thickness: 100 nm to 800 nm compressive residual stress: 300 MPa to 425 MPa
	151 ± 6 [67]	thickness: 470 nm, compressive residual stress of as-deposited film: 350 MPa ±12 MPa, which reduces to 20 MPa after annealing (600 °C–1100 °C) for 2 hours
	160 [54]	Tdep: 630 °C, thickness: 0.2 μm, compressive stress: 0.18 GPa
	150 ± 30 [68]	boron-doped, thickness: 2 μm to 10 μm
	174 ± 10 [60]	undoped at 610 °C then phosphorus diffusion at 1050 °C tensile residual stress: 10 ± 2 MPa
	130 ± 5 [69] (-42 ± 2 ppm/K)	in situ phosphorus-doped, Tdep: 610 °C, annealed at 1050 °C
	147 ± 6 [69] (-32.1 ± 1 ppm/K)	Tdep: 560 °C
Diamond	1143 [70], 1050 [87]	CVD polycrystalline
	891 [63]	PECVD
a-Si:H (8–10% H_2)	134 ± 5 [71]	PECVD, modulus highest at 8–10% H_2 where film density is highest: 2300 ± 20 kg/m^3
a-Si$_x$C$_{1-x}$:H ($27\% H_2$)	Biaxial: 200 ± 25 [72]	PECVD, modulus highest around $x = 0.4$–0.5 and decreases as film becomes Si rich ($x > 0.5$)
a-SiC	Biaxial: 340–420 [73]	increased Si-C bonding without hydrogen
SiC	414–462 [74], 331 [88]	CVD, Tdep: 1200 °C–1800 °C, moduli: averaged values
Al	42.7 [75]	Value 40% lower than for bulk (70 GPa), thickness: 75 nm
ZnO	Biaxial: 260–370 [65]	RF magnetron sputtering; values depend on sputtering power, substrate temperature, and gas pressures
	310 [76]	Tdep: 230 °C, tensile stress: 82.9 MPa
SiO$_2$	72–75 [55]	Thermal oxide, compressive stress: 240 MPa
	20 ± 10 [39]	2 μm CMOS contact oxide, compressive stress: 40 ± 10 MPa
	65 ± 5 [39]	2 μm CMOS intermetal oxide, compressive stress: 37 ± 6 MPa
Si-B-O glass	4.5–7.2 [77]	Flame hydrolysis deposition, linear decrease in values with increasing B (0 to 20% B_2O_3)
Silicon nitride	260 [78]	LPCVD, Tdep: 850 °C tensile stress of as-deposited Si-rich nitride: ~ 980 MPa
	320 [78]	after annealing at 1150 °C
	210 ± 42 [79]	
	290 [54]	Tdep: 790 °C, tensile stress: 1 GPa
	276 ± 2 [76]	Tdep: 835 °C, tensile stress: 226 ± 2 MPa
	195 ± 6 [39]	2 μm CMOS process, LPCVD, tensile stress: 1040 ± 30 MPa
	210 [54]	PECVD, Tdep: 300 °C, tensile stress: 110 MPa
	97 ± 6 [39]	2 μm CMOS process, PECVD, tensile stress: 82 ± 5 MPa
Oxynitride	180 [55]	LPCVD, O/N ratio: 0.7, tensile stress: 550 MPa
Polyimide	Biaxial: 6.0–8.2 [80]	Values dependent on thickness, thickness range: 10 μm to 30 μm
	3.2 ± 0.16 [62]	spin cast Dupont PI2525 (thickness: 5.2 μm), tensile stress: 32.2 MPa Hitachi PIQ13 (thickness: 11.4 μm), tensile stress: 35.2 MPa

Table 7.3. *Values of relevant material coefficients for various semiconductor, metal, and insulator thin film materials. Besides indicated sources, values are also drawn from [82–86] and references therein. Values for expansion coefficient are specified for room temperature*

Material	Modulus (GPa)	Poisson's ratio	Expansion coefficient ($10^{-6}/°C$)	Mass density (kg/m^3)	Other
Si*	See Table 7.1	See Table 7.1	2.33	2300	ε_{lim} : 28, σ_y : 7
Poly-Si	See Table 7.2	0.2–0.3	2.3–2.6		σ_f : 2–3 [81]
Diamond*	See Table 7.1	See Table 7.1	0.89–1.0	3500	σ_y : 53
			Polycrystalline: 1–1.5 [87]		
			0.07 [70]		
a-Si : H [71]	See Table 7.2	0.2 ± 0.05	2.3–2.6	2300 ± 20	
(8–10% H_2)		Increases with decreasing density		highest at 8–10% H_2	
SiC*	450	0.24	4.2	3200	σ_y : 21
	See Table 7.2				
Al*	See Tables 7.1 & 7.2	0.33	23–25	2700	ε_{lim} : 2, σ_y : 0.17
	70				
Au*	80	0.42	14.3	1900	
Cr*	140–180		7	7200	ε_{lim} : 2
Mo*	343		5	10 300	σ_y : 2.1
Ni*	See Table 7.1	0.31	12.9–13.3	8900	
	210				
Pt*	170	0.39	11.1	21 450	ε_{lim} : 1
Ti*	110	0.34	7.6–9.8	4510	ε_{lim} : 4
W*	See Table 7.1	0.28	4.5	19 300	ε_{lim} : 10, σ_y : 4
	410				
AlN	300	0.25	2.85–4.4		
Al_2O_3	300–380	0.22	5.6–7.1	4000*	σ_y : 15.4*
	530*		5.4*		
Quartz	See Table 8.9	0.169	14.3 (x/y cut)	2650	
		(Fused quartz)	7.8 (z cut)		
SiO_2	See Table 7.2	0.17–0.22	0.4–0.55	2220	ε_{lim} : 112
		0.2 (intermetal oxide) [39]			σ_y : 8.4 (fiber)
		0.5 (contact oxide) [39]			
7740 glass	64	0.2	2.9–3.3	2230	
Si-B-O glass		0.26 [77]			
Si_3N_4	See Table 7.2	0.10 ± 0.05 [79]	1.1–3.8	3100*	ε_{lim} : 37*
	385*				
		0.13 ± 0.02, 0.13 ± 0.07 [39]	0.8*		σ_y : 14*
		0.22*			
Polyimides	See Table 7.2		0.33–2.42 [80]		
PI2525,	3.04, 2.7 [53]	0.4 [62]	thickness		σ_f : 0.071, 0.112
PIQ13	at break point		dependent		[53]

* denotes value for single crystal or bulk material, σ_y the yield strength (GPa), ε_{lim} the limiting strain (%), and σ_f the stress (GPa) at fracture/break point

microstructure of the material [67]. Thermal annealing decreases the residual stress and increases Young's modulus. The latter is most likely due to an increase in the mass density of the film. Effects of density variation on mechanical properties are notable for low temperature processes such as low pressure chemical vapor deposition (LPCVD) or plasma enhanced chemical vapor deposition (PECVD). For example, Young's modulus of silicon nitride can increase from 260 GPa, at the as-deposited temperature of 850 °C, to 320 GPa after annealing at 1150 °C [78]. This is most pronounced with PECVD films in which the hydrogen content has a strong influence on the mechanical properties. For example, hydrogenated amorphous silicon (a-Si:H) films have the best mechanical properties when the atomic percent of hydrogen is 8–10%. Here, the density reaches a maximum value (see Tables 7.2 and 7.3) along with Young's modulus, which correlates with density [71]. In any case, regardless of the thin film deposition process, the values of material coefficients are considerably different from bulk. For example, Young's modulus of a 75 nm Al thin film is 40% lower than the value of 70 Ga for the bulk material [75].

7.3 Model Equations for Analysis of Fluid Transport

In contrast to the solid medium, fluid media (liquids and gases) offer very little resistance to a change in shape. While deformation in the solid, as seen in Sect. 7.2, is characterized by a displacement field, in the fluid media it is characterized by a velocity field. However, similar interpretations hold. The concept of infinitesimal strain, used in analysis of solid structures, can be extended to fluids to yield a rate of change of strain, whereby the displacement components are now replaced with velocity components; see (7.13) and (7.17).

Computation of the velocity field is crucial for the prediction of output response and sensitivity of microflow sensors [89, 90] and design optimization of the associated flow channel geometry/structure which may be an integral part of the sensor packaging [91]. It is also crucial for optimization of micromechanical devices, where damping of the structure by the surrounding fluid has a direct bearing on the dynamic performance of micromechanical devices (see Sect. 7.3.3). Additionally, other areas where fluids play an important role include cooling and thermal management of electronic systems (see [92]) and mass/heat transport in microfabrication process equipment, e.g., CVD and etching [93].

In what follows, we begin with a brief description of constitutive behavior to aid our discussion on fluid classification before moving to the governing field equations and boundary conditions.

7.3.1 Constitutive Properties

Depending on their constitutive properties, fluids can be broadly classified as being either Newtonian or non-Newtonian. Our discussion will be restricted to the former class of fluids. In Newtonian fluids, the shear stress components are linearly related to the strain-rate components. If the fluid is isotropic, which it is in most cases, the constitutive relations for a viscous, compressible medium can be written in terms of velocity components v_i as [3]

$$\sigma_{ij} = -p\,\delta_{ij} + \mu\left(\frac{\partial v_i}{\partial x_j} + \frac{\partial v_j}{\partial x_i}\right) + \lambda\,\delta_{ij}\frac{\partial v_k}{\partial x_k} \qquad \text{with} \quad i,j,k = 1,2,3.$$

(7.98)

Here, σ_{ij} denote stress components, p is a scalar which represents hydrostatic pressure, δ_{ij} the Kronecker delta, μ the coefficient of shear viscosity, and λ the coefficient of bulk viscosity. Relation (7.98) is also known as the Navier-Poisson law for Newtonian fluids. Employing the Stokes hypothesis, $\lambda + 2\mu/3 = 0$, which holds for most fluids [3], we can express the constitutive relation (7.98) in terms of a single material coefficient, viz.,

$$\sigma_{ij} = -p\,\delta_{ij} + \mu\left(\frac{\partial v_i}{\partial x_j} + \frac{\partial v_j}{\partial x_i}\right) - \frac{2}{3}\mu\,\delta_{ij}\frac{\partial v_k}{\partial x_k} \qquad \text{with} \quad i,j,k = 1,2,3.$$

(7.99)

Fluids that satisfy the constitutive property (7.99) are referred to as Stokesian fluids. In the case of an incompressible viscous fluid, the fluid density ρ is a constant and following (7.24), the last term in (7.99) vanishes to yield

$$\sigma_{ij} = -p\,\delta_{ij} + \mu\left(\frac{\partial v_i}{\partial x_j} + \frac{\partial v_j}{\partial x_i}\right) \qquad \text{with} \quad i,j = 1,2,3. \qquad (7.100)$$

For a non-viscous fluid, $\mu = 0$ and (7.100) yields the constitutive property that is characteristic of a perfect (or ideal) fluid, which is a special case of the Newtonian fluid:

$$\sigma_{ij} = -p\,\delta_{ij} \qquad \text{with} \quad i,j = 1,2,3. \qquad (7.101)$$

Here, the fluid is free of shearing stress and possesses only normal stress components which are equal and constitute negative pressure at any point in the fluid. The fluid pressure in such a hydrostatic stress system is then the arithmetic mean of the negative of normal stress components.

Non-Newtonian fluids do not satisfy the linear relationship between shear stress and strain-rate. Such behavior is characteristic of fluids such as pastes, blood, polymers. There is no accurate, yet simple, theory that describes the relation between stress and the strain rate. Models for the constitutive behavior proposed hitherto are empirical (see [3]).

7.3.2 Governing Equations

The equations governing fluid flow stem from the conservation laws of mass, momentum, and energy. The equation governing mass conservation, also referred to as the equation of continuity, expresses a balance between the masses entering and leaving a unit volume per unit time, and the local change in density. The equation of continuity was employed earlier in Sect. 7.2.1 in relation to analysis of static and dynamic behavior of solid structures. For convenience, we restate Eq. (7.24),

$$\frac{\partial \rho}{\partial t} + \frac{\partial (\rho v_i)}{\partial x_i} = 0 \quad \text{with} \quad i = 1, 2, 3. \tag{7.102}$$

Here, ρ denotes the mass density of the fluid and v_i the velocity components. In the case of an incompressible fluid, the density is independent of space and time, and the equation for continuity reduces to

$$\frac{\partial v_i}{\partial x_i} = 0 \quad \text{with} \quad i = 1, 2, 3, \tag{7.103}$$

which was employed earlier to simplify (7.99). The equations of fluid motion are derived from Newton's second law, which states that the mass-acceleration product must equal the sum of forces acting on the body; the general case is given by (7.22). In fluid motion, there are two classes of forces: forces acting throughout the mass of the body and forces acting on the surface. If f_i and s_i represent the components of the body and surface force densities, respectively, the equations of motion, following (7.23), can be stated as [3]:

$$\rho \left(\frac{\partial v_i}{\partial t} + v_j \frac{\partial v_i}{\partial x_j} \right) = f_i + s_i \quad \text{with} \quad i, j = 1, 2, 3. \tag{7.104}$$

Here, the s_i stems from the divergence of the stress tensor given by the first term on the left-hand side of (7.23). The surface force density can be resolved into nine stress components (see Fig. 7.2) that form the stress tensor. Symmetry considerations reduce the number of components to six due to vanishing moments. To introduce velocity components into the right-hand side of (7.104), we need to relate the stress components to the strain rates. This can be achieved by invoking the constitutive relations (7.99), which are linear for an isotropic Newtonian fluid. These considerations applied to (7.104) yield the celebrated Navier-Stokes equations [3]:

$$\rho\left(\frac{\partial v_i}{\partial t} + v_j \frac{\partial v_i}{\partial x_j}\right) = f_i - \frac{\partial p}{\partial x_i} + \frac{\partial}{\partial x_j}\left[\mu\left(\frac{\partial v_i}{\partial x_j} + \frac{\partial v_j}{\partial x_i} - \frac{2}{3}\delta_{ij}\frac{\partial v_k}{\partial x_k}\right)\right]$$

$$\text{with} \quad i,j,k = 1,2,3.$$

$$(7.105)$$

In the case of incompressible fluids, we can reduce (7.105) following the assumption of constant mass density. Also, for small variations in temperature, we can assume the viscosity μ to be a constant. In this case

$$\rho\left(\frac{\partial v_i}{\partial t} + v_j \frac{\partial v_i}{\partial x_j}\right) = f_i - \frac{\partial p}{\partial x_i} + \mu\frac{\partial^2 v_i}{\partial x_k \partial x_k} \qquad \text{with} \quad i,j,k = 1,2,3.$$

$$(7.106)$$

System (7.106), with the prescribed forces, along with the equation for continuity for an incompressible fluid, (7.103), yields four equations for the four unknowns; the three velocity components and the pressure. In the case of two-dimensional flows, the number of unknowns can be reduced to two by casting (7.106) in the vorticity stream function formulation [3, 89]. This is discussed in the example given in Sect. 7.4.

The equations we have dealt with so far are still incomplete for modeling fluid transport, since they do not account for effects of temperature variations on mass density and pressure [3]. Associated with fluid motion is a flow of heat. To account for the interaction of the two fields, the equations of motion have to be coupled with those of heat conduction. The energy balance for a fluid is determined by its internal energy, the generation of heat through friction, and heat loss by conduction and convection. In addition, there may be energy expended with volume changes and there may be heat loss by radiation. The former applies only to compressible fluids, but the latter applies to all cases. However, radiative heat loss may be neglected for moderate temperatures. Taking into account the various factors contributing to energy balance, and utilizing the constitutive relation, (7.99), we obtain the following energy equations of flow

$$\rho\left[\frac{\partial E}{\partial t} + v_i\frac{\partial E}{\partial x_i}\right] + p\frac{\partial v_i}{\partial x_i} = \frac{\partial}{\partial x_i}\left(\kappa\frac{\partial T}{\partial x_i}\right) + \mu\Phi \qquad \text{with} \quad i = 1,2,3.$$

$$(7.107)$$

Here, E denotes the internal energy, κ the thermal conductivity, T the temperature, and Φ the dissipation function, which, for convenience, is expressed in terms of Cartesian notation

$$\Phi = 2\left[\left(\frac{\partial u}{\partial x}\right)^2 + \left(\frac{\partial v}{\partial y}\right)^2 + \left(\frac{\partial w}{\partial z}\right)^2\right]$$
$$+ \left(\frac{\partial v}{\partial x} + \frac{\partial u}{\partial y}\right)^2 + \left(\frac{\partial w}{\partial y} + \frac{\partial v}{\partial z}\right)^2 + \left(\frac{\partial u}{\partial z} + \frac{\partial w}{\partial x}\right)^2 \qquad (7.108)$$
$$- \frac{2}{3}\left(\frac{\partial u}{\partial x} + \frac{\partial v}{\partial y} + \frac{\partial w}{\partial z}\right)^2,$$

where u, v, and w denote the components of the velocity vector. Following thermodynamic considerations, Eq. (7.107) can be simplified to the following forms. For an ideal fluid,

$$\rho c_p\left[\frac{\partial T}{\partial t} + v_i\frac{\partial T}{\partial x_i}\right] = \frac{\partial p}{\partial t} + v_i\frac{\partial p}{\partial x_i} + \frac{\partial}{\partial x_i}\left(\kappa\frac{\partial T}{\partial x_i}\right) + \mu\Phi \qquad (7.109)$$

$$\text{with} \quad i = 1, 2, 3$$

and for an incompressible fluid, since the density is independent of space and time,

$$\rho c\left[\frac{\partial T}{\partial t} + v_i\frac{\partial T}{\partial x_i}\right] = \frac{\partial}{\partial x_i}\left(\kappa\frac{\partial T}{\partial x_i}\right) + \mu\Phi \qquad \text{with} \quad i = 1, 2, 3$$

$$(7.110)$$

where c_p and c denote the respective specific heat which are dependent on temperature.

Equations (7.102), (7.105), and (7.109) or (7.110) constitute a coupled system of non-linear partial differential equations which can only be solved numerically; no general closed form solutions have been reported hitherto. In non-steady state, the equations are parabolic. In steady state, they become elliptic. In particular, the character of the equations change completely with vanishing viscosity. The system gives rise to five equations, but there are six unknown flow field variables, *viz.*, density (ρ), pressure (p), three velocity components (v_i), and temperature (T). Thus we need an additional equation. This can be drawn from the equation of state of the fluid, which for an ideal gas is

$$p = \rho RT \qquad (7.111)$$

where R denotes the specific gas constant. Values of relevant fluid parameters are given in Table 7.4 [94].

Although the system of coupled differential equations are applicable to a wide variety of problems, ranging from fluid flow over an aircraft to flow microsensors (see Chapt. 5), the contrasting nature of the flow field in each

Table 7.4. *Values of relevant parameters for selected gases; μ denotes the viscosity at $0°C$ and 1 atm, ρ_0 the density at $0°C$ and 1 atm, c the velocity of sound, κ the thermal conductivity at $0°C$, c_p the specific heat at constant pressure, and c_v the specific heat at constant volume. Densities at other temperatures can be calculated as $\rho = \rho_0 \, [p/(1\,atm)]$ $(273.16/T)$ where 1 atm $= 101.32\,kPa$. The gas constant, $R = 8.3144\,J/mol\text{-}K$*

Gas	μ $(10^{-6}\,Pa/s)$	ρ_0 (kg/m^3)	c (m/s)	κ $(W/m\text{-}K)$	c_p $(kJ/kg\text{-}K)$	c_p/c_v
Air	181.94	1.2929	331.45	0.0241	1.006 (0°C, 1 atm)	1.403 (0°C, 1 atm)
Argon	222.86	1.7837	307.8	0.0160	0.5252 (15°C, 1 atm)	1.668 (15°C, 1 atm)
Helium	196	0.1785	970	0.1436	5.239 (−180°C, 1 atm)	1.66 (−180°C, 1 atm)
Nitrogen	175.69	1.2506	337	0.0244	1.038 (15°C, 1 atm)	1.404 (15°C, 1 atm)

case is determined by the boundary and initial conditions. In the case of a viscous fluid, its interaction with a moving solid surface obeys the no-slip condition. Here, there is no relative velocity between the solid surface and the fluid and thus both the normal and tangential components of fluid velocity vanish, *viz.*,

$$v_i n_i = 0 \quad \text{with} \quad i = 1, 2, 3, \tag{7.112}$$

$$v_i t_i = 0 \quad \text{with} \quad i = 1, 2, 3, \tag{7.113}$$

where n_i and t_i denote the components of the unit vector normal and tangential to the surface, respectively.

For an inviscid (non-viscous) fluid, there is no friction to promote the sticking of the fluid to the surface. Thus the fluid slips over the surface resulting in a flow that is tangential to the surface. Hence, only the normal component of velocity vanishes on the surface as given by (7.112). Elsewhere at the non-physical boundaries the choice of boundary conditions depends on the nature of the problem considered; in fact, the proper selection of conditions at these boundaries still remains an art. The boundary conditions generally pertain to either inflow and outflow boundaries at a finite distance from the surface, or "infinity" boundary conditions at regions far from the surface.

7.3.3 Fluidic Damping

Damping of the microactuator by the surrounding fluid is an important design consideration which affects the amplitude response and stability,

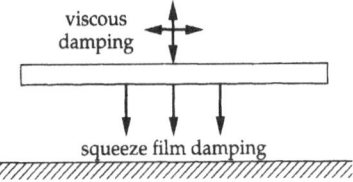

Fig. 7.6 Schematic representation of fluidic damping. Arrows denote actuation direction

and the quality factor of the micromechanical structure. In some structures (e.g. accelerometers [95, 96]), some degree of fluidic damping is crucial for stability of operation while in others (e.g. lateral resonators [97, 98]), damping degrades the quality factor. Thus for the desired application, it is crucial, by structural design, to eliminate or optimize energy dissipation so as to achieve the desired performance.

We consider two kinds of fluidic damping forces (see Fig. 7.6); forces due to viscous drag of the ambient fluid and squeeze film forces. The former is the major source of energy dissipation in laterally driven structures [97, 98] while the latter is the dominant dissipative source in vertically driven structures.

Viscous drag damping forces can be computed from distribution of the velocity field of the fluid surrounding the microactuator. The velocity field associated with the motion of an actuator in an incompressible viscous fluid ambient, in the absence of body forces, is governed by the Navier-Stokes equation (7.105). Restated in terms of vector notation, it reads,

$$\rho \frac{\partial \mathbf{v}}{\partial t} + \rho(\mathbf{v} \cdot \nabla)\mathbf{v} = -\operatorname{grad} p + \mu \operatorname{div} \operatorname{grad} \mathbf{v}, \tag{7.114}$$

where ρ, $\mathbf{v}$, p, and μ denote the density, velocity, pressure, and coefficient of viscosity of the ambient fluid, respectively, and ∇ the Nabla operator. The second term on the left-hand side accounts for non-linear convection. This term can be neglected for small Reynolds numbers (Re) and for relatively small oscillation amplitudes (see [97, 99]) to yield

$$\rho \frac{\partial \mathbf{v}}{\partial t} = -\operatorname{grad} p + \mu \operatorname{div} \operatorname{grad} \mathbf{v}. \tag{7.115}$$

Eq. (7.115) can be solved for the velocity, along with the conservation equation for mass, under suitable boundary conditions. If the viscosity of the surrounding fluid is high, we obtain a no-slip condition due to the large friction forces. This yields a vanishing normal and tangential component of velocity on the surface of the microactuator, as given by (7.112) and (7.113). In contrast, for a low viscosity fluid, slipping occurs and the flow is tangential. In this case, only the normal component of the velocity vanishes on the surface, as given by (7.112). The components of velocity in the fluid field at distances remote from the surface are determined by the fluid's free stream velocity. Based on the computed velocity distribution, we determine the viscous drag damping force f^D acting on the surface of the moving microactuator [3, 100] as

$$f_i^D = \int_\Gamma \sigma_{ij} n_j \, d\Gamma \quad \text{with} \quad i,j = 1,2,3, \tag{7.116}$$

where σ_{ij} are the stress components (which are related to the pressure and velocity components as given by (7.100) for an incompressible fluid), n_j denotes the jth component of the unit vector normal to the surface, Γ the integration surface, and $d\Gamma$ the surface element.

Equation (7.115) has been employed in a reduced form to investigate degradation in quality factor in lateral interdigital structures. Here, the pressure gradients and the velocity components orthogonal to the direction of actuator motion have been assumed negligible to yield a so-called "Stokes-type" damping model [97]. By further neglecting the time-dependent term at low oscillation frequencies, the "Couette-type" damping model has been employed [101]. However, these reduced forms of model equations have yet to yield good agreement with measurements.

Squeeze film damping in microactuators can be described by the generalized Reynold's equation [102, 103] that governs the pressure distribution p in thin fluid films. The Reynold's equation constitutes a limiting case of very slow motion or motion in highly viscous media. In the latter, the viscous forces are much larger than inertial forces thereby permitting a simplification of the Navier-Stokes equation. The reduced Navier-Stokes equation also provides valid description for motion with very small Reynolds numbers (Re $\rightarrow$ 0), since Re represents the ratio of inertial to friction forces. The Navier-Stokes equation coupled with the continuity equation of mass and the equation of state, yields the following description for the pressure distribution in a thin isothermal film:

$$\frac{\partial}{\partial x_i}\left(h^3 p \frac{\partial p}{\partial x_i}\right) = 6\mu\left\{2\frac{\partial(hp)}{\partial t} + \frac{\partial\left(hp\left[v_i^a + v_i^b\right]\right)}{\partial x_i}\right\} \tag{7.117}$$

$$\text{with} \quad i = 1,2.$$

Here, h denotes the thickness of the fluid thin film and v_i^a and v_i^b the respective velocity components on its bounding surfaces (a and b). In most cases of interest, p is the only dependent variable and h and $v_i^{a,b}$ are specified in space and time; otherwise Reynold's equation needs to be coupled to other equations describing the dynamics of the system. In the absence of lateral surface motion, we neglect the terms containing the velocity and the generalized Reynold's equation, (7.117), reduces to the following squeeze film equation [102, 103] which is non-linear and parabolic

$$\frac{\partial}{\partial x_i}\left(h^3 p \frac{\partial p}{\partial x_i}\right) = \frac{\partial}{\partial t}(hp) \qquad \text{with} \quad i = 1, 2. \tag{7.118}$$

The equation is valid when $\omega h^2 \rho \ll \mu$, where ω is the oscillation frequency of the micromechanical structure. From a computational standpoint, it may be more useful to employ a dimensionless form of (7.118), viz.,

$$\frac{\partial}{\partial X_i}\left(H^3 P \frac{\partial P}{\partial X_i}\right) = \sigma \frac{\partial(HP)}{\partial t} \qquad \text{with} \quad i = 1, 2, \tag{7.119}$$

where the dimensionless parameters are given as (see [102–104]),

$$X_i = \frac{x_i}{L}, \qquad H = \frac{h}{h_0}, \qquad P = \frac{p}{p_a}, \qquad \sigma = \frac{12 \mu L^2 \omega}{p_a h_0^2}. \tag{7.120}$$

Here, L denotes the plate length, h_0 the nominal film thickness, and p_a the ambient pressure. The squeeze film equation can be reduced to a simple linear Poisson-type equation if the following simplifying assumptions are employed: the fluid is incompressible; film thickness (h_0) is uniform; displacement of bounding surfaces is small relative to film thickness; and pressure variation is small. In terms of the original variables, it reads [95]

$$\frac{\partial}{\partial x_i}\left(\frac{\partial p}{\partial x_i}\right) = \left(12 \mu/h_0^3\right) \frac{\partial h}{\partial t} \qquad \text{with} \quad i = 1, 2 \tag{7.121}$$

which can be readily solved for the pressure using standard numerical routines employed for solution of linear Poisson's equation. At the non-bounding surfaces (boundaries), we set the value of p to the ambient pressure, viz.,

$$p = p_a. \tag{7.122}$$

At bounding surfaces, we assume a vanishing normal component of the pressure gradient,

$$\left(\frac{\partial p}{\partial x_i}\right) n_i = 0 \qquad \text{with} \quad i = 1, 2. \tag{7.123}$$

The squeeze film damping force f^D acting on the surface of the microactuator, following (7.101), can be computed as:

$$f_i^D = -\int_\Gamma p \, \delta_{ij} \, n_j \, d\Gamma \qquad \text{with} \quad i, j = 1, 2. \tag{7.124}$$

The resulting distribution of squeeze force is valid only for small displacements, incompressible fluids, and for large squeeze film dimensions. Damping behavior at large displacements of bounding surfaces can be accounted for by multiplying a displacement function to the solution of (7.121) [95, 105]. Compressibility effects can be accounted for by the squeeze number σ in (7.119) [102, 104]. At low $\omega, \sigma \to 0$, the film behaves like an incompressible viscous fluid. At high $\omega, \sigma \to \infty$, the film behaves like a spring. When the squeeze film thickness becomes significant relative to the mean free path of molecules in the fluid, continuum theory may no longer hold [95]. For example, when the thickness of the squeeze film (h) reduces to about two orders of the mean free path (λ), the Knudsen number (Kn $= \lambda/h$) increases and slip flow takes place at bounding surfaces effectively reducing damping forces.

7.4 Illustrative Simulation Example – Analysis of Flow Channels

The output response and sensitivity of microflow sensors are determined by the nature of fluid flow (laminar or turbulent) the sensor detects in a duct or a flow channel. The flow can create a recirculation (or vortex) region whose length depends on inlet velocity, viscosity of the fluid and channel geometry. Thus, the ability to predict reliably the character of flow for a given fluid viscosity and range of inlet velocities is critical to the geometrical/structural design of the flow channel, which is often an integral part of the microsensor packaging. Likewise if the flow channel suffers from dimensional constraints, knowledge of the exact velocity distribution in the channel can aid in calibration/compensation of the microsensor output response.

Fig. 7.7 illustrates the geometry of the flow channel considered in this example. Of practical importance are flow channels with a small enlargement angle, whereby the channel's cross-sectional area changes smoothly along the axial direction so as to delay the onset of fluid recirculation. The objective here is to determine the reattachment length (L_a) as a function of inlet velocity. Here, L_a is defined as the length for the flow stream to reattach to the surface (wall) of the flow channel. Considerable numerical and experimental work has been reported [106–109] for the reattachment lengths. In simple structures, e.g., circular cylindrical geometries, it is observed that the reattachment length varies linearly with the Reynolds number (Re) for low flows (Re $\leq$ 100). In more complex structures, e.g., "backward facing steps", the reattachment length is observed to be strongly dependent on the numerical method used in discretizing the governing equations. For Re of the order 500, the efficiency of the numerical scheme depends on the type of discretization scheme used for the convective derivative term. Many upwind methods have been proposed to deal with this issue. Depending upon the discretization, the

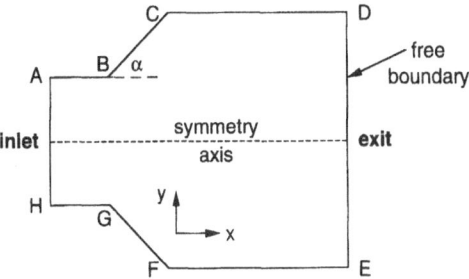

Fig. 7.7 Geometry of flow channel. AH denotes the inlet, DE the exit, α the enlargement angle, and ABCD and EFGH are no-slip boundaries $(u = v = 0)$ [89]

schemes may add numerical dissipation in the direction of the streamline or perpendicular to it. In this example, a rotated finite difference scheme, which takes into account the local characteristics of the streamline, is employed to simulate the two-dimensional, steady-state, incompressible viscous fluid motion in a flow channel.

7.4.1 Model Equations in Vorticity-Stream Function Formulation

The flow behavior is governed by the equations for continuity of mass, (7.102), and the Navier-Stokes equations of fluid motion, (7.105). The system is non-linear and virtually impossible to solve analytically. Various numerical methods have been proposed [106–112]. In the case of two-dimensional, steady-state, incompressible, viscous flow, the system of equations, (7.102) and (7.105), reduce to the following form

$$\frac{\partial u}{\partial x} + \frac{\partial v}{\partial y} = 0, \tag{7.125}$$

$$u\frac{\partial u}{\partial x} + v\frac{\partial u}{\partial y} = -\frac{1}{\rho}\frac{\partial p}{\partial x} + \nu\left[\frac{\partial^2 u}{\partial x^2} + \frac{\partial^2 u}{\partial y^2}\right], \tag{7.126}$$

$$u\frac{\partial v}{\partial x} + v\frac{\partial v}{\partial y} = -\frac{1}{\rho}\frac{\partial p}{\partial y} + \nu\left[\frac{\partial^2 v}{\partial x^2} + \frac{\partial^2 v}{\partial y^2}\right], \tag{7.127}$$

expressed in terms of Cartesian notation. Here, u and v denote the velocity components in the x- and y-directions, respectively, p the pressure, ρ the mass density, and ν the kinematic viscosity of the fluid (air in this case), which is given as μ/ρ. A linear velocity profile is assumed at the inlet of the flow channel (see Fig. 7.7). On the surface of the channel the no-slip boundary condition, $u = v = 0$, is applied. This is further elaborated in Sect. 7.4.2. The exit of the channel is left free, i.e., no boundary condition is imposed but the equations of motion are applied in the vorticity-stream function form. Since the geometry is symmetric with respect to the center-line of the channel (referred to as the symmetry axis in Fig. 7.7), the simulation has been confined to the upper portion of the flow channel and the symmetry boundary conditions are implemented on the center-line.

The number of unknown variables can be reduced to two by casting the governing equations in the vorticity-stream function formulation. The stream function ψ is defined as

$$u = \frac{\partial \psi}{\partial y}, \qquad v = -\frac{\partial \psi}{\partial x} \tag{7.128}$$

so that the mass conservation equation is satisfied automatically. Defining the vorticity ω as

$$\omega = \frac{\partial v}{\partial x} - \frac{\partial u}{\partial y}, \tag{7.129}$$

Eqs. (7.125) to (7.127) become

$$\frac{\partial^2 \psi}{\partial x^2} + \frac{\partial^2 \psi}{\partial y^2} = -\omega, \tag{7.130}$$

$$u \frac{\partial \omega}{\partial x} + v \frac{\partial \omega}{\partial y} = \nu \left[\frac{\partial^2 \omega}{\partial x^2} + \frac{\partial^2 \omega}{\partial y^2} \right]. \tag{7.131}$$

The boundary conditions have to be suitably applied in terms of the stream function and the vorticity.

7.4.2 Rotated Finite Difference Numerical Scheme

The governing equations (7.130) and (7.131) are discretized using the rotated finite difference scheme [113]. The nodal connectivity for an interior node I is shown in Figs. 7.8(a) to 7.8(d), where $N_1, N_2, N_3, \ldots, N_8$ are the surrounding nodes. For purposes of discretization, the vorticity transport equation (7.131) is recast in the form

$$\sqrt{u^2 + v^2} \frac{\partial \omega}{\partial s} = \nu \left[\frac{\partial^2 \omega}{\partial x^2} + \frac{\partial^2 \omega}{\partial y^2} \right], \tag{7.132}$$

where $\partial \omega / \partial s$ is the stream derivative of ω in the direction of the streamline given by $\partial x / u = \partial y / v$. The streamline IP (where P is an arbitrary point denoting the intersection of the local streamline with the boundary of the cell) is assumed to be linear within the cell connecting the eight surrounding nodes. The various possible choices for P are shown in Figs. 7.8(a) to 7.8(d). The stream derivative $\partial \omega / \partial s$ at I, for all the possible

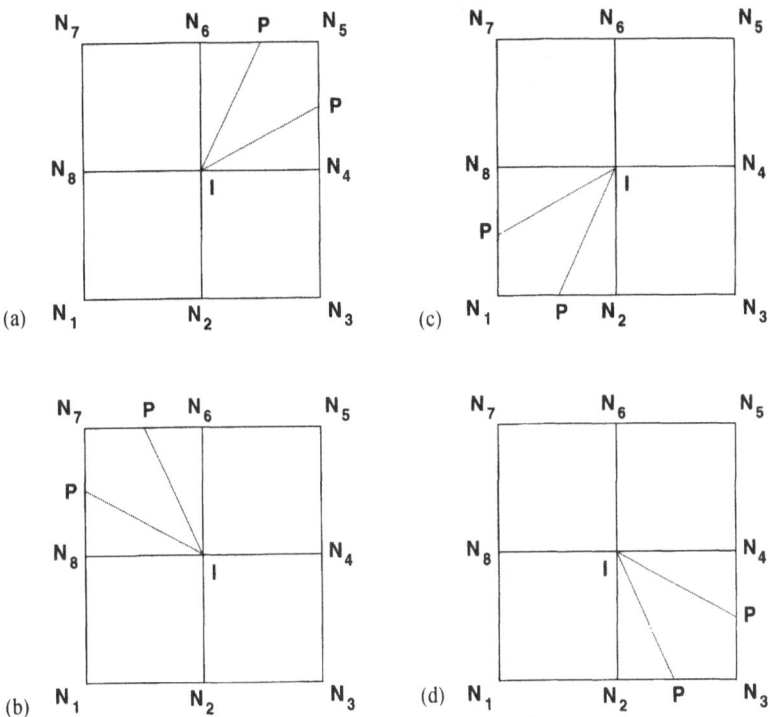

Fig. 7.8 Nodal connectivity for the interior node I for (a) $u \leq 0$ and $v \leq 0$, (b) $u \geq 0$ and $v \leq 0$, (c) $u \geq 0$ and $v \geq 0$, (d) $u \leq 0$ and $v \geq 0$. The two possible positions of P shown depend on the relative magnitudes of u and v [89]

choices shown, is then discretized by

$$\frac{\partial \omega}{\partial s} = \frac{\omega(I) - \omega(P)}{d_{IP}}, \tag{7.133}$$

where d_{IP} denotes the distance between I and P. When the streamline does not coincide with the coordinate axes, P must be interpolated between the neighboring nodes and this can be done in many ways. Here, a linear interpolation is employed. When the interior node is near the boundary, there will not be eight neighboring nodes, if the domain is not rectangular, and $\omega(P)$ has to be interpolated accordingly. When the velocity vector is in the direction of one of the coordinate axes, the scheme reduces to the usual first-order upwinding. This discretization results in a marginally diagonally dominant discrete set of equations. The scheme is closely related to the skew-upstream differencing technique of [114] but the diagonal dominance for the latter cannot be guaranteed. In many respects the discretization employed here is along the lines proposed by [115] but differs in the interpolation of point P. The diffusive operator is discretized by

$$\frac{\partial^2 \omega}{\partial x^2} = 2\left[\frac{d_{08}\omega_4 - (d_{08} + d_{04})\omega_I + d_{04}\omega_8}{d_{04}d_{08}(d_{04} + d_{08})}\right], \tag{7.134}$$

$$\frac{\partial^2 \omega}{\partial y^2} = 2\left[\frac{d_{02}\omega_6 - (d_{06} + d_{02})\omega_I + d_{06}\omega_2}{d_{02}d_{06}(d_{02} + d_{06})}\right], \tag{7.135}$$

where d_{02}, d_{04}, d_{06}, and d_{08} are the distances of the nodes N_2, N_4, N_6, and N_8 from node I, respectively.

Poisson's equation for the stream function and the transport equation for the vorticity are decoupled and solved as follows.

- Assuming an initial guess vorticity distribution ω^0, Poisson's equation

$$\frac{\partial^2 \psi}{\partial x^2} + \frac{\partial^2 \psi}{\partial y^2} = -\omega^0 \tag{7.136}$$

  is solved along with the boundary condition for ψ to obtain a new ψ^1. A linear velocity profile is assumed at the inlet of the duct. On the surface of the channel, ψ is set to zero. At the exit of the channel, Poisson's equation for the stream function is applied as a boundary condition.

- With the available stream function distribution ψ^1, the velocity components at nodes are calculated from

$$u = \frac{\partial \psi^1}{\partial y},$$
$$v = -\frac{\partial \psi^1}{\partial x}. \tag{7.137}$$

- With the calculated velocity distribution, the vorticity transport equation, (7.132), is solved along with the "updated" boundary condition for the vorticity to obtain a new ω^1. On the surface of the channel, the vorticity is unknown and it has to be calculated as a part of the iterative procedure. This wall vorticity is the source of disturbance and is advected and diffused at the interior fluid points. However, we are given only the no-slip boundary conditions for the velocity components. Noting that the governing non-linear equations are coupled and are of fourth order in stream function, the vorticity on the no-slip boundary can be calculated consistent with Poisson's equation for the stream function. An excellent detailed treatment of this boundary condition is elucidated in [110, 116], although it corresponds to the pressure boundary condition. On the no-slip boundary, $\partial^2 \psi / \partial^2 t = 0$, where t is a direction vector along the boundary and the vorticity,

$$\omega = -\frac{\partial^2 \psi}{\partial n^2},$$ (7.138)

where n is a unit normal to the boundary, can be calculated consistently with the no-slip boundary condition.

A remark must be made regarding the previously mentioned free boundary condition for the stream function and the vorticity. There has been significant interest [117, 118] in determining an optimal open boundary condition at the "exit" so as to reduce computational requirements. A detailed summary of all conditions at the free boundary is given in [118]; however, there appears to be no unanimous conclusion. In the calculations here, the vorticity transport equation itself, in the absence of any other information, is applied as a condition on the boundary.

• The three steps above are repeated until self-consistent solutions are achieved. As is well-known, it is crucial to under-relax the flow variables during the iteration cycle particularly for flows with high Reynolds number to avoid divergence.

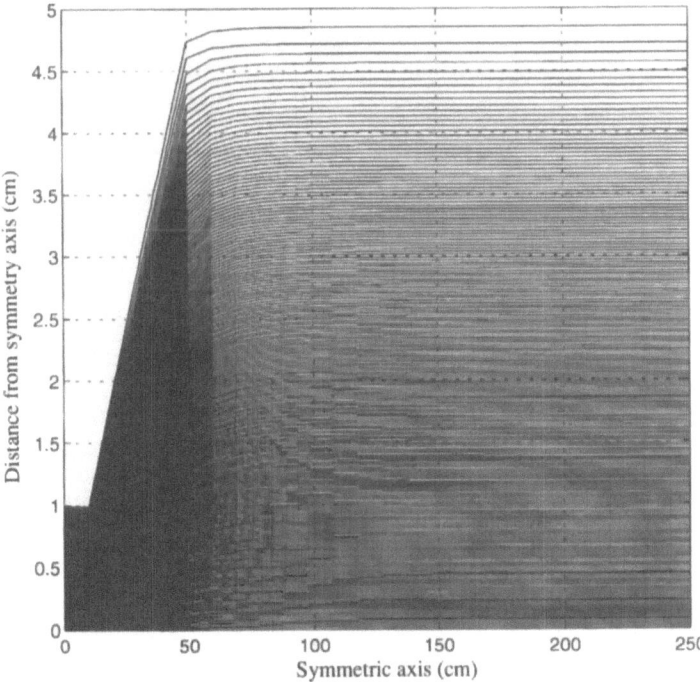

Fig. 7.9 Streamlines in the flow channel for an inlet velocity of 10 cm/s [89]

7.4.3 Computed Flow Profiles

In order to validate the accuracy of the numerical solutions, the scheme has been applied successfully to extreme or limiting conditions, such as pure diffusion and pure convection, for which the analytical solutions are known. In the pure convection case, with a known velocity along the diagonals of the unit cell, the scheme is able to predict the exact solution even when the analytical solution is discontinuous across the diagonal. Similarly, for the pure diffusion case, the scheme is able to predict the exact vorticity distribution, which is biquadratic in x and y when there is no flow.

The flow inside the channel depends on three geometrical quantities besides the inlet velocity and the kinematic viscosity of the fluid: enlargement angle α, the duct depth DE and duct length FE (Fig. 7.7). The numerical values for these quantities used are HG $= 10$ cm, GFcosα $= 40$ cm, FE $= 200$ cm, GFsin$\alpha = 4$ cm, DE $= 10$ cm, and $\alpha = 5.7°$. The kinematic viscosity of air is taken to be $\nu = 0.1785$ cm^2/s and Re is varied by increasing the inlet velocity. The computations are based on a non-uniform grid.

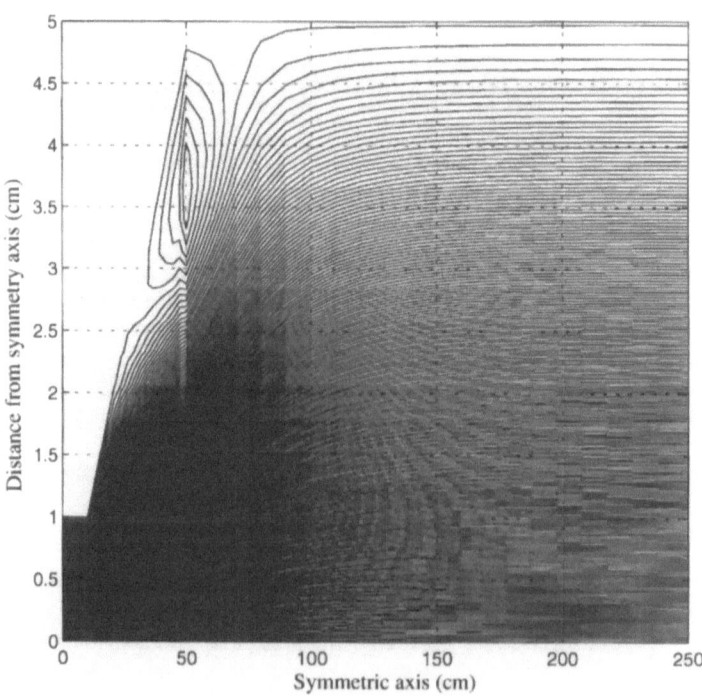

Fig. 7.10 Streamlines in the flow channel for an inlet velocity of 30 cm/s [89]

In the numerical experiments considered, the small enlargement angle means that there is no vortex formation even for high Reynolds numbers ($\text{Re} \leq 400$), for the given characteristic length DE/2. Fig. 7.9 depicts streamlines for an inlet velocity of 10 cm/s. Because of the relatively small velocity and the given channel geometry, no vortex formation is observed

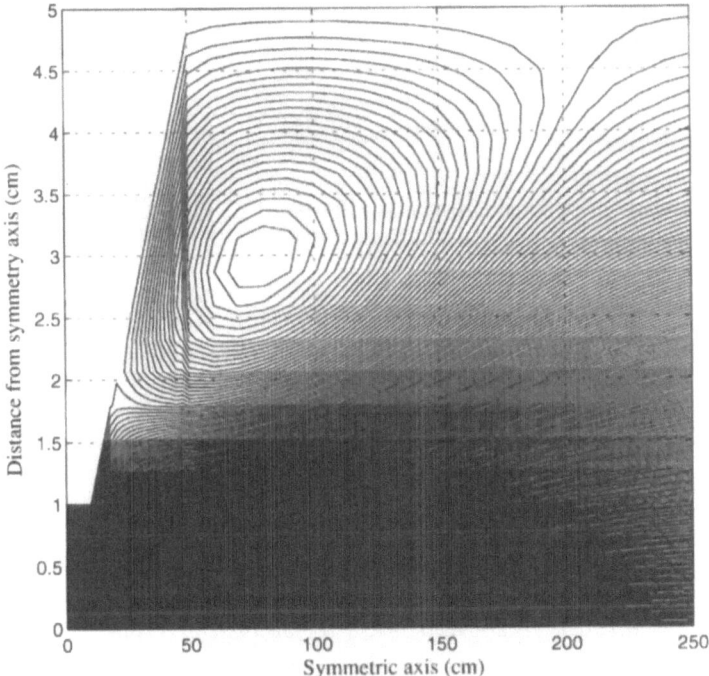

Fig. 7.11 Streamlines in the flow channel for an inlet velocity of 1 m/s [89]

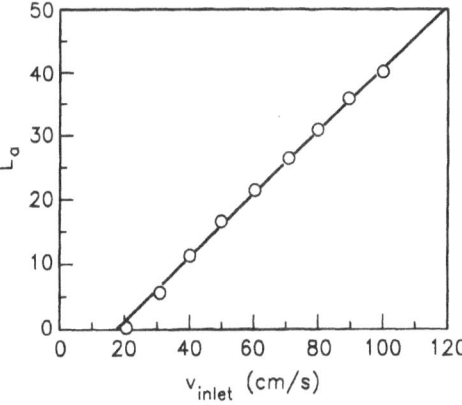

Fig. 7.12 The normalized reattachment length as a function of inlet velocity [89]

and the reattachment length is virtually zero. At higher velocities, 30 cm/s and 1 m/s, a vortex forms (Figs. 7.10 and 7.11) and moves downstream as the inlet velocity is increased, thus increasing the reattachment length. The width of the vortex region increases with Re. The reattachment length measured along the axis of the duct from point B (see Fig. 7.7) and normalized to $DE/2$ is shown as a function of the inlet velocity (Fig. 7.12). The reattachment length for low velocities (low Re) varies linearly with the Reynolds number.

7.5 References

[1] McDonald, P., *Continuum Mechanics*, Boston: PWS Publishing Co., 1996.
[2] Timoshenko, S. P., Goodier, J. N., *Theory of Elasticity*, 3rd Ed., New York: McGraw-Hill, 1970.
[3] Schlichting, H., *Boundary Layer Theory*, New York: McGraw-Hill, 1968.
[4] Nye, J. F., *Physical Properties of Crystals*, Oxford: Oxford University Press, 1957.
[5] Fung, Y. C., *A First Course in Continuum Mechanics*, 3rd Ed., New Jersey: Prentice Hall, 1994.
[6] Sokolnikoff, I. S., *Tensor Analysis, Theory and Applications to Geometry and Mechanics of Continua*, 2nd Ed., New York: Wiley, 1964.
[7] Landau, L. D., Lifshitz, E. M., *Fluid Mechanics*, 2nd Ed., New York: Pergamon, 1989.
[8] Middelhoek, S., Audet, S. A., *Silicon Sensors*, New York: Academic Press, 1989.
[9] Sze, S. M., (Ed.), *Semiconductor Sensors*, Wiley, New York, 1994.
[10] Bin, T. Y., Huang, R. S., CAPSS: A Thin Diaphragm Capacitive Pressure Sensor Simulator, *Sensors and Actuators*, 11 (1987), 1–22.
[11] Bouwstra, S., Geijselaers, B., On the Resonance Frequencies of Microbridges, *Digest of Technical Papers*, Transducers '91, San Francisco, 1991, pp. 538–542.
[12] Elgamel, H. E., Closed-Form Expressions for the Relationships Between Stress, Diaphragm Deflection, and Resisitance with Pressure in Silicon Piezoresistive Pressure Sensors, *Sensors and Actuators A*, 50 (1995) 17–22.
[13] Steinmann, R., Friemann, H., Prescher, C., Schellin, R., Mechanical Behaviour of Micromachined Sensor Membranes Under Uniform External Pressure Affected by In-Plane Stresses Using a Ritz Method and Hermite Polynomials, *Sensors and Actuators A*, 48 (1995), 37–46.
[14] Meng, Q., Mehregany, M., Theoretical Modeling of Microfabricated Beams with Elastically Restrained Supports, *J. of Microelectromechanical Systems*, 2 (1993), 128–137.
[15] Gerlach, G., Schroth, A., Pertsch, P., Influence of Clamping Conditions on Microstructure Compliance, *Sensors and Materials*, 8 (1996), 79–98.
[16] Lee, K. W., *Modeling and Simulation of Solid-State Pressure Sensors*, Ph.D. Dissertation, University of Michigan, Ann Arbor, USA, 1982.
[17] Korvink, J., *An Implementation of the Adaptive Finite Element Method for Semiconductor Sensor Simulation*, Ph.D. Dissertation, ETH Zurich, No. 10143, Switzerland, 1993.
[18] Zhang, Y., Wise, K. D., Performance of Non-Planar Silicon Diaphragms Under Large Deflections, *J. of Microelectromechanical Systems*, 3 (1994), 59–68.
[19] Mallon Jr., J. R., Pourahmadi, F., Petersen, K., Barth, P., Vermeulen, T., Bryzek, J., Low-Pressure Sensors Employing Bossed Diaphragms and Precision Etch-Stopping, *Sensors and Actuators*, A21–A23 (1990), 89–95.

[20] Bergqvist, J., Finite-Element Modelling and Characterization of a Silicon Condenser Microphone with a Highly Perforated Backplate, *Sensors and Actuators A*, 39 (1993), 191–200.

[21] Pourahmadi, F., Barth, P., Petersen, K., Modeling of Thermal and Mechanical Stresses in Silicon Microstructures, *Sensors and Actuators*, A21–A23 (1990), 850–855.

[22] Bessho, M., Tsuru, Y., Horiike, H., Jinmon, M., Yamagami, K., Wataya, S., High Reliability Absolute Semiconductor Pressure Sensor, *SAE Special Publication*, 536 (1983), 55–59.

[23] Suzuki, S., Yamada, K., Nishihara, M., Hachino, H., Minorikawa, S., Structural Analysis of a Semiconductor Pressure Sensor, *Proc., The 1st Sensor Symp.*, Japan, 1981, pp. 131–133.

[24] Suzuki, S., Yagi, Y., Optimum Design of Silicon Pressure Sensor by Nonlinear Finite Element Method, *Proc., The 2nd Sensor Symp.*, Japan, 1982, pp. 163–165.

[25] Barth, P. W., Pourahmadi, F., Mayer, R., Poydock, J., Petersen, K., A Monolithic Silicon Accelerometer with Integral Air Damping and Overrange Protection, *Technical Digest*, IEEE Solid-State Sensor and Actuator Workshop, Hilton Head Is., 1988, pp. 35–38.

[26] Puers, B., Peeters, E., Sansen, W., CAD Tools in Mechanical Sensor Design, *Sensors and Actuators*, 17 (1989), 423–429.

[27] Tschan, T., de Rooij, N., Characterization and Modelling of Silicon Piezoresistive Accelerometers Fabricated by a Bipolar-Compatible Process, *Sensors and Actuators A*, 25–27 (1991), 605–609.

[28] Tschan, T., de Rooij, N., Bezinge, A., Analytical and FEM Modeling of Piezoresistive Silicon Accelerometers: Predictions and Limitations Compared to Experiments, *Sensors and Materials*, 3 (1992), 189–203.

[29] Kovács, A., Stoffel, A., Mechanical Analysis of Polycrystalline and Single-Crystalline Silicon Microstructures, *Sensors and Actuators A*, Vol. 41–42 (1994), 672–679.

[30] Yamada, K., Kuriyama, T., A Novel Degree of Freedom Separation Technique in a Multi-Axis Accelerometer, *Sensors and Actuators A*, 43 (1994), 120–127.

[31] Lee, Y.-T., Seo, H.-D., Takano, R., Matsumoto, Y., Ishida, M., Nakamura, T., Design Consideration for Silicon Rectangular Diaphragm Pressure Sensor with Single-Element Four-Terminal Strain Gauge, *Sensors and Materials*, 7 (1995), 53–63.

[32] Marco, S., Samitier, J., Morante, J. R., Gotz, A., Esteve, J., Novel Structures for Miniature Pressure Transducers Obtained by Electrochemical Etch-Stop on Diffused Membranes, *Sensors and Materials*, 7 (1995), 331–345.

[33] Hein, S., Schlichting, V., Obermeier, S. E., Piezoresistive Silicon for Very Low Pressures Based on the Concept of Stress Concentration, *Digest of Technical Papers*, Transducers '93, Yokohama, 1993, pp. 628–631.

[34] Fotheringham, G., Simulation Methods for Multi-Chip Modules, *Sensors and Actuators A*, 30 (1992), 157–165.

[35] Lin, Y.-C., Hesketh, P. J., Schuster, J. P., Finite-Element Analysis of Thermal Stresses in a Silicon Pressure Sensor for Various Die-Mount Materials, *Sensors and Actuators A*, 44 (1994), 145–149.

[36] Pourahmadi, F., Petersen, K., Package Design of Silicon Micromachined Sensors Using Finite Element Modeling, *Digest of Technical Papers*, Transducers '93, Yokohama, 1993, pp. 774–778.

[37] Koen, E., Pourahmadi, F., Terry, S., A Multilayer Ceramic Package for Silicon Micromachined Accelerometers, *Digest of Technical Papers*, Vol. 1, Transducers '95, Stockholm, 1995, pp. 273–276.

[38] Washizu, K., Note on the Principle of Stationary Complimentary Energy Applied to Free Vibration of an Elastic Body, *Int. J. Solids and Structures*, 2 (1969), 27–35.

[39] Baltes, H., Korvink, J. G., Paul, O., Numerical Modelling and Materials Characterization for Integrated Micro Electro Mechanical Systems, *Simulation of Semiconductor Devices and Processes*, Vol. 6, Ryssel, H., Pichler, P. (Eds.), Wien-New York: Springer-Verlag, 1995, pp. 1–9.

[40] Hodge, Jr., P. G., *Plastic Analysis of Structures*, New York: McGraw-Hill, 1959.

[41] Korvink, J. G., Baltes, H., Microsystem Modelling, Chapt. 6, *Sensors Update*, Baltes, H., Göpel, W., Hesse, J., (Eds.), Weinheim: VCH, 1996, pp. 181–209.

[42] Washizu, K., *Variational Methods in Elasticity and Plasticity*, 3rd Ed., Oxford: Pergamon Press, 1982.

[43] Chau, K., Allegretto, W., Ristic, L., Simulation of Silicon Microstructures, *Sensors and Materials*, 2 (1991), 253–264.

[44] Timoshenko, S., Woinowsky-Krieger, S., *Theory of Plates and Shells*, New York: McGraw-Hill, 1959.

[45] Clark, S. K., Wise, K. D., Pressure Sensitivity in Anisotropically Etched Thin-Diaphragm Pressure Sensors, *IEEE Trans. Electron Devices*, ED-26 (1979), 1887–1896.

[46] Benaissa, K., *Integrated Silicon Opto-Mechanical Sensors*, Ph.D. Dissertation, Electrical and Computer Engineering, University of Waterloo, Waterloo, Ontario N2L 3G1, Canada, 1996.

[47] Benaissa, K., Nathan, A., IC Compatible Optomechanical Pressure Sensors Using Mach-Zender Interferometry, *IEEE Trans. Electron Devices*, 43 (1996), 1571–1582.

[48] Gorman, D. J., *Free Vibration Analysis of Rectangular Plates*, New York: Elsevier, 1982.

[49] Lee, K. W., Wise, K. D., SENSIM: A Simulation Program for Solid-State Pressure Sensors, *IEEE Trans. Electron Devices*, ED-29 (1982), 34–41.

[50] Stavsky, Y., Hoff, N. J., Mechanics of Composite Structures, *Composite Engineering Laminates*, Dietz, A. G. H, (Ed.), Cambridge: MIT Press, 1969.

[51] Senturia, S. D., Microfabricated Structures for the Measurement of Mechanical Properties and Adhesion of Thin Films, *Digest of Technical Papers*, Transducers '87, Tokyo, 1987, pp. 11–16.

[52] Allen, M. G., Mehregany, M., Howe, R. T., Senturia, S. D., Microfabricated Structures for the in situ Measurement of Residual Stress, Young's Modulus, and Ultimate Strain of Thin Films, *Appl. Phys. Letts.*, 51 (1987), 241–243.

[53] Maseeh, F., Schmidt, M. A., Allen, M. G., Senturia, S. D., Calibrated Measurements of Elastic Limit, Modulus, and the Residual Stress of Thin Films Using Micromachined Suspended Structures, *Technical Digest*, IEEE Solid-State Sensor and Actuator Workshop, Hilton Head Is., 1988, pp. 84–87.

[54] Tabata, O., Kawahata, K., Sugiyama, S., Igarashi, I., Mechanical Property Measurement of Composite Rectangular Membrane, *Sensors and Actuators A*, 20 (1989), 135–141.

[55] Puers, B., Vergote, S., A Subminiature Capacitive Movement Detector Using a Composite Membrane Suspension, *Sensors and Actuators A*, 31 (1992), 90–96.

[56] Guckel, H., Randazzo, T., Burns, D. W., A Simple Technique for the Determination of Residual Stress in Thin Films with Application to Polysilicon, *J. Appl. Phys.*, 57 (1985), 1671–1675.

[57] Campbell, D. S., *Handbook of Thin Films Technology*, New York: McGraw Hill, 1970.

[58] Peterson, K. E., Guarnieri, C. R., Young's Modulus Measurements of Thin Films Using Micromechanics, *J. Appl. Phys.*, 50 (1979), 6761–6766.

[59] Zhang, L. M., Uttamchandani, D., Culshaw, W., Measurement of the Mechanical Properties of Silicon Microresonators, *Sensors and Actuators A*, 29 (1991), 79–84.

[60] Pratt, R. I., Johnson, G. C., Howe, R. T., Chang, J. C., Micromechanical Structures for Thin Film Characterization, *Digest of Technical Papers*, Transducers '91, San Francisco, 1991, pp. 205–208.

[61] Osterberg, P. M., Gupta, R. K., Gilbert, J. R., Senturia, S. D., Quantitative Models for the Measurement of Residual Stress, Poisson Ratio and Young's Modulus Using Electrostatic Pull-In of Beams and Diaphragms, *Technical Digest*, IEEE Solid-State Sensor and Actuator Workshop, Hilton Head Is., 1994, pp. 184–188.

[62] Pan, J. Y., Lin, P., Maseeh, F., Senturia, S. D., Verification of FEM Analysis of Load-Deflection Methods for Measuring Mechanical Properties of Thin Films, *Technical Digest*, IEEE Solid-State Sensor and Actuator Workshop, Hilton Head Is., 1990, pp. 70–73.

[63] Seino, Y., Nagai, S., Temperature Dependence of the Young's Modulus of Diamond Thin Film Prepared by Microwave Plasma Chemical Vapor Deposition, *J. Mat. Sc. Lett.*, 12 (1993), 324–325.

[64] Bourgeois, C., Hermann, J., Blanc, N., de Rooij, N. F., Rudolf, F., Determination of the Elastic Temperature Coefficients of Monocrystalline Silicon, *Digest of Technical Papers*, Vol. 2, Transducers '95, Stockholm, 1995, pp. 92–95.

[65] Han, M. Y, Jou, J. H., Determination of the Mechanical Properties of RF-Magnetron-Sputtered Zinc Oxide Thin Films on Substrates, *Thin Solid Films*, 260 (1995), 58–64.

[66] Klein, C. A., Anisotropy of Young's Modulus and Poisson's Ratio in Diamond, *Mat. Res. Bull.*, 27 (1992), 1407–1414.

[67] Maier-Schneider, D., Köprülülü, A., Obermeier, E., Elastic Properties and Micro-structure of LPCVD Polysilicon Films, *J. of Micromechanics and Microengineering*, 5 (1995), 121.

[68] Kahn, H., Stemmer, S., Nandakumar, K., Hever, A. H., Mullen, R. L., Ballarini, R., Huff, M. A., Mechanical Properties of Thick, Surface Micromachined Polysilicon Films, *Proc. IEEE MEMS*, San Diego, 1996, pp. 343–348.

[69] Biebl, M., Brandl, G., Howe, R. T., Young's Modulus of in situ Phosphorus-Doped Polysilicon, *Digest of Technical Papers*, Vol. 2, Transducers '95, Stockholm, 1995, pp. 80–83.

[70] Obermeier, E., High Temperature Microsensors Based on Polycrystalline Diamond Thin Films, *Digest of Technical Papers*, Vol. 2, Transducers '95, Stockholm, 1995, pp. 178–181.

[71] Kuschnereit, R., Fath, H., Kolomenskii, A. A., Szabadi, M., Hess, P., Mechanical and Elastic Properties of Amorphous Hydrogenated Silicon Films Studied by Broad Band Surface Acoustic Wave Spectroscopy, *Appl. Phys. A*, 61 (1995), 269–276.

[72] Jean, A., El Khakani, M. A., Chaker, M., Boily, S., Gat, E., Kieffer, J. C., Pepin, H., Biaxial Young's Modulus of Silicon Carbide Thin Films, *Appl. Phys. Lett.*, 62 (1993), 2200–2202.

[73] Windischmann, H., Intrinsic Stress and Mechanical Properties of Hydrogenated Silicon Carbide Produced by Plasma-Enhanced Chemical Vapor Deposition, *J. Vac. Sci. Tech.*, A9 (1991), 2459–2463.

[74] Watkins, T. R., Green, D. J., Ryba, E. R., Determination of Young's Modulus in Chemically Vapor-Deposited SiC Coatings, *J. Am. Ceram. Soc.*, 76 (1993) 1965–1968.

[75] Walsh, D., Culshaw, B., Optically Activated Silicon Microresonator Transducers: An Assessment of Material Properties, *Sensors and Actuators A*, 25–27 (1991), 711–716.

[76] Stewart, R. A., Kim, J., Kim, E. S., White, R. M., Muller, R. S., Young Modulus and Residual Stress of LPCVD Silicon-Rich Silicon Nitride Determined from Membrane Deflection, *Sensors and Materials*, 2 (1991), 285–298.

[77] Tsukahara, Y., Ohira, K., Yanaka, M., Inaba, M., Satoh, A., Elastic Properties Measurement of Glass Layers Fabricated on Silicon Wafers for Microelectronics and Micromachines, *IEEE Trans. Ultrasonics, Ferroelectrics, and Frequency Control*, 42 (1995), 387–391.

[78] Maier-Schneider, D., Ersoy, A., Maibach, J., Schneider, D., Obermeier, E., Influence of Annealing on Elastic Properties of LPCVD Silicon Nitride and LPCVD Polysilicon, *Sensors and Materials*, 7 (1995), 121–129.

[79] Wells, G. M., Chen, H. T. H., Wallace, J. P., Engelstad, R. L., Cerrina, F., Radiation Damage-Induced Changes in Silicon Nitride Membrane Mechanical Properties, *J. Vac. Sci. Technol. B*, 13 (1995), 3075–3077.

[80] Jou, J., Chen, L., Relaxation and Thermal Expansion Coefficient of Polyimide Films Coated on Substrates, *Appl. Phys. Lett.*, 59 (1991), 46–47.

[81] Fan, L.-S., Tai, Y.-C., Muller, R. S., Integrated Movable Micromechanical Structures for Sensors and Actuators, *IEEE Trans. Electron Devices*, ED-35 (1988), 724–730.

[82] Lin, Y.-C., Hesketh, P. J., Schuster, J. P., Finite-Element Analysis of Thermal Stresses in a Silicon Pressure Sensor for Various Die-Mount Materials, *Sensors and Actuators A*, 44 (1994), 145–149.

[83] Reichl, H., Packaging and Interconnection of Sensors, *Sensors and Actuators A*, 25–27 (1991), 63–71.

[84] Peterson, K. E., Silicon as a Mechanical Material, *Proc. IEEE*, 70 (1982), 420–457.

[85] Hälg, B., On a Nonvolatile Memory Cell Based on Micro-Electro-Mechanics, *Proc. IEEE MEMS*, Napa Valley, 1990, pp. 172–176.

[86] Lide, D. R., *Handbook of Chemistry and Physics*, 72nd Ed., Boston: Chemical Rubber Publishing Co., 1992.

[87] Wur, D. R., Davidson, J. L., Kang, W. P., Kuiser, D. L., Polycrystalline Diamond Pressure Sensor, *IEEE J. of Microelectromechanical Systems*, 4 (1995), 34–41.

[88] Mehregany, M., Tong, L., Matus, L. G., Larkin, D. J., Internal Stress and Elastic Modulus Measurements on Micromachined 3C-SiC Thin Films, *IEEE Trans. Electron Devices*, 44 (1997), 74–79.

[89] Thangaraj, D., Nathan, A., Two Dimensional Analysis of Incompressible Viscous Flow in Ducts Using a Rotated Difference Scheme, *Sensors and Materials*, 8 (1996), 13–22.

[90] Nagata, M., Swart, N., Stevens, M., Nathan, A., Thermal Based Micro Flow Sensor Optimization Using Coupled Electrothermal Numerical Simulations, *Digest of Technical Papers*, Vol. 2, Transducers '95, Stockholm, 1995, pp. 447–450.

[91] Mastrangelo, C. H., Muller, R. S., A Constant-Temperature Gas Flowmeter with a Silicon Micromachines Package, *Technical Digest*, IEEE Solid-State Sensor and Actuator Workshop, Hilton Head Is., 1988, pp. 43–46.

[92] Engel, P. A., Chen, W. T., (Eds.), *Advances in Electronic Packaging*, Proc. ASME Int. Electro Packaging Conf., Vols. 1 and 2, 1993.

[93] Middleman, S., Hochberg, A. K., *Process Engineering Analysis in Semiconductor Device Fabrication*, New York: McGraw-Hill, 1993.

[94] *American Institute of Physics Handbook*, 3rd Ed., New York: McGraw-Hill, 1972.

[95] Starr, J.B., Squeeze-Film Damping in Solid-State Accelerometers, *Technical Digest*, IEEE Solid-State Sensor and Actuator Workshop, Hilton Head Is., 1990, pp. 44–47.

[96] van Kampen, R. P., Vellekoop, M. J., Sarro, P. M., Wolffenbuttel, R. F., Application of Electrostatic Feedback to Critical Damping of an Integrated Silicon Capacitive Accelerometer, *Digest of Technical Papers*, Transducers '93, Yokohama, 1993, pp. 818–821.

[97] Cho, Y.-H., Pisano, A. P., Howe, R. T., Viscous Damping Model for Laterally Oscillating Microstructures, *J. of Microelectromechanical Systems*, 3 (1994), 81–87.

[98] Zhang. X., Tang, W. C., Viscous Air Damping in Laterally Driven Microresonators, *Sensors and Materials*, 27 (1995), 415–430.

[99] Hosaka, H., Itao, K., Kuroda, S., Evaluation of Energy Dissipation Mechanisms in Vibrational Microactuators, *Proc. IEEE MEMS*, 1994, pp. 193–198.

[100] Reuther, H. M., Weinmann, M., Fischer, M., von Münch, W., Aßmus, F., Modeling Electrostatically Deflectable Microstructures and Air Damping Effects, *Sensors and Materials*, 8 (1996), 251–269.

[101] Tang, W. C., Lim, M. G., Howe, R. T., Electrostatic Comb Drive Levitation and Control Method, *J. Microelectromechanical Systems*, 1 (1992), 170–178.

[102] Langlois, W. E., Isothermal Squeeze Films, *Quart. Appl. Math.*, XX (1962), 131–150.

[103] Langlois, W. E., *Slow Viscous Flow*, New York: Macmillan, 1964.

[104] Yang, Y.-J., Senturia, S. D., Numerical Simulation of Compressible Squeezed-Film Damping, *Technical Digest*, IEEE Solid-State Sensor and Actuator Workshop, Hilton Head Is., 1996, pp. 76–79.

[105] Sadd, M. H., Stiffler, A. K., Squeeze Film Dampers: Amplitude Effects at Low Squeeze Numbers, *J. Eng. Indust.*, Trans. of the ASME, B97, (1975), 1366–1370.

[106] Morgan, K., Periaux, J., Thomasset, F. (Eds.), *A GAMM Workshop, Notes on Numerical Fluid Dynamics*, Braunschweig: Vieweg, 1984.

[107] Denis, S. C. R., Chang, G.-Z., Numerical Solutions for Steady Flow Past a Circular Cylinder at Reynolds Numbers up to 100, *J. Fluid Mech.*, 42 (1970), 471–489.

[108] Hamielec, A. E., Raal, J. D., Numerical Studies of Viscous Flow Around Circular Cylinders, *Phys. of Fluids*, 12 (1969), 11–17.

[109] Acrivos, A., Leal, L. G., Snowden, D. D., Pan, F., Further Experiments on Steady Separated Flows Past Bluff Objects, *J. Fluid Mech.*, 34 (1970), 25–48.

[110] Roache, P. J., *Computational Fluid Dynamics*, Albuquerque: Hermosa, 1976.

[111] Peyret, R. T., Taylor, T. D., *Computational Methods for Fluid Flow*, New York: Springer-Verlag, 1983.

[112] Patankar, S. V., *Numerical Heat Transfer and Fluid Flow*, New York: Hemisphere Publishing Co., 1980.

[113] Thangaraj, D., Wu, H., Jayaram, S., Stream Function Distribution of Petroleum Liquids in Relaxation Tanks, *Proc. IEEE-IAS 27th Meeting*, Denver, 1994, pp. 1676–1681.

[114] Raithby, G. D., Skew Upstream Differencing Schemes for Problems Involving Fluid Flow, *Computer Methods in Applied Mechanics and Engineering*, 9 (1976), 153–164.

[115] Rice, J. G., Schnipke, R. J., A Monotone Streamline Upwind Finite Element Method for Convection-Dominated Flows, *Computer Methods in Applied Mechanics and Engineering*, 48 (1985), 313–327.

[116] Roache, P. J., A Comment on the Paper "Finite Difference Methods for the Stokes and Navier-Stokes Equations" by J. C. Strickwerda, *Int. J. Num. Meth. in Fluids*, 8 (1988), 1459–1463.

[117] Gresho, P. M., Sani, R. L., Introducing Four Benchmark Solutions, *Int. J. Num. Meth. in Fluids*, 11 (1990), 951–952.

[118] Sani, R. L., Gresho, P. M., Résumé and Remarks on the Open Boundary Condition Minisymposium, *Int. J. Num. Meth. in Fluids*, 18 (1994), 983–1008.

[119] Small, M. K., Vlassak, J. J., Powell, S. F., Daniels, B. J., Nix, W. D., Accuracy and Reliability of Bulge Test Experiments, *Proc. MRS*, 308 (1993), 159–164.

[120] Ziebart, V., Paul, O., Münch, U., Baltes, H., A Novel Method to Measure Poisson's Ratio of Thin Films, *Proc. MRS*, 505 (1998), 27–32.

8 Micro-Actuation

Microactuators are miniaturized output transducers which convert an electrical input signal into a non-electrical output signal in the radiant, magnetic, thermal, mechanical, or chemical domains [1, 2]. Integrated silicon microactuators are realized using integrated circuit (IC) microfabrication techniques coupled with application-specific thin film deposition and micromachining technologies [3–11]. Central to current research is microactuation in the mechanical domain. Mechanical microactuators are three-dimensional structures with physical dimensions ranging from micrometers to millimeters. Progress in the field is rapid with evolution of new thin film actuation materials [12–14] and proliferation of increasingly complex micromechanical systems. Mechanical microactuators are part of Micro Electro Mechanical Systems (MEMS), a field which has grown to encompass a broad family of micromachined sensors, actuators, and systems that exploit coupled electrical, mechanical, radiant, thermal, magnetic, and selected chemical effects [15, 16]. The simulation of mechanical microactuators is the topic of this chapter.

Systems based on co-integration of MEMS with electronic driver and signal processing circuitry along with pertinent detection elements on the same chip are referred to as ICMEMS or IMEMS or, more specifically, CMOS MEMS [10, 11]. Co-integration of the actuator, detector, and circuitry is essential whenever the actuator, in a variety of applications, is a standalone closed loop control system. This is illustrated in terms of the following example of a microactuator in resonant configuration [17, 18]. An input electrical signal is converted into an electrostatic force which excites the microactuator into mechanical resonance. Its oscillation frequency is modulated by a physical or chemical signal. To maintain the oscillation frequency, the mechanical signal is converted to an electrical signal, by using modulation of capacitance in the air gap, and is fed back to the input. Thus the detector, which forms part of the feedback loop, must be an integral part of the actuator. Apart from use of microactuators in resonator configurations, further promising application areas of integrated actuators are emerging (see [15, 16]). These include optical signal processing, microfluidics, microrobotics, and material (including chemical and biological) analysis microinstruments. Specific applications will be discussed below when we address modeling and simulation issues pertinent to electrostatic, thermal, magnetic, piezoelectric, and electroacoustic actuation.

8.1 Transduction Principles

Like microsensors [1, 2], mechanical microactuators can be classified by either transduction principle, primary input signal, material and/or technology, application, cost, or accuracy. From a modeling and simulation standpoint, a pragmatic classification is in terms of the primary input signal domain and the associated excitation field(s) and/or effect(s). This is depicted in Table 8.1 illustrating the coupling of the different energy fields along with pertinent actuation devices and detection mechanisms for use in a closed loop configuration. The excitation and detection mechanisms given in Table 8.1 stem from either reversible or irreversible processes [18].

Table 8.1. *Transduction principles for mechanical micro-actuation. "Microstructures" include beams, bridges, and membranes. In addition, mechanical-to-electrical signal conversion effects for closed loop operation include capacitive, piezoresistive, direct piezoelectric, and inductive detection or sensor effects*

Primary or tandem input energy	Excitation or transduction effect	Microactuator or MEMS
Electrical	Inverse piezoelectric	Electrically excited microstructures
	Electrostatic	Electroacoustic devices
	Electrostrictive	Electrostatic micromirror
	Electrothermal	Resonators
	Electrodynamic I	Electrostatic ciliary motion arrays
Radiant and thermal	Thermal expansion	Bimorph microstructures
	Shape memory	Thermo-mechanical ciliary motion
	Thermopneumatic	Shape memory alloy devices
	Electrothermal	Thermal microfluidic devices
	Optothermal	Thermally excited resonators
		Golay cell
		Crooks radiometer
Magnetic	Inductive	Microcoils
	Magnetostrictive	Magnetic microfluidic devices
	Electrodynamic II	Magnetic micropositioners
		Magnetic resonators
Mechanical	Pneumatic	Microgears
	Hydraulic	Microfluidic devices
	Acoustic	Microresonators

Reversible excitation implies reversible energy exchange and storage. Thus the same transduction effect can be used for both excitation and detection. Irreversible excitation implies energy dissipation.

Revisiting our earlier resonant microactuator example, the excitation signal, which actuates the structure into resonance, is derived from the conversion of one or several combined input signals in the radiant, magnetic, thermal, mechanical, or electrical domain. Here, actuation signals such as electrostatic excitation with an air gap, piezoelectric excitation, and magnetic excitation are reversible, while electrothermal excitation and radiant (infrared or optothermal) excitation are irreversible. The resonant frequency of the microactuator is modulated by the variation of physical parameters and environmental factors, or by the presence of a measurand. The detection signal is converted from the mechanical to the electrical domain and fed back to the input to maintain resonance. Here, the signal conversion mechanisms based on capacitance modulation with an air gap, inductive, and piezoelectric effects are reversible. In particular, the piezoelectric effect is reciprocal and thus facilitates reverse transduction by the same principle, *viz.*, an input mechanical signal (e.g., stress) generates an output electrical signal and vice versa (direct and inverse piezoelectric effects). On the other hand, detection based on stress induced resistivity modulation (piezoresistance) is an irreversible process.

Although most of the effects and principles given in Table 8.1 are conceptually simple, some clarification may be necessary in view of apparent ambiguities associated with tandem transduction and multiple input signal domains. For example, an input radiant signal in the infrared (IR) range gives rise to absorption-induced heat generation causing a differential thermal strain in a bimorph structure to yield thermomechanical actuation. Thus radiant signals in the IR range, while of electromagnetic nature, can also be considered as thermal signals. Other ambiguous cases are electrothermal excitation (*viz.*, generation of Joule heat in a resistive element by an electrical current), electromagnetic excitation (*viz.*, generation of a magnetomotive force in a magnetic circuit by an electrical current), and electrodynamic excitation which requires presence of both electrical and magnetic signal inputs. In the latter case, a beam placed in an external bias magnetic field displaces as a result of the action of the Lorentz force on the current flowing through the beam. Since the input signals involve electrical and magnetic domains, the term "electrodynamic effect" applies in both cases.

8.2 State-of-the-Art and Preview

In this section, we summarize the simulation challenges, progress in the field, and the organization of this chapter.

Even the conceptually most simple microactuator involves transduction principles that stem from the interaction of different energy fields; the coupling is even more pronounced with microactuators based on tandem transduction. For example, thermomechanical actuation of a bimorph structure relies on electrothermal excitation for mechanical displacement whereby the thermal excitation is induced by an electrical input. Here, an electrical input signal into a resistive element upsets the temperature distribution due to Joule heating. This generates a differential thermal strain by mismatch in thermal expansion coefficients of the materials involved leading to mechanical displacement of the bimorph microstructure.

Thus simulation of thermomechanical micro-actuation involves the coupled system of partial differential equations (PDEs) that govern electrical transport, heat transport, and mechanical displacement [19]. The coupling in PDEs stem from three sources. The first comes from the presence of the direct coupling terms in a given PDE that contain the state variables of the other PDE, e.g., Seebeck, Peltier, and Joule heating effects (see Chapt. 5) and thermal strain (see Chapt. 7). The second source of coupling arises from the dependence of material and transport coefficients on the underlying state variables, e.g., temperature- and strain-dependence of electrical and thermal conductivites, and of elastic coefficients of the materials involved (see Chapts. 6 and 7). These coefficients are not only non-uniform (position-dependent), but can also turn out to be anisotropic (tensors). Third, as identified in all previous chapters, the PDEs are coupled through boundary and interface conditions. Of particular importance is the changing structural shape of the microactuator during operation. In thermomechanical actuation, the influence of displaced boundaries is significant in terms of heat transport (see Chapt. 5). All of the above considerations exclude modeling by analytical solutions as virtually impossible and make numerical microactuator modeling a compelling task (see [19–25]). This is even more so with IMEMS as the materials associated with IC technologies impose a narrow design latitude, thus requiring extensive optimization with respect to transduction efficiency and reliability [24].

Obtaining numerical solutions of the system of coupled PDEs is challenged by numerical convergence problems due to strong non-linearities. This is particularly true with electrostatic actuation since the electrostatic excitation forces (which are inversely proportional to the square of the displacement) vary with the change in shape of the device structure [25]. The degree of numerical difficulty in microactuator computer-aided-design (CAD) can become comparable to that of *p-n* junction based microsensors where the space charge, on the right-hand side of Poisson's equation, is an exponential function of the electric and Fermi potentials [26] (see Chapts. 2 to 6). Returning to our thermomechanical example, non-linearities in the system of PDEs stem from residual (intrinsic) stresses in the materials involved, non-linear dependence of material coefficients on temperature, and non-linear boundary conditions arising from possible convective and radiative heat transport. Other sources of non-linearity include large deflection, surface contact, creep phenomena, time-dependent masses, and non-linear damping effects. The latter three are associated with transient simulations [19].

Progress in the field of microactuator CAD can be classified in terms of six broad categories of micro-actuation principles: electrostatic, thermal, magnetic, piezoelectric, electroacoustic, and fluidic. This is summarized in Table 8.2 along with a list of relevant general-purpose and application-specific software tools developed in industry and academia. Our description of pertinent equation systems will be restricted to the first five categories of microactuators listed above with emphasis on the various coupling mechanisms and associated terms, along with sources of non-linearity in the system. Models for the underlying material and transport coefficients associated with thermal, mechanical, and fluidic signal domains are discussed in Chaps. 5, 6, and 7, respectively. A summary of pertinent

Table 8.2. *Summary of tools for different simulation modules (see Fig. 8.9)*

Simulation Modules	Tools
Environments	MEMCAD [25, 131], SOLIDIS [37], INTELLICAD [132], SESES [133], CAEMEMS [134], ALECSIS [135]
Interfaces and geometry	PATRAN [136], Geomview [137], OYSTER [138], Pro/Engineer [139], DESIGNBASE [140], MemBuilder [141]
Parameter	ICMAT [24], INTELLICAD [132], MemBase [142]
Electrostatic	SOLIDIS [37], FASTCAP [57, 143], ANSYS [144], Maxwell [145], PHI3D, ELECTRO, COULOMB [146], EFCREL, EFDYN [161], BEMMODULE [41], WATCAP [166]
Thermal	SOLIDIS [37], INTELLICAD [132], SESES [133], ANSYS [144], FLUX2D/3D [146], ADINA [147], KELVIN [148].
Magnetic and electromagnetic	ANSYS [144], FLUX2D/3D & ATILA [146], MAGNETO, AMPERES, OERSTED, FARADAY [148], EFCAD [161]
Piezoelectric and acoustic	ANSYS [144], ATILA [146]
Fluidic damping/transport	FIDAP [149], FLUENT [150], FLOTRAN [151], FLOTHERM [152], PUSI [162],
Mechanical structural	SOLIDIS [37], ANSYS [144], ADINA [147], I-DEAS [153], ABAQUS [154], MSC/NASTRAN [155], COSMOS/M [156], FLOWERS [157], SENSIM [158], TPS10 [159], MARC [160].

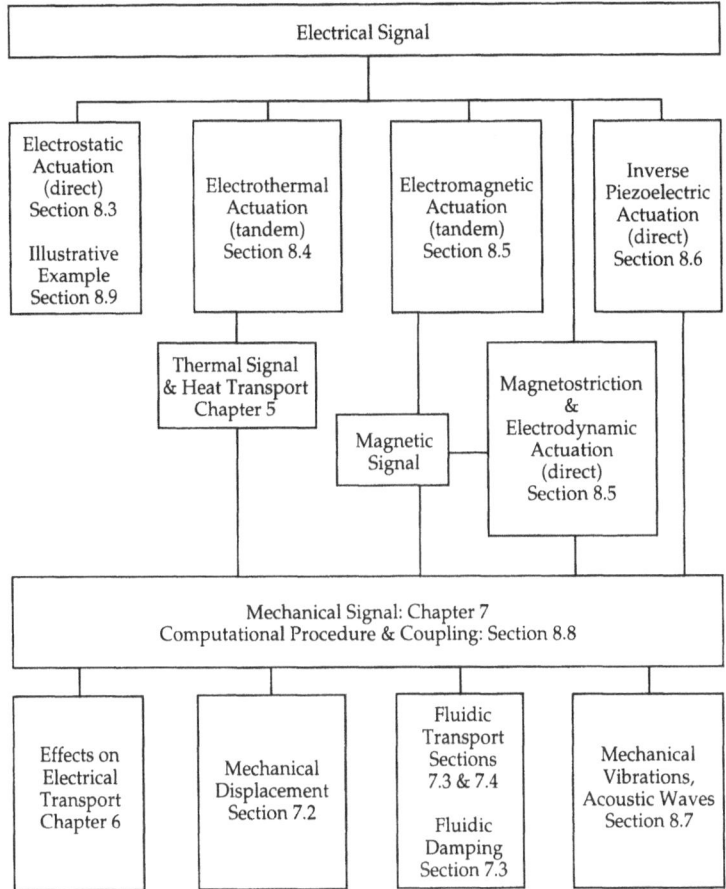

Fig. 8.1 Actuator overview and organization of Chapt. 8 in relation to other chapters

equations, including the computational procedure and coupling mechanisms, is provided in Sect. 8.8. Sect. 8.9 concludes the chapter with a numerical simulation example of a CMOS micromirror. Discussion of the rather broad family of microfluidic devices, which are rapidly emerging for use in chemical and biological analyses instruments, is beyond our scope, although an elementary treatment of fluidic transport modeling is given in Sects. 7.3 and 7.4. The organization of this chapter and relation to other chapters is shown in Fig. 8.1.

8.3 Electrostatic Actuation

The Coulomb force between two charged electrodes of a capacitor forms the basis of electrostatic microactuation. The one electrode is part of a

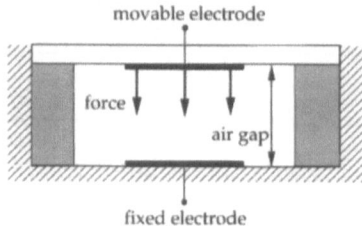

Fig. 8.2 Schematic representation of electrostatic actuation

movable beam while the other electrode is fixed and separated by an air gap that is of the order of micrometers. The related actuation principle is illustrated in Fig. 8.2. Actuation by electrostatic forces has become attractive due to the favorable scaling of electric field with miniaturization, since the electrostatic force for a fixed dc voltage varies inversely with the square of the distance between electrodes [27]. Electrostatic microactuators operate at low voltage, consume very little power, and have a high switching speeds. To date, there are two basic configurations of electrostatic microactuators: the parallel plate structure, where the direction of actuation is perpendicular to the chip surface, and the (more efficient) lateral interdigital or comb structure [28] where the direction of actuation is parallel to the chip surface. In comparison, the latter has high vibration amplitude (over $10\,\mu m$), high quality factor at atmospheric pressure, and its drive capacitance varies linearly with beam displacement [29]. Examples of electrostatic actuators include accelerometers, rate gyroscopes, resonant microactuators, microgrippers, microtweezers, micropositioners, rotational microstages, mechanical shutters, relays, micromirrors, micromotors, and valves [15, 16].

Due to their structural complexity, realistic design of electrostatic microactuators requires numerical solutions, in two, if not three, dimensions, of the system of differential equations governing electro-mechanical interactions (see [23–25, 30–41]); the latter being particularly true for comb structures. Electromechanical interactions are strongly coupled since the electrically induced mechanical loading is significant relative to the internal mechanical forces in the structure. Simple analytical and one-dimensional numerical models of electrostatic actuation are useful only for a first design or to establish quick estimates of device geometry and bias limits [42]. Such models rely on simple parallel plate approximations that neglect effects of the non-trivial fringing capacitance [43] stemming from propagation of the electric field to regions exterior to the actual physical structure. In addition, these models neglect the effects of distributed capacitance associated with beam deformation. In particular, the design of electrostatic microactuators must account for the effects of elastic deformation; here the electric field in the air gap becomes non-uniform leading to a redistribution of charges and in the electrically-induced mechanical loading conditions. This places limits on use of such approaches for design verification because of the restricted range of solution validity; if acceptable, they may be applied only to simple designs which have sufficient structural symmetry.

Critical issues pertinent to design of electrostatic microactuators include:

- Minimization or elimination of electrically induced levitation forces through appropriate optimization of device design and/or bias conditions without compromising drive forces [29].
- Prediction of non-linearity stemming from large electrostatic forces and large displacements, non-linear material properties, and large initial or residual stresses (or stiffening) [44, 45].
- Effects of geometrical non-idealities, induced by residual stress gradients leading to beam deformation, including effects of over-etching and mask misalignment [46].
- Accurate prediction of transition voltages and contact forces associated with sticking/release of the actuation beam and subsequent control of associated hysteresis characteristics by structural design [47–49].
- Effects of air damping on amplitude response, stability, and quality factor [50, 51].

8.3.1 Electrostatic Analysis

The equation governing electrostatics stems directly from Maxwell's equations under static conditions. Here, "static" implies that the charge distribution does not vary as a function of time. Thus we set all the time derivatives to zero. We also assume vanishing electric current and magnetic field. Thus the electric field vector can be represented as a gradient of the scalar electrostatic potential (see Sect. 2.1) and Poisson's equation, which relates the electrostatic potential (ψ) to the space charge density (ρ), in terms of index notation, reads

$$\frac{\partial}{\partial x_i}\left(\varepsilon_r \varepsilon_0 \frac{\partial \psi}{\partial x_i}\right) = -\rho \qquad \text{with} \quad i = 1, 2, 3. \tag{8.1}$$

Here, ε_r denotes the relative permittivity or dielectric constant and ε_0 is the permittivity of free space. The effect of the electric field on isotropic dielectric or insulating materials, such as silicon dioxide (SiO_2) and silicon nitride (Si_3N_4), is already accounted for through ε_r in Eq. (8.1). The bound positive and negative charges internal to the material displace with respect to each other leading to creation of atomic dipoles which yield a uniform electric polarization, the divergence of which vanishes [52]. In such materials as well as in any cubic crystalline material, the relative permittivity is a scalar, although it may be position-dependent in an inhomogeneous system. However, in anisotropic materials, such as piezoelectric, pyroelectric, or strained materials, the orientation of the polarization and electric field vectors differs, and the permittivity becomes a tensor of rank two [53]. Values of the relative permittivity in homogeneous isotropic and anisotropic materials are given in Sect. 2.1 and Sect. 8.6, respectively.

The space charge term on the right-hand side of (8.1) depends on the material-type, and in particular, on its electrical conductivity, since the internal charge distribution may be affected by the external electrostatic field. Insulating materials such as SiO_2 and Si_3N_4 contain no mobile charge, and if there is no trapped charge, $\rho = 0$. In piezoelectric and pyroelectric materials, a polarization may already be present in the material that is induced by stress or temperature. Unlike before, the polarization vector $\mathbf{P}$ may be non-uniform, yielding a non-vanishing divergence [52] and hence, an additional space-charge-like term, *viz.*,

$$\frac{\partial}{\partial x_i}\left(\varepsilon_r \varepsilon_0 \frac{\partial \psi}{\partial x_i}\right) = -\rho + \frac{\partial P_i}{\partial x_i} \qquad \text{with} \quad i = 1, 2, 3. \tag{8.2}$$

In highly conductive materials (e.g., metals and highly doped semiconductors) subject to an electric field, mobile carriers very quickly reach their equilibrium positions and re-establish charge neutrality to yield a zero electric field inside the conductor. Thus the conductor is at an equipotential. Its interior space charge density (ρ) is zero and any static charge induced locates within a few atomic layers from the surface. The electric field is normal to the surface of the conductor if there is no charge flow along the surface. Just outside the conductor surface, the electric field, and hence the electrostatic force, can be expressed solely in terms of the local surface charge density despite its dependence on the magnitude and position of all charges whether internal, or external, to the conductor [54].

In low doped semiconductors, due to the limited number of mobile carriers, the space charge, $\rho = q(p - n + N_D - N_A)$, and hence, the electric field, in the material does not necessarily vanish. Here, p and n denote mobile holes and electrons concentrations, respectively, N_D the ionized donor density, and N_A the ionized acceptor density. The concentration of mobile charge carriers can be modeled using Boltzmann statistics. They are highly non-linear functions of the electrostatic potential in the material and need to be obtained from solutions to the carrier continuity equations [55]. The dopant density distributions depend on process conditions. These distributions can be predicted using process simulators, e.g., SUPREM [56]. The magnitude of the space charge in the semiconductor depends on the level of doping. At large doping levels, the semiconductor behaves more like a metal and the charge induced by the external field resides on the surface of the semiconductor. However, unlike metals, where the induced charge resides within a few atomic layers, the thickness of the charge layer in semiconductors is extended and dependent on doping level.

Eq. (8.2) governs the static electrical behavior valid for a broad range of materials, and operating voltages and frequencies normally encountered in electrostatic microactuation. In numerical simulations using, e.g., the finite element method, Eq. (8.2) is solved for the electrostatic potential ψ under suitable Dirichlet, Neumann, and mixed boundary conditions, with the relevant specification of the permittivity $\varepsilon(= \varepsilon_r \varepsilon_0)$ and the space charge distribution ρ. If the simulation domain involves active electronics (e.g., p-n junctions and related devices), ρ needs to be constructed from a self-consistent solution of the complete system of equations governing charge conservation and carrier continuity (see Chapt. 2). In the case of materials with non-uniform polarization induced by stress or otherwise, we need to account for the additional term, div $\mathbf{P}$. This is discussed in Sect. 8.6.

The Dirichlet boundaries constitute the electrode or other conductor regions in the electrostatic microactuator where the potential is prescribed by a fixed or applied voltage, V_a:

$$\psi = V_a. \tag{8.3}$$

At Neumann boundaries, the electric field (E) component normal to the boundary has a defined value, E_0

$$E_i n_i = E_0 \quad \text{with} \quad i = 1, 2, 3, \tag{8.4}$$

which can be a function of position along the boundary. Here, n_i denotes the component of the unit vector normal to the boundary. If $E_0 = 0$, then (8.4) becomes a homogeneous Neumann condition which is applied at the symmetry planes of the microactuator. At floating boundaries, such as a conductor at equipotential with a prescribed total charge Q_0, we impose the condition,

$$\int_{\Gamma} (\varepsilon E_i n_i) d\Gamma = Q_0 \quad \text{with} \quad i = 1, 2, 3, \tag{8.5}$$

where Γ denotes the integration surface and $d\Gamma$ the surface element on the conductor. The potential ψ on the conductor is spatially-invariant but its value is unknown. At ideal interfaces comprising two different materials, a and b, we impose continuity conditions on the electric displacement,

$$(\varepsilon E_i n_i)_a = (\varepsilon E_i n_i)_b \quad \text{with} \quad i = 1, 2, 3. \tag{8.6}$$

In simple microactuator structures comprising an air gap sandwiched between two equipotentials (electrodes), ε is uniform and $\rho = 0$, yielding Laplace's equation in the electrostatic potential. Relevant boundary conditions for this case are (8.3) and (8.4).

With the computed spatial distribution of electrostatic potential in the air gap, we next determine the induced surface charge density, and hence, the electrostatic force, at the air gap-material interface. The surface charge density is simply the electric displacement and the electrostatic force is obtained from the product of the charge and electric field. The electrostatic force (f^E) acts on the surface of the actuator and can be computed from the Maxwell stress tensor σ_{ij}^M whose divergence, using indicial notation, reads [23]:

$$f_i^E = \int_\Gamma \sigma_{ij}^M n_j \, d\Gamma \qquad \text{with} \quad i,j = 1,2,3. \tag{8.7}$$

Here, Γ denotes the integration surface and n_j is the jth component of the unit vector normal to the surface element $d\Gamma$. The stress tensor is defined as

$$\sigma_{ij}^M = E_i D_j - \frac{1}{2}\delta_{ij} E_k D_k \qquad \text{with} \quad i,j,k = 1,2,3, \tag{8.8}$$

where E denotes the electric field, D the electric displacement, and δ_{ij} the Kronecker delta. The second term on the right-hand side is representative of the electrostatic pressure.

An alternate, and perhaps even more suitable, approach to electrostatic analysis employs a specific class of boundary element methods, referred to as the panel method, which relies on the free space Green's function. The solution variable is the surface charge density (σ) on the electrode, which satisfies Laplace's equation for the potential over the (free space) charge free region and the boundary conditions, (8.3), on the electrodes. The computation can be accelerated using multipole expansion [41, 57] or exponential expansion [165]. With the computed distribution of surface charge (or normal displacement), the electrostatic pressure (or equivalently the surface force density) on a surface element, $d\Gamma$, following (8.8), can be computed as

$$p^E = \frac{\sigma^2}{2\varepsilon}, \tag{8.9}$$

where σ is uniform over $d\Gamma$ for an ideal (or perfect) conductor.

With the above description for the electrostatic force, we now have the complete system of model equations governing electrostatic microactuation. The equations are highly coupled; mechanical deformation of the structure alters the distribution of the electrostatic potential, thus changing the distribution of the surface charge density, and hence, the electrostatic forces acting on the structure. For self-consistent solutions of the field variables, the equation governing dynamic mechanical behavior, along with the constitutive relations including residual stresses and strains, is coupled with the equations governing electrostatics and fluidic damping. Pertinent model equations, governing electrostatic microactuation, drawn from this section and Chapt. 7 are summarized in Sect. 8.8.

In the simple case of a system in the absence of body forces, residual stresses and strains, space charge, and dielectric anisotropy, the system of equations reduce to the following form governing static operation:

$$\frac{\partial}{\partial x_i}\left(C_{ijkl}\varepsilon_{kl} + \varepsilon_0 E_i E_j - \frac{1}{2}\delta_{ij}\varepsilon_0 E_k E_k\right) = 0 \tag{8.10}$$

$$\text{with} \quad i,j,k,l = 1,2,3,$$

$$\frac{\partial}{\partial x_i}\left(\frac{\partial \psi}{\partial x_i}\right) = 0 \quad \text{with} \quad i = 1,2,3. \tag{8.11}$$

Here, C denotes the tensor of stiffness coefficients, ε is the strain tensor whose notation is similar to the permittivity, ε_0, and the rest in usual notation. Eq. (8.10) is solved along with the strain-displacement relations (7.13) and boundary conditions (7.44) to (7.47). Boundary conditions for (8.11) are given by (8.3) to (8.6).

8.4 Thermal Actuation

Thermal microactuators rely on thermal energy as the source of excitation. They are capable of large displacements with relatively high driving power and from a thermal standpoint they scale favorably with miniaturization, leading to reduced input power and long thermal response time. Devices most common to the thermal microactuator family include electrothermal [58] or optothermal [59] structures and shape memory alloys (SMAs) [60]. Other device structures include lossy dielectrics using radio frequency (RF) heating [61], thermally-induced (low boiling point) liquid-to-vapor phase transformation (see, e.g., [62]), and pneumatic structures.

The operation of electrothermal and optothermal microactuators can be described in terms of the simple bimorph theory derived from the theory of bimetals; the bimorph structure comprises a sandwich of two materials with different expansion coefficients, $\alpha_{1,2}$ (Fig. 8.3). Thermal excitation leads to differential thermal strain (see Sect. 7.1.4) in the materials, and hence to beam deformation due to mismatch in expansion coefficients; the extent of mechanical displacement is governed by the difference in expansion coefficients. Thermal excitation is derived from Joule heating with an in situ resistive element or by absorption of IR radiation by a

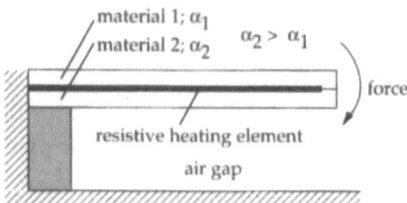

Fig. 8.3 Schematic representation of thermomechanical actuation with a bimorph beam

dielectric layer (e.g., SiO_2). In contrast to electrothermal bimorph structures, SMAs have very high output forces and very large mechanical displacements. In particular, for thin film SMA, large force-to-weight ratio with mechanical characterisitics similar to their bulk counterparts [60] have been demonstrated. Actuation by shape memory alloys relies on the deformation induced by a bias force at low (e.g., room) temperature. The SMA remains deformed, even upon removal of bias force, until it is subject to thermal excitation (e.g., Joule heat). When the temperature exceeds a certain phase transition temperature of the material, the SMA recovers its intial undeformed shape.

Thermal microactuators are intrinsically tandem transducers which rely on the interaction of electrical, thermal, and mechanical fields. Meaningful design of these microactuators can only be accomplished through numerical solutions to the system of PDEs that govern electrical transport, heat transport, and mechanical displacement. This must be accompanied with a suitable description of mechanical material properties and boundary conditions. Indeed thermo-mechanical behavior (e.g., stress-strain relationship), particularly in SMAs, can be relatively complex and accurate specification of boundary conditions is essential for realistic simulation of heat transfer, and hence, thermal behavior.

These considerations, not to mention the complexities associated with transducer geometry and structure, render analytical modeling as virtually impossible. This is even more so with IMEMS as the materials associated with IC technologies impose a narrow design latitude, thus requiring extensive optimization with respect to transduction efficiency and reliability [24]. Critical design issues pertinent to thermal micro-actuation include:

- Prediction of non-linearity stemming from non-linear material properties, temperature- and/or fabrication-process-dependent electrical, thermal, and mechanical material parameters, large initial stresses, large thermal strains, and displacement-dependent radiative and convective heat losses [19].
- Structural and geometrical design optimization to limit escalations in temperature (e.g., in hot spots) which undermine device reliability or

lead to irreversible thermally-induced changes in the material's microstructure (see, e.g., [63]).

- Control of strain levels, particularly, in SMAs to avoid introduction of mechanical fatigue [60].
- Minimization of temperature asymmetry induced by electrothermal (e.g., Peltier and Thomson) effects at conductor interfaces [19].
- Maximizing actuation efficiency by reducing/eliminating parasitic radiative and fluidic (e.g., conduction, forced and natural convection) heat loss components through optimization of device structure, geometry, and operating conditions [21].

8.4.1 Electrothermal Analysis

Electrical and heat transport in the material under non-isothermal conditions are tightly coupled and give rise to a number of electrothermal effects which can be effectively utilized in microtransducers (see Chapt. 5). Because the associated material systems invariably involve thin film conductors and dielectrics, several assumptions can be invoked to simplify the underlying constitutive (or transport) relations. First we review the assumptions, then provide the reduced form of constitutive relations, and finally proceed to obtain the electrothermal model equations using the associated conservation laws. To avoid possible confusion with material-property-related subscripts, we will refrain from use of the index notation where possible in the model equations.

We consider a unipolar system and assume transport to be governed only by electrons, such as in an n-type material (the general bipolar case is treated in Sect. 5.2). This automatically precludes carrier generation and recombination processes and associated energy (heat) exchanges with the lattice. In addition, we make the following assumptions: Electrons are in thermal equilibrium with the lattice which implies identical carrier and lattice temperatures; negligible carrier concentration gradients implying no diffusion current components; and no interaction of the carriers with optical or magnetic fields. Despite the interaction of electrical and heat transport with the strain field, we will maintain, for simplicity, a scalar description for the various electrothermal coefficients.

Following the above assumptions, the laws of irreversible thermo-dynamics, notably Onsager's theorem, lead to the following system of electrical and heat transport equations to include temperature gradient a driving force [64]:

$$\mathbf{J}_n^e = -\sigma_n(T)\,\mathrm{grad}\,\psi + \sigma_n(T)\alpha_{sn}(T)\,\mathrm{grad}\,T, \tag{8.12}$$

$$\mathbf{J}_n^q = -\alpha_{sn}(T)T\,\mathbf{J}_n^e - \kappa_{\mathrm{tot}}(T)\,\mathrm{grad}\,T. \tag{8.13}$$

Here, $\mathbf{J}_n^e$ and $\mathbf{J}_n^q$ denote the electrical and heat current densities, respectively, ψ denotes the electrostatic potential, $\sigma_n(T)$ the temperature-dependent electrical conductivity, $\alpha_{sn}(T)$ the temperature-dependent Seebeck coefficient (also referred to as the thermoelectric power), and $\kappa_{tot}(T)$ the total (*viz.*, measured) temperature-dependent thermal conductivity which accounts for both electron and lattice (phonon) contributions to thermal conduction. The latter dominates in semiconductors as well as insulators while the former dominates in metals (see Sect. 5.3.2). Eqs. (8.12) and (8.13) are valid for *n*-type material; the relations for a *p*-type material are analogous. Note that because of Onsager's theorem there are only three independent coefficients in (8.12) and (8.13); all three depend on temperature. The Seebeck coefficient plays a crucial role as it determines electrical transport in the presence of a temperature gradient and heat transport in the presence of a potential gradient. For convenience, the equations have been expressed in terms of measurable material coefficients, values for which can be retrieved through the use of dedicated test structures (see [65]). Values and associated temperature coefficients (TC) for selected IC materials relevant to microactuation are given in Chapts. 2 and 5. Descriptions for these coefficients based on microscopic solid-state theory of ideal materials are not recommended for simulation purposes since the electrical and thermal conductivities and the Seebeck coefficient of a given material can strongly depend on microfabrication process conditions.

In most thermally isolated (micromachined) device structures, electro-thermal operation in the transient regime is determined by heat transport since the electrical time constants are several orders of magnitude smaller than corresponding thermal time constants. Thus we only need to consider the steady-state form of the carrier continuity equations. With negligible recombination, the divergence of the electrical current density vanishes to yield

$$\text{div}\left[\text{grad}\,\psi - \alpha_{sn}(T)\,\text{grad}\,T\right] = 0. \tag{8.14}$$

To arrive at an equivalent relation for heat transport, we compute the total energy current density (cf (5.20)),

$$\mathbf{J}_{tot}^u = \psi\,\mathbf{J}_n^e + \mathbf{J}_n^q \tag{8.15}$$

the divergence of which yields the total energy balance:

$$\partial u/\partial t + \text{div}\,\mathbf{J}_{tot}^u = 0. \tag{8.16}$$

Substituting for the various terms yields the following well-known form of the heat conduction equation

$$\text{div}\left[\kappa_{tot}(T)\,\text{grad}\,T\right] = -H + c_{tot}(T)\partial T/\partial t, \tag{8.17}$$

where

$$H = \frac{\mathbf{J_n^2}}{\sigma_n(T)} + T\mathbf{J_n} \operatorname{grad} \alpha_{sn}(T), \tag{8.18}$$

and c_{tot} denotes the total heat capacity which, like the thermal conductivity, comprises both carrier and phonon contributions [64]. Here, we have reverted back to the standard notation $\mathbf{J_n}$ for the electrical current density. In (8.18), the first term corresponds to the well-known Joule heat and the second term accounts for both Peltier and Thomson heat. The Peltier heat (associated coefficient, $\Pi = \alpha_{sn}T$) describes the heat flux that goes along with an electrical current and thus constitutes a heat current discontinuity at an interface of materials with differing Seebeck coefficients. The Thomson heat (associated coefficient, $\gamma = T\partial\alpha_{sn}/\partial T$) describes the exchange of heat when an electrical current flows in a temperature gradient, and is significant only when the Seebeck coefficient is strongly temperature-dependent.

Equations (8.14) and (8.17), along with relation (8.18), are subject to appropriate boundary conditions at contact and interface regions. Conditions for electrical transport are relatively straightforward. At contacts, the electrostatic potential is prescribed by the applied voltage (V_a), viz., $\psi = V_a$. If biased by a constant current source (I_0), the contact is treated as an equipotential whose value is constant but unknown. The potential of the floating contact is determined by solving

$$\int_\Gamma \sigma_n(T) \left(\operatorname{grad}\psi \cdot \mathbf{n}\right) d\Gamma = I_0, \tag{8.19}$$

where Γ denotes the integration surface, $d\Gamma$ a surface element, and $\mathbf{n}$ is a unit normal vector. At conductor-insulator interfaces, $\mathbf{J_n} \cdot \mathbf{n} = qR^{\text{surf}}$, where R^{surf} denotes the surface recombination rate. This rate can be assumed negligible in most cases of practical interest.

As for the heat transport, either the temperature or the heat flux (q_0) can be prescribed at isothermal regions. Similar to the electrostatic case, the former leads to the condition $T = T_0$, where T_0 denotes the temperature of the isothermal surface, while for the latter, the condition reads

$$\int_\Gamma \kappa_{\text{tot}}(T) \left(\operatorname{grad} T \cdot \mathbf{n}\right) d\Gamma = q_0 \tag{8.20}$$

in the usual notation. At interfaces, including device boundaries, the conditions on the temperature must be consistent with the total energy balance. This is satisfied by imposing continuity of the energy current density (8.15), viz., $\mathbf{J}_{\text{tot}}^{\text{u1}} \cdot \mathbf{n} = \mathbf{J}_{\text{tot}}^{\text{u2}} \cdot \mathbf{n}$, where the superscripts 1 and 2 refer to the associated materials at the interface. At a conductor-insulator interface, this yields

$$\kappa_c(T) \left[\operatorname{grad} T \cdot \mathbf{n}\right]_c - \kappa_i(T) \left[\operatorname{grad} T \cdot \mathbf{n}\right]_i = \left[\psi - T\alpha_{sn}(T)\right]\mathbf{J_n} \cdot \mathbf{n}, \tag{8.21}$$

where the subscripts c and i denote conductor and insulator regions, respectively. At the interface between two conducting materials ($c1$ and $c2$), we obtain for an ideal interface

$$\kappa_{c1}(T) \left[\text{grad}\, T \cdot \mathbf{n} \right]_{c1} - \kappa_{c2}(T) \left[\text{grad}\, T \cdot \mathbf{n} \right]_{c2}$$
$$= T[\alpha_{c2}(T) - \alpha_{c1}(T)] \mathbf{J_n} \cdot \mathbf{n}. \tag{8.22}$$

Here, the potential is continuous across the interface. At device boundaries interfaced with an ambient fluid, one can choose between the following conditions; the choice depends on the respective (conductor or insulator) material at the interface:

$$\kappa_c(T) \left[\text{grad}\, T \cdot \mathbf{n} \right]_c = h(T_\infty - T) + [\psi - T\alpha_{sn}(T)] \mathbf{J_n} \cdot \mathbf{n} + q_{\text{rad}}, \tag{8.23}$$

$$\kappa_i(T) \left[\text{grad}\, T \cdot \mathbf{n} \right]_i = h(T_\infty - T) + q_{\text{rad}}. \tag{8.24}$$

Here, h denotes an effective heat transfer coefficient, T_∞ the fluid free-stream temperature, and q_{rad} the heat loss or gain by radiation. As seen in Sect. 5.5, the heat loss to the ambient fluid takes place by conduction through the fluid, forced convection, and radiation. The former two components can be accounted for through h; conduction or forced convection predominate in the absence or presence of a flow stream, respectively. Heat loss by natural convection can be neglected since the buoyancy forces associated with most microstructures, even when heated to temperatures of 400 °C, are not sufficient to produce any significant natural convective currents [66, 67]. Methods for computing h range from lumped analytical to numerical, including boundary element, techniques. The choice depends on the device structure and its temperature distribution.

Solution of the electrothermal model equations under pertinent boundary and interface conditions, collectively described by (8.12) to (8.24), yields the temperature distribution. Based on this temperature distribution, we compute the resulting thermal strain (see Sect. 7.1.4) for subsequent mechanical analysis, viz., $\partial/\partial x_i [C_{ijkl}(\varepsilon_{kl} + \alpha_{kl}\Delta T)] = 0$, where $i, j, k, l = 1, 2, 3$. A summary of pertinent model equations is given in Sect. 8.8.

8.4.2 Shape Memory Actuation

The principle underlying this actuation mechanism involves thermally-induced reversible crystalline transformation in a SMA from the low-temperature martensitic (ductile) phase to the high-temperature austenitic (high strength or parent) phase. In the martensitic phase, the SMA is deformed by a bias force, which if sufficiently large can lead to plastic deformation of the alloy. However, unlike most regular materials where plasticity, which is induced by dislocations, is irreversible, the plastic behavior in SMAs is caused by twin deformation and is reversible. When the SMA is externally or resistively heated to induce a phase transformation, the bias force is overcome and the alloy regains its original annealed shape. A commonly used material in the SMA family is nitinol which is a binary alloy with approximately equal atomic weight of nickel (Ni) and titanium (Ti) [68]. Because shape memory behavior and associated phase transformation temperatures are very sensitive to composition, control of the film composition is crucial. This makes the SMA deposition and subsequent annealing processes very challenging, oxygen contamination being a danger with the latter. Recent reports [69] show that addition of copper makes the alloy (NiTiCu) much more tolerant with respect to sensitivity of thermomechanical properties to variations in composition.

Shape memory behavior is present only in the crystalline alloy. The alloy in thin film form has thermomechanical properties that are comparable to their bulk counterparts. The thermomechanical properties can be characterized by the alloy's phase transformation hysteresis loop (see Fig. 8.4). The transformation temperatures can be reasonably well estimated by measurement of the temperature-dependent stress-strain behavior or electrical resistivity [69]. At the (high temperature) austenitic phase, a thin film SMA has a high tensile stress. Upon cooling, the transformation phase begins at the martensite start temperature (M_s) and completes at the martensite finish temperature (M_f). The residual stress is referred to as the martensitic yield stress (σ_{ym}). Reverse transformation takes place when the thin film is heated. The transformation begins at the austenite start temperature (A_s), completes at the austenite finish temperature (A_f), and the film recovers the high tensile stress (σ_{rec}). This forms the basis for microactuation. The hysteresis width can be defined as

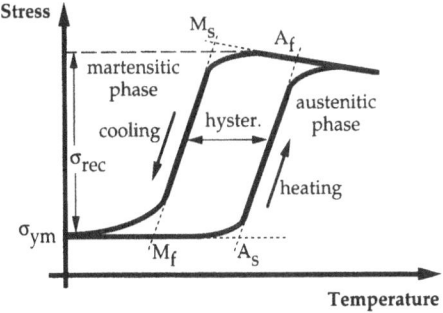

Fig. 8.4 Sketch of thermomechanical behavior of an SMA illustrating the major hysteresis loop. Depending on the conditions of thermal treatment, minor loops may arise

Table 8.3. *Electrical, thermal, and mechanical properties of the NiTi SMA family;*
transformation temperatures are sensitive to composition as seen by the values given in the
last two rows containing Cu and Pd

SMA Material Parameter	Value
Density	$6.5 \, \text{g/cm}^3$
Specific heat	$\sim 48.2 \, \text{J/(°C mol)}$
Thermal expansion coefficient	$6.6 \times 10^{-6}/°\text{C}$ (martensite)
	$11 \times 10^{-6}/°\text{C}$ (austenite)
Thermal conductivity	$0.3 \, \text{W/(cm °C)}$
Electrical resistivity	$75 \, \mu\Omega \, \text{cm}$ (martensite)
	$60 \, \mu\Omega \, \text{cm}$ (austenite)
Yield strength	$0.21 \, \text{GPa}$
Young's modulus [70]	$29 \, \text{GPa}$ (martensite)
	$62 \, \text{GPa}$ (austenite)
Poisson's ratio	~ 0.3
NiTi [70]: $M_s, M_f, A_s, A_f, \sigma_{ym}$	$32 °\text{C}, -3 °\text{C}, 44 °\text{C}, 72 °\text{C}, 100 \, \text{MPa}$
$\text{Ni}_{42}\text{Ti}_{51}\text{Cu}_7$ [69]: M_s, M_f, A_s, A_f	$53 °\text{C}, 36 °\text{C}, 47 °\text{C}, 62 °\text{C}$
hysteresis width, $\sigma_{ym}, \sigma_{rec}$	$10 °\text{C}, 120 \, \text{MPa}, 330 \, \text{MPa}$
$\text{Ni}_{37}\text{T}_{54}\text{Pd}_9$ [72]: M_s, M_f, A_s, A_f	$26 °\text{C}, 14 °\text{C}, 32 °\text{C}, 45 °\text{C}$
$\text{Ni}_{18}\text{T}_{54}\text{Pd}_{28}$ [72]: M_s, M_f, A_s, A_f	$192 °\text{C}, 179 °\text{C}, 203 °\text{C}, 220 °\text{C}$

$(A_f - M_s)$ or $(A_s - M_f)$. Values of electrical, thermal, and mechanical
parameters for the nitinol SMA family are given in Table 8.3. The values
given vary with film fatigue, which stems from prolonged thermal cycling
as well as extreme mechanical loading conditions whereby very large
stresses lead to irreversible plastic deformation. For example, a NiTiCu thin
film SMA with a recovery stress of 500 MPa ($\sim 5\%$ strain) reduces to
300 MPa ($\sim 2\%$ strain) after thousands of cycles [69]. Apart from the
martensitic and austenitic phases, the SMA is known to also exhibit an
intermediate rhombohedral phase. The presence of this phase depends on
material composition and process conditions [70].

Mechanical microactuation of the thin film SMA following transformation from the martensitic to austenitic phases is based on a coupling of
electrothermal and thermomechanical effects. While the former is modeled
along the lines discussed in Sect. 8.4.1, with incorporation of relevant
material data, the latter can be modeled using the standard equation for
thermoelasticity as we saw in Sect. 7.1.4. Here, the thermal strain term is
suitably modified to account for the recoverable strain (ε_{rec}) to yield the
following relation for the net thermal strain

$$\varepsilon_{ij}^T = (\alpha \Delta T + \varepsilon_{rec})\delta_{ij} \qquad \text{with} \quad i,j = 1,2,3, \qquad (8.25)$$

where α is the linear (isotropic) thermal expansion coefficient and δ_{ij} is the
Kronecker delta. Treatment of this thermal strain is analogous to that

shown previously in Sect. 8.4.1. Eq. (8.25) is strictly valid only when ΔT exceeds the phase transformation temperature. The thermal expansion term can become dominant when temperatures exceed A_f. Here, this component of thermal strain opposes the recoverable strain thus reducing the net thermal strain, and hence, the overall displacement. The recoverable strain in (8.25) associated with the SMA in the austenitic phase can be described as [69]:

$$\varepsilon_{rec} = \sigma_{rec}/E_a. \tag{8.26}$$

Here, σ_{rec} denotes the recoverable stress and E_a is the Young's modulus of the austenitic alloy. Eq. (8.26) predicts the recoverable displacement in the austenitic phase but does not account for the SMA's mechanical behavior during phase transformation or mechanical hysteresis. The material properties of the SMA are complex. Not only is the alloy a non-linear material with hysteresis, its phase transformation parameters are sensitive to film stresses, process conditions, temperature, and thermal cycling. Thus modeling of its thermomechanical properties is rather complex, although descriptions based on variable sub-layer models comprising parallelly connected layers of different mechanical characterisitics, along with a thermodynamic description of phase transformations, have been reported recently [70, 71]. From a simulation standpoint, it may be more convenient to employ values of electrical and thermomechanical parameters retrieved from measurements of the phase-dependent stress-strain behavior or electrical resistance using substrate curvature methods, bulge tests, or dedicated test microstructures (see [69]). Measured values of relevant SMA parameters are given in Table 8.3.

The computational process for self-consistent electro-thermo-mechanical analysis is very similar to that of the electrostatic case in Sect. 8.3. Essential changes lie in replacing the electrostatic analysis and associated forces by electrothermal analysis and associated thermal strains, and in the constitutive relations and boundary conditions. This is summarized in Sect. 8.8.

8.5 Magnetic Actuation

This class of microactuators are magnetically driven. The excitation field is generated either externally or in situ through integrated inductor microcoils. Although magnetic microactuators are inferior, in terms of actuation distances, to phase transformation actuators (e.g., SMAs), they are superior to their electrostatic counterparts. Magnetic microactuators are only in their infancy, but are rapidly proliferating with the evolution of new thin film magnetic, including magnetostrictive, materials and associated fabrication processes. The latter include vacuum deposition or electro-

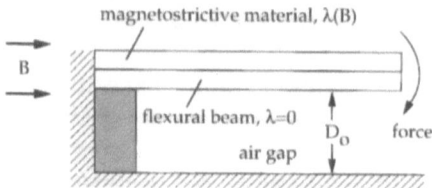

Fig. 8.5a Schematic representation of magnetostrictive actuation with a bimorph beam; λ denotes the magnetostrictive strain and D_0 the zero-field separation

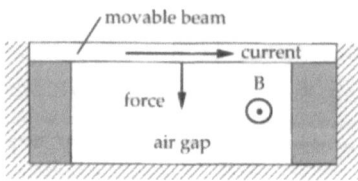

Fig. 8.5b Schematic representation of electrodynamic actuation

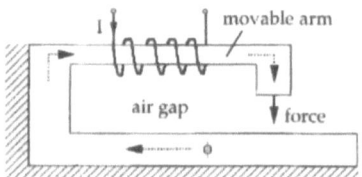

Fig. 8.5c Schematic representation of electromagnetic actuation

deposition (using LIGA [73] or competing techniques [74]) in conjunction with IC microfabrication technology, to yield actuation elements that are stand-alone or integrated with IC dielectric (e.g., Si_3N_4 [14]) or conductor (e.g., poly-Si [75]) flexural structures. Applications of magnetic micro-actuator systems, whether fully integrated or manually micro-assembled, include micromirrors for microoptic systems [75] and holographic data storage [76], tunable IR filters [77], microrelays [78], micromotors [79–81], and microfluidic systems (see, e.g., [82, 83]).

Magnetic microactuators rely on magnetostrictive, electrodynamic, or electromagnetic transduction principles for mechanical displacement. The underlying operating principles are illustrated in Fig. 8.5. Magnetostriction, similar to piezoelectricity, is a reciprocal effect. The direct effect, known as the Joule effect, describes a positive or negative change in linear displacement (strain) in ferromagnetic materials in response to an in-plane magnetic field. The inverse effect, also referred to as the Villari effect, describes the change in magnetic properties, and hence, in the induced magnetization, by stress or strain. In unimorph configuration, where the magnetostrictive material is an integral part of a flexural beam, the field-induced strain results in beam deflection (see Fig. 8.5a). The electro-dynamic principle relies on the presence of both electrical and magnetic input signals. Here, a conductor with a current in a magnetic field orthogonal to current flow displaces as a result of the Lorentz force. This constitutes the magnetic driving force (Fig. 8.5b). The electromagnetic principle, considered to be the analog of electrostatics, forms the basis of microrelay operation (Fig. 8.5c). Here, a current through a coil wound around an insulated highly permeable soft magnetic core (to form a driving electromagnet) generates a magnetic flux ϕ which flows through the air gap in a well-defined magnetic circuit. The flux in the movable arm (rotor) creates a force causing it to contact the stationary arm (stator). Contact is disabled when the driving coil current is switched off.

As with electrostatic or thermal microactuation reviewed previously, there are several design issues associated with magnetic microactuation, for which numerical simulations play an essential role. The simulation of magnetic microactuators involves electromagnetic and mechanical ana-lyses coupled through strain or actuation forces induced by the magnetic field. Design issues include:

- Prediction of non-linearity stemming from large magnetic forces or magnetostrictive strains; non-linear dependence of material coefficients (such as the magnetostrictive coefficient (λ) and permeability (μ)) on magnetic field and associated hysteresis effects; and temperature dependence of material coefficients arising from Joule heat at high drive currents or through induced eddy currents, particularly in mechanically, and hence, thermally isolated structures.
- Control of drive currents in integrated structures to prevent irreversible damage (e.g., due to conductor electromigation).
- Optimization of structure/geometry and relevant bias conditions to restrict device operation to the linear magnetic regime (where λ, μ are approximately constant).
- Investigation of actuator packaging/shielding effects on magnetic field distribution and associated minimization or elimination of flux leakage for maximum transduction efficiency.

These considerations along with structural/geometrical complexities and associated material inhomogeneities, renders meaningful modeling by analytical solutions an impossible task. Thus numerical simulations are mandatory.

In what follows, we present the differential equations, boundary conditions, constitutive relations, and associated magnetic material models and/or data pertinent to the analysis of the different actuation mechanisms. Without recourse to the rather involved science of magnetic materials, we

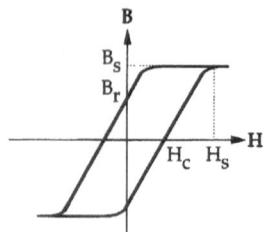

Fig. 8.6 Magnetization of ferromagnetic materials; B_r denotes the remanent flux density, B_s the saturation flux density, H_c the coercivity, and H_s the saturation field

Table 8.4. *Relations for magnetic materials. Notation used is standard (see, e.g., [54])*

Magnetic relations		Electrostatic	
Anisotropic	Isotropic	equivalent	Description
$M_i = \chi_{mij} H_j$	$\mathbf{M} = \chi_m \mathbf{H}$	$\mathbf{P} = \varepsilon_0 \chi_e \mathbf{E}$	M = magnetization
$\chi_{mij} = \chi_{mji}$			χ_m = magnetic susceptibility
			$\chi_m > 0$; paramagnetic (Al, Cr)
			$\chi_m < 0$; diamagnetic (Cu, Zn)
			$\chi_m \gg 1$; ferromagnetic (Fe, Co, Ni)
$B_i = \mu_{ij} H_j$	$\mathbf{B} = \mu_0(\mathbf{H} + \mathbf{M})$	$\mathbf{D} = \varepsilon_0 \mathbf{E} + \mathbf{P}$	ν = reluctivity = $1/\mu$
$\mu_{ij} = \mu_{ji}$	$\mathbf{B} = \mu \mathbf{H}$	$\mathbf{D} = \varepsilon \mathbf{E}$	ν_0 = vacuum reluctivity = $1/\mu_0$
	$\mathbf{H} = \nu \mathbf{B}$		
μ_{ij}	$\mu = \mu_0(1 + \chi_m)$	$\varepsilon = \varepsilon_0(1 + \chi_e)$	$\mu_0 = 4\pi \times 10^{-7}$ H/m
$= \mu_0(\delta_{ij} + \chi_{mij})$	$\mu = \mu_0 \mu_r$	$\varepsilon = \varepsilon_0 \varepsilon_r$	$\varepsilon_0 = 8.85 \times 10^{-12}$ F/m
			(vacuum permeability and permittivity)
$\mu_{rij} = \mu_{ij}/\mu_0$	$\mu_r = 1 + \chi_m$	$\varepsilon_r = 1 + \chi_e$	$\mu_r \sim 1$; para- and diamagnetic materials
$\mu_{rij} = \mu_{rji}$			

provide in Fig. 8.6 and Tables 8.4 to 8.7 the necessary relations and terminology as well as an overview of ferromagnetic material properties. These properties are highly sensitive to fabrication, and in many cases, magnetic-processing histories. This is particularly true in the case of thin films, whose properties can be quite different from bulk. Thus for accurate simulations, experimentally retrieved values of material data are preferred. However, even with a well-defined material behavior, modeling of the $B–H$ curve, from a numerical standpoint, is not a trivial task even if hysteresis and material anisotropies are ignored. The non-linear dependence of permeability on the field variable (B or H) and its wide range of values can give rise to convergence problems [84]. Depending on the choice of the solution variable, the behavior can be modeled using curve representations of $\nu = \nu(B^2)$, $H = H(B)$, or $\mu = \mu(H^2)$ expressed in terms of the square of the field variable. Here, $\nu = 1/\mu$ is called the reluctivity. A hard magnetic material (e.g., permanent magnet) may be treated as soft by shifting the

Table 8.5. *Summary of magnetic properties related to Fig. 8.6; T_{curie} denotes the Curie temperature and λ the magnetostrictive coefficient*

Magnetic material	Property
Soft (e.g., transformers)	Easy to magnetize and demagnetize
	Magnetization M induced by external field
	$B(H) \Rightarrow 0$ as $H \Rightarrow 0$
	Small H_c (usually < 1 Oe and $H_c \ll H_s$)
	Large B_s (~ 1–$2\,\mathrm{T}$), small B_r
	Large permeability $\mu = \Delta B / \Delta H$
	μ constant at low field
Soft (e.g., magnetic storage)	$H_c \sim H_s$, $B_r \sim B_s$

In soft magnetic materials, spin alignment exists only in small domains where net magnetization can only be maintained by external field. Examples of soft materials: Fe and Fe-Ni alloys.

Hard (e.g., permanent magnet)	Hard to magnetize and demagnetize
	Very large H_c and B_r and $\mu_0 M \geqslant 1\,\mathrm{T}$
	$B(H) \Rightarrow B_r$ as $H \Rightarrow 0$ and $H = H_c$ when $B = 0$

In hard materials, the spontaneous alignment of spins (at $T < T_{\text{curie}}$) yields a remanent magnetization $\mathbf{M_0}$ in the absence of external field; $\mathbf{B} = \mu\mathbf{H} + \mu_0\mathbf{M_0}$ or $\mathbf{H} = \nu\mathbf{B} - (\nu/\nu_0)\mathbf{M_0}$. For $T > T_{\text{curie}}$, material becomes paramagnetic. Examples of hard materials: tungsten steel, Cu-Ni-Fe alloys, and Al-Ni-Co alloys.

Magnetostriction	Partly responsible for "hardness"
	μ enhanced through reduced λ (e.g. $\lambda \sim 0$ in
	$Fe_{81}Ni_{19}$, since $\lambda > 0$ in Fe and $\lambda < 0$ in Ni)
Giant magnetostriction	Large H_c and H_s, large λ, small μ
Weak magnetostriction	Low H_c and H_s, small λ, high μ

B–H curve to the origin and introducing a magnetic pole or an electrical current distribution that is equivalent to the displaced coercivity (H_c) or remanence (B_r). With anisotropic materials, μ or ν become tensors.

8.5.1 Magnetostriction Analysis

Magnetostriction is a direct result of the rotation of magnetic domains in the amorphous or crystalline material, from the easy to hard magnetization axes, when subject to a magnetic field parallel to the latter. Magnetostriction is present in isotropic and anisotropic materials; but generally, the larger the material anisotropy, the larger the effect. The original orientation (or pre-alignment) of domains is very important; domains parallel or anti-parallel to the direction of magnetization do not rotate, and thus do not contribute to dimension change, while the domains aligned orthogonally contribute the most.

Table 8.6. *Magnetic properties of selected magnetostrictive alloys; values are sensitive to composition and process conditions*

Magnetostrictive material parameter	Value/process conditions
$Fe_{78}Si_9B_{13}$ [91] (tradename: Metglas2605S2)	dc magnetron sputtering
λ_s, H_s, coercivity (H_c)	$+30 \times 10^{-6}$, 15 Oe, 1 Oe
Young's modulus, Poisson's ratio	170 GPa, 0.3
Tb-Fe alloys [92]	rf magnetron sputtering
λ (for 28–46 at.% Tb)	$> 250 \times 10^{-6}$ at 16 kOe
Sm-Fe alloys	dc magnetron sputtering
λ	-220×10^{-6} at 1 kOe, -300×10^{-6} at 5 kOe [93]
(for 30–40 at.% Sm)	rf magnetron sputtering
	$-(250–300) \times 10^{-6}$ at 1 kOe [92]
	$-(300–400) \times 10^{-6}$ at 16 kOe [92]
$(Tb_{0.3}Dy_{0.7})_{0.42}Fe_{0.58}$ [93] (tradename: Terfenol)	dc magnetron sputtering
λ	$+250 \times 10^{-6}$ at 1 kOe, $+400 \times 10^{-6}$ at 5 kOe
Young's modulus	50 GPa

Table 8.7. *Properties of selected ferromagnetic materials* [98]

Material	Curie temperature (°C)	Relative permeability $\times 10^4$	Electrical resistivity $(10^{-8} \Omega\,m)$	Density (kg/m^3)	Saturation field (Tesla)
Fe (purified)	770	1–20	10	7800	2.15
Fe-Si (3% Si)	740	0.75–5.5	47	7670	2.0
Ni	360	0.011–0.06	7	8900	0.61
45 Permalloy (45% Ni)	400	0.25–2.5	45	8170	1.6
78 Permalloy (78.5% Ni)	600	0.8–10	16	8600	1.08
Supermalloy (5% Mo, 79% Ni)	400	10–100	60	8770	0.79
Mumetal		3–10	42	8900	0.8

Magnetostrictive materials are characterized by two key material figures of merit [85]: the saturation magnetostrictive strain λ_s and the saturation magnetic field H_s. In giant magnetostrictive materials, such as Tb-Fe, Sm-Fe, and Tb-Dy-Fe (Terfenol) alloys, λ_s can be in excess of 1000 ppm (strain $\sim 0.1\%$) but with a high saturation field, $H_s > 1000$ Oe. The units of Oersteds (Oe) can be converted as follows: 1 Oe = 1 gauss = 10^{-4} tesla in vacuum, where 1 tesla = $1\,Vs/m^2 = 10^4$ gauss is unit of B and 1 Oe = 79.6 A/m is unit of H. Giant magnetostrictive materials are strong candidates for magnetic actuation because of their large actuation distances. Another class of materials is the Fe-based amorphous alloy family referred to as Metglas. These alloys have a combination of relatively large λ_s (~ 20–60 ppm) and low H_s(< 20 Oe). However, they are capable of only small actuation distances and are generally more suited for sensing applications.

The generalized description of magnetostriction, valid for operation in the linear elastic and linear magnetic regimes, takes the form [86]

$$\sigma_{ij} = J_{ijkl}B_k B_l \qquad \text{with} \quad i,j,k,l = 1,2,3, \tag{8.27}$$

$$H_i H_j = V_{ijkl}\varepsilon_{kl} \qquad \text{with} \quad i,j,k,l = 1,2,3 \tag{8.28}$$

where σ_{ij} and ε_{kl} denote the stress and strain tensors, respectively, $B_{k,l}$ denote the components of magnetic induction, and $H_{i,j}$ the components of magnetic field. The coefficients J_{ijkl} and V_{ijkl} are fourth rank tensors associated with the direct (Joule) and indirect (Villari) effects, respectively. The exact form of these tensors depend on the crystal structure of the material. Relations (8.27) and (8.28) can also be cast in various other forms (see [86]) following use of the stress-strain and field-flux constitutive relations: σ-ε (see Sect. 7.2) and **B-H** (see Table 8.4). Relations (8.27) and (8.28) only hold for the idealized (or perfect) material. Often, the materials used in micro-actuation applications are either amorphous or polycrystalline, and the associated magnetostrictive coefficient λ can be regarded as being isotropic with no preferred direction for the easy axis of magnetization (**M**$_{\text{easy}}$). In the low field regime, the magnetostrictive strain can be related to the magnetic field H through this simple approximation [87]

$$\lambda = CH^2, \tag{8.29}$$

where C is a so-called piezomagnetic coefficient, *viz.*,

$$C = \frac{3}{2}\frac{\lambda_s}{H_a^2}. \tag{8.30}$$

Here, H_a is the anisotropic field, related to the anisotropic energy, associated with the easy axis. In metallic glasses (e.g., Metglas), H_a is equivalent to the saturation field H_s. The strain, for a perfectly oriented material, is largest when the magnetic field is parallel to the hard axis. This is also referred to as the longitudinal strain λ_l. (For convenience, as it will soon be apparent, we will refer to this as the maximum attainable strain since the value of λ_l heavily depends on the thermal and magnetic processing histories.) The strain is otherwise reduced, and its relative decrease using the constant volume approximation can be computed as [85]:

$$\frac{\lambda_\theta}{\lambda_l} = \frac{3}{2}\left(\cos^2\theta - \frac{1}{3}\right). \tag{8.31}$$

Here, λ_θ denotes the magnetostrictive strain at an angle θ to the magnetization direction. From (8.31), we see that the strain in the transverse direction $(\theta = 90°)$ is half the value of the longitudinal counterpart and negative in sign. Relations (8.29) and (8.30) follow from the assumption of perfect alignment of all magnetic domains in the material. In practice, such ideal conditions are not completely satisfied leading to a reduction in the piezomagnetic coefficient, and in the effective magneto-mechanical coupling. This is particularly so with thin magnetic films. Here, the desired domain orientation and magnetic material parameters (e.g., μ_r, λ_s, H_s) are not just material related, but strongly depend on the thermal and magnetic processing histories as well as on the mechanical and chemical states [88, 89]. For example, residual stress in the thin film disturbs the in-plane domain orientation thus reducing λ_l, and the film's material properties strongly depend on the chemical composition, particularly in binary and tertiary alloys. Thus, for accurate simulations, parameter values retrieved using experimental test structures are recommended.

For example, the field-dependent magnetostrictive strain of a thin film can be retrieved from capacitance measurements of cantilever beam deflection in a dc magnetic field (see [90]). For a tip deflection (D) much smaller than the zero field separation (D_0), the Kolkholm formula [163] for tip deflection reads

$$D = 3\lambda_l L^2 \frac{E_f}{E_s} \frac{t_f}{t_s^2} \frac{1 - \nu_s}{1 + \nu_f}, \tag{8.32}$$

which according to recent reports [164] has an error in the sign of Poisson's ratio of the substrate (ν_s), i.e., values retrieved using (8.32) are overestimated by a factor of $(1 + \nu_s)/(1 - \nu_s)$. In (8.32), $E_{f,s}$ are the Young's moduli of the film and flexural substrate (thickness t_s), respectively, ν_f denotes the Poisson's ratio of the thin film, L the beam length, and t_f the thickness of the magnetostrictive thin film. Here, we have assumed that magnetization effects have negligible influence on mechanical properties. This assumption is reasonable since the change in elastic moduli of ferromagnetic materials such as Fe or Ni is about 1% and 6%, respectively.

Assuming a parabolic profile for the deflection, the change in capacitance can be retrieved as $\Delta C = DC_0/(3D_0)$, where C_0 denotes the zero-field capacitance. Values of the strain retrieved for thin film $Fe_{78}Si_9B_{13}$ (Metglas2605S2) are given in Fig. 8.7 [91]. In the case considered, in view of the substrate being glass $(\nu_s = 0.17)$, the indicated values of λ_l (Fig. 8.7) are overestimated by a factor of 1.4. Values for other alloys, gathered from selected sources [92, 93], are given in Table 8.6. If needed, the associated piezomagnetic coefficient C can be retrieved from

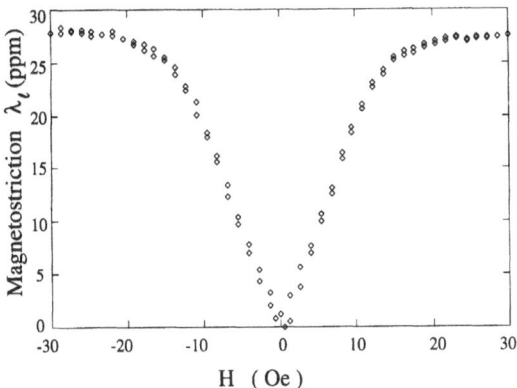

Fig. 8.7 Longitudinal magnetostriction λ_l of Fe$_{78}$Si$_9$B$_{13}$ as a function of magnetic field retrieved using the Kolkholm formula (8.32)

measured data using (8.29). Note that instead of using the analytical relation (8.32), the magnetostrictive strain can also be obtained numerically, for arbitrary test geometries, by fitting measured and simulated values of beam deflection; along the lines of the load-deflection procedure reported for retrieval of mechanical properties of thin films [94]. This is discussed in Sect. 7.2.3.

Simulation of magneto-mechanical coupling based on magnetostriction is similar to the thermal bimorph problem discussed in Sect. 8.4. For convenience of simulation, the magnetostrictive strain term ε^M for the thin film can be cast in the following form:

$$\varepsilon_i^M = \frac{|\mathbf{H} \times \mathbf{M}_{\text{easy}}|}{|\mathbf{H}||\mathbf{M}_{\text{easy}}|}\lambda_i \qquad \text{with} \quad i = 1, 2, \ldots, 6, \tag{8.33}$$

where $\mathbf{H}$ and $\mathbf{M}_{\text{easy}}$ are in-plane, $\lambda_1 = \lambda_l$, $\lambda_2 = -\lambda_l/2$, and $\lambda_{3,4,5,6}$ are assumed zero. However, since the magnetostrictive strain is non-linear with respect to the magnetic field H, for small variations in H we can employ a Taylor expansion around a bias field H_0. Retaining the linear terms in H, the magnetostrictive strain can be approximated as

$$\lambda \approx \lambda(H_0) + \left.\frac{\partial \lambda}{\partial H}\right|_{H_0} (H - H_0). \tag{8.34}$$

Eq. (8.34) is valid only in the non-saturation magnetic region, where $H \ll H_s$, and in the elastic regime where the stress-strain relation is linear. Alternatively, if the piezomagnetic coefficient is known, following (8.29), we can write (8.34) as:

$$\lambda \approx CH_0[1 + 2(H - H_0)]. \tag{8.35}$$

8.5.2 Electrodynamic Analysis

Mobile charge in a conductor in the presence of an orthogonal magnetic induction (**B**) is subject to a magnetic force that depends on charge velocity and acts orthogonal to the velocity vector (**v**). The force on each charge is $\mathbf{f} = q\mathbf{v} \times \mathbf{B}$, where q denotes the elementary charge. Since the force acts on all mobile charges in the conductor, the resultant magnetic body force density $\mathbf{f}^M$ (N/m^3) in terms of the current density (**J**) through the conductor is then given as

$$\mathbf{f}^M = \mathbf{J} \times \mathbf{B}. \tag{8.36}$$

The force can be exploited to yield bi-directional mechanical displacement through suitable arrangement (routing) of conductors (e.g., poly-Si, IC metallization) that are sandwiched in flexural dielectric materials (e.g., SiO$_2$, Si$_3$N$_4$) [95]. Relation (8.36) can also be expressed in terms of a distributed surface force. This is discussed in Sect. 8.5.3.

The conductors, depending on their relative placement, can exert forces on each other since the current in each conductor generates a magnetic field. For example, two parallel conductors attract when currents are in the same direction and repel when currents are in opposite directions. The generated field, and hence the forces, can be computed from Maxwell's equations under static conditions. Together with Stokes' theorem, we obtain the generalized Ampere's law

$$\int_C \mathbf{B} \cdot \mathbf{dl} = \mu_0 \int_\Gamma \mathbf{J} \cdot \mathbf{n} \, d\Gamma \tag{8.37}$$

noting that div $\mathbf{B} = 0$. Here, μ_0 denotes the permeability of free space. The left-hand side integral is taken around a cross-sectional closed loop contour (C), $\mathbf{dl}$ is a length segment on the contour, $d\Gamma$ is an elemental surface area, and Γ is the area bounded by C, through which current flows. The force between two (electrically closed loop) conductors (α and β) can be expressed as [54]

$$\mathbf{f}^M_{\alpha,\beta} = -\frac{\mu_o}{4\pi} \int_\alpha \int_\beta \frac{(\mathbf{J}_\alpha \cdot \mathbf{n} d\Gamma_\alpha)(\mathbf{J}_\beta \cdot \mathbf{n} d\Gamma_\beta)}{r^2_{\alpha,\beta}} (\mathbf{dl}_\alpha \cdot \mathbf{dl}_\beta)\mathbf{u}, \tag{8.38}$$

where the integrals are taken over the complete circuit. Here, $\mathbf{J}_{\alpha,\beta}$ denote the respective current densities, $\mathbf{dl}_{\alpha,\beta}$ are the elemental conductor lengths, $r_{\alpha,\beta}$ is the separation distance between conductors, and $\mathbf{u}$ is a unit vector that points at the conductor on which the force acts.

8.5.3 Electromagnetic Drive Analysis

Forces associated with actuation of either permanent magnets or materials magnetized through solenoidal excitation (Fig. 8.5c) are computed from the distribution of the magnetizing field $\mathbf{H}$ or the associated flux density $\mathbf{B}$. This follows from solution of Maxwell's equations that govern dynamic behavior of electric and magnetic fields. In differential form, the equations are [96]:

$$\text{curl } \mathbf{H} = \mathbf{J}, \tag{8.39}$$

$$\text{curl } \mathbf{E} = -\partial \mathbf{B}/\partial t, \tag{8.40}$$

where $\mathbf{J}$ is the current density and $\mathbf{E}$ the electric field. Following Ohm's law for a linear isotropic material, these quantities are related as $\mathbf{J} = \sigma \mathbf{E}$ where σ denotes the isotropic electrical conductivity. Intrinsic to (8.39) and (8.40) is the quasi-static approximation. Here, we neglect displacement currents thus transforming the PDEs from hyperbolic (with propagating wave solutions) to diffusion or elliptic character. The quasi-static approximation is justified since for the range of frequencies of interest, the associated wavelengths are much larger than practical microactuator dimensions. For example, the wavelength at a 1 MHz frequency in air is 300 m, and frequencies in this range are considered very large from a structural analysis viewpoint. The field-flux constitutive relations relevant to (8.39) and (8.40) is given in Table 8.4, along with their electrostatic equivalents. The flux density and electric field can be expressed in terms of the associated potentials:

$$\mathbf{B} = \text{curl } \mathbf{A}, \tag{8.41}$$

$$\mathbf{E} = -\partial \mathbf{A}/\partial t - \text{grad } \psi, \tag{8.42}$$

to gain mathematical advantage in prescribing the relevant boundary conditions. Here, $\mathbf{A}$ denotes the magnetic vector potential, which satisfies the requirement of a vanishing divergence of flux density, and ψ the electric scalar potential. Additionally, to ensure uniqueness of the vector potential, we employ the Coulomb gauge condition,

$$\text{div } \mathbf{A} = 0. \tag{8.43}$$

Eqs. (8.39) to (8.43), along with the field-flux constitutive relations, yield a PDE in terms of the magnetic vector potential,

$$\text{div grad } \mathbf{A} - \mu\sigma \, \partial^2 \mathbf{A}/\partial t^2 = \mu\sigma \text{ grad } \psi. \tag{8.44}$$

In its given form, (8.44) is strictly valid only for an isotropic and linear ferromagnetic material. It does not account for anisotropies in material permeability μ (or reluctivity ν) as well as (hard) materials whereby remanent magnetization exists (see Table 8.5). The second term on the left-hand side accounts for eddy currents in the conducting material, induced by time variations in $\mathbf{B}$, and hence, $\mathbf{A}$. The eddy currents, which decay exponentially over the conductor depth reaching $(1/e)$ of the surface value in one skin depth, can locally augment or oppose the source current density $\sigma\nabla\psi$.

Thus equation (8.44), or a modified form of it to account for material anisotropy and non-linearity, can also be employed for simulation of sensors based on the eddy current principle. Solving (8.44), for a prescribed electric potential, yields the spatial distribution of the vector potential, and hence, flux density. The form of (8.44) differs over the simulation domain due to material inhomogeneities. For example, free space regions are non-permeable and non-conducting. Stator and rotor regions are permeable but non-conducting if they are made of insulating materials, and depending on the type of material (i.e. hard or soft), we have to specify its remanent magnetization ($\mathbf{M_0}$). At solenoidal and other conducting regions, we need to specify the permeability and conductivity.

At interfaces between different regions, appropriate conditions must be specified to account for changes in the field behavior, since the change in device boundaries with actuation constitutes one of the coupling mechanisms between electromagnetic and structural analyses. Interface conditions are obtained from the integral forms of Maxwell's equations applied at the interface between two media. Assuming there exists an interface charge density ρ_s (couls/m^2) and an interface current $\mathbf{J_s}$ (A/m), we obtain the following conditions which, for convenience, are expressed in terms of the field variables [96]:

$$(\mathbf{E}_1 - \mathbf{E}_2) \times \mathbf{n} = 0, \tag{8.45}$$

$$(\mathbf{H}_1 - \mathbf{H}_2) \times \mathbf{n} = \mathbf{J_s}, \tag{8.46}$$

$$(\mathbf{D}_1 - \mathbf{D}_2) \cdot \mathbf{n} = \rho_s, \tag{8.47}$$

$$(\mathbf{B}_1 - \mathbf{B}_2) \cdot \mathbf{n} = 0. \tag{8.48}$$

Here, $\mathbf{n}$ is the surface normal directed from material 1 to material 2. Conditions (8.45) and (8.46), in the absence of $\mathbf{J_s}$, ensure continuity of the tranverse field components while (8.47) and (8.48) ensure flux conservation. These conditions can also be expressed in terms of the magnetic vector potential using relations (8.39) to (8.43) supplemented with the field-flux constitutive relations (Table 8.4). Eqs. (8.45) to (8.48) are valid for a vanishing remanent magnetization.

At regions distant from rotor and stator, we can assume a vanishing magnetic vector potential $\mathbf{A}$ since most of the flux is concentrated in the high permeability rotor and stator. As a result, there is rapid spatial decay in the field strength. Nevertheless, the extension of the outer simulation boundaries have to be judiciously chosen to ensure correctness of the field distribution, and hence forces, in the device active region.

In the time invariant case, we can drop the $\partial A / \partial t$ term and (8.44) reduces to the well-known Poisson's equation for magnetostatics (see, e.g., [97]):

$$\text{div grad } \mathbf{A} = -\mu \mathbf{J}. \tag{8.49}$$

As before, we must specify the source term $\mathbf{J}$ for solution of $\mathbf{A}$. In the case of solenoidal excitation, we have to account for the contribution of each winding in the solenoid. Eq. (8.49) is mathematically identical to Poisson's equation for electrostatics, given in Sect. 8.3, the difference being its expression in terms of a vector potential.

Computation of magnetostatic forces is similar to that of electrostatic forces produced by induced charges. We can view the magnetostatic forces as acting on the poles of the magnetized material, whose magnitude depends on pole strength and the associated field strength. However, unlike electrostatic forces, magnetostatic forces are considered to be body forces [79, 83, 96]:

$$\mathbf{f}^{\mathbf{M}} = (\mathbf{M} \cdot \nabla)\mathbf{B}, \tag{8.50}$$

where, $\mathbf{f}^{\mathbf{M}}$ denotes the body force density (N/m^3), $\mathbf{M}$ the magnetization, $\mathbf{B}$ the flux density, and ∇ the Nabla operator. Relation (8.50) provides a valid description for forces stemming from an electromagnet or permanent magnet. In the case of a permanent magnet in an applied field, the forces on the poles form a couple, $\tau^{\mathbf{M}}$ (N/m^3), that is given by

$$\tau^{\mathbf{M}} = \mathbf{M} \times \mathbf{B} \tag{8.51}$$

which induces rotational motion for alignment of its magnetization with the applied field. The force or couple is computed by integrating the respective densities, (8.50) or (8.51), over the volume of the magnet. Values of relevant magnetic parameters are given in Table 8.7 [98].

The body force (8.50) can also be represented in terms of a distributed surface force. Since the tractions on the surface of the solid are related to internal stresses through Gauss' theorem, assuming the latter is spatially continuous and differentiable, the total force acting on the surface is obtained by integrating the traction over the surface. Reverting to indicial notation, it reads

$$f_i^M = \int_\Gamma \sigma_{ij}^M n_j \, d\Gamma \qquad \text{with} \quad i,j = 1,2,3. \tag{8.52}$$

Here, Γ denotes the integration surface, n_j is the jth component of the unit vector normal to the surface element $d\Gamma$, and σ^M is the Maxwell stress tensor defined as [96]

$$\sigma_{ij}^M = B_i H_j - \frac{1}{2}\delta_{ij}B_k H_k \qquad \text{with} \quad i,j,k = 1,2,3, \tag{8.53}$$

where $B_{i,k}$ denote the components of the flux density, $H_{j,k}$ the components of the magnetizing field, and δ_{ij} is the Kronecker delta. Eqs. (8.52) and (8.53) are analogous to relations (8.7) and (8.8) obtained for electrostatics. The second term on the right-hand side of (8.53) is representative of the magnetic pressure. Relation (8.53) is only valid in soft magnetic materials, i.e. zero remanent magnetization, $\mathbf{M}_0 = 0$. Note that relations (8.52) and (8.53) also hold for the case of electrodynamic actuation by Lorentz forces discussed in Sect. 8.5.2. When dealing with a conducting material in air, then $\mathbf{B} = \mu_0\mathbf{H}$, and the total force acting on the conductor is reduced simply to an expression in terms of the field value at its surface.

The computational process for self-consistent magnetic micro-actuation is similar to the previous electrostatic and thermomechanical cases, discussed in Sects. 8.3 and 8.4, but with appropriate replacement of strain and force terms. For example, with magnetostrictive micro-actuation, we replace the thermal strain by the magnetostrictive strain. With electro-dynamic and electromagnetic actuation, the computation of electrostatic forces is replaced by the respective Lorentz force and the magnetostatic drive force. In all of these cases, these relations are coupled with the PDEs, boundary conditions, and constitutive relations governing structural behavior. Depending on device structure and operation, effects of fluidic damping have to be taken into consideration. A summary of pertinent equations and coupling terms is given in Sect. 8.8.

8.6 Piezoelectric Actuation

Piezoelectric micro-actuation relies on the inverse piezoelectric effect whereby electrical excitation of a ferroelectric material leads to strain-induced deformation [99]. The nature of the strain, i.e., compressive or tensile, and the resulting displacement (or actuation) depend on the type of material and the magnitude of the underlying coefficient. The latter is determined by the direction of the electric field vector relative to the material's polarization vector. When the two vectors are parallel, there is a tensile strain in the direction of the electric field. If the vectors are anti-parallel, the strain is compressive. Thus the strain dependence on the electric field can be exploited for operation in push-pull mode to yield increased displacement [100]; a schematic of a push-pull piezoelectric bimorph is illustrated in Fig. 8.8.

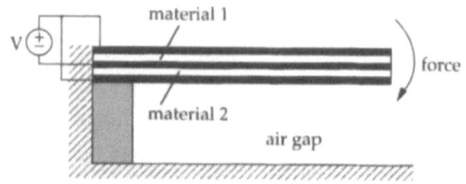

Fig. 8.8 Schematic representation of piezoelectric actuation with a bimorph beam

The bimorph constitutes just one class of piezoelectric microactuators. Various other structures have been reported for a number of applications which include [99–103]: frequency modulation with mass loading in (quartz) resonant structures; microbalances; crystal resonators for frequency control in consumer and other electronics; surface acoustic wave filters; optical signal processing elements; ultrasonic transducers for solids, liquids, and fluids; and more recently, ultrasonic micromotors and force or displacement sensors for microscopy. Device design requirements for these applications, again, can be meaningfully addressed only with numerical simulations. Here, optimization of electro-mechanical coupling requires solutions to the system of equations governing the electrical and mechanical displacement fields. Simple mathematical models enjoy a limited range of applicability. This is particularly true for piezoelectric actuation since the transduction efficiency is governed by material orientation and crystal cut, thus making the coupling coefficients anisotropic.

The piezoelectric and related effects, associated materials, and application areas have been extensively treated in dedicated texts and review articles [53, 99, 100, 102]. Our brief review on the subject will focus on the constitutive relations and material properties/data pertinent to simulation of piezoelectric micro-actuation, including its interaction with the thermal field and the consequence on associated coefficients. The various interactions and their relationships, in the broad class of ferroelectric and pyroelectric materials, are summarized in Table 8.8.

Table 8.8. *Matrix of mechanical-electrical-thermal interactions and coupling coefficients in ferroelectric and pyroelectric materials (adapted from [99]). Diagonal entries reflect primary effects (only the associated variables and coefficients are shown) and the off-diagonal entries describe the coupling effect and associated relation. Notation: σ = stress, ε = strain, C = elastic stiffness, S = elastic compliance, P = electric polarization, d = piezoelectric coefficient, α = thermal expansion coefficient, E = electric field, D = electric displacement, ε = electric permittivity (same notation as strain, but can be distinguished since permittivity appears with the electric field), β = dielectric impermeability, p = pyroelectric coefficient, c = heat capacity/volume, T = temperature, and s = entropy*

	Output signal		
Input signal	Mechanical	Electrical	Thermal
Mechanical	$(\sigma, \varepsilon, c, s)$ $\sigma_{ij} = C_{ijkl}\varepsilon_{kl}$ $\varepsilon_{ij} = S_{ijkl}\sigma_{kl}$	Piezoelectric (d) $P_i = d_{ijk}\sigma_{jk}$	Piezocaloric (α) $\Delta s = \alpha_{ij}\sigma_{ij}$
Electrical	Inverse piezoelectric (d) $\varepsilon_{jk} = d_{ijk}E_i$	$(E, D, \varepsilon, \beta)$ $D_i = \varepsilon_{ij}E_j$ $E_i = \beta_{ij}D_j$	Electrocaloric (p) $\Delta s = p_i E_i$
Thermal	Thermal expansion (α) $\varepsilon_{ij} = \alpha_{ij}\Delta T$	Pyroelectric (p) $\Delta P_i = p_i\Delta T$	$(c/T, T/c, T, s)$ $\Delta s = (c/T)\Delta T$

Here, optical and magnetic interactions have been excluded. Entries on the diagonal reflect primary effects where signal conversion takes place within the same domain. The off-diagonal entries describe coupled interactions. For example, the piezoelectric effect, present only in ferroelectric materials, arises from mechanical energy associated with stress or strain. Conversely, the inverse piezoelectric effect arises from the electric field or electric displacement. The effect is reciprocal; the latter is utilized for actuators, the former for sensors, or both are utilized to provide excitation and sensing functions for closed loop operation. Similarly, the piezocaloric effect and thermal expansion are reciprocal effects. They describe the stress induced change in entropy (heat) or thermally induced change in strain, respectively. The remaining off-diagonal terms are the pyroelectric and electrocaloric effects. These describe the electric polarization induced by a temperature change and generation of entropy by an electric field, respectively.

Ferroelectric materials behave analogously to ferromagnetic materials [99]. Individual domain states may be reoriented by applied electric and/or stress fields. Thus by analogy, ferroelectric materials exhibit spontaneous polarization, remnant polarization, and a coercive field (E_c). However, unlike ferromagnetic materials, ferroelectric materials must have crystallinity. Polycrystalline ferroelectric materials exhibit no piezoelectric effect because of the isotropic orientation of spontaneous polarization in the various grains. Alignment can only be achieved by a so-called "poling process" which involves subjecting the sample to a large electric field ($E > E_c$) at high temperature and subsequently cooling the sample with the field maintained. The resulting attainable piezoelectric strain thus depends on the degree of orientation in individual grains. Relevant crystalline, ceramic, and polymer materials include [104–108] quartz (α-SiO_2), lithium niobate ($LiNbO_3$), barium titanante ($BaTiO_3$), lead titanate ($PbTiO_3$), lead zirconium titanate ($PbZrTiO_3$ or PZT), lead niobate ($PbNbO_3$), zinc oxide (ZnO), and polyvinylidene fluoride (PVF_2 or PVDF). Although the field of piezoelectrics is well-established in terms of bulk material data and applications, the evolution of new ceramic and polymer thin film micro-actuation materials [109–111] is posing new challenges. A key challenge lies in extraction of the various electro-mechanical coupling coefficients, including their influence by thermal interactions. Numerical simulations, and in particular, inverse modeling, can greatly facilitate the extraction process.

8.6.1 Piezoelectric Analysis

According to the linear theory of piezoelectricity coupled with thermodynamic considerations [53], the components of the electric field vector in the material and the components of the strain tensor are linearly related. The coefficients underlying direct and inverse piezoelectric effects are the same. In index notation, both effects can be stated as

$$P_i = d_{ijk}\sigma_{jk} \qquad \text{with} \quad i,j,k = 1,2,3, \tag{8.54}$$

$$\varepsilon_{jk} = d_{ijk}E_i \qquad \text{with} \quad i,j,k = 1,2,3, \tag{8.55}$$

respectively. Here, P_i, E_i, σ_{jk}, and ε_{jk} denote the the components of polarization, electric field, stress, and strain, respectively, and d_{ijk} is a third rank tensor of twenty-seven piezoelectric coefficients. These coefficients provide the means to couple the equations of elasticity to electrostatics. Using the piezoelectric strain (8.55), the associated displacement can be computed using the stress-strain constitutive relations and the equation governing mechanical equilibrium. In the elastic regime, the stress in the material can be stated as

$$\sigma_{ij} = \sigma_{ij}^0 + C_{ijkl}(\varepsilon_{kl} - \varepsilon_{0kl} + d_{mkl}E_m) \qquad \text{with} \quad i,j,k,l,m = 1,2,3, \tag{8.56}$$

which includes the initial stresses (σ_{0ij}) and strains (ε_{0kl}). The divergence of (8.56) yields the force balance from which we compute the displacement field.

We recall that for an anisotropic material in the absence of stress, the electric displacement $D_i = \varepsilon_{ij}E_j$, where ε denotes the permittivity. In the presence of stress, due to the induced polarization P_i, the displacement is modified. Following (8.54), it reads

$$D_i = \varepsilon_{ij}E_j + P_i = \varepsilon_{ij}E_j + d_{ijk}\sigma_{jk} \qquad \text{with} \quad i,j,k = 1,2,3. \tag{8.57}$$

The divergence of (8.57) yields Poisson's equation in the electrostatic potential ψ, where $E_i = -\partial\psi/\partial x_i$, by virtue of the quasi-static approximation. The effective space charge on the right-hand side is modified to include an additional charge term stemming from the divergence of the induced polarization. The divergence of (8.56) and (8.57), taking into consideration the respective source terms, are solved subject to suitable boundary and interface conditions [112]. At electrode regions, the electrostatic potential ψ is prescribed by the applied voltage. At traction-free and displacement-free surfaces, $\sigma_{ij}n_j = 0$ and $u_j = 0$, respectively, where u_j denote the components of mechanical displacement. At the interface between two materials (a and b), we impose continuity conditions on the stress. Additionally, continuity conditions are required for the mechanical displacement, electric displacement, and the electrostatic potential:

$$u_{ja} = u_{jb}; \qquad (D_j n_j)_a = (D_j n_j)_b; \qquad \psi_a = \psi_b,$$
$$\text{with} \quad j = 1, 2, 3, \tag{8.58}$$

where we have assumed an absence of interface charges. Here, n_j denote the components of the surface unit normal and the subscripts a and b denote the respective material regions. At material-air interfaces, for a large value of material permittivity compared to air, the condition $D_j n_j = 0$ is approximately satisfied, where D_j is the electric displacement in the material.

The various terms in relations (8.54) to (8.58) are defined with respect to the crystallographic axes; for simulation purposes, they have to be transformed to the coordinate system of interest. The coefficients d_{ijk} are symmetric in j and k [53], $d_{ijk} = d_{ikj}$. Thus the number of independent coefficients is reduced to eighteen, and (8.54) and (8.55) can be expressed in terms of the engineering (matrix) notation: $P_i = d_{ij}\sigma_j$ and $\varepsilon_j = d_{ij}E_i$, where $i = 1, 2, 3$ and $j = 1, 2, \ldots, 6$. The matrix of piezoelectric coefficients take the form

$$d_{ij} = \begin{bmatrix} d_{11} & d_{12} & d_{13} & d_{14} & d_{15} & d_{16} \\ d_{21} & d_{22} & d_{23} & d_{24} & d_{25} & d_{26} \\ d_{31} & d_{32} & d_{33} & d_{34} & d_{35} & d_{36} \end{bmatrix}, \tag{8.59}$$

which is valid for a general anisotropic material. The number of independent coefficients can be further reduced through symmetry considerations stemming from the nature of crystal structure. With centrosymmetric crystals (e.g., Si), all the d_{ij} vanish. In the case of a class 32 trigonal crystal (e.g., quartz), there are two independent coefficients:

$$d_{ij} = \begin{bmatrix} d_{11} & -d_{11} & 0 & d_{14} & 0 & 0 \\ 0 & 0 & 0 & 0 & -d_{14} & -2d_{11} \\ 0 & 0 & 0 & 0 & 0 & 0 \end{bmatrix}. \tag{8.60}$$

Values of d_{11} and d_{14} and their temperature coefficients along with values of various other coefficients for selected crystalline, ceramic, and polymer bulk and thin film materials, compiled from various sources [104–111, 113–115], are given in Table 8.9. Values of coefficients for thin film materials are, in most cases, notably different from bulk counterparts. In general, the values depend on processing history including the electric field strength, temperature, and time duration of the poling process [99]. In some cases, values are given for an alternate form of the coefficients (e) based on the product of piezoelectric and stiffness coefficients. The elasticity coefficients given for quartz have been obtained under constant electric

Table 8.9. *Values of relevant coefficients gathered from various sources for bulk and thin film materials. Values shown are drawn directly from the references cited or from sources given therein*

Material	Relative permittivity	Piezoelectric coefficients d (pC/N), e (C/m^2)	Elasticity coefficients (GN/m^2)
α-SiO$_2$ [104]	$\varepsilon_{r11} = 4.5$ $\varepsilon_{r22} = 4.5$ $\varepsilon_{r33} = 4.6$	$d_{11} = -2.31$; -200 to -350 ppm/°C $d_{14} = -0.727$; 1290 to 1770 ppm/°C	$C_{11} = 86.80 \pm 0.04$; -46.4 ± 2.1 ppm/°C $C_{33} = 106.2 \pm 0.8$; $\quad \cdot \quad -172 \pm 24$ ppm/°C $C_{12} = 7.10 \pm 0.09$; -2901 ± 183 ppm/°C $C_{13} = 11.91 \pm 0.01$; -476 ± 111 ppm/°C $C_{44} = 58.17 \pm 0.13$; -170 ± 13 ppm/°C $C_{66} = 39.85 \pm 0.03$; 177 ± 8 ppm/°C $C_{14} = -18.02 \pm 0.07$; 105 ± 8 ppm/°C
LiNbO$_3$ [105]	$\varepsilon_{r11} = 44.3$ $\varepsilon_{r33} = 27.9$	$e_{15} = 3.76$, $e_{22} = 2.43$ $e_{31} = 0.23$, $e_{33} = 1.33$	$C_{11} = 203$, $C_{33} = 242.4$ $C_{12} = 57.3$, $C_{13} = 75.2$ $C_{44} = 59.5$, $C_{14} = 8.5$
BaTiO$_3$ [106]	$\varepsilon_{r11} = 1268$ $\varepsilon_{r33} = 1419$	$d_{15} = 270$, $d_{31} = -79$, $d_{33} = 191$	$C_{11} = 166$, $C_{33} = 162$ $C_{12} = 76.6$, $C_{13} = 77.5$ $C_{44} = 42.9$, $C_{66} = 44.8$
PZT	$\varepsilon_{r33} = 1800$ [113] $\varepsilon_{r33} = 1000$ [114]	$d_{31} = -205$, $d_{33} = 450$ [113] $d_{33} = 282$ [114]	80 [107]
sol-gel	1300 [103] 800–1100 [109]	$d_{31} = -88.7$, $d_{33} = 220$ [103] $d_{33} = 190$–250 [109]	$C_{33} = 10$–40 [109]
sputtered	930 [110]	$d_{31} = -49$ [110] $d_{31} = -100$ [111]	75 [111]
ZnO [109] sputtered	10.8–11	$d_{33} = 10.5$–11.5	$C_{33} = 140$–230
PVDF	10–15 [107], [115] 10–13 [108]	$d_{31} = 18$–30, $d_{32} = 2$–3, $d_{33} = -30$ [107] $d_{33} = -18$ [108]	1–3 [107] 2–5 [108]

field and adiabatic conditions. The conditions of measurement, in principle, have an influence on the retrieved values, although this is significant only for the thermal coupling coefficients. Thus no distinction is made in values obtained for different measurement conditions. This is further discussed in Sect. 8.6.2. Unless stated otherwise, all values quoted are for $T = 300$ K.

Also given are their first order temperature coefficients. In addition to the data given in Table 8.9, a comprehensive list of values related to the thermophysical (e.g., specific heat, thermal expansion, thermal conductivity) and optical properties of quartz can be found in [104].

8.6.2 Electro-Thermo-Mechanical Interactions and Coupling Coefficients

In the previous Section, we dealt with electrical-mechanical interactions without coupling to the thermal field. However, due to the nature of associated materials, the thermal field has to be considered as an integral part of the analysis for precise formulation of the coefficients and their interrelations for different conditions of measurement. This has a bearing on the accuracy of retrieved values, and hence, on the simulation results. For example, the values of elastic coefficients can vary when retrieved under constant entropy (adiabatic) or under isothermal conditions. In this case, the difference is less than 1%, which is why no distinction was made between measurement conditions in the values given in Table 8.9. We next summarize the key results that illustrate the coupling between electrical, thermal, and mechanical signals. A comprehensive treatment of the coupling and associated accuracy in retrieved coefficients is given in [53].

Following the notation used in Table 8.8, if we choose the input (or independent) variables as (σ_{ij}, E_i, T) and the output (or dependent) variables as $(\varepsilon_{ij}, D_i, s,$ etc.), equilibrium thermodynamics provides us with the following relations:

$$d\varepsilon_{ij} = \left(\frac{\partial \varepsilon_{ij}}{\partial \sigma_{kl}}\right)_{E,T} d\sigma_{kl} + \left(\frac{\partial \varepsilon_{ij}}{\partial E_k}\right)_{\sigma,T} dE_k + \left(\frac{\partial \varepsilon_{ij}}{\partial T}\right)_{\sigma,E} dT, \qquad (8.61)$$

$$dD_i = \left(\frac{\partial D_i}{\partial \sigma_{jk}}\right)_{E,T} d\sigma_{jk} + \left(\frac{\partial D_i}{\partial E_j}\right)_{\sigma,T} dE_j + \left(\frac{\partial D_i}{\partial T}\right)_{\sigma,E} dT, \qquad (8.62)$$

$$ds = \left(\frac{\partial s}{\partial \sigma_{ij}}\right)_{E,T} d\sigma_{ij} + \left(\frac{\partial s}{\partial E_i}\right)_{\sigma,T} dE_i + \left(\frac{\partial s}{\partial T}\right)_{\sigma,E} dT, \qquad (8.63)$$

which give rise to differential coefficients that are not independent. Each of the Eqs. (8.61), (8.62), and (8.63) corresponds to the effects listed in the second, third, and fourth columns of Table 8.8, respectively. For example, the coefficients on the right-hand side of (8.61) describe, in order, the elastic compliance, the inverse piezoelectric effect, and thermal expansion. Similarly, (8.62) describes the direct piezoelectric effect, the permittivity, and the pyroelectric effect. Finally, the piezocaloric effect, electrocaloric effect, and heat capacity are described by (8.63). Since the coefficients of the respective off-diagonal terms are the same, we can now assemble the complete list of coupled constitutive relations. Following (8.61) to (8.63), we obtain

$$\varepsilon_{ij} = S_{ijkl}^{E,T} \sigma_{kl} + d_{ijk}^T E_k + \alpha_{ij}^E \Delta T, \tag{8.64}$$

$$D_i = d_{ijk}^T \sigma_{jk} + \varepsilon_{ij}^{\sigma,T} E_j + p_i^\sigma \Delta T, \tag{8.65}$$

$$\Delta s = \alpha_{ij}^E \sigma_{ij} + p_i^\sigma E_i + \frac{c^{\sigma,E}}{T} \Delta T, \tag{8.66}$$

where ε_{ij} in (8.64) denotes the strain tensor and $\varepsilon_{ij}^{\sigma,T}$ in (8.65) the permittivity tensor. The superscripts denote the independent variable that is held constant. Thus we clearly see from (8.64) to (8.66), how the coefficients depend on measurement conditions. For example, a constant T implies isothermal conditions, a constant E and in particular, $E = 0$ implies an equipotential condition, and a constant σ implies a mechanically stress free state. The relative difference in values of most coupling coefficients under the different measurement conditions, viz.,

adiabatic and isothermal: $\quad d_{ijk}^s - d_{ijk}^T = -p_i^\sigma \alpha_{jk}^E (T/c^{\sigma,e}),$

$$\tag{8.67}$$

electrically free and clamped: $\quad \alpha_{ij}^D - \alpha_{ij}^E = -d_{kij}^T \beta_{kl}^{\sigma,T} p_l^\sigma,$

$$\tag{8.68}$$

mechanically free and clamped: $\quad p_i^\varepsilon - p_i^\sigma = -\alpha_{jk}^E C_{jklm}^{E,T} d_{ilm}^T,$

$$\tag{8.69}$$

are generally of the order of, or less than, 1%. However, a relative difference of the order of 100% can exist for pyroelectric coefficients, (8.69), retrieved under constant strain and constant stress conditions. The pyroelectric effect for a mechanically free material under constant E can be stated in the following form [53]

$$\left(\frac{\partial D}{\partial T}\right)_\sigma = \left(\frac{\partial D}{\partial T}\right)_\varepsilon + \left(\frac{\partial D}{\partial \sigma}\right)_T \left(\frac{\partial \sigma}{\partial \varepsilon}\right)_T \left(\frac{\partial \varepsilon}{\partial T}\right)_\sigma, \tag{8.70}$$

where the first and second term on the right-hand side denote the primary and secondary pyroelectric effects, respectively. The latter, which a function of the piezoelectric, elastic stiffness, and thermal expansion coefficients, shows that an electric displacement can also be generated by virtue of a combination of thermal expansion and the direct piezoelectric effects. However, the pyroelectric effect via this route would depend on material properties. For example, because of crystal symmetry quartz does not exhibit a net pyroelectric behavior under an isothermal temperature

Table 8.10. *Compilation of pyroelectric coefficients for bulk and thin film materials. Values shown are drawn from the references given or from sources cited within*

Material	Pyroelectric coefficients ($\mu C\ K^{-1}\ m^{-2}$)
BaTiO$_3$ [107]	200
PZT [107]	50–300
sol-gel PZT [109]	500–700
PVDF	30–40 [107], 40 [108], 30 [116], 24–28 [115]
ZnO [116]	$p^\sigma = -9.4,\ p^\varepsilon = -6.9$
thin film ZnO [109]	9.5–10.5
CdS [116]	$p^\sigma = -4,\ p^\varepsilon = -2.97$
CdSe [116]	$p^\sigma = -3.5,\ p^\varepsilon = -2.94$

change. Values of pyroelectric coefficients for selected materials, gathered from various sources [107–109, 115, 116], are given in Table 8.10.

8.7 Electroacoustic Transducers

Electroacoustic sensors and actuators rely on the piezoelectric effect to transform mechanical (acoustic) energy into electrical energy and vice versa. The acoustic energy from an actuator, with proper impedance matching, can be transmitted through solids and fluids. Depending on frequency and energy of the emitted acoustic field, electroacoustic actuators are applied in a variety of areas which include ultrasonic cleaning, welding, soldering, biomedical probing including microsurgery, sonar, and sonochemistry (see, e.g., [99]). The acoustic wave is extremely sensitive to changes in material properties, such as mass, elasticity, viscosity, permittivity and conductivity, of the propagation medium. Thus changes in material properties, induced by a physical or chemical signal, lead to a modulation of the amplitude and/or velocity of the acoustic wave, which makes it very useful for measurement of physical and chemical variables, such as temperature, pressure, acceleration, phase transitions, and gas concentrations, including realization of chemical and biological measurement systems [117, 118].

Design approaches based on analytical [119] or semi-numerical models [120] of electroacoustic interactions have restricted applicability; they are generally valid only for simple (limiting) device geometries, and in most cases, they are unable to account for the influence of higher order acoustic interactions on device performance [121, 122]. Thus design by numerical simulations, in at least two spatial dimensions, is necessary. In particular, critical issues which have a strong bearing on device optimization include:

Table 8.11. *Characteristics and associated material orientation and cut for various types of acoustic waves propagating in solids*

Type of wave		Characteristic	Material and orientation/cut
Bulk		Longitudinal or shear wave	AT-cut quartz: stable shear mode over wide temperatures
Surface generated	Surface acoustic waves (SAW)	Shear horizontal mode or Bleustein-Gulyaev wave	Exists only in selected piezoelectric materials (e.g. $KTiOPO_4$ or KTP)
	Love modes	Conversion of shear surface skimming bulk waves (SSBW) to waveguide modes in a layer	Shear wave velocity in layer smaller than shear velocity of substrate; SSBW are most notable in 90° rotated Y-cut quartz
	Leaky SAW	Attenuates in the direction of propagation with growth of a bulk wave	Selected Y-rotated X-propagating cut in $LiNbO_3$ and $LiTaO_3$; predominantly a shear displacement component in the plane of the crystal
	Acoustic plate modes (APM)	Shear horizontal, shear vertical, or longitudinal displacements	Shear horizontal APM with displacement parallel to the crystal plane are predominant in ST-cut quartz and ZX cut $LiNbO_3$
	Lamb waves	Excited in finite plate thicknesses	Waves on both plate surfaces interact to yield two Lamb plate modes; a symmetric mode which behaves like a SAW and an antisymmetric mode whose velocity vanishes as plate thickness is decreased

- Interactions of different acoustic modes (e.g., bulk and surface waves) [123]; anisotropies in the piezoelectric material (as seen in Sect. 8.6), which can introduce unusual diffraction and beam steering effects [124], thus requiring optimization with respect to orientation or crystal cut.
- Acoustic feedback stemming from interactions of the transducer with the surrounding media (e.g., liquids and gases) [121, 122].
- Minimization or elimination of non-useful acoustic modes and associated power radiation to maximize transduction efficiency.

We begin Sect. 8.7.1 with a short recapitulation of the various modes associated with acoustic wave propagation in solids summarized in Table 8.11. Next we establish the model equations, including constitutive relations and boundary conditions pertinent to the analysis of acoustic wave propagation in solids. Interactions of the acoustic wave with the surrounding media are reviewed in Sect. 8.7.2.

8.7.1 Acoustic Wave Propagation in Solids

Various types of acoustic waves can be distinguished and depending on their nature of propagation through a solid, they can be broadly classified as bulk waves or surface generated waves [117]. The latter are electrically excited, e.g., through interdigital electrodes (IDT), on the surface of a piezoelectric material. While bulk acoustic waves (BAW) propagate through the bulk of a solid, surface generated waves travel, with relatively lower acoustic velocities, on or near the surface of the material. Most of the acoustic energy lies within a depth of one wavelength from the surface and the wave amplitude decays exponentially with depth from the surface. There are many forms of surface generated waves [117] (see Table 8.11). These include surface acoustic waves (SAW), Love modes, leaky SAW, acoustic plate modes, and Lamb waves; their relative significance is determined by electrode geometry and material-type, orientation, cut, and geometry of the substrate.

Mechanical displacements stemming from the compressive and dilational stress field associated with acoustic wave propagation in the solid are coupled to the electromagnetic excitation fields by the piezoelectric coefficients. Since the sonic velocities associated with the acoustic wave are about five orders of magnitude smaller than the corresponding electromagnetic wave velocity, the governing electric field variable can be treated as quasi-static even in the dynamical sense. Thus, displacement currents can be neglected and the electric field can be represented as the gradient of the scalar potential, $E_i = -\partial \psi / \partial x_i$. In what follows, we adopt the procedures used in Sect. 7.2 in modeling the acoustic wave propagation in a solid. Denoting the mechanical displacement of a point in the solid by u_j, the velocity of the point in continuum is given by $v_j = \partial u_j / \partial t$. If the density ($\rho$) of the solid is assumed constant, the force density acting on the point is given as $\rho \partial^2 u_j / \partial t^2$, and following the conservation of momentum, we obtain, in the absence of external forces,

$$\frac{\partial \sigma_{ij}}{\partial x_i} = \rho \frac{\partial^2 u_j}{\partial t^2} \quad \text{with} \quad i, j = 1, 2, 3, \tag{8.71}$$

where σ_{ij} is the second rank symmetric stress tensor. The constitutive relations governing the coupling between elasticity and electrostatics, as given by (8.56) and (8.57) in Sect. 8.6.1, can be restated in a slightly modified form

$$\sigma_{ij} = C_{ijkl} \varepsilon_{kl} - e_{kij} E_k \quad \text{with} \quad i, j, k, l = 1, 2, 3, \tag{8.72}$$

$$D_i = \varepsilon_{ij} E_j + e_{ikl} \varepsilon_{kl} \quad \text{with} \quad i, j, k, l = 1, 2, 3, \tag{8.73}$$

where ε_{kl} in (8.72) denotes the strain tensor and ε_{ij} in (8.73) the permittivity tensor. In (8.72) and (8.73), we use an alternate form of the piezoelectric tensor (e) which stems from the product of the stiffness (C) and piezoelectric coefficients (d), and we have omitted, for convenience, the terms containing initial stresses and strains. The strain tensor is determined by the spatial gradient of mechanical displacement. Restating (7.13)

$$\varepsilon_{ij} = \frac{1}{2}\left(\frac{\partial u_i}{\partial x_j} + \frac{\partial u_j}{\partial x_i}\right) \qquad \text{with} \quad i,j = 1,2,3, \tag{8.74}$$

which is based on the infinitesimal strain approximation. Relations (8.71) to (8.74), in addition to Gauss' law, $\partial D_i/\partial x_i = 0$, yield the following systems of equations that govern acoustic wave propagation in a solid [125]:

$$C_{ijkl}\frac{\partial^2 u_k}{\partial x_l \partial x_i} + e_{kij}\frac{\partial^2 \psi}{\partial x_k \partial x_i} = \rho\frac{\partial^2 u_j}{\partial t^2} \qquad \text{with} \quad i,j,k,l = 1,2,3, \tag{8.75}$$

$$\varepsilon_{ij}\frac{\partial^2 \psi}{\partial x_i \partial x_j} = e_{kij}\frac{\partial^2 u_i}{\partial x_j \partial x_k} \qquad \text{with} \quad i,j,k = 1,2,3, \tag{8.76}$$

where ε_{ij} in (8.76) denotes the permittivity tensor.

The boundary conditions associated with the system of Eqs. (8.75) and (8.76) are identical to those stated in Sect. 8.6.1 for the electrostatic potential ψ and Sect. 7.2.1 for the stress and displacement. At traction-free and displacement-free surfaces, we assume a vanishing normal component of force at the surface; $\sigma_{ij}n_j = 0$ and $u_j = 0$. Similarly, a vanishing electrical displacement at the surface yields $D_j n_j = 0$. At material interfaces, continuity conditions on the stress and mechanical displacement, as described by relation (8.58) in Sect. 8.6.1, hold.

8.7.2 Interactions with Ambient Fluid

The interaction of the transducer with the ambient acoustic media can have a strong bearing on the dynamic performance of the transducer, due to mass loading and damping effects. To account for the acoustic feedback by the surrounding fluid, the electroacoustic equations, (8.75) and (8.76), in the solid have to be solved in conjunction with the equation governing propagation of the acoustic wave in the fluid. For an inviscid (i.e., non-viscous) fluid, the acoustic wave propagation can be described in terms of a scalar velocity potential ϕ, viz., $v_i = \partial\phi/\partial x_i$. The associated acoustic pressure is then $p = \rho_0 \partial\phi/\partial t$, where ρ_0 is the static density of the fluid. The momentum equation for inviscid fluids along with the continuity equation for mass (see Sect. 7.3) yields the following equation which governs acoustic wave propagation in the fluid [126]

$$\frac{\partial^2 \phi}{\partial t^2} + \frac{\partial}{\partial t}\left(\frac{\partial \phi}{\partial x_i}\right)^2 + \frac{1}{2}\left(\frac{\partial \phi}{\partial x_i}\right)\frac{\partial}{\partial x_i}\left(\frac{\partial \phi}{\partial x_j}\right)^2 - c^2\frac{\partial^2 \phi}{\partial x_i \partial x_i} = 0 \qquad (8.77)$$

with $\quad i,j = 1,2,3.$

Here, c denotes the sound velocity in the fluid.

Under certain conditions, we can reduce Eq. (8.77) to well-known forms. For small wave amplitudes, terms in $(\partial \phi/\partial x_i)^2$ vanish and we obtain

$$\frac{\partial^2 \phi}{\partial x_i \partial x_i} = \frac{1}{c^2}\frac{\partial^2 \phi}{\partial t^2} \qquad \text{with} \quad i = 1,2,3. \qquad (8.78)$$

For incompressible fluids, $c \rightarrow \infty$ and we obtain Laplace's equation in ϕ;

$$\frac{\partial^2 \phi}{\partial x_i \partial x_i} = 0 \qquad \text{with} \quad i = 1,2,3. \qquad (8.79)$$

In the case where wave propagation is steady, (8.77) reduces to

$$v_i v_j \frac{\partial^2 \phi}{\partial x_i \partial x_j} = c^2 \frac{\partial^2 \phi}{\partial x_i \partial x_i} \qquad \text{with} \quad i,j = 1,2,3, \qquad (8.80)$$

where, for convenience, we have reintroduced the velocity components. Equation (8.80) can have a very distinct character depending on the value of v relative to c [126]. For $v < c$ (Mach number, $M < 1$), the equation is elliptic with diffusive character, and for $v > c$ ($M > 1$), the equation becomes hyperbolic with wave character. Here, v denotes the magnitude of the velocity vector.

Eq. (8.78), which serves as the case of interest, together with (8.75) and (8.76) constitute the system that governs acoustic interactions between the solid and fluid [121] for an inviscid medium; air being a possible example. At the solid-fluid interface, we assume that the force is continuous. Thus the normal stress components at the solid surface are equated to the pressure of the fluid at the interface. At the outer regions of the fluid, the value of the scalar acoustic potential has to be judiciously chosen so as to not exclude possible damping by Sommerfield radiation and mass loading effects [121].

8.8 Computational Procedure and Coupling

The computational process for self-consistent solutions of the various mixed-signal interactions described in Sects. 8.3 to 8.7, including fluidic

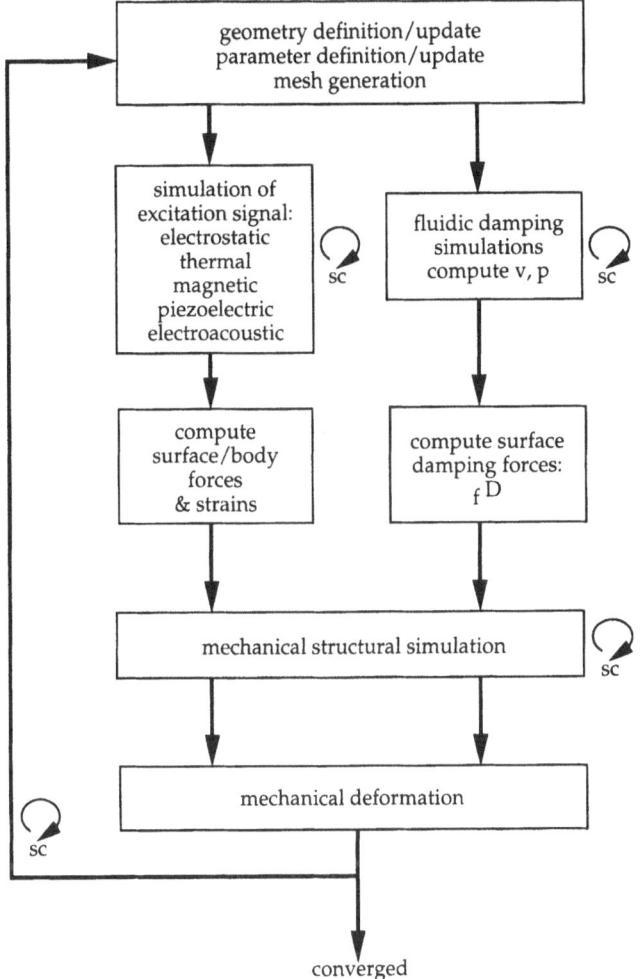

Fig. 8.9 Simulation procedure for mixed-signal interactions; *sc* denotes iterations for self-consistent solutions. Corresponding equation systems and coupling terms are given in Tables 8.12 and 8.13

damping effects, is summarized in Fig. 8.9. This is accompanied with Tables 8.12 and 8.13 which draw reference to model equations, constitutive relations, boundary conditions, and coupling terms pertinent to the different types of micro-actuation. Here, we can adopt one of two simulation approaches. The first lies in linking well supported commercial software packages that are highly specialized in the one or the other task, i.e., geometry definition and mesh generation, simulation of the excitation signal, simulation of fluidic damping, and mechanical structural simulation, with the coupling between variables performed externally [25]. Specialized

Table 8.12. *Summary of model equations pertinent to the various excitation signals corresponding to Fig. 8.9*

Excitation	Governing equation	Constitutive relations/material data	Boundary conditions	Solution variable(s)	Coupling terms
Electrostatic Sect. 8.3	(8.1)	Table 2.2	(8.3)–(8.6)	Electrostatic potential ψ Surface charge density σ	Electrostatic force (8.7) Electrostatic pressure (8.9)
Thermal Sect. 8.4	(8.14) (8.17)	Electrical: Chapt. 2 Thermal: Chapts. 5 and 7 Table 8.3	(8.3), (8.19)–(8.24) See Sect. 5.5	Electrostatic potential ψ Temperature T	Temperature T Thermal strain (7.15), (8.25)
Magnetic Sect. 8.5	Magnetostatics: (8.44) (8.49)	Table 8.4 Table 8.6 Table 8.7	Magnetostatics: (8.45)–(8.48)	Magnetostatics: Magnetic vector potential $\mathbf{A}$	Magnetostrictive strain (8.33) Lorentz force (8.36), (8.38), or (8.52) Magnetostatic force (8.50) or (8.52) Magnetostatic couple (8.51)
Piezoelectric Sect. 8.6	(8.2)	(8.57) Table 8.9	(8.58)	Electrostatic potential ψ	Piezoelectric strain (8.55)
Electroacoustic and fluid interactions	(8.75) (8.76) (8.78)	Table 2.2 Chapt. 7 Table 8.9	(8.58) See Section 8.7.2	Electrostatic potential ψ Mechanical displacement u Wave velocity in fluid v	

Table 8.13. *Summary of model equations pertinent to damping and structural analysis corresponding to Fig. 8.9*

	Governing equation	Constitutive relations/ material data	Boundary conditions	Solution variables(s)	Coupling terms
Fluidic damping	Viscous drag: (7.103), (7.115)	Table 7.4	(7.112), (7.113)	Velocity v Pressure p	Viscous drag damping force: (7.100), (7.116)
	Squeeze film: (7.118) or (7.121)	Table 7.4	(7.122), (7.123)	Velocity v Pressure p	Squeezed film damping force: (7.124)
Mechanical structural simulation	(7.29) (7.32)	(7.13) (7.33), (7.35), or (7.43) Tables 7.1, 7.2, 7.3	(7.44)–(7.50)	Displacement u Natural frequency ω	Stress σ Strain ε Displacement u Natural frequency ω

tools for the various simulation modules are given in Table 8.2. Such an approach appears attractive since each tool is independently optimized in terms of computational efficiency for the given simulation task. For example, mechanical structural analysis is based on finite elements while electrostatic simulations are based on boundary elements. In particular, the latter technique avoids numerical computations in the relative large charge-free bulk regions. Afterall to compute electrostatic coupling forces, only the field at the boundaries is relevant. However, several limitations are apparent with linking different simulation packages. The input/output data structures and numerical meshes employed in the different tools may be incompatible, which necessitates use of suitable data conversion routines. Since the coupling takes place external to each package, this could lead to a large number of iterations for self-consistent solutions, and in some cases, due to the strong intrinsic non-linearity, convergence may not even be possible.

These limitations can be overcome with a single integrated software tool (e.g., [37]) that is tailored specifically to deal with non-linearities in such coupled systems and associated convergence issues, yet benefiting from the use of computational algorithms highly optimised for the one or other energy domain. For example, in electrostatic micro-actuation, the coupling can be implemented internally through use of a linearized form of the force-displacement relation [41, 127] and iterated till convergence within an inner loop. This type of a Newton update for the surface displacement can dramatically decrease the number of non-linear iterations necessary for a self-consistent solution.

8.9 Illustrative Example – CMOS Micromirror

By way of example, we illustrate the electrostatically deflectable micromirror device fabricated in CMOS technology [40, 128, 129]. The more advanced version of the micromirror [130] enjoys demonstrated success in video projection displays; it is capable of switching at high speeds and lends itself to high integration density with over two million micromirrors on a single chip. Fig. 8.10 shows a scanning electron micrograph (SEM) of CMOS aluminium (Al) micromirror along with a device schematic illustrating its operating principle. The free standing micromirror, supported by two thin Al hinges, is suspended over two independently addressable electrodes located on the chip surface.

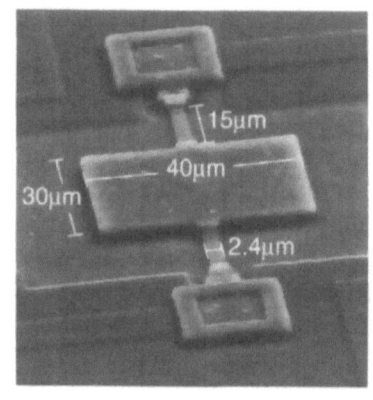

(a)

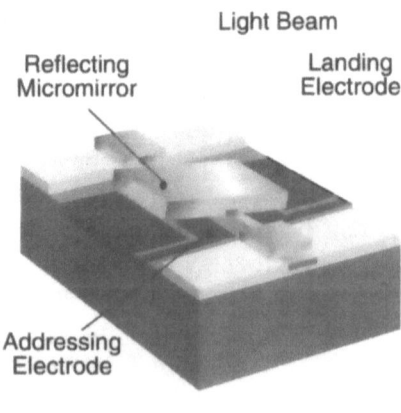

(b)

Fig. 8.10 (a) Scanning electron micrograph of CMOS micromirror and (b) schematic illustrating its operating principle. Source [40, 128]

Depending on the bias state of electrodes, the micromirror deflects, in a controlled manner, to reflect incident light.

Micromirror design for efficient, stable, and reliable operation requires numerical optimization of device structure and geometry to accommodate the large number of critical design requirements. These include: high optical-quality-flatness of mirror regardless of deflection state to avoid divergence in the reflected optical beam; high fill factor for high projection quality; high deflection angles without establishing contact to landing electrodes; low threshold voltage to minimize dissipative energy; assurance that maximum stresses in the structure do not exceed the yield stress limit;

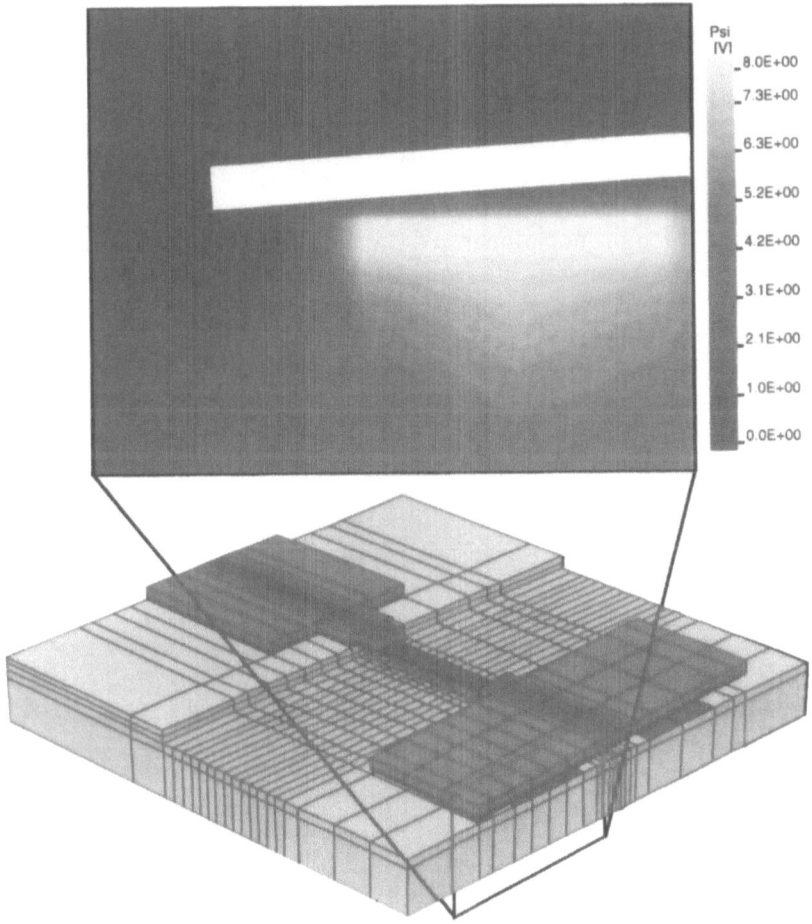

Fig. 8.11 Numerical mesh and simulated electrostatic potential distribution in air gap region. Source [40]

and low voltage operation to maintain compatibility with CMOS technology.

The three-dimensional simulation mesh and the computed electrostatic potential distribution (viewed at the cross section of the micromirror in Fig. 8.10) are shown in Fig. 8.11 for a bias voltage of 8 V on one addressing electrode. Due to device symmetry, only half of the micromirror is shown. We see a significant drop in electrostatic potential within the air gap region between the mirror and electrode. For the given structure, fringing fields are notable only at the vicinity of the hinges. The simulations are based on a finite element scheme for both electrostatic and mechanical domains along with a smoothing procedure for the electric field to avoid possible numerical artifacts of spurious charges, and hence forces, at element boundaries. This is due to the inherent approximation of piecewise constant electric fields associated with linear finite elements. Comparison of simulations with measurements for mirror deflection as a function of bias voltage is shown in Fig. 8.12; the worst case discrepancy is approximately 10%. Here, the threshold limit signifies the voltage at which the electrostatic force exceeds the reaction force associated with torsional rigidity of hinges, leading to contact of the micromirror with the landing electrode. Although not illustrated here, simulations indicate that the maximum stress in support hinges at full deflection is well below the yield stress value for bulk Al and that a minimum Al thickness of $1.0\,\mu m$, which is readily available with multi-level CMOS technology, provides the necessary optical-quality flatness of the micromirror even in the deflected state. The simulation results shown do not include the effects of fluidic damping.

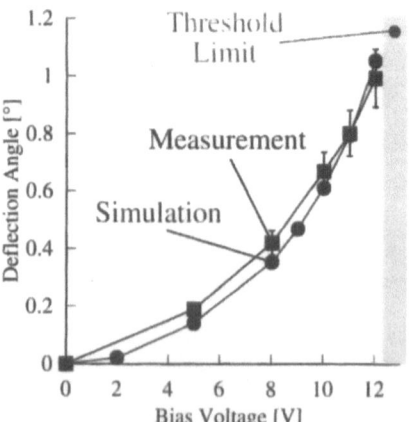

Fig. 8.12 Comparison of simulated and measured micromirror deflection angle as a function of bias voltage. Source [40]

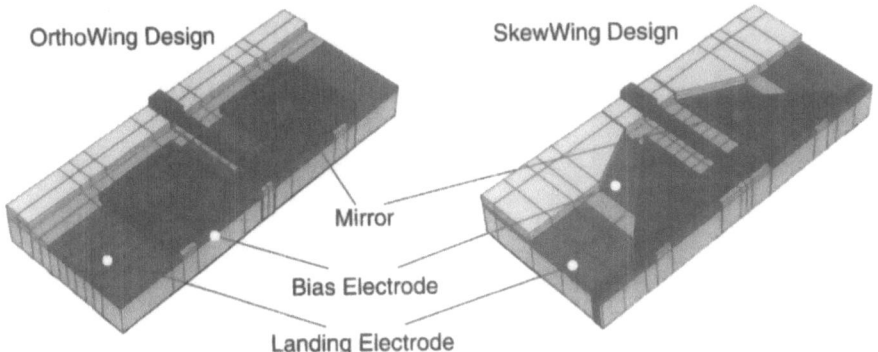

Fig. 8.13 Structural variants of the regular wing micromirror design in Fig. 8.10 used in fill factor and threshold voltage optimization. Source [40]

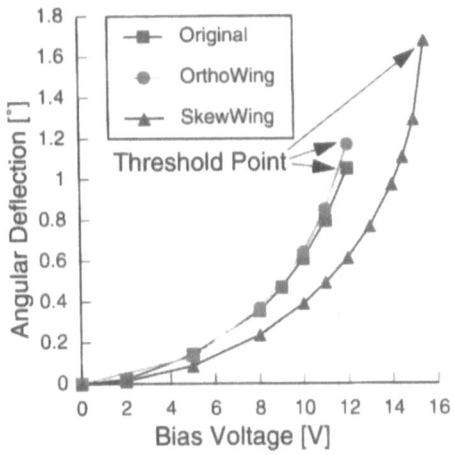

Fig. 8.14 Simulated comparison of deflection behavior for the various micromirror designs. Values of fill factor and threshold voltage are given in Table 8.14. Source [40]

Fill factor, defined as the ratio of the optically active to total device area, is an important design consideration as it determines the quality of the projected image. Selected structural variants (Fig. 8.13) of the micromirror shown in Fig. 8.10 were simulated for the fill factor, along with the threshold voltage. The values are given in Table 8.14 along with the deflection-voltage behavior in Fig. 8.14. The numerical study shows a clear preference for the OrthoWing design in view of its high fill factor and low threshold voltage. These, and other considerations, clearly demonstrate how numerical simulations can benefit the designer, particularly, when dealing with standard IC technologies in view of a constrained design space

Table 8.14. *Simulated values for different CMOS micro-mirror structures [40]*

Design	Fill factor (%)	Threshold voltage (V)
Original	53	11.8
OrthoWing	73	11.5
SkewWing	72	15.0

arising from the restricted selection in materials, material thicknesses and associated physical properties.

8.10 References

[1] Middelhoek, S., Audet, S. A., *Silicon Sensors*, New York: Academic Press, 1989.
[2] Grandke, T., Hesse, J., Introduction, in: *Sensors*, Vol. 1, Grandke, T., Ko, W. H. (Eds.), Weinheim: VCH, 1989, pp. 1–16.
[3] Trimmer, W. S. N., Microrobots and Micromechanical Systems, *Sensors and Actuators*, 19 (1989), 267–287.
[4] Elwenspoek, M., Blom, F. R., Bouwstra, S., Lammerink, T. S. J., Popma, Th. J. A., Fluitman, J. H. J., Transduction Mechanisms and Their Applications in Micromechanical Devices, *Digest of Technical Papers*, Transducers '89, Montreux, 1989, pp. 126–132.
[5] Muller, R. S., Microdynamics, *Sensors and Actuators*, A21–A23 (1990), 1–8.
[6] Muller, R. S., Howe, R. T., Senturia, S. D., Smith, R. L., White, R. M. (Eds.), *Microsensors*, New York: IEEE Press, 1991.
[7] Fujita, H., Gabriel, K. J., New Opportunities for Micro Actuators, *Digest of Technical Papers*, Transducers '91, San Francisco, 1991, pp. 14–20.
[8] Benecke, W., Silicon-Microactuators: Activation Mechanisms and Scaling Problems, *Digest of Technical Papers*, Transducers '91, San Francisco, 1991, pp. 46–50.
[9] Sze, S. M. (Ed.), *Semiconductor Sensors*, New York: Wiley, 1994.
[10] Baltes, H., Future of IC Microtransducers, *Sensors and Actuators A*, 56 (1996), 179–192.
[11] Baltes, H., Paul, O., Korvink, J. G., Schneider, M., Bühler, J., Schneeberger, N., Jaeggi, D., Malcovati, P., Hornung, M., Häberli, A., von Arx, M., Mayer, F., Funk, J., IC MEMS Microtransducers, *Technical Digest*, IEEE IEDM, San Francisco, 1996, pp. 521–524.
[12] Hunter, I. W., Lafontaine, S., A Comparison of Muscle with Artificial Actuators, *Technical Digest*, IEEE Solid-State Sensor and Actuator Workshop, Hilton Head Is., 1992, pp. 178–185.
[13] Quandt, E., Holleck, H., Materials Development for Thin Film Actuators, *Microsystem Technologies*, 1 (1995), 178–184.
[14] Lu, Y., Nathan, A., Manku, T., Ning, Y., Thin Film Magnetostrictive Sensor with On-Chip Readout and attoFarad Capacitance Resolution, *Technical Digest*, IEEE IEDM, San Francisco, 1996, pp. 777–780.
[15] *Proceedings, IEEE Micro Electro Mechanical Systems Conference*, Nagoya, 1997.
[16] *Digest of Technical Papers*, Transducers '97, Chicago, 1997.

[17] Prak, A., Elwenspoek, M., Fluitman, J. H. J., Selective Mode Excitation and Detection of Micromachined Resonators, *IEEE J. Microelectromechanical Systems*, 1 (1992), 179–186.

[18] Prak, A., Lammerink, T. S. J., Fluitman, J. H. J., Review of Excitation and Detection Mechanisms for Micromechanical Resonators, *Sensors and Materials*, 5 (1993), 143–181.

[19] Korvink, J. G., Funk, J., Baltes, H., IMEMS Modeling, *Sensors and Materials*, 6 (1994), 235–243.

[20] A. Nathan, (Ed.), *Special Issue on Microsensor Modeling, Sensors and Materials*, Vol. 6, Nos. 2–4 (1994).

[21] Nathan, A., Microtransducer CAD, *Proc. ESSDERC '96*, Baccarani, G., Rudan, M. (Eds.), Bologna, 1996, pp. 707–715.

[22] Korvink, J. G., Bächtold, M., Emmenegger, M., Paganini, R., Ruehl, R., Funk, J., Baltes, H., TCAD for MEMS, *Proc. ESSDERC '96*, Baccarani, G., Rudan, M. (Eds.), Bologna, 1996, pp. A5–A7.

[23] Korvink, J. G., Baltes, H., Microsystem Modeling, in: *Sensors Update*, Baltes, H., Göpel, W., Hesse, J. (Eds.), Chapt. 6, Weinheim: VCH, 1996, pp. 181–209.

[24] Baltes, H., Korvink, J. G., Paul, O., Numerical Modelling and Materials Characterization for Integrated Micro Electro Mechanical Systems, in: *Simulation of Semiconductor Devices and Processes*, Vol. 6, Ryssel, H., Pichler, P., (Eds.), Wien-New York: Springer-Verlag, 1995, pp. 1-9.

[25] Senturia, S. D., Harris, R. M., Johnson, B. P., Kim, S., Nabors, K., Shulman, M. A., White, J. K., A Computer-Aided Design System for Microelectromechanical Systems (MEMCAD), *IEEE J. of Microelectromechanical Systems*, 1 (1992), 3–14.

[26] Allegretto, W., Nathan, A., Baltes, H., Numerical Analysis of Magnetic-Field-Sensitive Bipolar Devices, *IEEE Trans. CAD of ICAS*, 10 (1991), 501–511.

[27] Trimmer, W. S. N., Gabriel, K. J., Design Considerations for a Practical Electrostatic Micro-Motor, *Sensors and Actuators*, 11 (1987), 189–206.

[28] Tang, W. C., Nguyen, T.-C. H., Howe, R. T., Laterally-Driven Polysilicon Resonant Microstructures, *Sensors and Actuators*, 20 (1989), 25–32.

[29] Tang, W. C., Lim, M. G., Howe, R. T., Electrostatic Comb Drive Levitation and Control Method, *IEEE J. Microelectromechanical Systems*, 1 (1992), 170–178.

[30] Price, R. H., Wood, J. E., Jacobsen, S. C., Modelling Considerations for Electrostatic Forces in Electrostatic Microactuators, *Sensors and Actuators A*, 20 (1989), 107–114.

[31] Schwarzenbach, H. U., Korvink, J. G., Roos, M., Sartoris, G., Anderheggen, E., A Micro Electro Mechanical CAD Extension for SESES, *J. Micromech. Microeng.*, 3 (1993), 118–122.

[32] Korvink, J., *An Implementation of the Adaptive Finite Element Method for Semiconductor Sensor Simulation*, Ph. D. Dissertation, ETH Zurich, No. 10143, Switzerland, 1993.

[33] Fischer, M., Graef, H., von Münch, W., Electrostatically Deflectable Polysilicon Torsional Mirrors, *Sensors and Actuators A*, 44 (1994), 83–89.

[34] Cai, X., Osterberg, P., Yie, H., Gilbert, J., Senturia, S., White, J., Self-Consistent Electromechanical Analysis of Complex 3-D Microelectromechanical Structures Using Relaxation/Multipole-Accelerated Method, *Sensors and Materials*, 6 (1994), 85–99.

[35] Pourahmadi, F., Review of Modeling Silicon Microsensors and Actuators, *Sensors and Materials*, 6 (1994), 193–209.

[36] Yamada, K., Kuriyama, T., FEM Analysis for Single-Chip Multiaxial Servo Accelerometer, *Sensors and Materials*, 6 (1994), 211–223.

[37] Korvink, J. G., *SOLIDIS Reference Manual 1.0*, Internal Report No. 95/01, Physical Electronics Laboratory, ETH Zurich, 1995. ISE Integrated Systems Engineering AG, Technopark Zürich, Technoparkstrasse 1, CH-8005 Zürich, Switzerland.

[38] Lee, J. S., Yoshimura, S., Yagawa, G., Shibaike, N., A CAE System for Micromachines: Its Application to Electrostatic Micro Wobble Actuator, *Sensors and Actuators A*, 50 (1995), 209–221.

[39] Senturia, S. D., CAD for Microelectromechanical Systems, *Digest of Technical Papers*, Vol. 2, Transducers '95, Stockholm, 1995, pp. 5–8.

[40] Funk, J., *Modeling and Simulation of IMEMS*, Ph. D. Dissertation, ETH Zürich, No. 11378, Switzerland, 1996.

[41] Bächtold, M., *Efficient 3D Computation of Electrostatic Fields and Forces in Microsystems*, Ph. D. Dissertation, ETH Zürich, No. 12165, Switzerland, 1997.

[42] Osterberg, P., Yie, H., Cai, X., White, J., Senturia, S., Self-Consistent Simulation and Modeling of Electrostatically Deformed Diaphragms, *Proc. IEEE MEMS*, Oiso, 1994, pp. 28–32.

[43] Johnson, B. P., Kim, S., Senturia, S. D., White, J., MEMCAD Capacitance Calculations for Mechanically Deformed Square Diaphragm and Beam Microstructures, *Digest of Technical Papers*, Transducers '91, San Francisco, 1991, pp. 494–497.

[44] Korvink, J. G., Funk, J., Roos, M., Wachutka, G., Baltes, H., SESES: A Comprehensive MEMS Modelling System, *Proc. IEEE MEMS*, Oiso, 1994, pp. 22–27.

[45] Wang, P. K. C., Hadaegh, F. Y., Computation of Static Shapes and Voltages for Micromachined Deformable Mirrors with Nonlinear Electrostatic Actuators, *IEEE J. of Microelectromechanical Systems*, 5 (1996), 205–220.

[46] Yie, H., Bart, S. F., White, J., Senturia, S. D., A Computationally Practical Approach to Simulating Complex Surface-Micromachined Structures with Fabrication Non-Idealities, *Proc. IEEE MEMS*, Amsterdam, 1995, pp. 128–133.

[47] Gilbert, J. R., Legtenberg, R., Senturia, S. D., 3D Coupled Electro-Mechanics for MEMS: Applications of CoSolve-EM, *Proc. IEEE MEMS*, Amsterdam, 1995, pp. 122–127.

[48] Gilbert, J. R., Ananthasuresh, G. K., Senturia, S. D., 3D Modeling of Contact Problems and Hysteresis in Coupled Electro-Mechanics, *Proc. IEEE MEMS*, San Diego, 1996, pp. 127–132.

[49] Jaecklin, V. P., Linder, C., de Rooij, N. F., Moret, J. M., Micromechanical Comb Actuators with Low Driving Voltage, *J. Micromech. Microeng.*, 2 (1992), 250–255.

[50] Cho, Y.-H., Pisano, A. P., Howe, R. T., Viscous Damping Model for Laterally Oscillating Microstructures, *IEEE J. of Microelectromechanical Systems*, 3 (1994), 81–87.

[51] Zhang. X., Tang, W. C., Viscous Air Damping in Laterally Driven Microresonators, *Sensors and Materials*, 27 (1995), 415–430.

[52] Kittel, C., *Introduction to Solid State Physics*, 6th Ed., New York: Wiley, 1986.

[53] Nye, J. F., *Physical Properties of Crystals*, Oxford: Oxford University Press, 1957.

[54] Lorrain, P., Corson, D. R., Lorrain, F., *Electromagnetic Fields and Waves*, 3rd Ed., New York: Freeman, 1987.

[55] Selberherr, S., *Analysis and Simulation of Semiconductor Devices*, Wien-New York: Springer-Verlag, 1984.

[56] *SUPREM*, Integrated Circuits Laboratory (ICL), Department of Electrical Engineering, Stanford University, CA, USA. http://www-tcad.stanford.edu/tcad/org.html

[57] Nabors, K., White, J., FastCap: A Multipole-Accelerated 3-D Capacitance Extraction Program, *IEEE Trans. CAD of ICAS*, 10 (1991), 1447–1459.

[58] Riethmüller, W., Benecke, W., Schnakenberg, U., Heuberger, A., Micromechanical Silicon Actuators Based on Thermal Expansion Effects, *Digest of Technical Papers*, Transducers '87, Tokyo, 1987, pp. 834–837.

[59] Lammerink, T. S. J., Elwenspoek, M., Fluitman, J. H. J., Optical Excitation of Micro-Mechanical Resonators, *Proc. IEEE MEMS*, Nara, 1991, pp. 160–165.

[60] Johnson, A. D., Vacuum-Deposited TiNi Shape Memory Film: Characterization and Applications in Microdevices, *J. Micromech. Microeng.*, 1 (1991), 34–41.

[61] Rashidian, B., Allen, M. G., Electrothermal Microactuators Based on Dielectric Loss Heating, *Proc. IEEE MEMS*, Fort Lauderdale, 1993, pp. 24–29.

[62] Bergstrom, P. L., Ji, J., Liu, Y.-N., Kaviany, M., Wise, K. D., Thermally-Driven Phase-Change Microactuation, *IEEE J. Microelectromechanical Systems*, 4 (1995), 10–17.

[63] Swart, N. R., Nathan, A., Reliability Study of Polysilicon for Microhotplates, *Technical Digest*, IEEE Solid-State Sensors and Actuators Workshop, Hilton Head Is., 1994, pp. 119–122.

[64] Wachutka, G. K., Rigorous Thermodynamic Treatment of Heat Generation and Conduction in Semiconductor Device Modeling, *IEEE Trans. on CAD of ICAS*, 9 (1990), 1141–1149.

[65] Paul, O., von Arx, M., Baltes, H., CMOS IC Layers: Complete Set of Thermal Conductivities, in: *Semiconductor Characterization: Present and Future Needs*, Bullis, W. M., Seiler, D. G., Diebold, A. C. (Eds.), New York: AIP, 1995, pp. 197–201.

[66] Nathan, A., Swart, N. R., Quasi-three-Dimensional Simulation of Heat Transfer in Thermal-Based Microsensors, in: *Simulation of Semiconductor Devices and Processes*, Vol. 6, Ryssel, H., Pichler, P., (Eds.), Wien-New York: Springer-Verlag, 1995, pp. 30–33.

[67] Nagata, M., Swart, N. R., Stevens, M., Nathan, A., Thermal Based Microflow Sensor Optimization Using Coupled Electrothermal Numerical Simulations, *Digest of Technical Papers*, Vol. 2, Transducers '95, Stockholm, 1995, pp. 447–450.

[68] Busch, J. D., Johnson, A. D., Shape-Memory Properties in Ni-Ti Sputter-Deposited Film, *J. Appl. Phys.*, 68 (1990), 6224–6228.

[69] Krulevitch, P., Lee, A. P., Ramsey, P. B., Trevino, J. C., Hamilton, J., Northrup, M. A., Thin Film Shape Memory Alloy Microactuators, *IEEE J. of Microelectromechanical Systems*, 5 (1996), 270–282.

[70] Ikuta, K., Shimizu, H., Two-Dimensional Mathematical Model of Shape Memory Alloy and Intelligent SMA-CAD, *Proc. IEEE MEMS*, Fort Lauderdale, 1993, pp. 87–91.

[71] Madill, D. R., Wang, D., The Modeling and L_2-Stability of a Shape Memory Alloy Position Control System, *Proc. IEEE Int. Conf. on Robotics and Automation*, San Diego, 1994, pp. 293–299.

[72] Quandt, E., Halene, C., Holleck, H., Feit, K., Kohl, M., Schloßmacher, P., Skokan, A., Skrobanek, K. D., Sputter Deposition of TiNi, TiNiPd and TiPd Films Displaying the Two-Way Shape-Memory Effect, *Sensors and Actuators A*, 53 (1996), 434–439.

[73] Becker, E. W., Ehrfeld, W., Hagmann, P., Maner, A., Münchmeyer, D., Fabrication of Microstructures with High Aspect Ratios and Great Structural Heights by Synchrotron Radiation Lithography, Galvanoformung, and Plastic Moulding (LIGA Process), *Microelectronic Engineering*, 4 (1986), 35–56.

[74] Allen, M. G., Polyimide-Based Processes for the Fabrication of Thick Electroplated Microstructures, *Digest of Technical Papers*, Transducers '93, Yokohama, 1993, pp. 60–65.

[75] Judy, J. W., Muller, R. S., Zappe, H. H., Magnetic Microactuation of Polysilicon Flexure Structures, *IEEE J. of Microelectromechancial Systems*, 4 (1995), 162–169.

[76] Miller, R. A., Burr, G. W., Tai, Y.-C., Psaltis, D., Ho, C.-M., Katti, R. R., Electromagnetic MEMS Scanning Mirrors for Holographic Data Storage, *Technical*

Digest, IEEE Solid-State Sensor and Actuator Workshop, Hilton Head Is., 1996, pp. 183–186.

[77] Ohnstein, T. R., Zook, J. D., French, H. B., Guckel, H., Earles, T., Klein, J., Mangat, P., Tunable IR Filters with Integral Electromagnetic Actuators, *Technical Digest*, IEEE Solid-State Sensor and Actuator Workshop, Hilton Head Is., 1996, pp. 196–199.

[78] Taylor, W. P., Allen, M. G., Dauwalter, C. R., A Fully Integrated Magnetically Actuated Micromachined Relay, *Technical Digest*, IEEE Solid-State Sensor and Actuator Workshop, Hilton Head Is., 1996, pp. 231–234.

[79] Wagner, B., Kreutzer, M., Benecke, W., Linear and Rotational Magnetic Micromotors Fabricated Using Silicon Technology, *Proc. IEEE MEMS*, Travemünde, 1992, pp. 183–189.

[80] Ahn, C. H., Kim, Y. J., Allen, M. G., A Planar Variable Reluctance Magnetic Micromotor with Fully Integrated Stator and Coils, *IEEE J. of Microelectromechanical Systems*, 2 (1993), 165–173.

[81] Guckel, H., Christenson, T. R., Skobris, K. J., Klein, J., Karnowsky, M., Design and Testing of Planar Magnetic Micromotor Fabricated by Deep X-Ray Lithography and Electroplating, *Digest of Technical Papers*, Transducers '93, Yokohama, 1993, pp. 76–79.

[82] Zhang, W., Ahn, C. H., A Bi-Directional Magnetic Micropump on a Silicon Wafer, *Technical Digest*, IEEE Solid-State Sensor and Actuator Workshop, Hilton Head Is., 1996, pp. 94–97.

[83] Kruusing, A., Mikli, V., Flow Sensing and Pumping Using Flexible Permanent Magnet Beams, *Digest of Technical Papers*, Vol. 2, Transducers '95, Stockholm, 1995, pp. 299–302.

[84] Lowther, D. A., Silvester, P. P., *Computer-Aided Design in Magnetics*, Berlin: Springer-Verlag, 1985.

[85] Cullity, B. D., *Introduction to Magnetic Materials*, London: Addison-Wesley, 1972.

[86] McDonald, P. H., *Continuum Mechanics*, Boston: PWS Publishing Co., 1996.

[87] Livingstone, J. D., Magnetomechanical Properties of Amorphous Metal, *Phys. Stat. Sol. A*, 70 (1982), 591–596.

[88] Carr, W. J., Magnetostriction, in: *Magnetic Properties of Metals and Alloys*, American Society of Metals, Cleveland, Ohio, 1959.

[89] Chin, G. Y., Processing Control of Magnetic Properties for Magnetostrictive Transducer Applications, *J. Metals*, 23, No. 1 (1971), 42–45.

[90] Lu, Y., *Magnetostrictive Sensors with On-Chip Readout*, Ph. D. Dissertation, Electrical and Computer Engineering, University of Waterloo, Waterloo, Ontario, N2L 3G1, Canada, 1997.

[91] Lu, Y., Nathan, A., Metglass Thin Film with as-Deposited Domain Alignment for Smart Sensor and Actuator Applications, *Appl. Phys. Letts.*, 70 (1997), 526–528. (Erratum: Appl. Phys. Letts., Vol. 12, No. 18, 1998, in press).

[92] Honda, T., Arai, K. I., Yamaguchi, M., Fabrication of Actuators Using Magnetostrictive Thin Films, *Proc. IEEE MEMS*, Oiso, 1994, pp. 51–56.

[93] Quandt, E., Seeman, K., Fabrication and Simulation of Magnetostrictive Thin Film Actuators, *Sensors and Actuators A*, 50 (1995), 105–109.

[94] Pan, J. Y., Lin, P., Maseeh, F., Senturia, S. D., Verification of FEM Analysis of Load-Deflection Methods for Measuring Mechanical Properties of Thin Films, *Technical Digest*, IEEE Solid-State Sensor and Actuator Workshop, Hilton Head Is., 1990, pp. 70–73.

[95] Shen, B., Allegretto, W., Ma, Y., Yu, B., Hu, M., Robinson, A. M., Cantilever Micromachined Structures in CMOS Technology with Magnetic Actuation, *Sensors and Materials*, 9 (1997) 347–362.

[96] Moon, F. C., *Magneto-Solid Mechanics*, New York: Wiley, 1984.

[97] Affane, W., Gibbs, M. R. J., Powell, A. L., Performance Modeling of Micromachined Sensor Membranes Coated with Piezomagnetic Material, *Sensors and Actuators A*, 51 (1996), 219–224.

[98] Stanley, J. K., *Electrical and Magnetic Properties of Metals*, American Society of Metals, Metals Park, Ohio, 1963.

[99] *Piezoelectricity*, Rosen, C. Z., Hiremath, B. V., Newnham, R. (Eds.), New York: American Institute of Physics, 1992.

[100] Smits, J. G., Dalke, S. I., Cooney, T. K., The Constituent Equations of Piezoelectric Bimorphs, *Sensors and Actuators A*, 28 (1991), 41–61.

[101] Smits, J. G, Ballato, A., Dynamic Admittance Matrix of Piezoelectric Cantilever Bimorphs, *IEEE J. of Microelectromechanical Systems*, 3 (1994), 105–112.

[102] Benes, E., Gröschl, M., Burger, W., Schmid, M., Sensors Based on Piezoelectric Resonators, *Sensors and Actuators A*, vol. 48 (1995), 1–21.

[103] Flynn, A. M., Tavrow, L. S., Bart, S. F., Brooks, R. A., Ehrlich, D. J., Udayakumar, K. R., Cross, L. E., Piezoelectric Micromotors for Microrobots, *IEEE J. of Microelectromechanical Systems*, 1 (1992), 44–51.

[104] Brice, J. C., Crystals for Quartz Resonators, *Rev. Mod. Phys.* 57 (1985), 105–146 (also in [99], pp. 35–76).

[105] Koos, G. L., Wolfe, J. P., Phonon Focusing in Piezoelectric Crystals: Quartz and Lithium Niobate, *Phys. Rev. B.*, 30 (1984), 3470–3481 (also in [99], pp. 77–88).

[106] Bechmann, R., Elastic, Piezoelectric, and Dielectric Constants of Polarized Barium Titanate Ceramics and Some Applications of the Piezoelectric Equations, *Jour. of the Acoustical Soc. of America*, 28 (1956), 347–350 (also in [99], pp. 155–158).

[107] Lovinger, A. J., Ferroelectric Polymers, *Science*, 220 (1983), 1115–1121 (also in [99], pp. 182–188).

[108] Gallantree, H. R., Review of Transducer Applications of Polyvinylidene Fluoride, *IEE Proc.*, 130 (1983), 219–224 (also in [99], pp. 189–194).

[109] Tjhen, W., Tamagawa, T., Ye, C.-P., Hsueh, C.-C., Schiller, P., Polla, D. L., Properties of Piezoelectric Thin Films for Micromechanical Devices and Systems, *Proc. IEEE MEMS*, Nara, 1991, pp. 114–119.

[110] Abe, T., Reed, M. L., RF-Magnetron Sputtering of Piezoelectric Lead-Zirconate-Titanate Actuator Films Using Composite Targets, *Proc. IEEE MEMS*, Oiso, 1994, pp. 164–169.

[111] Sakata, M., Wakabayashi, S., Goto, H., Totani, H., Takeuchi, M., Yada, T., Sputtered High $|d_{31}|$ Coefficient PZT Thin Film for Micro Actuators, *Proc. IEEE MEMS*, San Diego, 1996, pp. 263–266.

[112] Tiersten, H. F., *Linear Piezoelectric Plate Vibrations*, New York: Plenum Press, 1969.

[113] Safari, A., Sa-gong, G., Giniewicz, J., Newnham, R. E., Composite Piezoelectric Sensors, *Proc. 21st Univ. Conf. Ceramic Sci.*, Vol. 20, 1986, pp. 445–454 (also in [99], pp. 195–204).

[114] Ohara, Y., Miyayama, M., Kuomoto, K., Yanagida, H., PZT-Polymer Piezoelectric Composites: A Design for an Acceleration Sensor, *Sensors and Actuators A*, 36 (1993), 121–126.

[115] Leaver, P., Cunningham, M. J., Jones, B. E., Piezoelectric Polymer Pressure Sensors, *Sensors and Actuators*, 12 (1987), 225–233.

[116] Liu, S. T., Long, D., Pyroelectric Detectors and Materials, *Proc. IEEE*, 66 (1978), 14–26 (also in [99], pp. 310–322).

[117] Andle, J. C., Vetelino, J. F., Acoustic Wave Biosensors, *Sensors and Actuators A*, 44 (1994), 167–176.

[118] Venema, A., (Ed.), Acoustic-Wave-Based Microsensors, *Sensors and Actuators A*, 44 (1994).

[119] Mason, W. P., *Electro-Mechanical Transducers and Wave Filters*, 3rd Ed., New Jersey: D. van Nostrand, 1948.

[120] Schwarzenbach, H. U., Lechner, H., Steinle, B., Baltes, H. P., Schwendimann, P., Calculation of Vibrations of Thick Piezoceramic Disk Resonators, *Appl. Phys. Lett.*, 38 (1981), 854–855.

[121] Lerch, R., Finite Element Analysis of Piezoelectric Transducers, *Proc. IEEE Ultrasonics Symp.*, 1988, pp. 643–654.

[122] Lerch, R., Piezoelectric and Acoustic Finite Elements as Tools for the Development of Electroacoustic Transducers, *Siemens Forsch.-u. Entwickl.-Ber.*, Bd. 17, Nr. 6 (1988), pp. 283–290.

[123] Langer, E., Selberherr, S., Morkowich, P. A., Ringhofer, C. A., Numerical Analysis of Acoustic Wave Generation in Anisotropic Piezoelectric Materials, *Sensors and Actuators A*, 4 (1983), 71–76.

[124] Farnell, G. W., SAW Propagation in Piezoelectric Solids, in: *Computer-Aided Design of Surface Acoustic Wave Devices*, Collins, J. H., Masotti, L. (Eds.), Amsterdam: Elsevier Scientific, 1976, pp. 1–24.

[125] *IEEE Standard on Piezoelectricity*, Std 176-1987 (also in [99], pp. 235–280).

[126] Thompson, P. A., *Compressible-Fluid Dynamics*, New York: McGraw-Hill, 1972.

[127] Bächtold, M., Korvink, J. G., Funk, J., Baltes, H., New Convergence Scheme for Self-Consistent Electromechanical Analysis of iMEMS, *Technical Digest*, IEEE IEDM, Washington, 1995, pp. 605–608.

[128] Bühler, J., Funk, J., Steiner, F.-P., Sarro, P. M., Baltes, H., Double Pass Metallization for CMOS Aluminum Actuators, *Digest of Technical Papers*, Vol. 2, Transducers '95, Stockholm, 1995, pp. 360–363.

[129] Funk, J. M., Korvink, J. G., Bühler, J., Bächtold, M., Baltes, H., SOLIDIS: A Tool for Microactuator Simulation in 3-D, *J. of Microelectromechanical Systems*, 6 (1997), 70–82.

[130] Hornbeck, L. J., Current Status of the Digital Micromirror Device (DMD), for Projection Television Applications, *Technical Digest*, IEEE IEDM, Washington, 1993, pp. 381–384.

[131] *MICROCOSM*, 201 Willesden Dr., Cary, NC 27513, USA.

[132] *IntelliSense Corp.*, 16 Upton Dr., Wilmington, MA 01887, USA.

[133] Anderheggen, E., Korvink, J. G., Roos, M., Sartoris, G. E., Schwarzenbach, H. U., *SESES User Manual*, NM Numerical Modelling GmbH, Thalwil, Switzerland, 1993.

[134] Crary, S. B., Zhang, Y., CAEMEMS: An Integrated Computer-Aided Engineering Workbench for Micro-Electro-Mechanical Systems, *Proc. IEEE MEMS*, 1990, pp. 113–114.

[135] *ALECSIS*, Inst. of Prec. Eng., TU Vienna, Floragasse 7, A-1040 Vienna, Austria.

[136] *PATRAN*, PDA Engineering, Costa Mesta, CA, USA.

[137] *Geomview*, Software Development Group, Geometry Center, University of Minnesota, 1300 South Second Street, Suite 500, Minneapolis, MN 55454, USA. http://www.geom.umn.edu/welcome.html.

[138] Koppelman, G. M., OYSTER, A Three-Dimensional Structural Simulator for Microelectromechanical Design, *Sensors and Actuators*, 20 (1989), 179–185.

[139] *Pro/ENGINEER*, Parametric Technology, Waltham, MA, USA.

[140] Chiyokura, H., *Solid Modeling with DESIGNBASE: Theory and Implementation*, Reading, MA.: Addison-Wesley, 1988.

[141] Osterberg, P., Senturia, S., MEMBUILDER: An Automated 3D Solid Model Construction Program for Microelectromechanical Structures, *Digest of Technical Papers*, Vol. 2, Transducers '95, Stockholm, 1995, pp. 21–24.

[142] Shulman, M., Ramaswamy, M., Heytens, M., Senturia, S. D., An Object-Oriented Material-Property Database Architecture for Microelectromechanical CAD, *Digest of Technical Papers*, Transducers '91, San Francisco, 1991, pp. 486–489.

[143] Nabors, K., Kim, S., White, J., Senturia, S., *FastCap User's Guide*, Research Laboratory of Electronics, Department of Electrical Engineering and Computer Science, MIT, Cambridge, MA 02139, USA.

[144] *ANSYS*, Inc., 275 Technology Drive, Canonsburg, PA 15317, USA.

[145] *Maxwell Solver*, Ansoft Corp., 4 Station Square, 660 Commerce Court Bldg., Pittsburgh, PA, USA.

[146] *CEDRAT S. A.*, 10 Chemin du Pré Carré, 38240 Meylan, France.

[147] *ADINA*, Adina R & D, Inc., 71 Elton Ave., Watertown, MA 02172, USA.

[148] *IES*, Integrated Engineering Software, 46-1313 Border Place, Winnipeg, Manitoba, R3H 0X4, Canada.

[149] *FIDAP*, Fluid Dynamics International, Evanston, Illinois, USA.

[150] *FLUENT*, FLUENT Inc., Centerra Resource Park, 10 Cavendish Court, Lebanon, N. H. 03766-1442, USA.

[151] *FLOTRAN*, see: Ulrich, J., Zengerle, R., Static and Dynamic Flow Simulation of a KOH-Etched Microvalve Using the Finite Element Method, *Sensors and Actuators A*, 53 (1996), 379–385.

[152] *FLOTHERM*, see, Fotheringham, G., Simulation Methods for Multi-Chip Modules, *Sensors and Actuators A*, 30 (1992), 157–165.

[153] *I-DEAS*, Structural Dynamics Research Corp, Milford, OH., USA.

[154] *ABAQUS*, Hibbit, Karlsson, and Sorenson, Inc., 1080 Main Street, Pawtucket, RI 02860, USA.

[155] *MSC/NASTRAN*, McNeal-Schwendler Corp., Los Angeles, CA, USA.

[156] *COSMOS/M*, Structural Research Analysis Corp., Santa Monica, CA, USA.

[157] *FLOWERS*, Inst. für Informatik, ETH, CH-8093 Zürich, Switzerland.

[158] Lee, K. W., Wise, K. D., SENSIM: A Simulation Program for Solid State Pressure Sensors, *IEEE Trans. Electron Devices*, 29 (1982), 34–41.

[159] *TPS10 Benutzerhandbuch*, 11th Ed., Reutlingen: T-Programm GmbH, 1989.

[160] *MARC*, MARC Analysis Research Corp., (see 38]).

[161] *EFCREL, EFDYN, EFCAD*, see: Lefèvre, Y., Lajoie-Mazenc, M., Sarraute, E., Lamon, H., First Stop Towards Design, Simulation, Modeling and Fabrication of Electrostatic Micromotors, *Sensors and Actuators A*, 46–47 (1995), pp. 645–648.

[162] *PUSI*, see: Zengerle, R., Richter, M., Brosinger, F., Richter, A., Sandmaier, H., Performance Simulation of Microminiaturized Membrane Pumps, *Digest of Technical Papers*, Transducers '93, Yokohama, 1993, pp. 106–109.

[163] Klokholm, E., The Measurement of Magnetostriction in Ferromagnetic Thin Films, *IEEE Trans. Magn.*, 12 (1976), 819–821.

[164] de Lacheisserie, E. du T., Peuzin, J. C., Magnetostriction and Internal Stresses in Thin Films: The Cantilever Method Revisited, *J. of Magnetism and Magnetic Materials*, 136 (1994), 189–196.

[165] Pham, H. H., Nathan, A., A New Approach for Rapid Evaluation of The Potential Field in Three Dimensions, Proc. Royal Society London A, 455 (1999), 1–39.

[166] Pham, H. H., Nathan, A., WATCAP: A New Simulation Engine for Interconnect Capacitance Extraction, 1st Canadian Workshop on RF IC Research and Development, Nov. 16, Ottawa, Canada, 1998.

9 Microsystem Simulation

Support electronics has become an essential part of the microtransducer providing necessary control and signal processing/conversion functions for improved accuracy, reliability, and functionality (see [1, 2]). Integration of these elements on a single chip constitutes the first step towards realization of microsystems [3, 4]. The efficient design of successful microsystems critically rests on accommodating the interaction of mixed electrical, thermal, mechanical, magnetic, radiant, and chemical signals. Most importantly, since the microtransducer is central to control and feedback operation, it cannot be isolated from circuitry in the design process. For example, in the design process for an integrated accelerometer micro-system (see Fig. 9.1a) one has to: evaluate the system response to transient electrical and mechanical signals; optimize operating bias and self-test procedures with respect to accelerometer reliability associated with electrostatic pull-in; minimize the influence of read-out operation on accelerometer performance; and evaluate the sensitivity of electrical and mechanical system performance to variations in accelerometer or circuit parameters. Thus, it is crucial that the simulation tool or environment accounts for the mixed-signal microtransducer-circuit interactions and yet provides reasonably accurate functional descriptions for both.

In the integrated circuit (IC) world, merged device-circuit simulations at the numerical level have been reported for analog and digital circuits [5–7], high power devices and circuits [8], including coupling of electrical and thermal effects [9–11], as well as for general-purpose device analysis [12, 13]. In all of these cases, device-circuit interactions are handled either in a device simulation environment or a circuit simulation environment. The latter is through use of lumped or distributed circuit models. Here, we define the terms "lumped" or "distributed" as referring to a coarse or fine discretization of the device continuum, respectively. The use of either simulation environment has its advantages and disadvantages and the choice is ultimately driven by the application and nature of the problem. Our treatment of mixed-signal microsystem simulations will focus on use

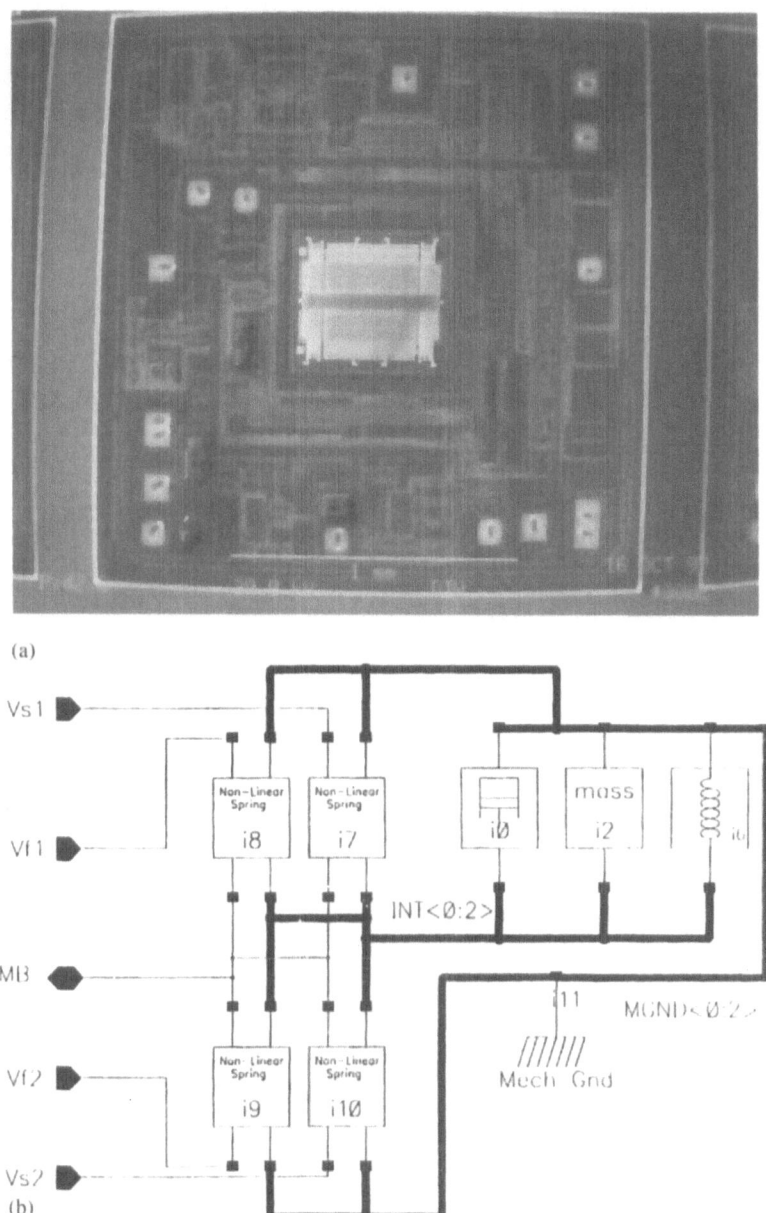

(a)

(b)

Fig. 9.1 (a) Photograph of integrated accelerometer (ADXL 76B) and (b) schematic of lumped equivalent SPICE circuit model. Courtesy of Analog Devices, Wilmington, MA, USA

of the latter environment [14]. By exploiting the historically well-known analogues between different physical and chemical variables, merged microtransducer-circuit simulation in a circuit environment can provide, with sufficient accuracy and detail, a low cost and expedient means of designing high performance and highly reliable microsystems. An example of this is illustrated in Fig. 9.1b which depicts the equivalent electrical lumped circuit model for the integrated accelerometer for SPICE simulation of the microsystem.

From an engineering standpoint, microsystem design in a circuit simulation environment has the following advantages:

• It provides a means of interfacing to different (mixed) signal types, through use of passive circuit elements such as dependent sources, in an environment that is already well-established in industry.
• The models and boundary conditions are simple and fast to implement with reduced complexity of computer code, thus isolating the designer from the complexities involved in a full numerical analysis.
• It allows for simulation of complicated microtransducer structures by using a combination of lumped and distributed equivalent circuits through device partitioning.
• It is highly modular in that the microtransducer, which is suitably discretized using a complex network of circuit elements, is treated simply as a subcircuit along with the other circuits that provide, e.g., the necessary control and feedback operation.

Additionally, the circuit simulator offers the potential of performing mixed-signal sensitivity analysis within the same environment. Assuming that layout extraction tools can be extended to also include automatic generation of microtransducer model description files based on material data made available in the technology file [15], the sensitivity of the mixed-signal microsystem performance to statistical and deterministic variations of geometry- and process-dependent microtransducer and circuit parameters can be estimated (see, e.g., [16]). An example of the data flow in an automated extraction and simulation environment is shown in Fig. 9.2 for the thermal-based flow microsensor [14].

Despite the convenience of a circuit simulation environment, the physical plausibility of solutions obtained is strongly dependent on the technique used in synthesis of the circuit model. The resulting equivalent circuit must provide an accurate description of device behavior which requires insight into the device operating principles. But this can be gained by numerical simulation, which, if necessary, can also serve to provide the macroscopic model of device functionality. Also, standard circuit simulators suffer from non-optimal numerical algorithms which could limit their ability to handle large algebraic systems of equations stemming from a fine discretization with distributed circuits. However, this may

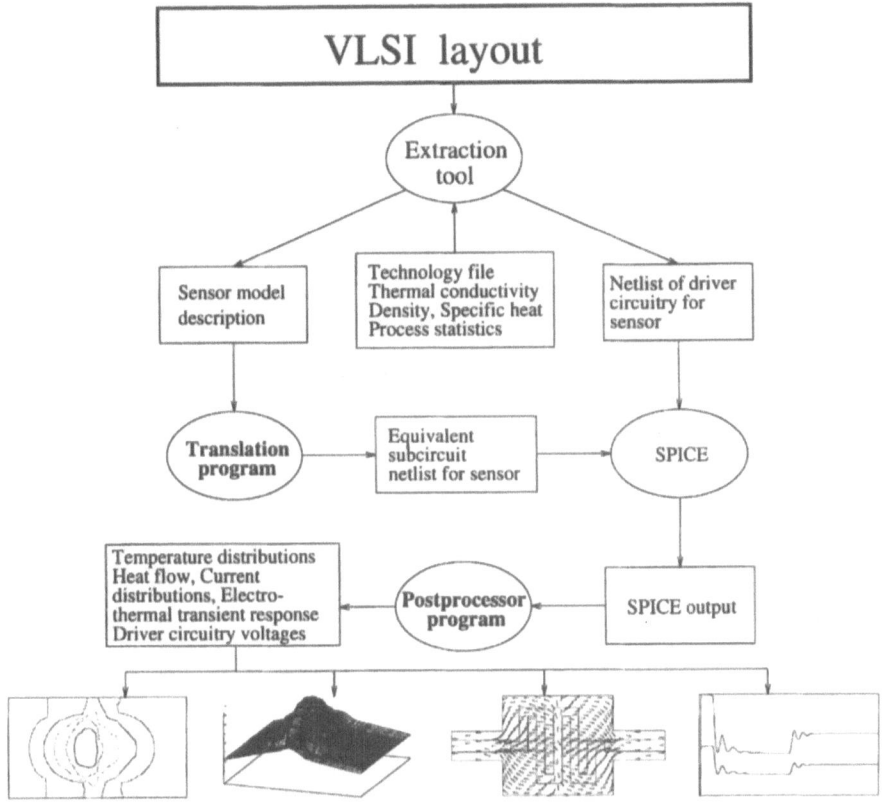

Fig. 9.2 Data flow in an automated extraction/simulation environment for thermal microsensors

improve with future design tools, needed to cope with the growing complexity of ICs.

The simulation of non-electrical systems using circuit techniques is not new. Early work in this area dates back to the forties. The subject has been extensively treated by Kron [17] and others (see [18] and Sect. 9.1). Our brief treatment of the subject is presented in the form of examples. We focus on selected aspects of mixed-signal circuit model synthesis and implementation in standard circuit simulators. These are described in Sect. 9.2. Sects. 9.3 and 9.4 follow with examples based on lumped and distributed analysis, respectively, using SPICE. Our choice in SPICE, as opposed to other tools based on high-level functional description languages (e.g., VHDL), stems from it being the current industry standard. Moreover, SPICE is a low-level (or primitive) tool requiring circuit descriptions that are simple and transparent; these are attributes essential for insight into microsystem operation.

9.1 Electrical Analogues for Mixed-Signals and Historical Developments

First formal attempts to develop electrical equivalent circuit models for various mixed-signal systems date back to the 1940's with circuit synthesis for elastic fields [19]. Models were developed to solve problems related to propagation of elastic waves, natural frequencies of vibration of elastic bodies, non-linear stress-strain analysis, analysis of electromagnetic fields [20], and compressible and incompressible fluid flow [21]. Circuit synthesis in all these cases was based on establishing suitable analogues between pertinent system variables with voltage and current, and system components with two-terminal electrical components. Circuit analogues had been in use long before then; an extensive list of applications of circuit models is given in [22]. Table 9.1 illustrates the analogies between different physical variables encountered in selected engineering disciplines. For convenience, we have also provided in Tables 9.2 and 9.3 the electrical analogues for thermal and selected chemical variables, respectively. The latter illustrates the isomorphisms between the semiconductor *p-n* junction and the liquid junction; when two solutions of different compositions are brought into contact, a junction is formed leading to ionic charge separation [23].

Without digital computing resources, the numerical evaluation of complex equivalent circuits was not feasible and analog computers, composed of racks of electrical elements such as resistors, transformers, and current sources, were assembled to simulate solutions to partial differential equations (PDEs) (see [24]). Shortly after, linear graph theory and operational approaches were introduced into the discipline [25]. The introduction of linear graph theory led to the more general concept of multi-terminal components [26], whose values were determined by measurement or calibration. The operational approach treats components as black boxes, concerned only with terminal behavior rather than the intricate workings of the black box interior. These developments led to the abandonment of distributed analysis, stemming from the governing PDEs, in favor of lumped parameter systems aimed at reducing the number of multi-terminal components used to describe the system.

The above ideas can be extended to mixed-signal microsystem simulations. The physical behavior of the microtransducer can be simulated using either lumped or distributed models. The former is based on reducing the physical properties of the microtransducer (e.g, capacitance, electrical/thermal resistances, mass, stiffness, etc.) to single elements. Values and functional behavior of elements can be constructed from the underlying models equations, the measured terminal behavior and/or numerical simulations (see Sect. 9.3). However, this involves a coarse linearization of

Table 9.1. *Analogies between physical variables for mixed-signal simulation. t denotes time. Alternate sets of analogies for the mechanical domain are also possible*

Generic	Electrical	Hydraulic	Thermal	Translational mechanics	Rotational mechanics
Across variable $x(t)$	Voltage $v(t)$	Pressure $p(t)$	Temperature $T(t)$	Force $F(t)$	Torque $\tau(t)$
Through variable $y(t)$	Current $i(t)$	Fluid flow $q(t)$	Heat flow $q(t)$	Velocity $v(t)$	Angular velocity $\omega(t)$
Across driver $x(t) = x_0(t)$	Voltage source $v(t) = v_0(t)$	Pressure driver $p(t) = p_0(t)$	Temperature driver	Force driver	Torque driver
Through driver $y(t) = y_0(t)$	Current source $i(t) = i_0(t)$	Flow driver $q(t) = q_0(t)$	Heat source $q(t) = q_0(t)$	Velocity driver	Angular velocity driver
Algebraic $x(t) = r\,y(t)$	Resistance $v(t) = Ri(t)$	Hydraulic resistance (viscosity, leak, valve)	Thermal resistance	Friction/ damping	Friction/ damping
1st order across $y(t) = s[dx(t)/dt]$	Capacitance $i(t) = C\,dv(t)/dt$	Hydraulic capacitance (compressibility, accumulation)	Thermal capacitance	Spring	Torsion
1st order through $x(t) = i[dy(t)/dt]$	Inductance $v(t) = L\,di(t)/dt$	Hydraulic inertance (inertia)	Thermal inductance	Mass	Moment of Inertia
Constitutive $\begin{bmatrix} x_1 \\ y_2 \end{bmatrix} = \begin{bmatrix} a & b \\ c & d \end{bmatrix}\begin{bmatrix} y_1 \\ x_2 \end{bmatrix}$	Ideal transformer, transistor	3-Terminal reservoir, free flow			
Parametric drivers	VCVS, VCCS CCVS, CCCS non-uniform and tensor coefficients			Dimensional coupling	Dimensional coupling

system behavior, and care must be taken not to overlook intricacies in the continuum pertinent to the analysis. Distributed models can be synthesized either by extending lumped elements or self-consistently from the governing differential equations to yield a network of multi-terminal components that stretch across the microtransducer domain (see Sect. 9.4). Both modeling approaches have been demonstrated for simulation of

Table 9.2. *Illustration of electrical and thermal duals. Here, $\mathbf{J}^{e,q}$ denote electrical and heat current densities, respectively, R the recombination rate, and H the heat density; remaining notation described in text*

Electrical	Thermal
Current density (A/cm^2)	Power density (W/cm^2)
Voltage (V)	Temperature (K)
Electric field (V/cm)	Temperature gradient (K/cm)
Electrical conductivity $(\Omega\text{-cm})^{-1}$	Thermal conductivity (W/cm-K)
Charge concentration (couls/cm^3)	Energy density (J/cm^3)
Resistance (Ω)	Thermal resistance $R_{th} = \frac{L}{\kappa A}$ (K/W)
Capacitance (F)	Thermal capacitance $C_{th} = \rho c_p.\text{vol}$ (J/K)
Voltage source (V)	Temperature source (K)
Current source (A)	Power source (W)
$\mathbf{J}^e = -\sigma \,\mathrm{grad}\,\psi$	$\mathbf{J}^q = -\kappa \,\mathrm{grad}\,T$
$\mathrm{div}\,\mathbf{J}^e = qR + q\partial n/\partial t$	$\mathrm{div}\,\mathbf{J}^q = -H + \rho c_p(\partial T/\partial t)$
Kirchhoff's Voltage Law (KVL)	$\int \mathrm{grad}\,T \cdot \mathbf{dl} = 0$
Kirchhoff's Current Law (KCL)	$\int \mathbf{J}^q \cdot \mathbf{n} d\Gamma = 0$

Table 9.3. *Analogy between the semiconductor junction and the liquid junction. $N_{D,A}$ denote donor and acceptor doping densities, respectively, ψ_j and π_j denote junction potentials, and λ_i the equivalent ionic conductivity of univalent solutions A and B*

Semiconductor junction		Liquid junction	
Symbol	Quantity	Symbol	Quantity
n	Electron carrier density	a_i	Activity
ψ	Electric potential	ϕ	Electrochemical potential
J_n	Electron current density	J_i	Mass flux density of species i
μ_n	Electron mobility	U_i	Absolute mobility
q	Charge on electron	$Z_i F$	Available charge
k	Boltzmann constant	R	Rydberg constant
$J_n = +qD_n\dfrac{dn}{dx} - qn\mu_n\dfrac{d\psi}{dx}$		$J_i = -U_i RT\dfrac{da_i}{dx} - Z_i FU_i a_i\dfrac{d\phi}{dx}$	
$\dfrac{D_n}{\mu_n} = \phi_T = \dfrac{kT}{q}$		$\dfrac{D}{U_i} = \dfrac{RT}{Z_i F}$	
		Simplified Case – Henderson Junction	
		• constant activity coefficients	
		• linear concentration profile	
$\psi_j = \phi_T \ln\dfrac{N_D}{N_A}$		$\pi_j = \pm\dfrac{RT}{F} \ln\dfrac{\lambda_i(B)}{\lambda_i(A)}$	

Table 9.4. *Selected publications on mixed-signal simulations in a circuit simulation environment based on lumped (L) or distributed (D) analysis*

Signal-type	Description
Magnetic	MOS Hall devices (*D*) [27, 28], p-i-n magnetodiodes (*D*) [29]
	Bipolar magnetotransistors (*L & D*: device partitioning) [30]
	Noise analysis including correlation (*D*) [31]
	Magnetostatics (*L*) [32]
Thermal	Electrothermal and flow microsensors (*D*) [14, 33, 34], (*L*) [35, 36]
	Flow microsensors (*L*) [37]
	Microbolometers (*L*) [38]
Mechanical and fluidic	Thermal stress analysis (*D*) [39]
	Accelerometers & damping (*L*) [40]
	Squeeze film damping (*D*) [41]
	Gyroscope (*L*) [42]
	Electromechanical microsystems (*L*) [43]
	Accelerometer packaging (*L*) [44]
	Electrostatic, electromagnetic, electrodynamic, and electroacoustic microactuators (*L*) [45, 46]
	Electrostatic micro-actuation (*L*) [35, 36, 47]
	Acoustic resonators (*D*) [48]
	Electro-fluidic (*L*) [49]
Chemical	ISFETs (*L*) [50]
Synthesis techniques	Lumped (*L*): consistent with numerical simulations [51]
	Distributed (*D*): consistent with PDE discretization [52, 53]

various types of microsystems; Table 9.4 provides a cross section of recent progress in the field drawn from the microsystems literature.

9.2 Circuit Modeling and Implementation Considerations

Interactions between mixed-signals may be strong and in many cases, the coupling can be highly non-linear. The coupling often manifests itself in some elements (in bond graph terminology [54] they are referred to as multi-port elements), whose values are dependent on signals from a different energy domain. Thus an important step in mixed-signal circuit synthesis lies in establishing a reasonably accurate description of the coupling elements and associated variables. The resulting circuit should be simple and comprehensible to allow insight into system behavior under diffferent geometric and operating conditions as well as compatible with standard circuit simulators. Most importantly, the circuit should be effective in handling non-linear coupling. Regardless of the analysis technique, *viz.*, lumped or distributed, one can employ several approaches to deal with mixed-signal coupling; each varies in terms of the degree of simplicity and effectiveness [36].

- Analog computer techniques [55, 56]. These involve the use of basic computing elements (circuits) such as integrators, adders, and multipliers. This can lead to complex circuits involving a large number of elements such as integrators. Analog computing techniques place more emphasis on computation rather than the intended purpose of providing insight into system behavior.
- Direct synthesis from numerical simulators [51]. This can be achieved by reduction of the numerical stiffness matrix to yield a minimum number of elements and interconnections between terminal nodes of the microtransducer. The procedure is analogous to network reduction techniques employed for transformation of a network to yield only terminal nodes or to erase internal nodes. An example of the latter is the Y-Δ transformation. Direct synthesis, although as accurate as numerical simulations, has a computational overhead associated with matrix reduction, and the handling of non-linear coupling requires the use of small signal models.
- Domain decoupling. Each signal-type is treated by an equivalent circuit in its native domain and coupling is realized through exchange/update of values upon convergence within each domain. The scheme requires a netlist regeneration at each iteration in the outer loop and shares similar concerns as the sequential (decoupled) numerical algorithms whereby convergence slows down in the presence of strong coupling.
- Dependent passive elements [36]. The background for this approach stems from the theory of bond graphs whereby coupling elements can be described in terms of dependent passive elements such as dependent resistors (R), dependent capacitors (C), or dependent inductors (L). The bond graph representation is an approach to circuit modeling that unifies systems of mixed-signal domains. Once the representation is established, it can be transformed to a detailed mathematical model suitable for simulation. The formulation of coupling between signals is based on the exchange of energy across the domains. Thus any system with a notion of "energy" exchange can be described by a bond graph. But how big a role does the bond graph play in the synthesis process? It is only useful at the very early stages of synthesis since it helps identify the nature of coupling and provides useful hints about possible constitutive relations. But for a specific constitutive relation, it is up to the design engineer to construct the coupling circuit that is most effective and efficient.

Given the various approaches, the choice is dictated by the following factors:

- Circuit simplicity in terms of realization and verification, and its ability to provide direct insight into system behavior.
- Effectiveness in handling non-linear coupling without use of small signal models which restrict the range of analysis.

In particular, the macroscopic model of the mixed-signal system can have different equivalent forms and care is needed to identify the form that is most appropriate in dealing with the above issues.

Our treatment of equivalent circuit synthesis techniques will be based on the last two approaches, *viz.*, domain decoupling and dependent passive elements. First we describe, in Sect. 9.2.1, a method for synthesis of coupling elements using multi-variate polynomial dependent sources [36]. This is followed, in Sect. 9.2.2, by a procedure for self-consistent synthesis of circuits from the governing multi-dimensional field equations [52]. The former is used in the illustrative example based on lumped analysis given in Sect. 9.3 and the latter in an example on distributed analysis given in Sect. 9.4.

9.2.1 Multi-Variate Polynomial Dependent Sources

In most mixed-signal systems, passive elements (e.g., R, C, L) are dependent on some variable excitation or detection signal (x) and serve as a source of coupling. Some circuit simulators do not support dynamically-varying passive elements; their values are not allowed to change during the course of simulation. To deal with this problem, we need to find an alternate means of expressing the dependence. Fortunately, multi-variate polynomial dependent sources have become a standard feature in state-of-the-art SPICE-like simulators.

Let us denote a dependent passive element, $P \in \{R, C, L\}$, whose value $P(x)$ is dependent on some signal x such as temperature or mechanical displacement. We need an approximate or possibly exact polynomial form for either $P(x)$ or $1/P(x)$, *viz.*,

$$P(x) = P_0[1 + f(x)], \tag{9.1}$$

$$\frac{1}{P(x)} = \frac{1}{P_0}[1 + g(x)], \tag{9.2}$$

where $f(x)$ and $g(x)$ are some polynomial functions of the external signal x and P_0 is the unperturbed $(x = 0)$ value of the element. What is important with the above polynomial forms is that either $P(x)$ or $1/P(x)$ can be decomposed into two components; one which is independent of x and the other which can be realized using a polynomial dependent source. Either form, (9.1) or (9.2), will suffice in describing the coupling element; the choice depends on the behavior of the physical system. If there is a tendency for $P(x)$ to become large, then the choice lies in (9.2). This will become evident in the examples that follow.

A. The Dependent Resistor

In this case, $P(x)$ in (9.1) and (9.2) denotes a resistor, $R \equiv R(x)$. With the polynomial form (9.1), we choose the constitutive relation $V = iR(x)$. This yields, following (9.1),

$$iR_0[1 + f(x)] = iR_0 + iR_0 f(x), \tag{9.3}$$

where we have an independent resistor R_0 whose voltage drop is V_0. Thus

$$V = V_0 + V_0 f(x), \tag{9.4}$$

where the dependence of R on x can now be described by an independent resistor R_0 and a dependent voltage-controlled voltage source (VCVS) whose value is $V_0 f(x)$. The circuit realization is depicted in Fig. 9.3.

In the case of the polynomial form (9.2), we choose the constitutive relation $i = V/R(x)$. This yields, following (9.2),

$$i = \frac{V[1 + g(x)]}{R_0}. \tag{9.5}$$

Here, i is the current through $R(x)$ and V is the associated voltage drop. This current can also be viewed as the current through the resistor R_0, the voltage across which is $V[1 + g(x)]$. Thus we can put in series an independent resistor R_0 and a VCVS of value $-Vg(x)$. The circuit realization is depicted in Fig. 9.4. Note the minus sign for the VCVS; the voltage across the independent resistor R_0 is $V - [-Vg(x)] = V[1 + g(x)]$, as expected.

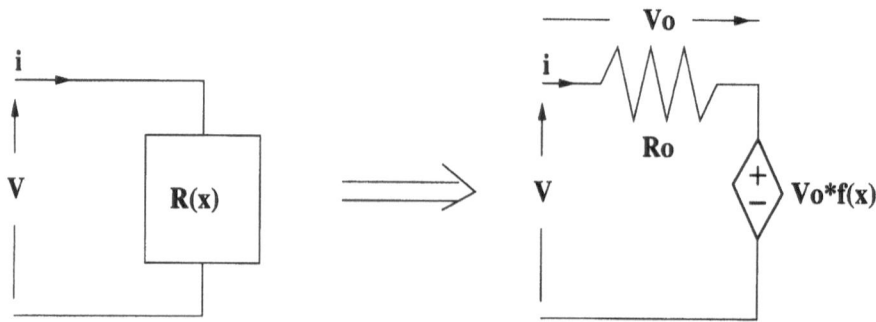

Fig. 9.3 Dependent source realization of coupling resistor, $R(x) = R_0[1 + f(x)]$

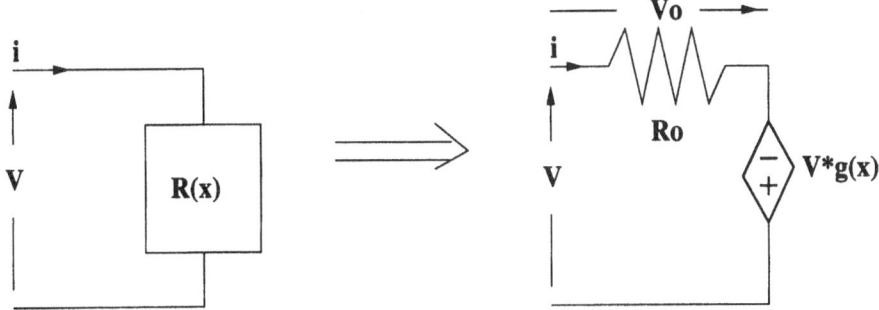

Fig. 9.4 As in Fig. 9.3, but for coupling resistor $1/R(x) = (1/R_0)[1 + g(x)]$

Alternatively, we can also express relation (9.5) as

$$i = \frac{V}{R(x)} = \frac{V}{R_0}[1 + g(x)] = \frac{V}{R_0} + \frac{V}{R_0}g(x),$$ (9.6)

where the current i can also be viewed as the sum of two currents, V/R_0 and $(V/R_0)g(x)$. This suggests an alternate realization of $R(x)$, whereby the independent resistor R_0 can be placed in parallel. If i_0 denotes the current through resistor R_0, we have

$$i = i_0 + i_0 g(x).$$ (9.7)

Thus the dependence of R on x can now be described by a dependent current-controlled current source (CCCS), whose value is $i_0 g(x)$, and an independent resistor R_0. The resulting circuit is depicted in Fig. 9.5. In some circuit simulators, the controlling current in a CCCS must be through a voltage source. In this case, we re-express (9.7) as $i = i_0 + (V/R_0)g(x)$, where $i_0 = V/R_0$ and the CCCS can now be turned into a voltage-controlled current source (VCCS).

We illustrate the synthesis scheme, based on polynomial dependent sources, with an example drawn from thermal analysis whereby the electrothermal coupling stems from the temperature dependence of

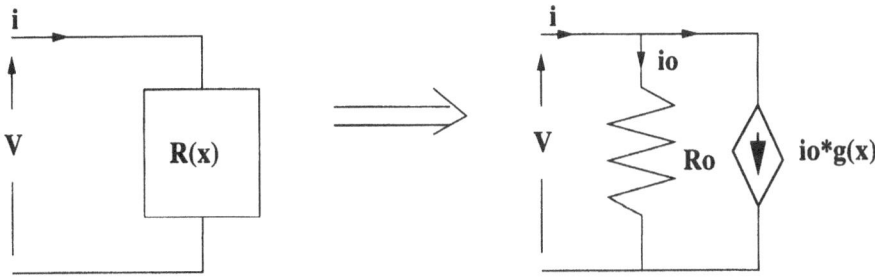

Fig. 9.5 An alternate circuit realization for the coupling resistor in Fig. 9.4

electrical resistance and Joule heat. For not too large variation in temperature T, the electrical resistance R can be modeled in terms of its temperature coefficient α

$$R(T) = R_0[1 + \alpha T].\tag{9.8}$$

Assuming that the resistor is isothermal with heat exchange to the environment (T_∞) described with a heat transfer coefficient (h), the model equations for the electrothermal system read

$$V = iR_0[1 + \alpha T] \qquad\qquad \text{electrical domain,} \tag{9.9}$$

$$Vi = C\frac{dT}{dt} + h[T - T_\infty] \qquad \text{thermal domain,} \tag{9.10}$$

where C denotes the thermal capacitance, V the voltage across the electrothermal resistor, and i the current through it. Eq. (9.10) describes the dissipation of the input electrical power Vi to the thermal capacitor (C) for heat energy storage and to the thermal resistor $(R_{th} = 1/h)$ for heat exchange with the environment.

The temperature-dependence of the electrical resistance $R(T)$, described by (9.8), is of the exact polynomial form, Eq. (9.1), where $f(x) = f(T) = \alpha T$. Therefore, the dependent resistor $R(T)$ can be realized using the circuit topology shown in Fig. 9.3. If V_{12} (instead of V_0) denotes the voltage across the resistor R_0 placed in series, the current through the resistor $R(T)$ is identical to the current through the independent resistor R_0. Therefore,

$$i = \frac{V_{12}}{R_0},\tag{9.11}$$

where the subscripts $(1, 2)$ denote the nodes in the circuit. The Joule heat is then

$$Vi = \frac{VV_{12}}{R_0},\tag{9.12}$$

which can be realized using a bi-variate polynomial VCCS of value VV_{12}/R_0. Hence, the model equations for the electrothermal system can now be expressed as

$$V = V_{12} + V_{12}\alpha T,\tag{9.13}$$

$$V\frac{V_{12}}{R_0} = C\frac{dT}{dt} + h(T - T_\infty).\tag{9.14}$$

Fig. 9.6 shows the resulting equivalent circuit for the electrothermal system. Here, the voltage V_3 represents the temperature T. The model equations (9.13) and (9.14), and the associated equivalent circuit can be employed to simulate transient electrothermal behavior; the SPICE

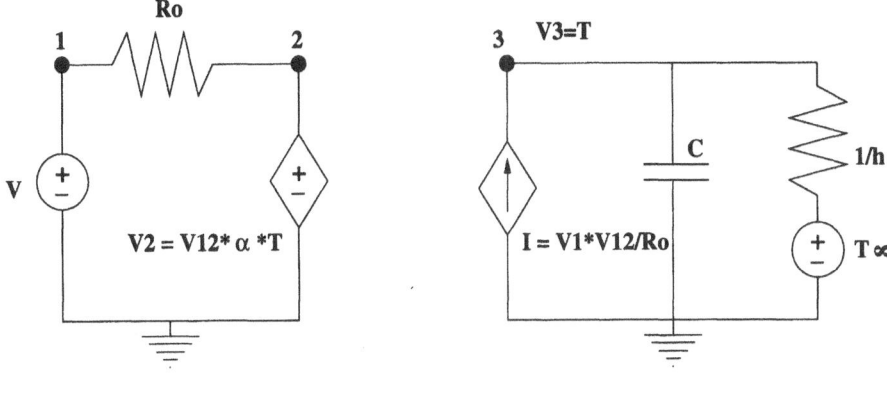

```
*temperature as a function of input voltage
.OPTION POST=2
.PARAM Alpha=3.7e-3 * Temperature coefficient
.PARAM Ro=1e4 *10K ohm Independent resistor
.PARAM InvRo='1/Ro'
.PARAM Rth=1e5 *1M  ohm Thermal resistor for heat convection
.PARAM Cth=1e-4  *100 uF Thermal capacitor
.PARAM Tambient=300 *300 K Ambient temperature

** Electric side
 * Input voltage Vin is of pulse form
 * Independent resistor Ro
 * Dependent voltage source (E2): V2=V12*Alpha*T
Vin 1 0 pulse(0 5 50 2 2 100 200 )
R1  1 2 Ro
E2  2 0 poly(2) 1 2  3 0    0 0 0 0 Alpha

** Thermal side
 * Temperature is the voltage V(3)
 * Dependent current source (G3): I=Vin*V23/Ro
G3  0 3 poly(2) 1 0  1 2    0 0 0 0  InvRo
Cth 3 0 Cth IC=Tambient
Rth 3 4 Rth
Vt  4 0 Tambient

** Simulation starting at 0s and ends at 400s with step 1s
 * Recording input voltage V=v(1), temperature in C=v(3)-273,
.tran 1 400
.print tran vin=v(1) Temp=PAR('v(3)-273')
.end
```

Fig. 9.6 Equivalent circuit for electrothermal resistor $R(T)$ and SPICE netlist

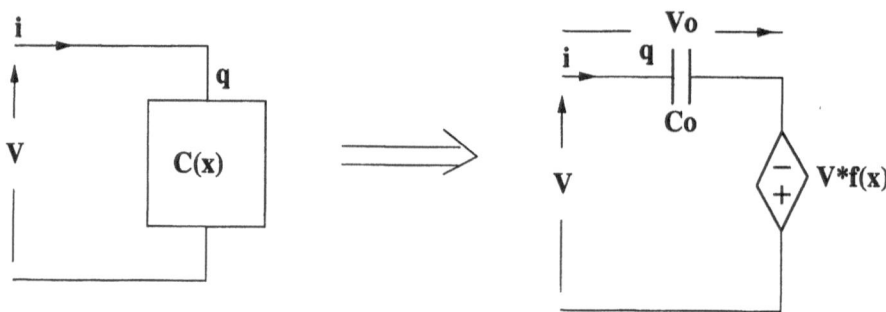

Fig. 9.7 Dependent source realization of coupling capacitor, $C(x) = C_0[1 + f(x)]$

implementation of the circuit is described in the accompanying netlist shown in Fig. 9.6.

B. The Dependent Capacitor

In this case, $P(x)$ in (9.1) and (9.2) denotes a capacitor $C \equiv C(x)$. By replacing i with q and R_0 with $1/C_0$, we can identify the correspondences between the resistor and capacitor; $(V = iR) \leftrightarrow (V = q/C)$. With polynomial form (9.1), we obtain

$$q = CV = C_0[1 + f(x)]V, \tag{9.15}$$

where q is the charge on the plate of the capacitor C_0, the voltage across which is $V[1 + f(x)]$. This is analogous to the case of the dependent resistor associated with the polynomial form (9.2). A realization of $C(x)$ is shown in Fig. 9.7. The circuit is similar in structure to the one shown in Fig. 9.4. Alternatively, if (9.15) was expressed as

$$q = C_0V + C_0Vf(x) = q_0 + q_0f(x), \tag{9.16}$$

where $q_0 = C_0V$, the resulting circuit for $C(x)$ could turn out to be complex. The current through C is

$$i = \frac{dq}{dt} = \frac{dq_0}{dt} + \frac{dq_0}{dt}f(x) + q_0\frac{df(x)}{dt}, \tag{9.17}$$

which is the sum of three currents in parallel; the current i_0 through the independent capacitor C_0, a CCCS of value $i_0 f(x)$, and VCVS of value $C_0Vf(x)$, and the resulting circuit that is synthesized is inferior to the one shown in Fig. 9.7.

In the case of polynomial form (9.2), by choosing the constitutive relation $V = q/C$, we obtain

$$V = \frac{q}{C_0}[1 + g(x)] = V_0 + V_0g(x), \tag{9.18}$$

where $V_0 = (q/C_0)$. The voltage V across the dependent capacitor C can be

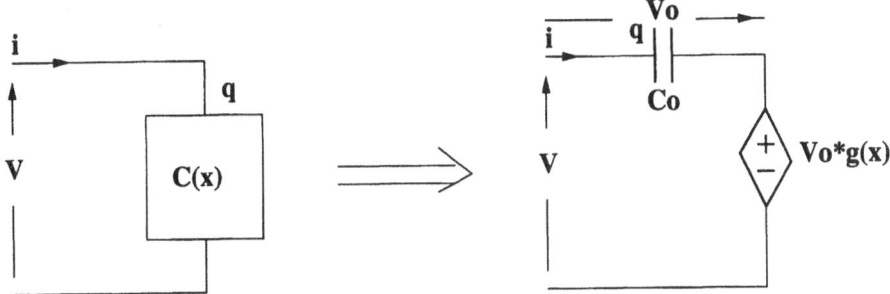

Fig. 9.8 As in Fig. 9.7, but for coupling capacitor, $1/C(x) = (1/C_0)[1 + g(x)]$

viewed as a sum of two voltages: the voltage V_0 of the independent C_0 and the VCVS of value $V_0g(x)$. A realization of $C(x)$ in this case is shown in Fig. 9.8. The circuit has a structure similar to that shown in Fig. 9.3.

For illustrative purposes, we apply the synthesis scheme to an electrostatically actuated parallel plate capacitor whereby the coupling stems from the dependence of capacitance on mechanical displacement. Fig. 9.9 shows a schematic of the actuation arrangement. The capacitor plate is constrained by a spring (coefficient k) and a damper (b) which models the internal friction. With a voltage (V_p) across the plate, a charge (q) is induced, creating an electrostatic force (F) which causes plate displacement (x). This in turn changes the capacitance. In bond graph terminology, the actuator is a two-port C-field. In modeling the electromechanical system, the constitutive relationship between the variables F, x, V_p, and q, is constructed from expressing C in terms of x, and F in terms of V_p and x.

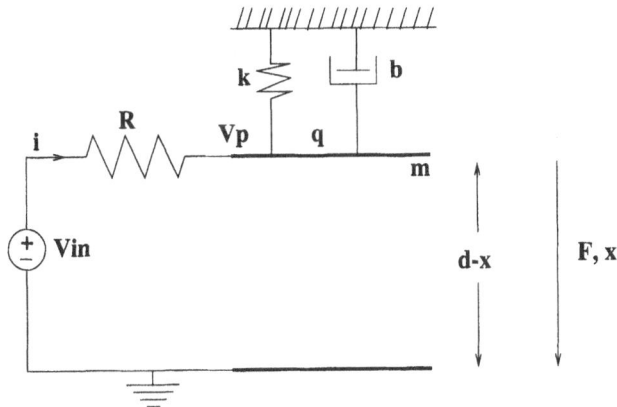

Fig. 9.9 Schematic of vertically-actuated parallel plate capacitor

The model equations for electrical and mechanical operation are:

$$V_p = V_{in} - iR \qquad \text{with} \quad i = dq/dt, \tag{9.19}$$

$$F = m\frac{d^2x}{dt^2} + b\frac{dx}{dt} + kx. \tag{9.20}$$

The capacitance C can be expressed as

$$C = \frac{\varepsilon A}{d-x} = \frac{\varepsilon A}{d}\frac{d}{d-x} = C_0\frac{d}{d-x}, \tag{9.21}$$

where C_0 denotes the initial capacitance (when $x = 0$), d the initial distance between the two plates, ε the permittivity of the associated dielectric (assumed to be air), and A the plate area.

For the electrostatic force F, we employ

$$F = \frac{\varepsilon A}{2(d-x)^2}V_p^2. \tag{9.22}$$

Thus the complete electromechanical system is modeled using Eqs. (9.19) to (9.22).

The main difficulty in using the above forms stems from the high non-linearity associated with the plate displacement, which requires use of a time-varying capacitor model [43], small signal approximation [45], or Taylor expansion [47], since

$$\lim_{x \to d} C = \lim_{x \to d} \varepsilon A/(d-x) = \infty. \tag{9.23}$$

This limitation can be overcome by employing an alternate form of the constitutive relation, viz.,

$$\frac{1}{C} = \frac{1}{C_0}\left(1 - \frac{x}{d}\right), \tag{9.24}$$

which is of the form given by Eq. (9.2) whereby, $g(x) \equiv -x/d$. Thus although C does not have an exact polynomial form, see Eq. (9.21), $1/C$ as described as (9.24) does, to allow use of a multi-variate polynomial dependent source. Furthermore, by replacing V_p in Eq. (9.22) with $V_p = q/C$, we can express the electrostatic force in terms of charge using a simpler form

$$F = \frac{q^2}{2C_0 d}. \tag{9.25}$$

The form (9.25) in terms of charge constitutes a more convenient form, compared to (9.22), to deal with non-linearity.

The form (9.24) for the capacitance can be realized using the circuit topology shown in Fig. 9.7. If V_{23} (instead of V_0) denotes the voltage

Table 9.5. *Set of analogies between electrical and mechanical domains employed here for synthesis of electromechanical equivalent circuits*

Electrical	Translational mechanics
Voltage V	Force F
Charge q	Displacement x
Current i	Velocity $v = dx/dt$
di/dt	Acceleration dv/dt

across C_0, the charge $q = \int i\, dt$ on the dependent capacitor C, is also the charge q on the independent capacitor C_0. Therefore

$$q = C_0 V_{23}. \tag{9.26}$$

The voltage across the dependent capacitor and the force given by (9.25) can be expressed as

$$V_p = V_{23} - V_{23}\frac{x}{d}, \tag{9.27}$$

$$F = \frac{C_0}{2d}V_{23}^2. \tag{9.28}$$

Thanks to the capacitor C_0, we have also managed to eliminate the need to integrate dq/dt. In fact, q is totally eliminated and replaced by $C_0 V_{23}$ as given by (9.26). Eq. (9.27) suggests the realization of the force using a polynomial VCVS of value $(C_0/2d)V_{23}^2$. The next step lies in obtaining the displacement x without use of an integrator. Employing an alternate set of analogies (see Table 9.5), we can synthesize mechanical operation, Eq. (9.20), using a VCVS and a series RLC circuit in which kx is the voltage across the capacitor, whose capacitance is $1/k$. Denoting this voltage by V_6, we obtain $x = V_6/k$, and Eq. (9.27) becomes

$$V_p = V_{23} - V_{23}\frac{V_6}{kd}. \tag{9.29}$$

The model equations for the complete electromechanical system now become (9.19), (9.20), (9.28), and (9.29). Fig. 9.10 shows the resulting equivalent circuit for the electromechanical system. Here, the voltage V_4 represents the electrostatic force F, and V_2 denotes the plate voltage V_p. The implementation of the circuit in SPICE is described in the accompanying netlist.

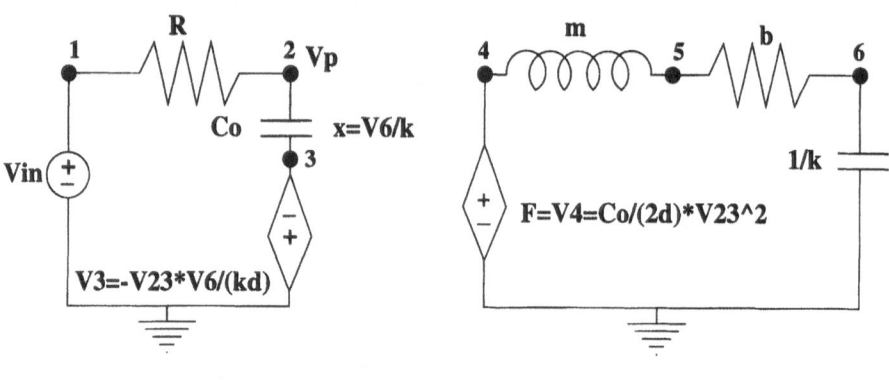

```
*constant charge
.OPTION post=2
.PARAM Rin=1e3
.PARAM Co=3.41e-12                          *Co
.PARAM Gap=1e-6                             *d
.PARAM SpringCoeff=2.4049e4                 *k
.PARAM SpringCap='1/SpringCoeff'
.PARAM InvKD='-1/(Gap*SpringCoeff)'         *1/kd
.PARAM CoOverGap='Co/(2*Gap)'
.PARAM Mass=1e-6                            *m
.PARAM Damper=1e-1                          *b


** Electrical operation
 * Dependent voltage source E3: V3=-V23*V6/kd (node 3 and 0)
 * Constant charge is simulated by applying a voltage source across Co
R1 0 2 Rin
Vin 2 3
C2 2 3 Co IC=0
E3 3 0 poly(2) 2 3 6 0   0 0 0 0  InvKD


** Mechanical operation:
 * Force: Dependent voltage source E4: V4=Co/(2d)*V23^2
E4 4 0  poly(1)  2 3      0 0 CoOverGap
L4 4 5  Mass
R5 5 6  Damper
C6 6 0  SpringCap IC=0


** DC-Analysis
.dc Vin 0 118 1
.print dc Gap= PAR('Gap-V(6)*SpringCap')
.end
```

Fig. 9.10 Equivalent circuit for electromechanical capacitor and SPICE netlist for constant voltage operation

9.2.2 Synthesis from Multi-Dimensional Field Equations

An alternate approach to simulating mixed-signal microsystems is through use of distributed models. These can be constructed by extending the lumped elements, discussed in Sect. 9.2.1, or self-consistently from the governing differential equations. Circuit synthesis with the latter approach forms the subject of this section. We present a generalized network synthesis procedure based on suitable multi-dimensional discretization of the underlying partial differential equation(s) governing microtransducer behavior. The procedure is demonstrated with second order time-dependent partial differential equations, taking into account material inhomogeneities and anisotropic (tensorial) coefficients; the latter is of importance, for example, in magnetic and piezoresistive-based mechanical microsystems, microsystem packaging, and in the analysis of mechanically strained integrated circuits.

A relatively simple and reliable discretization scheme commonly employed in numerical device modeling is the control-region approximation, more commonly known as the box integration method [57]. Based on this scheme, we present a procedure for two-dimensional network synthesis maintaining consistency with the numerical discretization of second-order partial differential equations; the procedures for three-dimensional network synthesis are analogous. We first describe the procedure when material coefficients are isotropic, although spatially non-uniform. The synthesis procedure is then extended to the case when the coefficients become anisotropic (tensorial).

A. Isotropic Case

Consider a general time-dependent PDE of the form

$$\text{div}\,(a\,\text{grad}\,b) = c + d\frac{\partial b}{\partial t}, \tag{9.30}$$

where the coefficient a describes the material property (e.g., permittivity ε, electrical conductivity σ, thermal conductivity κ, permeability μ), grad b is a vector field (e.g., electric field, current density, temperature gradient, magnetic polarization), c denotes a source field (e.g., space charge ρ, optical generation G_{opt}), and the last term accounts for transient behavior. We consider the two-dimensional mesh arrangement shown in Fig. 9.11 commonly employed in the control-region approximation. Applying the divergence theorem to Eq. (9.30) yields for the left-hand side

$$\int_{\Omega_i} \text{div}\,(a\,\text{grad}\,b)d\Omega_i = \int_{\partial\Omega_i} (a\,\text{grad}\,b)\cdot\mathbf{n}\,d(\partial\Omega_i), \tag{9.31}$$

where Ω_i denotes the control area (cell) formed by the union of perpendicular bisectors to element edges and $\partial\Omega_i$ the boundary of Ω_i

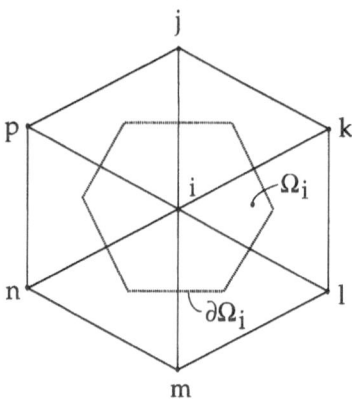

Fig. 9.11 Two-dimensional mesh arrangement used in control area approximation

(see Fig. 9.11). The integral on the right-hand side of Eq. (9.31) denotes the total flux leaving the cell, Ω_i. With (9.31), Eq. (9.30) now becomes

$$\int_{\partial\Omega_i} (a\,\text{grad}\,b) \cdot \mathbf{n}\,d(\partial\Omega_i) = \int_{\Omega_i} c\,d\Omega_i + \int_{\Omega_i} d\frac{\partial b}{\partial t}\,d\Omega_i. \tag{9.32}$$

To approximate the various contributions of flux, we consider an element ijk and assume that the coefficient "a" can be approximated by an arithmetic average, $viz.$, $a_{ijk} \equiv [a_i(T) + a_j(T) + a_k(T)]/3$, where we have included its possible dependence on temperature.

For an element ijk in the mesh arrangement (see Fig. 9.11), the left-hand side of (9.32) becomes:

$$a_{ijk}\left[\frac{b_j - b_i}{l_{ij}}w_{ij} + \frac{b_k - b_i}{l_{ik}}w_{ik}\right] = \left(a_{ijk}\frac{w_{ij}}{l_{ij}}\right)(b_j - b_i)$$
$$+ \left(a_{ijk}\frac{w_{ik}}{l_{ik}}\right)(b_k - b_i). \tag{9.33}$$

Upon close inspection of (9.33), we see that terms of the form $[a(w/l)]$ denote, for example, the equivalent of a two-dimensional conductance per unit device thickness. Thus we can write for the flux contribution to node i

$$g_{ij}(b_j - b_i) + g_{ik}(b_k - b_i) = \frac{(b_j - b_i)}{r_{ij}} + \frac{(b_k - b_i)}{r_{ik}}, \tag{9.34}$$

where r denotes the equivalent of a resistance. The total flux contribution from all nodes surrounding node i then becomes

$$\frac{b_{ji}}{r_{ij}} + \frac{b_{ki}}{r_{ik}} + \frac{b_{li}}{r_{il}} + \frac{b_{mi}}{r_{im}} + \frac{b_{ni}}{r_{in}} + \frac{b_{pi}}{r_{ip}}, \tag{9.35}$$

where b_{ji} denotes the difference in potentials between nodes i and j. Eq. (9.35) represents the discrete form of the left-hand side of Eq. (9.30). Based

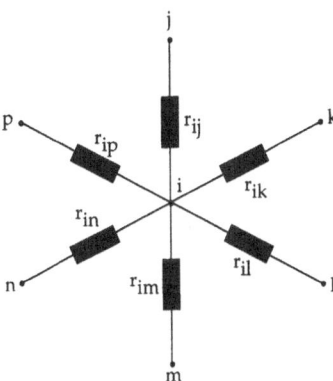

Fig. 9.12 Equivalent resistor network for isotropic transport coefficients based on mesh shown in Fig. 9.11

on (9.35), we can synthesize an equivalent circuit of the form shown in Fig. 9.12, which is consistent with the differential Eq. (9.30) with a right-hand side that is zero, *viz.*, div (a grad b) $= 0$. These lumped elements account for the physical properties of the two- (or three-)dimensional space to locally satisfy the flux conservation laws [58]. The discrete circuit arrangement shown can be employed for:

- Electrical problems with spatially varying conductivity; div $[\sigma(x, y) \operatorname{grad} \psi] = 0$. In this case, the circuit elements denote electrical resistances.
- Capacitance calculations; div $[\varepsilon(x, y) \operatorname{grad} \psi] = 0$, where the elements represent capacitances.
- Thermal analysis; div $[\kappa(x, y) \operatorname{grad} T] = 0$, where the elements represent thermal resistances.
- Magnetic field analysis; div $[\mu(x, y) \operatorname{grad} P] = 0$, where the elements denote reluctances.

In most practical problems, the right-hand side of (9.30) is non-zero. In such cases, the same circuit arrangement holds, but with the inclusion of an appropriate source element connected to node i. From Eq. (9.32), the value of the source element is simply the integral of c over the sub-domain, Ω_i. We illustrate the synthesis scheme with an example drawn from thermal analysis. Under steady state conditions, the heat conduction equation reads

$$\operatorname{div} [\kappa(T) \operatorname{grad} T] = -\sigma(T) (\operatorname{grad} \psi)^2, \tag{9.36}$$

where $\sigma(T) (\operatorname{grad} \psi)^2$ denotes Joule heat and ψ the electrostatic potential. The Joule heat generated in element ijk can be easily approximated as follows. Applying the divergence theorem to Eq. (9.36) yields

$$\int_{\partial \Omega_i} (\kappa(T) \operatorname{grad} T) \cdot \mathbf{n} \, d(\partial \Omega_i) = -\int_{\Omega_i} \sigma(T) (\operatorname{grad} \psi)^2 d\Omega_i. \tag{9.37}$$

For the element ijk, the right-hand side of (9.37) yields

$$\left(\frac{\psi_j - \psi_i}{l_{ij}}\right)^2 \sigma_{ijk} \frac{1}{2} l_{ik} w_{ij} + \left(\frac{\psi_k - \psi_i}{l_{ik}}\right)^2 \sigma_{ijk} \frac{1}{2} l_{ik} w_{ik} = \frac{\psi_{ji}^2}{2r_{ij}} + \frac{\psi_{ki}^2}{2r_{ik}},$$

$$(9.38)$$

where σ_{ijk} denotes the arithmetic average of the temperature-dependent thermal conductivity. The total Joule heat contribution q is then

$$\frac{\psi_{ji}^2}{2r_{ij}} + \frac{\psi_{ki}^2}{2r_{ik}} + \frac{\psi_{li}^2}{2r_{il}} + \frac{\psi_{mi}^2}{2r_{im}} + \frac{\psi_{ni}^2}{2r_{in}} + \frac{\psi_{pi}^2}{2r_{ip}} = q. \qquad (9.39)$$

We next discuss the implementation of q along with treatment of the transient term.

To model transient behavior, we include the time dependent term, $viz.$,

$$\int_{\partial\Omega_i} (\kappa(T)\mathrm{grad}\,T) \cdot \mathbf{n}\, d(\partial\Omega_i)$$

$$= -\int_{\Omega_i} \sigma(\,\mathrm{grad}\,\psi)^2 d\Omega_i + \int_{\Omega_i} \rho c_p \frac{\partial T}{\partial t} d\Omega_i$$

$$(9.40)$$

resulting in a shunt thermal capacitance, C_{th} (per unit device thickness):

$$C_{\mathrm{th}} = \rho_i c_{pi} A \qquad (9.41)$$

that is added to node i [59]. Here, A denotes the area of Ω_i, and ρ_i and c_{pi} denote the density and specific heat at node i, respectively. Thus our discrete equivalent circuit takes the form shown in Fig. 9.13. Here, q denotes Joule heat as described by Eq. (9.39), C_{th} the thermal capacitance as given by (9.41), and the resistive elements are thermal resistances described earlier in relation to (9.33).

B. Anisotropic Case

In the presence of strain or magnetic field, the conductivity becomes a tensor and the current density relation is described as

$$\mathbf{J} = -\sigma\,\mathrm{grad}\,\psi, \qquad (9.42)$$

where σ, in general, takes the following form

$$\sigma = \begin{bmatrix} \sigma_{xx} & \sigma_{xy} & \sigma_{xz} \\ \sigma_{yx} & \sigma_{yy} & \sigma_{yz} \\ \sigma_{zx} & \sigma_{zy} & \sigma_{zz} \end{bmatrix}. \qquad (9.43)$$

As an example, we will consider the synthesis of a discrete equivalent

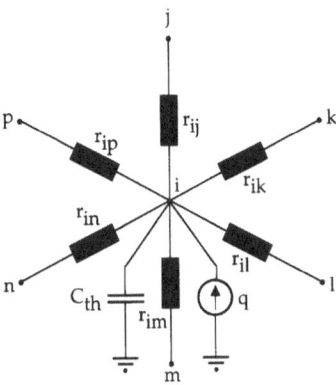

Fig. 9.13 Equivalent network for transient simulations

circuit for a unipolar n-type Hall vector probe. The current density in the presence of a magnetic field $\mathbf{B}$ (see Sect. 4.1), after a slight change of notation, reads

$$\mathbf{J} = -\sigma_n(1 + \mu_{Hn}\mathbf{B} \times + \mu_{Hn}^2\mathbf{B} \times \mathbf{B} \times)\,\text{grad}\,\psi \qquad (9.44)$$

where $\sigma_n = q\mu_n n$ denotes the $B = 0$ conductivity and $\mu_{Hn} = r_{Hn}\mu_n$. To arrive at a form suitable for a circuit model, we expand Eq. (9.44) in terms of components

$$\mathbf{B} \times \text{grad}\,\psi = \left(\frac{\partial\psi}{\partial z}B_y - \frac{\partial\psi}{\partial y}B_z\right)\mathbf{i} - \left(\frac{\partial\psi}{\partial z}B_x - \frac{\partial\psi}{\partial x}B_z\right)\mathbf{j}$$
$$+ \left(\frac{\partial\psi}{\partial y}B_x - \frac{\partial\psi}{\partial x}B_y\right)\mathbf{k}, \qquad (9.45)$$

where $\mathbf{i}$, $\mathbf{j}$, and $\mathbf{k}$ denote unit vectors, and

$$\mathbf{B} \times \mathbf{B} \times \text{grad}\,\psi = (\mathbf{B} \cdot \text{grad}\,\psi)\mathbf{B} - \mathbf{B}\mathbf{B}\,\text{grad}\,\psi. \qquad (9.46)$$

With (9.45) and (9.46), Eq. (9.44) can be expressed in terms of its components:

$$J_x = -\sigma_n\frac{\partial\psi}{\partial x}(1 + \mu_{Hn}^2 B_x^2 - \mu_{Hn}^2 B^2)$$
$$- \sigma_n\frac{\partial\psi}{\partial y}(\mu_{Hn}^2 B_x B_y - \mu_{Hn}B_z)$$
$$- \sigma_n\frac{\partial\psi}{\partial z}(\mu_{Hn}B_y + \mu_{Hn}^2 B_x B_z),$$

$$J_y = -\sigma_n \frac{\partial \psi}{\partial y} (1 + \mu_{Hn}^2 B_y^2 - \mu_{Hn}^2 B^2)$$

$$- \sigma_n \frac{\partial \psi}{\partial z} (\mu_{Hn}^2 B_y B_z - \mu_{Hn} B_x)$$

$$- \sigma_n \frac{\partial \psi}{\partial x} (\mu_{Hn} B_z + \mu_{Hn}^2 B_x B_y),$$

$$J_z = -\sigma_n \frac{\partial \psi}{\partial z} (1 + \mu_{Hn}^2 B_z^2 - \mu_{Hn}^2 B^2)$$

$$- \sigma_n \frac{\partial \psi}{\partial x} (\mu_{Hn}^2 B_x B_z - \mu_{Hn} B_y)$$

$$- \sigma_n \frac{\partial \psi}{\partial y} (\mu_{Hn} B_x + \mu_{Hn}^2 B_y B_z), \tag{9.47}$$

where $B = \sqrt{B_x^2 + B_y^2 + B_z^2}$. Eq. (9.47) can represented in terms of the form $\mathbf{J} = -\sigma \operatorname{grad} \psi$, with σ being a 3×3 magneto-conductivity matrix, viz.,

$$\sigma = \sigma_n \begin{bmatrix} 1 + \mu_{Hn}^2 B_x^2 - \mu_{Hn}^2 B^2 & \mu_{Hn}^2 B_x B_y - \mu_{Hn} B_z & \mu_{Hn} B_y + \mu_{Hn}^2 B_x B_z \\ \mu_{Hn} B_z + \mu_{Hn}^2 B_x B_y & 1 + \mu_{Hn}^2 B_y^2 - \mu_{Hn}^2 B^2 & \mu_{Hn}^2 B_y B_z - \mu_{Hn} B_x \\ \mu_{Hn}^2 B_x B_z - \mu_{Hn} B_y & \mu_{Hn} B_x + \mu_{Hn}^2 B_y B_z & 1 + \mu_{Hn}^2 B_z^2 - \mu_{Hn}^2 B^2 \end{bmatrix}. \tag{9.48}$$

For purposes of illustrating the equivalent circuit, we will assume that $B = (0, 0, B_z)$ and in two-dimensional analysis, (9.48) reduces to

$$\sigma = \sigma_n \begin{bmatrix} 1 - \mu_{Hn}^2 B_z^2 & -\mu_{Hn} B_z \\ \mu_{Hn} B_z & 1 - \mu_{Hn}^2 B_z^2 \end{bmatrix}. \tag{9.49}$$

For convenience, we write the current density components in the form:

$$J_x = -\sigma_{xx} \frac{\partial \psi}{\partial x} - \sigma_{xy} \frac{\partial \psi}{\partial y},$$

$$J_y = -\sigma_{yx} \frac{\partial \psi}{\partial x} - \sigma_{yy} \frac{\partial \psi}{\partial y}. \tag{9.50}$$

Note that each component of current density has an additional term that constitutes a driving force. These are the terms that account for the interaction of the magnetic field and current flow, i.e., $\sigma_{xy} = -\sigma_{yx} = -\mu_{Hn} B_z \sigma_n$. The other terms $\sigma_{xx} = \sigma_{yy} = \sigma_n$ are purely ohmic since $(\mu_{Hn} B_z)^2 \ll 1$ for practical values of magnetic field strength. Thus the diagonal terms yield purely ohmic elements while the off-diagonal terms

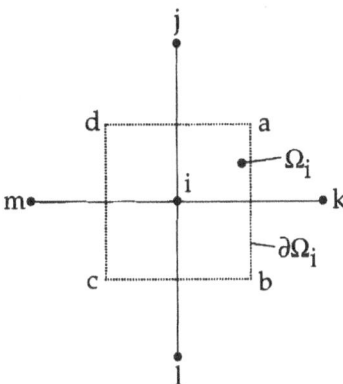

Fig. 9.14 Cell arrangement to illustrate circuit synthesis for anisotropic coefficients

account for the coupling. The former can be represented by a resistor while the latter by a field-controlled parametric element such as a voltage-controlled current source (VCCS). We formally show this using the procedures outlined earlier in Eqs. (9.30) through (9.35).

For simplicity of illustration, we choose the rectangular grid arrangement shown in Fig. 9.14 for the circuit synthesis. Here, i, j, k, l, and m denote nodes and $abcd$ the control area (cell) formed by union of perpendicular bisectors. Assuming no recombination in the device,

$$\text{div } \mathbf{J} = -\text{div} \left(\sigma \text{ grad } \psi \right) = 0. \tag{9.51}$$

Applying the divergence theorem to Eq. (9.51) yields

$$\int_{\partial \Omega_i} \sigma \text{ grad } \psi \cdot \mathbf{n} \, d(\partial \Omega_i) = 0, \tag{9.52}$$

which states that the algebraic sum of all fluxes entering node i is zero. Using system (9.50), we simplify (9.52) to read:

$$\frac{\psi_k - \psi_i}{l_{ik}} W_{ab} \sigma_{xx} + \frac{\psi_l - \psi_i}{l_{il}} W_{bc} \sigma_{xy}$$

$$+ \frac{\psi_m - \psi_i}{l_{im}} W_{cd} \sigma_{xx} + \frac{\psi_j - \psi_i}{l_{ij}} W_{ad} \sigma_{xy}$$

$$+ \frac{\psi_j - \psi_i}{l_{ij}} W_{ad} \sigma_{yy} + \frac{\psi_k - \psi_i}{l_{ik}} W_{ab} \sigma_{yx}$$

$$+ \frac{\psi_l - \psi_i}{l_{il}} W_{cb} \sigma_{yy} + \frac{\psi_m - \psi_i}{l_{im}} W_{cd} \sigma_{yx} = 0. \tag{9.53}$$

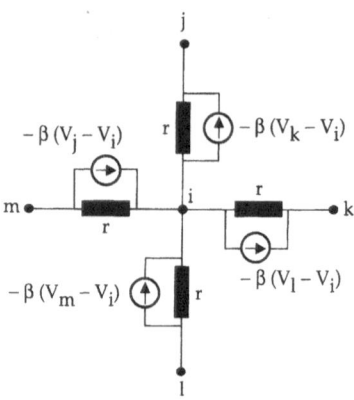

Fig. 9.15 Equivalent network for anisotropic transport coefficients

For convenience of interpretation, we employ a square grid arrangement ($w = l$) to simplify the various expressions for the circuit elements. With a slight change of notation ($\psi \rightarrow V$), Eq. (9.53) becomes

$$(V_k - V_i)\sigma_{xx} + (V_l - V_i)\sigma_{xy} + (V_m - V_i)\sigma_{xx} + (V_j - V_i)\sigma_{xy}$$
$$+ (V_j - V_i)\sigma_{yy} + (V_k - V_i)\sigma_{yx} + (V_l - V_i)\sigma_{yy}$$
$$+ (V_m - V_i)\sigma_{yx} = 0. \qquad (9.54)$$

Substituting for the σ's using (9.49), we get the equivalent circuit shown in Fig. 9.15. Here, the values associated with the resistor and VCCS are: $r = 1/[\sigma_n(1 - \beta^2)]$, where $\beta = \mu_{Hn}B_z$. The values of all circuit elements are in terms of per unit device thickness. In general, the circuit holds true for any possible spatial variation in σ_n, μ_{Hn}, or magnetic field, including magnetic domains. Note that the same equivalent circuit holds for stress-induced conductivity modulation (piezoresistance) in the device. In this case, instead of a magnetic-field-controlled VCCS, we would have a stress-dependent VCCS, whose value is computed from the piezo-resistance coefficients which are transformed from the crystal- to the circuit-axes.

As for the boundary conditions, at the fixed ohmic boundaries, we can either prescribe a voltage or current source. In terms of implementation, we simply short-circuit the nodes that are on these boundaries which forces an equipotential condition. At floating boundaries, the desired zero current flow condition is implemented by connecting a very high resistance between nodes at these boundaries and ground. Nodes on probe regions are treated similar to floating boundaries (i.e., by connecting a high resistance to ground) except that respective nodes are short-circuited to force an equipotential. Alternatively, depending on the value of current I_{leakage} drawn by the read-out circuit, the probes can be connected to a current source of that value, I_{leakage}.

9.3 Lumped Analysis: Illustrative Example – Electrostatic Micromirror

We illustrate SPICE simulation with the CMOS electrostatic micromirror [60] considered in Chapt. 8. Fig. 9.16 shows the schematic of the mirror used in the analysis [36]. The mirror consists of a rotating plate RP, two fixed electrode plates EP1 and EP2, and two landing plates LP1 and LP2. The rotating plate RP has a length $2l_0$, rotates about the pivot in the center, and is grounded. The fixed electrodes EP1 and EP2 have a length l, with $l < l_0/2$. When one of the electrodes is biased, RP rotates under the action of the electrostatic force. If the bias voltage is large enough, electrostatic pull-in occurs and RP contacts the closest LP.

Equivalent circuit modeling of the micromirror is similar to the vertically-actuated parallel plate capacitor (see Sect. 9.2.1), except that we now have rotational movement; the correspondence in associated variables is shown in Table 9.6. In what follows, we develop model equations for the capacitance C and torque τ, using both analytical approximation and numerical simulations, for verification, based on the boundary element method, and we investigate electrostatic pull-in behavior from the viewpoint of constant charge operation.

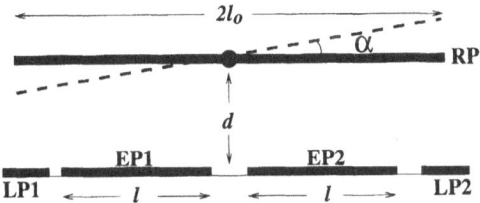

Fig. 9.16 Schematic for electrostatic micromirror used in SPICE simulations

Table 9.6. *Illustration of duals for transla-tional and rotational mechanics*

Translational mechanics	Rotational mechanics
Force F	Torque τ
Displacement x	Rotation angle α
Mass m	Moment of inertia I
Spring constant k	Torsion constant κ
Damper b	Damper b

9.3.1 Capacitance and Torque Modeling

There are three dominant mutual capacitors C_{01}, C_{02}, and C_{12} for the pair of plates (RP, EP1), (RP, EP2), and (EP1, EP2), respectively. Due to symmetry, $C_{01} = C_{02}$. Furthermore, we can safely assume that $C_{12} \ll C_{01}$; hence, we ignore the role of C_{12}. When one of the electrodes, e.g., EP1, is biased and the other (EP2) grounded, the electrostatic interaction is mainly between EP1 and the left-half part of RP. In this case, we are interested in the capacitor $C \equiv C_{01}$ which can be approximated as (see [36])

$$C \approx \varepsilon W \frac{\ln\left(1 - \frac{l}{d}\alpha\right)}{-\alpha},\tag{9.55}$$

where W denotes the plate width. By substituting $x = -(l/d)\alpha$, Eq. (9.55) can be expressed as

$$C \approx C_0 \frac{\ln(1+x)}{x},\tag{9.56}$$

where $C_0 = \varepsilon W l / d$ denotes the capacitance when $\alpha = 0$. Expression (9.56), in its given form, is not suitable for simulation since it has an indeterminate form at $x = 0$. We need to resort to Taylor-series expansion. We can either expand $\ln(1+x)/x$ or $x/\ln(1+x)$. The latter is preferable since it approaches 0 when x approaches -1. Therefore, we work with the reciprocal $(1/C)$ form:

$$\frac{1}{C} \approx \frac{1}{C_0} g(x),$$

$$\text{with} \quad g(x) \approx 1 + \frac{1}{2}x - \frac{1}{12}x^2 + \frac{1}{24}x^3 - \frac{19}{720}x^4 + \cdots + O(x^{n+1}).\tag{9.57}$$

With the expression for capacitance, we now obtain model equations for the torque τ

$$\tau = \frac{1}{2}V^2 \frac{dC}{d\alpha} \approx \begin{cases} \dfrac{1}{2C_0}\dfrac{l}{d}q^2 g'(x), & \text{constant charge,} \\[2ex] \dfrac{1}{2}C_0\dfrac{l}{d}V^2 \dfrac{g'(x)}{g^2(x)}, & \text{constant voltage,} \end{cases}\tag{9.58}$$

for the two operation modes. Depending on device structure, constant charge operation delays or eliminates electrostatic pull-in (see [36]). In particular, with vertically-actuated parallel plate type structures under constant voltage operation, regardless of the value of spring constant, electrostatic pull-in occurs when the capacitor plate displaces by one third of the initial separation, *viz.*, $d/3$.

To verify the above approximate models for the capacitance C and torque τ, we employ numerical simulation based on the panel method [61].

9.3.2 Verification Using the Panel Method

The panel method constitutes a special class of boundary element methods. Here, the planar surfaces of the mirror including the fixed and landing electrodes are meshed into small panels. On each panel, the potential is assumed constant. For a pair of panels i and j, the contribution of a unit charge on panel j to the potential at panel i is described by a coefficient p_{ij} which is part of a large and dense matrix P. The matrix can be made sparse using multipole expansion [64] or exponential expansion [65]. The relation between the charge q on the panels and the potential ψ at the center of each panel is described by

$$\psi = Pq, \tag{9.59}$$

where ψ and q are vectors. With the prescribed boundary condition for the potential on electrode surfaces, we solve Eq. (9.59) for the charge distribution, based on which we compute the capacitance and the torque acting on the mirror [36]. The simulations are based on mirror geometry data taken from [60]; here, $2l_0 = 40\,\mu\text{m}$, $W = 30\,\mu\text{m}$, $d = 1.4\,\mu\text{m}$, and l is chosen to be $18\,\mu\text{m}$. We can now compare the analytical approximation of C and τ described by (9.57) and (9.58) with the numerical simulation result.

Fig. 9.17 shows the capacitance C as a function of rotation angle α computed using numerical simulations, the logarithm form, and a third-order polynomial approximation; the latter two are associated with Eq. (9.57). While the third-order polynomial approximation appears to be sufficient for capacitance calculations, it is insufficient for accurate prediction of torque. However, this can be overcome by choosing a polynomial of higher order; see Fig. 9.18 which illustrates the comparison between analytical and numerical simulation results for the torque. Thus in the SPICE simulations that follow, the capacitance and torque are modeled using third- and seventh-order polynomial approximations, respectively.

9.3.3 SPICE Simulation

The model equations for capacitance and torque used in simulation of the coupled electromechanical circuit (Fig. 9.10) are summarized in Table 9.7. Fig. 9.19 shows the simulated rotation angle as a function of applied

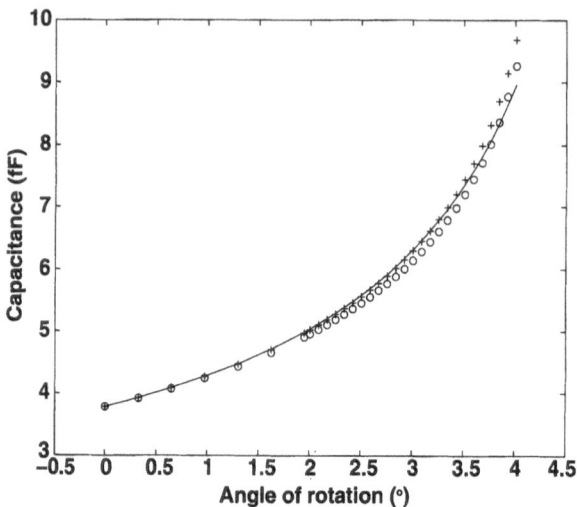

Fig. 9.17 Comparison of capacitance computed using approximate formulas with that by numerical simulation using the panel method: solid line = polynomial form ($n = 3$); + = log form; and ◯ = numerical simulation

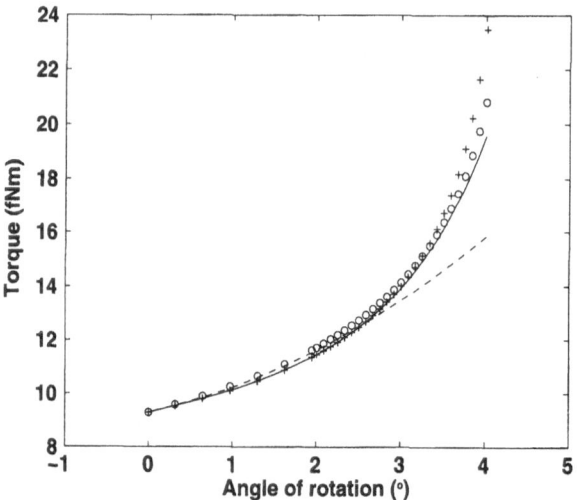

Fig. 9.18 Comparison of torque computed using approximate formulas with that of the panel method: solid line = polynomial form ($n = 7$); dashed line = polynomial form ($n = 3$), + = log form; and ◯ = numerical simulation

voltage. The data is in agreement with numerically obtained values [60]. The torsion constant κ needed in the SPICE simulations is extracted from measurement data [60] of the critical (pull-in) voltage V_c, which is taken to be 12.5 V.

Table 9.7. *Summary of model equations for capacitance and torque (under constant charge operation) used in micromirror simulation*

Capacitance	$\dfrac{1}{C(\alpha)} = \dfrac{1}{C_0}\left[1 - \dfrac{1}{2}\dfrac{l}{d}\alpha - \dfrac{1}{12}\left(\dfrac{l}{d}\right)^2\alpha^2 - 2\dfrac{1}{24}\left(\dfrac{l}{d}\right)^3\alpha^3\right]$
Torque	$\tau(\alpha) = \dfrac{1}{2C_0}\dfrac{l}{d}q^2\left[\dfrac{1}{2} + \dfrac{1}{6}\dfrac{l}{d}\alpha + \dfrac{1}{8}\left(\dfrac{l}{d}\right)^2\alpha^2 + \ldots + 2\dfrac{33953}{453600}\left(\dfrac{l}{d}\right)^7\alpha^7\right]$

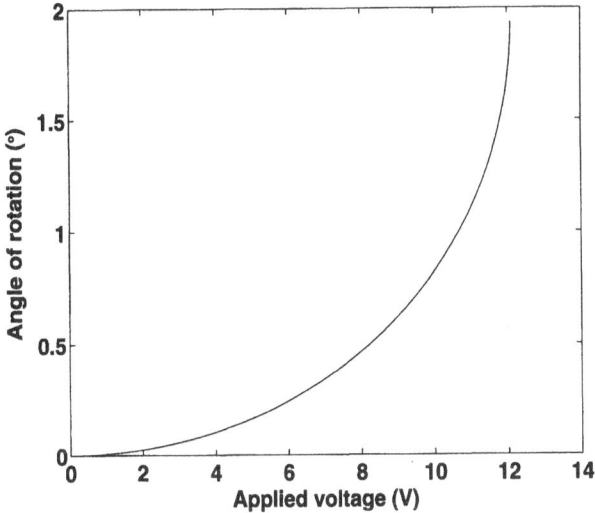

Fig. 9.19 SPICE simulation of the rotation angle as a function of applied voltage

To determine the critical angle and critical voltage at electrostatic pull-in, we solve for α using the following system of equations:

$$\tau(\alpha) = \kappa\alpha, \qquad \frac{\partial\tau(\alpha)}{\partial\alpha} = \kappa, \tag{9.60}$$

where τ is given in Table 9.7. System (9.60) can be reduced by eliminating κ to yield

$$\alpha\frac{\partial\tau(\alpha)}{\partial\alpha} - \tau(\alpha) = 0. \tag{9.61}$$

Using the relation $x = -(l/d)\alpha$ and Eq. (9.58), Eq. (9.61) can be rewritten as:

$$x_q g''(x) - g'(x) = 0, \qquad\qquad \text{constant charge,}$$
$$x_v\{g''(x)g(x) - 2[g'(x)]^2\} - g'(x)g(x) = 0, \qquad \text{constant voltage.}$$
$$\tag{9.62}$$

Solution of (9.62) yields $x_q = -0.71$ and $x_v = -0.44$, which in terms of the rotation angle α reads

$$\alpha_q = 0.71 \frac{d}{l}, \tag{9.63}$$

$$\alpha_v = 0.44 \frac{d}{l}, \tag{9.64}$$

for constant charge and constant voltage operation, respectively. Substituting these expressions into Eq. (9.58) and using the first equation in system (9.60), we obtain the corresponding critical voltages

$$V_{c,q} = 1.34 \frac{d}{l} \sqrt{\frac{\kappa}{C_0}}, \tag{9.65}$$

$$V_{c,v} = 0.91 \frac{d}{l} \sqrt{\frac{\kappa}{C_0}}. \tag{9.66}$$

Because the maximum angle of rotation is $\alpha_{max} = d/l_0$, the relative angle where no pull-in occurs is given as

$$r_q = \frac{\alpha_q}{\alpha_{max}} = 0.71 \frac{l_0}{l}, \tag{9.67}$$

$$r_v = \frac{\alpha_v}{\alpha_{max}} = 0.44 \frac{l_0}{l}. \tag{9.68}$$

Table 9.8 provides a summary of the above results, based on which we analyse the electrostatic pull-in behavior in what follows.

Unlike the vertically-actuated parallel plate capacitor where constant charge operation eliminates electrostatic pull-in, the micromirror is subject to pull-in even under constant charge operation, since the torque is not constant despite a constant charge. This is due to the non-constant term $g'(x)$ in Eq. (9.58). However, Eqs. (9.63), (9.64) and (9.67), (9.68) suggest that for a given geometry (l_0, l, and d), the onset of electrostatic pull-in could be delayed. Fig. 9.20 shows the angle of rotation as a function of

Table 9.8. *Summary of results of electrostatic pull-in associated with constant voltage and constant charge operation of micromirror*

Operation mode	Critical angle	Relative angle	Critical Voltage
Constant charge	$\alpha_q = 0.71 \frac{d}{l}$	$r_q = \frac{\alpha_q}{\alpha_{max}} = 0.71 \frac{l_0}{l}$	$V_{c,q} = 1.34 \frac{d}{l} \sqrt{\frac{\kappa}{C_0}}$
Constant voltage	$\alpha_v = 0.44 \frac{d}{l}$	$r_v = \frac{\alpha_v}{\alpha_{max}} = 0.44 \frac{l_0}{l}$	$V_{c,v} = 0.91 \frac{d}{l} \sqrt{\frac{\kappa}{C_0}}$

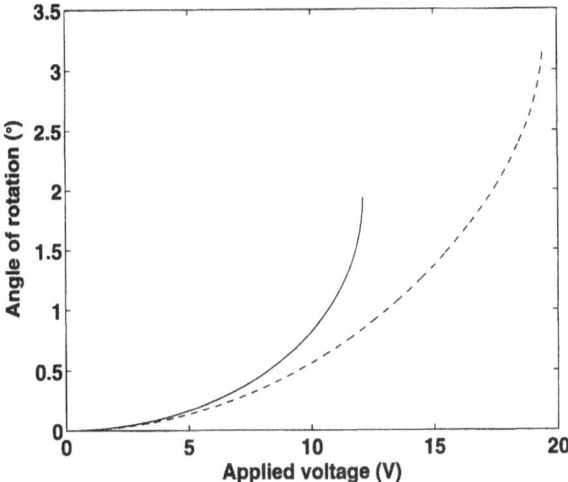

Fig. 9.20 A comparison of rotation behavior associated with electrostatic pull-in under constant voltage (solid curve) and constant charge (dashed curve) operation modes

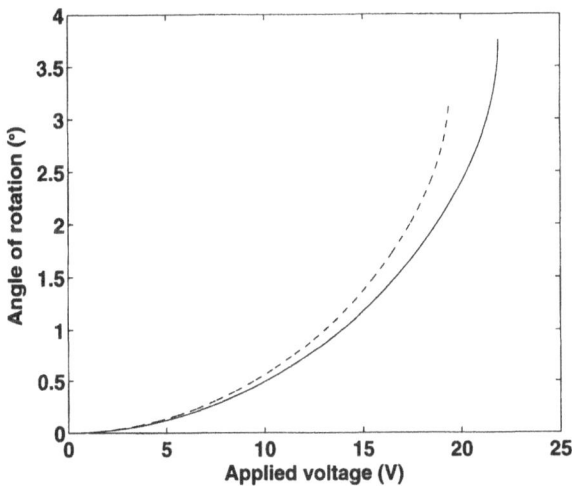

Fig. 9.21 A comparison of rotation behavior for different micromirror electrode geometries operated under constant charge mode; $l_0/l = 20\,\mu\text{m}/16\,\mu\text{m}$ (solid curve) and $l_0/l = 20\,\mu\text{m}/18\,\mu\text{m}$ (dashed curve)

applied bias under constant charge and constant voltage operation modes. As predicted by Eqs. (9.63) and (9.64), the critical angle under constant charge operation is larger than that of constant voltage operation.

Also, by changing the micromirror geometry, we can delay the onset of pull-in even under constant voltage operation. For both operation modes, this can be achieved by choosing l small enough so that the relative angle, r_q or r_v, in Eqs. (9.67) or (9.68) becomes unity. Fig. 9.21 shows how the rotation angle can be improved by increasing the ratio l_0/l. A smaller value of l or a larger value of d, increases the critical angle. However, it decreases the initial capacitance C_0 and results in an increase in the required applied voltage to achieve the same angle of rotation.

9.4 Distributed Analysis: Illustrative Example – Flow Microsensor

We illustrate the SPICE simulation of transient thermal behavior, circuit transient response, and system behavior under transient flow conditions for a flow microsensor integrated with control circuitry. Fig. 9.22 shows a simple control circuit for microheater operation at constant temperature [14]. In what follows, we review pertinent electrothermal modeling equations and provide a detailed description of the convective boundary conditions and their implementation on SPICE for distributed analysis.

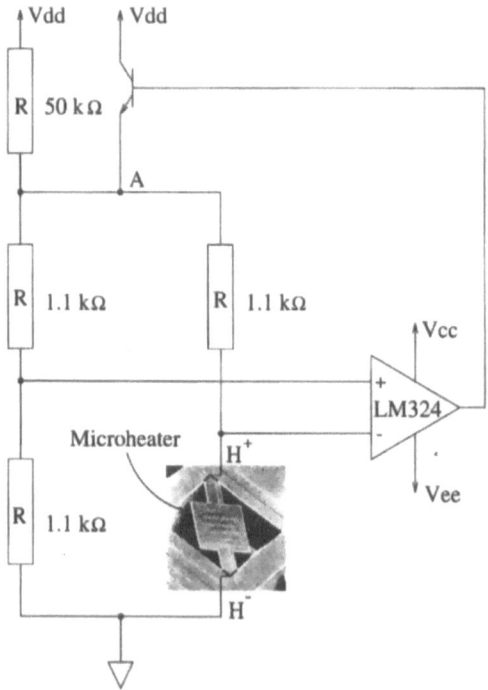

Fig. 9.22 A simple constant temperature circuit used in simulation. The SEM indicates the flow microsensor

9.4.1 Model Equations and Circuit Synthesis

The thermal balance of the sensor comprises conduction, heat generation, and surface heat loss, denoted as Q_{cond}, Q_{Joule}, and Q_{sur}, respectively. Neglecting thermoelectric effects, the differential equations governing the transient electrothermal behavior, in the thermally isolated active sensor region, take the following well-known forms:

$$\text{div}\,[\sigma(T)\,\text{grad}\,\psi] = 0, \tag{9.69}$$

$$\text{div}\,[\kappa(T)\,\text{grad}\,T] = \rho c_p \frac{\partial T}{\partial t} + \mathbf{J} \cdot \text{grad}\,\psi, \tag{9.70}$$

where $\sigma(T)$ and $\kappa(T)$ denote temperature-dependent electrical and thermal conductivities, respectively, ψ the electric potential, $\mathbf{J}$ the electrical current density, ρ the density, and c_p the specific heat. Eq. (9.69) governs the current flow in the heating element and Eq. (9.70) governs the temperature distribution in both the electrically conducting and insulating regions. In all simulations discussed, we assume the temperature and potential gradients in the z-direction (perpendicular to the sensor plane) are small in comparison to the gradients in the x-y plane, and any heat losses in the z-direction are treated as mixed boundary conditions. However, the simulation approach presented here is not limited to two dimensions.

Equations (9.69) and (9.70) are solved subject to appropriate boundary and interface conditions (see Chapt. 5). Fixed temperatures and potentials are assumed where pertinent. At the conductor/insulator boundaries the normal component of the current density is assumed to be zero, and at interfaces between two different conducting materials, current continuity is maintained. A similar condition for heat current continuity is maintained at different material boundaries.

The surface heat loss Q_{sur} which represents the heat lost to the surroundings, depends on the emissivity of the sensor material, the sensor surface temperature distribution and the thermophysical properties of the gas surrounding the device.

The net radiative heat loss from the sensor surface can be approximated using

$$Q_{rad} = \sigma \varepsilon \int_A [T^4(x, y) - T_\infty^4]dA, \tag{9.71}$$

where σ is the Stefan-Boltzmann constant, ε the surface emissivity, $T(x, y)$ the surface temperature distribution, and T_∞ the temperature of the environment.

The local heat transfer $q(x)$ resulting from a forced one-dimensional laminar flow over a non-isothermal plate can be approximated as [62]

$$q(x) = \int_0^x h(x, \xi) \frac{d\theta(\xi)}{d\xi} d\xi. \tag{9.72}$$

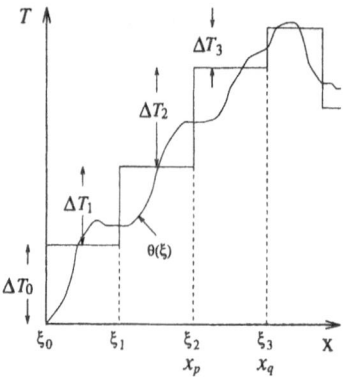

Fig. 9.23 Piecewise constant temperature approximation used in modeling convective heat
transfer

Here, $\theta(\xi)$ denotes the excess surface temperature variation, where ξ varies from 0 to x, and $h(x,\xi)$ the heat transfer coefficient (for laminar flow) which is defined as

$$h(x,\xi) = 0.332 \frac{\kappa Pr^{1/3}}{x} Re_x^{1.2} \frac{1}{[1 - (\xi/x)^{3/4}]^{1/3}}. \tag{9.73}$$

Here, κ, Pr and Re_x denote the thermal conductivity, Prandtl number and Reynolds number, respectively. Here, the Reynolds number $Re_x = u_\infty x/\nu$, where u_∞ is the free-stream velocity and ν the gas viscosity.

Assuming a piecewise constant surface temperature, the discrete form of the local heat transfer in Eq. (9.72) can be integrated between two stream points $(x_p$ and x_q, $x_p < x_q)$ on the plate to yield the net heat transfer between points (see Fig. 9.23), viz.,

$$Q_{p,q} = \sum_{i=0}^{m} h_a(x_q, \xi_i)\Delta T_i - \sum_{j=0}^{n} h_a(x_p, \xi_j)\Delta T_j, \tag{9.74}$$

where m, n denote the number of integration sub-intervals and h_a the average heat transfer coefficient obtained by integrating Eq. (9.73) with respect to x. The net heat transfer, Eq. (9.74), is now in a form amenable to SPICE implementation.

In the absence of flow, the heat loss to the surroundings is diffusive; microheaters, in view of their small dimensions, have a Grashof number that is typically Gr ≈ 0.01. This heat loss can be modeled using a variety of approaches, all of which are in a form

$$q = h\kappa(T_{sa})[T_{sa} - T_\infty]. \tag{9.75}$$

Here, $h = Nu/L_c$ denotes the heat transfer coefficient in terms of the Nusselt number (Nu) and the characteristic length (L_c) of the active sensing region. In Eq. (9.75), T_{sa} denotes the average surface temperature,

$\kappa(T_{sa})$ the thermal conductivity evaluated at T_{sa}, and T_∞ the temperature of the cavity walls, substrate, or fluid far from the surface. The drawback to this approach is the evaluation of Nu and L_c, which exist only for simple geometries. Moreover, with a non-linear surface temperature distribution, which is typically the case in realistic devices, use of the average temperature T_{sa} leads to inaccuracies in the calculated heat loss.

We employ a two-dimensional local heat-transfer coefficient $h_c(x, y)$ derived from a three-dimensional solution of Laplace's equation governing the temperature distribution in the fluid around the heated plate to accurately model this heat loss. This heat transfer coefficient represents the local heat loss per unit thermal conductivity and per unit difference in temperature between an isothermal plate (T_s) and its surrounding (T_∞). For an infinitesimally thin plate, the function can be expressed as

$$h_c(x, y) = h_{top}(x, y) + h_{bot}(x, y), \qquad (9.76)$$

where the heat loss from the top (h_{top}) and bottom (h_{bot}) surfaces are simultaneously calculated using the boundary element (panel) method [61, 63]. Hence, the total heat loss from the sensor surface, for an arbitrary temperature distribution $T(x, y)$, can be approximated using the following discrete form,

$$Q = \sum_{i,j} h_c(x_i, y_j)\kappa(T_a)[T(x_i, y_j) - T_\infty]\Delta x_i \Delta y_j. \qquad (9.77)$$

The thermal conductivity is evaluated at an average temperature, $T_a = [T(x_i, y_j) + T_\infty]/2$. The function $h_c(x, y)$ need be calculated only once for use in all subsequent simulations of the same structure. Eq. (9.77) is applied as a mixed boundary condition in the solution of Eq. (9.70).

9.4.2 SPICE Simulation

Following the procedures outlined in Sect. 9.2.2, Eqs. (9.69) and (9.70) yield two circuits (of the form shown in Fig. 9.13) denoted as C^e (electrical) and C^t (thermal), although a single grid is used to discretize the simulation domain. These two circuits are entered as one netlist which is then evaluated using SPICE.

The coupling between equations is accounted for by employing temperature-dependent resistor values in C^e and C^t, and by Joule heat generation in C^t. Both the Joule heat and temperature (voltage)-dependent resistors can be implemented in any circuit simulator which supports complex definitions of resistors and current sources.

With the above circuit formulation, C^e and C^t are coupled through behavioral elements, although no current flows between the two circuits. Relevant information is exchanged between the circuits, but the circuit topologies are distinct. We note that not all circuit simulators necessarily handle such coupled circuit systems efficiently. For example, the thermal circuit nodal voltage values (temperatures) can be more than two orders of magnitude greater than those in the electrical circuit. An efficient solution in this case (and particularly so in transient analysis) is to partition the problem into electrical and thermal networks, with a separate convergence criterion applied to each network. Also, for geometrically complex structures leading to large systems of equations, the current solution algorithms employed should be modified to improve the convergence rate and reduce the memory requirements.

As for boundary conditions, the radiative heat loss, Eq. (9.71), is implemented using an array of fourth-order voltage-controlled current sources (VCCS) distributed over the sensor surface. For the forced convective heat loss, Eq. (9.74), a behavioral VCCS is used where velocity is treated as a variable and the ΔT's are the controlling voltages. The implementation of this VCCS, however, may appear crude. But, this is due to the lack of appropriate elements in some circuit simulators to efficiently describe this convective heat loss. Diffusive heat loss due to natural convection is implemented using multi-dimensional VCCS with the current sources distributed over the entire sensor surface. The form of the VCCS follows directly from Eq. (9.77).

The equivalent circuit models are incorporated in a SPICE subcircuit which is synthesized from a model description file of the sensor generated

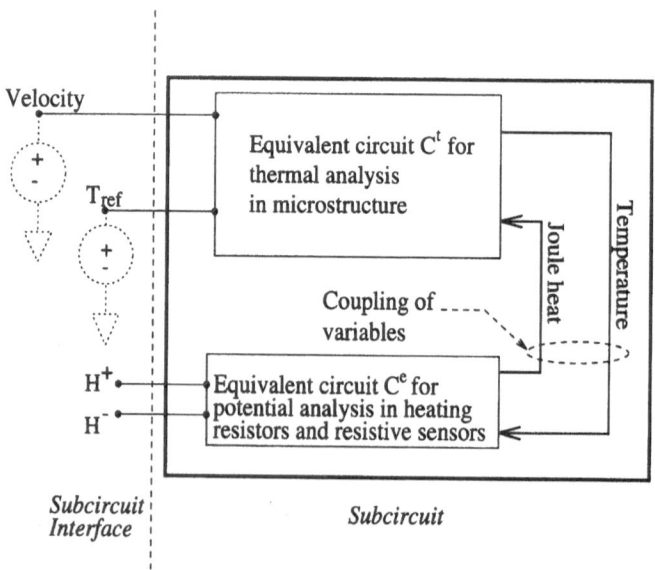

Fig. 9.24 Block diagram and subcircuit interface of equivalent coupled subcircuits used in SPICE simulation of flow microsensor

automatically or by the user [14]. The model description file for a thermal sensor is analogous, in terms of functionality, to a netlist for an electrical circuit. A block diagram of the resulting subcircuit and corresponding electrical-thermal interactions is depicted in Fig. 9.24. Internal to the subcircuit are the two coupled circuits, $C^{e,t}$. The subcircuit interface consists of four terminals. The terminals, H^+ and H^-, represent the heater connections while the terminal nodes, Velocity and T_{ref}, represent the gas velocity and gas temperature, respectively, values of which are controlled through the voltages (either DC or time-varying) at the corresponding nodes.

To illustrate device-circuit interactions under transient and steady-state conditions, we consider the control circuit shown in Fig. 9.22. The circuit has been designed for constant temperature operation; the feedback loop maintains the average temperature of the microheater at 95 °C. The SPICE netlist for the circuit is shown in Fig. 9.25. In particular, it illustrates the hierarchy and modularity of the underlying simulation approach, using the various subcircuit calls, *viz.*, the op-amp (xamp1), the coupled electro-thermal circuit (xsens1), and other circuit elements.

```
* Concurrent sensor simulation

* Subcircuit call for flow sensor
xsens1 bm 0 vel tref FlowSensor
Vtref tref 0 20.0

* Op-amp
xamp1 bp bm vcc vee opout nslm324
Vee vee 0 0.0

* Bridge
r1    bridge bp   1100
r2    bridge bm   1100
r3    bp     0    1100
Rbon bridge vdd 50k

* Driver transistor
xq1 vdd opout bridge t2n3904

* Power-on
Vel vel 0 (pulse 0.0 15.0 6ms 2ns 2ns 6ms  12ms)
Vdd vdd 0 (pulse 0.0  8.0 0ns 2ns 2ns 12ms 12ms)
Vcc vcc 0 (pulse 0.0  8.0 0ns 2ns 2ns 12ms 12ms)

.TRAN 0.01ms 12ms
.PRINT V(bp) V(bm) V(vcc) PAR('V(bm)/I(r2)') V(opout)
.end
```

Fig. 9.25 SPICE netlist for simulation of circuit and sensor shown in Fig. 9.22

Simulation results of the device transient thermal behavior, circuit transient response, and system behavior under transient flow conditions are illustrated in Fig. 9.26. Here, the voltage across the sensor and bridge (node A in Fig. 9.22), is shown in response to "power-on" of all supplies at $t = 0^+$s. The extent of ringing observed in the bridge and sensor voltages is influenced primarily by the op-amp. Also shown is the corresponding average sensor temperature which reaches its steady-state value of 95 °C in approximately 2 ms, with an overshoot of only a few degrees.

The simulations shown so far are based on zero flow conditions. The system response to a flow transient is simulated using a pulsed voltage source that is connected to the Velocity terminal of the sensor subcircuit (see SPICE netlist in Fig. 9.25). The simulated results are shown in Fig. 9.26. We note that the average sensor temperature is virtually constant despite the extent of ringing observed with the bridge voltage. As expected, the steady-state voltages across the sensor and bridge increase, since more power is required to maintain the sensor temperature at 95 °C with increased heat loss to the flow stream.

Although experimental results, and comparison with simulations, are not presented here, excellent agreement with measurement results has been achieved for similar device/circuit configurations [34] for both transient and steady-state conditions.

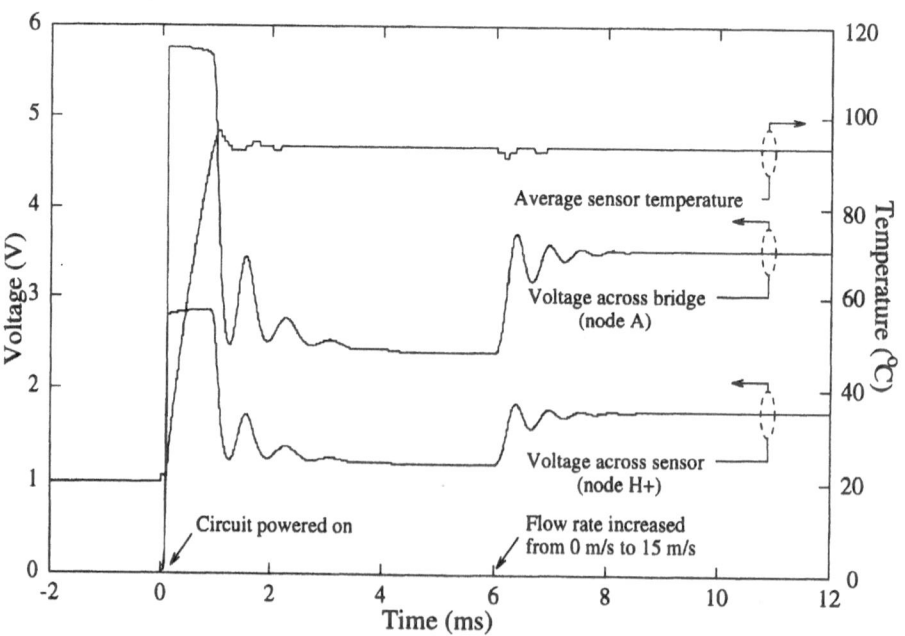

Fig. 9.26 Transient response of bridge voltages (Fig. 9.22) and average sensor temperature on "power-on" and to a velocity pulse

9.5 References

[1] Middelhoek, S., Audet, S. A., *Silicon Sensors*, New York: Academic Press, 1989.
[2] Wise, K. D., Microelectromechanical Systems: Interfacing Electronics to a Non-Electronic World, *Technical Digest*, IEEE IEDM, San Francisco, 1996, pp. 11–18.
[3] Baltes, H., Future of IC Microtransducers, *Sensors and Actuators A*, 56 (1996), 179–192.
[4] Baltes, H., Paul, O., Korvink, J. G., Schneider, M., Bühler, J., Schneeberger, N., Jaeggi, D., Malcovati, P., Hornung, M., Häberli, A., von Arx, M., Mayer, F., Funk, J., IC MEMS Microtransducers, in: *Technical Digest*, IEEE IEDM, San Francisco, 1996, pp. 521–524.
[5] Engl, W. L., Laur, R., Dirks, H. K., MEDUSA – A Simulator for Modular Circuits, *IEEE Trans. CAD of ICAS*, 1 (1982), 85–93.
[6] McMacken, J. R. F., Chamberlain, S. G., CHORD – A Modular Semiconductor Device Simulation Development Tool Incorporating External Network Models, *IEEE Trans. CAD of ICAS*, 8 (1989), 826–836.
[7] Gnudi, A., Ciampolini, P., Guerrieri, R., Rudan, M., Baccarani, G., Circuit Analysis by Using Device Simulator, *Proc. NASECODE V Conf.*, Dublin, Miller, J. J. H., (Ed.), 1987, pp. 201–206.
[8] Litsios, J., Muller, S., Fitchner, W., Mixed-Mode Multi-Dimensional Device and Circuit Simulation, in: *Simulation of Semiconductor Devices and Processes*, Vol. 5, Selberherr, S., Stippel, H., Strasser, E. (Eds.), Wien-New York: Springer-Verlag, 1993, pp. 129–132.
[9] Latif, M., Bryant, P. R., Network Analysis Approach to Multidimensional Modeling of Transistors Including Thermal Effects, *IEEE Trans. CAD of ICAS*, 1 (1982), 94–101.
[10] Krabbenborg, B. H., de Graaff, H. C., Mouthaan, A. J., Boezen, H., Bosma, A., Tekin, C., 3D Thermal/Electrical Simulation of Breakdown in a BJT Using a Circuit Simulator and a Layout-to-Circuit Extraction Tool, in: *Simulation of Semiconductor Devices and Processes*, Vol. 5, Selberherr, S., Stippel, H., Strasser, E. (Eds.), Wien-New York Springer-Verlag, 1993, pp. 57–60.
[11] Litsios, J., Schmithüsen, B., Fichtner, W., Large Scale Thermal Mixed Mode Device and Circuit Simulation, in: *Simulation of Semiconductor Devices and Processes*, Vol. 6, Ryssel, H., Pichler, P. (Eds.), Wien-New York: Springer-Verlag, 1995, pp. 368–371.
[12] Green, M. A., Shewchun, J., The General Transmission Line Equivalent Circuit Model for Degenerate and Non-Degenerate Carrier Concentrations in Semiconductors, *Solid-State Electronics*, 17 (1974), 717–723.
[13] Green, M. A., Shewchun, J., Application of the Small-Signal Transmission Line Equivalent Circuit Model to the a. c., d. c. and Transient Analysis of Semiconductor Devices, *Solid-State Electronics*, 17 (1974), 941–949.
[14] Swart, N. R., Nathan, A., Mixed-Mode Device-Circuit Simulation of Thermal-Based Microsensors, *Sensors and Materials*, 6 (1994), 179–192.
[15] Mouthaan, T. J., Krabbenborg, B. H., Thermodynamic Analysis of Semiconductor Structures Using a Device Simulator and Lumped Circuit Modelling, *Sensors and Materials*, 6 (1994), 125–137.
[16] Kuzmicz, W., Denisiuk, W., Gempel, J., Jaworski, Z., Niewczas, M., Pfitzner, A., Piworarska, E., Pleskazc, W., Wojtasik, A., Coupling a Statistical Process-Device Simulator with a Circuit Layout Extractor for a Realistic Circuit Simulation of VLSI Circuits, *Simulation of Semiconductor Devices and Processes*, Vol. 5, Selberherr, S., Stippel, H., Strasser, E. (Eds.), Wien-New York: Springer-Verlag, 1993, pp. 37–40.
[17] Kron, G., Electric Circuit Models of Partial Differential Equations, *Electrical Engineering*, 67 (1948), 672.
[18] Lynn, J. W., *Tensors in Electrical Engineering*, London: Edward Arnold Ltd., 1963.

[19] Kron, G., Equivalent Circuits for the Elastic Field, *J. Appl. Mech.*, 12 (1945), 149–161.
[20] Kron, G., Equivalent Circuits for the Field Equations of Maxwell, *Proc. I. R. E.*, 32 (1944), 289.
[21] Kron, G., Equivalent Circuits of Compressible and Incompressible Fluid Flow Fields, *J. Aero. Sci.*, 12 (1945), 221.
[22] Higgins, T. J., Electroanalogic Methods, *Appl. Mech. Rev.*, Jan. 1956, Feb., Aug., Oct., 1957, May 1958.
[23] Janata, J., *Principles of Chemical Sensors*, New York: Plenum Press, 1989.
[24] MacNeal, R. H., *Electric Circuit Analogies for Elastic Structures*, New York: Wiley, 1962.
[25] Koenig, H. E., Blackwell, W. A., Linear Graph Theory – A Fundamental Engineering Discipline, *IRE Transactions of Education*, 105 (1960), 42–49.
[26] Koenig, H. E., Tokad, Y., State Models of Systems of Multiterminal Linear Components, *IEEE Int. Convention Record*, 12, Part 1 (1964), 318–329.
[27] Arnold, E., Computer Simulation of Conductivity and Hall Effect in Inhomogeneous Inversion Layers, *Surface Sci.*, 113 (1982), 239–243.
[28] Popovic, R. S., Numerical Analysis of MOS Magnetic Field Sensors, *Solid-State Electronics*, 28 (1985), 711–716.
[29] Caverly, R., Peck, E., A Finite-Element Model and Characterization of the p-i-n Magnetodiode at Microwave Frequencies, *Solid-State Electronics*, 30 (1987), 473–477.
[30] Salim, A., Manku, T., Nathan, A., Modeling of Magnetic Field Sensitivity of Bipolar Magnetotransistors Using HSPICE, *IEEE Trans. CAD of ICAS*, 14 (1995), 464–469.
[31] Mohajerzadeh, S., Nathan, A., Modeling Noise Correlation Behaviour in Dual-Collector Magnetotransistors Using Small Signal Equivalent Circuit Analysis, *IEEE Trans. Electron Devices*, 43 (1996), 883–888.
[32] Rombach, P., Langheinrich, W., Modelling of a Micromachined Torque Sensor, *Sensors and Actuators A*, 46–47 (1995), 294–297.
[33] Swart, N. R., Nathan, A., Flow-Rate Microsensor Modelling and Optimization Using SPICE, *Sensors and Actuators A*, 34 (1992), 109–122.
[34] Swart, N. R., Nathan, A., Coupled Electrothermal Modeling of Microheaters Using SPICE, *IEEE Transactions on Electron Devices*, 41 (1994), 920–925.
[35] Pham, H. H., Nathan, A., Compact MEMS-SPICE Modeling, *Sensors and Materials*, 10 (1998), 63–75.
[36] Pham, H. H., Nathan, A., Circuit Modeling and SPICE Simulation of Mixed-Signal Microsystems, *Sensors and Materials*, Special Issue on CAD for MEMS, 10, No. 7 (1998), (to appear).
[37] Auerbach, F. J., Meiendres, G., Müller, R., Scheller, G. J. E., Simulation of the Thermal Behaviour of Thermal Flow Sensors by Equivalent Electrical Circuits, *Sensors and Actuators A*, 41–42 (1994), 275–278.
[38] Shie, J.-S., Chen, Y.-M., Ou-Yang, M., Chou, B. C. S., Characterization and Modeling of Metal-Film Microbolometer, *J. of Microelectromechanical Systems*, 5 (1996), 298–306.
[39] Reimer, D. E., Electrical Equivalent Method for Thermal Stress Analysis, *ECC* (1989), 869–874.
[40] Marco, S., Samitier, J., Ruiz, O., Herms, A., Morante, J. R., Analysis of Electrostatic-Damped Piezoresistive Silicon Accelerometers, *Sensors and Actuators A*, 37–38 (1993), 317–322.
[41] Veijola, T., Kuisma, H., Lahdenperä, J., Ryhänen, T., Equivalent-Circuit Model of the Squeezed Gas Film in a Silicon Accelerometer, *Sensors and Actuators A*, 48 (1995), 239–248.
[42] Burstein, A., Kaiser, W. J., The Microelectromechanical Gyroscope – Analysis and Simulation Using SPICE Electronic Simulator, *Proc. SPIE*, 2642 (1995), 225–232.

[43] Fedder, G. K., Howe, R. T., Multimode Digital Control of a Suspended Polysilicon Microstructure, *J. of Microelectromechanical Systems*, 5 (1996), 283–297.

[44] Pourahmadi, F., Review of Modeling Silicon Microsensors and Actuators, *Sensors and Materials*, 6 (1994), 193–209.

[45] Tilmans, H. A. C., Equivalent Circuit Representation of Electromechanical Transducers: I. Lumped-Parameter Systems, *J. Micromech. Microeng.*, 6 (1996), 157–176.

[46] Romanowicz, B., Lerch, Ph., Renaud, Ph., Fullin, E., de Coulon, Y., Simulation of Integrated Electromagnetic Device Systems, *Digest of Technical Papers*, Transducers '97, Chicago, 1997, pp. 1051–1054.

[47] Senturia, S. D., CAD for Microelectromechanical Systems, *Digest of Technical Papers*, Vol. 2, Transducers '95, Stockholm, June 25–29, 1995, pp. 5–8.

[48] Ando, S., Tanaka, K., Abe, M., Fishbone Architecture: An Equivalent Mechanical Model of Cochlea and Its Application to Sensors and Actuators, *Digest of Technical Papers*, Transducers '97, Chicago, 1997, pp. 1027–1030.

[49] Voigt, P., Wachutka, G., Electro-Fluidic Microsystem Modeling Based on Kirchhoffian Network Theory, *Digest of Technical Papers*, Transducers '97, Chicago, 1997, pp. 1019–1022.

[50] Massobrio, G. Martinoia, S., Grattarola, M., Use of SPICE for Modeling Silicon-Based Chemical Sensors, *Sensors and Materials*, 6 (1994), 101–123.

[51] Korvink, J. G., Bächtold, M., Emmenegger, M., Paganini, R., Ruehl, R., Funk, J., Baltes, H., TCAD for MEMS, *Proc. ESSDERC '96*, Baccarani, G., Rudan, M. (Eds.), Bologna, 1996, pp. A5–A7.

[52] Nathan, A., Self-Consistent Network Synthesis for Mixed-Signal Simulations, *Int. Rep. No. 95/06*, Physical Electronics Laboratory, ETH Zürich, Switzerland, 1995.

[53] Nathan, A., Microtransducer CAD, *Proc. ESSDERC '96*, Baccarani, G., Rudan, M. (Eds.), Bologna, 1996, pp. 707–715.

[54] Karnopp, D., Margolis, D., Rosenberg, R., *System Dynamics: A Unified Approach*, 2nd Ed., New York: Wiley, 1990.

[55] Giloi, W. K., *Principles of Continuous System Simulation*, Stuttgart: Teubner, 1975.

[56] Karayanakis, N. M., *Computer-Assisted Simulation of Dynamic Systems with Block Diagram Languages*, Boca Raton, FL: CRC Press, 1993.

[57] Rudan, M., Guerrieri, R., Ciampolini, P., Baccarani, G., Discretization Strategies and Software Implementation for a General-Purpose 2D-Device Simulator, in: *New Problems and New Solutions for Device and Process Modeling*, Miller, J. J. H. (Ed.), Dublin: Boole Press, 1985, pp. 110–121.

[58] Kron, G., Basic Concepts of Space Filters, *Trans. A. I. E. E.*, 28, Part I (1959), 554.

[59] Ackroyd, R. T., Houston, J., Lynn, J. W., Mann, E., An Electrical Analogue for Heat Waves in an Exothermic Medium, *Proc. I. E. E.*, 108, Part B, (1961), 33.

[60] Funk, J. M., Korvink, J. G., Bühler, J., Bächtold, M., Baltes, H., SOLIDIS: A Tool for Microactuator Simulation in 3-D, *IEEE J. of Microelectromechanical Systems*, 6 (1997), 70–82.

[61] Harrington, R. F., *Field Computation by Moment Methods*, New Jersey: IEEE Press, 1993.

[62] Rubesin, M. W., Inouye, M., Parikh, P. G., in: *Handbook of Heat Transfer Fundamentals*, Rohsenow, W. M., Hartnett, J. P., Ganic, E. N. (Eds.), New York: McGraw-Hill, 1985.

[63] Swart, N. R., *Heat Transport in Thermal-Based Microsensors*, Ph.D. Dissertation, University of Waterloo, Waterloo, Ontario N2L 3G1, Canada, 1994.

[64] Nabors, K., Kim, S., White, J., Fast Capacitance Extraction of General Three Dimensional Structures, *IEEE Trans. on Microwave Theory and Techniques*, 40 (1992), 1496–1506.

[65] Pham, H. H., Nathan, A., A New Approach for Rapid Evaluation of the Potential Field in Three Dimensions, *Proceedings Royal Society London A*, 455 (1999), 1–39.

Subject Index

SpringerEngineering

Andreas Schenk

Advanced Physical Models
for Silicon Device Simulation

1998. XVIII, 349 pages. 125 figures.
Hardcover DM 248,–, öS 1736,–
(recommended retail price)
ISBN 3-211-83052-9
Computational Microelectronics

The quality of physical models is decisive for the understanding of the physical processes in semiconductor devices and for a reliable prediction of the behavior of a new generation of devices. The first part of the book contains a critical review on models for silicon device simulators, which rely on moments of the Boltzmann equation. With reference to fundamental experimental and theoretical work, an extensive collection of widely used models is discussed in terms of physical accuracy and application results. The second part outlines the derivation of physics-based models for bulk mobility, band-to-band tunneling, defect-assisted tunneling, thermal recombination, non-ideal metal-semiconductor contact, and direct and multiphonon-assisted tunneling through insulating layers, all from a microscopic level. The models are compared with experimental data and applied to a number of simulation examples. This part also describes some new approaches of "taylored quantum mechanics" for deriving device models from "first principles" and the fundamental problems therein.

Contents
• Simulation of Silicon Devices: An Overview
• Mobility Model for Hydrodynamic Transport Equations
• Advanced Generation-Recombination Models
• Metal-Semiconductor Contact
• Modeling Transport Across Thin Dielectric Barriers
• Summary and Outlook

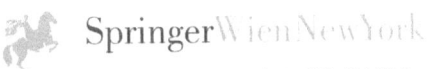

SpringerWienNewYork

Sachsenplatz 4–6, P.O.Box 89, A-1201 Wien, Fax +43-1-330 24 26, e-mail: books@springer.at, Internet: http://www.springer.at
New York, NY 10010, 175 Fifth Avenue • D-14197 Berlin, Heidelberger Platz 3 • Tokyo 113, 3–13, Hongo 3-chome, Bunkyo-ku

SpringerEngineering

Dietmar Schroeder

Modelling of Interface
Carrier Transport
for Device Simulation

1994. XI, 221 pages. 69 figures.
Hardcover DM 204,–, öS 1428,–
ISBN 3-211-82539-8
Computational Microelectronics

The book contains a comprehensive review of the physics, modelling and simulation of electron transport at interfaces in semiconductor devices. Particular emphasis is put on the consistent derivation of interface or boundary conditions for semiconductor device simulation. It combines a review of existing interface charge transport models with original developments. A unified representation of charge transport at semiconductor interfaces is introduced. Models for the most important interfaces are derived, classified within the unique modelling framework, and discussed in the context of device simulation. Discretization methods for numerical solution techniques are presented.

Contents
Introduction • Charge Transport in the Volume • General Electronic Model of the Interface • Charge Transport Across the Interface • Semiconductor-Insulator Interface • Metal-Semiconductor Contact • Semiconductor Heterojunction • MOSFET Gate • Discretization • Appendices

Wolfgang Joppich,
Slobodan Mijalković

Multigrid Methods
for Process Simulation

1993. XVII, 309 pages. 126 figures.
Hardcover DM 198,–, öS 1386,–
ISBN 3-211-82404-9
Computational Microelectronics

This book is the first one that combines both research in multigrid methods and a particular application field here - process simulation.
The introduction to multigrid is strictly directed towards the goal to provide the algorithmical overview one needs to compose optimal multigrid algorithms for evolution problems of process simulation and similar applications. The necessary explanation how and why multigrid works is derived from the roots. So the book preassumes no advanced familiarity with numerical analysis.
Additionally a complete strategy to implement different algorithmical components on an adaptive multilevel grid structure is presented. The outlined principle of grid definement and adaption is based on the control of errors and is reliable as well as general. Last but not least the described strategies are applied to "real life" problems of process simulation.
Consequently this book is an important contribution to the interdisciplinary challenge of improving numerical techniques for diffusion problems of process simulation.

All prices are recommended retail prices

SpringerWienNewYork

Sachsenplatz 4–6, P.O.Box 89, A-1201 Wien, Fax +43-1-330 24 26, e-mail: books@springer.at, **Internet: http://www.springer.at**
New York, NY 10010, 175 Fifth Avenue • D-14197 Berlin, Heidelberger Platz 3 • Tokyo 113, 3–13, Hongo 3-chome, Bunkyo-ku

SpringerEngineering

Narain Arora

MOSFET Models for VLSI Circuit Simulation

Theory and Practice

1993. XXII, 605 pages. 270 figures.
Hardcover DM 328,–, öS 2295,–
ISBN 3-211-82395-6
Computational Microelectronics

The book has 12 chapters. Starting from the overview of various aspects of device modeling for circuit simulators, a brief but complete review of semiconductor device physics and *pn* junction theory required for understanding MOSFET models is covered. The MOS transistor characteristics as applied to current MOS technologies are then discussed. Finally, the statistical variation of model parameters due to process variations are discussed.

Contents
• Overview • Review of Basic Semiconductor and *pn* Junction Theory • MOS Transistor Structure and Operation • MOS Capacitor • Threshold Voltage • MOSFET DC Model • Dynamic Model • Modeling Hot-Carrier Effects • Data Acquisition and Model Parameter Measurements • Model Parameter Extraction Using Optimization Method • SPICE Diode and MOSFET Models and Their Parameters • Statistical Modeling and Worst-Case Design Parameters • Appendices

Wilfried Hänsch

The Drift Diffusion Equation and Its Applications in MOSFET Modeling

1991. XII, 271 pages. 95 figures.
Hardcover DM 164,–, öS 1148,–
ISBN 3-211-82222-4
Computational Microelectronics

The drift diffusion equation and its applications in MOSFET modeling will bridge the gap between phe-nomenological modeling and a rigorous microscopic approach. The five chapters cover: Wigner's and Boltzmann's equation, the relaxation time approximation and the hydro dynamic equations for the case of strong non equilibrium, charge transport in an inversion channel, analytical approaches to determine the high energy distribution of carriers in high electric fields, and the spatial and temporal built up of charges in the gate ox-ide of a MOSFET device under electrical stress.

Contents
• Boltzmann's Equation • Hydrodynamic Model • Carrier Transport in an Inversion Channel • High Energetic Carriers • Degradation • Appendices: Perturbation Theory and Diagram Technique; Inversion Channel Particle-Density Distribution in Equilibrium

All prices are recommended retail prices

 SpringerWien New York

Sachsenplatz 4–6, P.O.Box 89, A-1201 Wien, Fax + 43-1-330 24 26, e-mail: books@springer.at. **Internet: http://www.springer.at**
New York, NY 10010, 175 Fifth Avenue • D-14197 Berlin, Heidelberger Platz 3 • Tokyo 113, 3–13, Hongo 3-chome, Bunkyo-ku

SpringerEngineering

Carlo Jacoboni, Paolo Lugli

The Monte Carlo Method

for Semiconductor

Device Simulation

1989. X, 356 pages. 228 figures.
Hardcover DM 204,–, öS 1430,–
ISBN 3-211-82110-4
Computational Microelectronics

The application of the Monte Carlo method to the simulation of semiconductor devices is presented. A review of the physics of transport in semiconductors is given, followed by an introduction to the physics of semiconductor devices. The Monte Carlo algorithm is discussed in great details, and specific applications to the modelling of semiconductor devices are given. A comparison with traditional simulators is also presented.

Contents
• Introduction
• Charge Transport in Semiconductors
• The Monte Carlo Simulation
• Review of Semiconductor Devices
• Monte Carlo Simulation
  of Semiconductor Devices
• Applications
• Appendices

Peter A. Markowich

The Stationary Semiconductor

Device Equations

1986. IX, 193 pages. 40 figures.
Hardcover DM 119,–, öS 836,–
ISBN 3-211-81892-8
Computational Microelectronics

The static semiconductor device problem is treated in an "applied mathematics" way. Qualitative properties, e.g. existence and uniqueness of solutions, and quantitative properties, particularly the structure of steady state solutions, are analysed. Physical interpretations of the mathematical results are given. Also, these results serve as a basis for the derivation and convergence analysis of numerical discretisation techniques.

Contents
• Introduction • Mathematical Modeling of Semiconductor Devices • Analysis of the Basic Stationary Semiconductor Device Equations • Singular Perturbation Analysis of the Stationary Semiconductor Device Problem • Discretisation of the Stationary Device Problem • Numerical Simulation – A Case Study • Appendix: Notation of Physical Quantities; Mathematical Notation

All prices are recommended retail prices

SpringerWienNewYork

Sachsenplatz 4–6, P.O.Box 89, A-1201 Wien, Fax +43-1-330 24 26, e-mail: books@springer.at, Internet: http://www.springer.at
New York, NY 10010, 175 Fifth Avenue • D-14197 Berlin, Heidelberger Platz 3 • Tokyo 113, 3–13, Hongo 3-chome, Bunkyo-ku

Springer-Verlag
and the Environment

WE AT SPRINGER-VERLAG FIRMLY BELIEVE THAT AN international science publisher has a special obligation to the environment, and our corporate policies consistently reflect this conviction.

WE ALSO EXPECT OUR BUSINESS PARTNERS – PRINTERS, paper mills, packaging manufacturers, etc. – to commit themselves to using environmentally friendly materials and production processes.

THE PAPER IN THIS BOOK IS MADE FROM NO-CHLORINE pulp and is acid free, in conformance with international standards for paper permanency.